Oxygen Ion and Mixed Conductors and their Technological Applications

NATO ASI Series

Advanced Science Institute Series

A Series presenting the results of activities sponsored by the NATO Science Committee, which aims at the dissemination of advanced scientific and technological knowledge, with a view to strengthening links between scientific communities.

The Series is published by an international board of publishers in conjunction with the NATO Scientific Affairs Division

A. Life Sciences **B. Physics**	Plenum Publishing Corporation London and New York
C. Mathematical and Physical Sciences **D. Behavioural and Social Sciences** **E. Applied Sciences**	Kluwer Academic Publishers Dordrecht, Boston and London
F. Computer and Systems Sciences **G. Ecological Sciences** **H. Cell Biology** **I. Global Environment Change**	Springer-Verlag Berlin, Heidelberg, New York, London, Paris and Tokyo

PARTNERSHIP SUB-SERIES

1. Disarmament Technologies	Kluwer Academic Publishers
2. Environment	Springer-Verlag / Kluwer Academic Publishers
3. High Technology	Kluwer Academic Publishers
4. Science and Technology Policy	Kluwer Academic Publishers
5. Computer Networking	Kluwer Academic Publishers

The Partnerschip Sub-Series incorporates activities undertaken in collaboration with NATO's Cooperation Partners, the countries of the CIS and Central and Eastern Europe, in Priority Areas of concern to those countries.

NATO-PCO-DATA BASE

The electronic index to the NATO ASI Series provides full bibliographical references (with keywords and/or abstracts) to about 50,000 contributions from international scientists published in all sections of the NATO ASI Series. Access to the NATO-PCO-DATA-BASE is possible via a CD-ROM "NATO Science and Technology Disk" with user-friendly retrieval software in English, French, and German (©WTV GmbH and DATAWARE Technologies, Inc. 1989). The CD-ROM contains the AGARD Aerospace Database.

The CD-ROM can be ordered through any member of the Board of Publishers or through NATO-PCO, Overijse, Belgium.

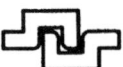

Series E: Applied Sciences – Vol. 368

Oxygen Ion and Mixed Conductors and their Technological Applications

edited by

Harry L. Tuller

Department of Materials Science and Engineering,
Massachusetts Institute of Technology,
Cambridge, MA, U.S.A.

Johannes Schoonman

Laboratory for Inorganic Chemistry,
Delft University of Technology,
Delft, The Netherlands

and

Ilan Riess

Physics Department,
Technion-Israel Institute of Technology,
Haifa, Israel

Kluwer Academic Publishers

Dordrecht / Boston / London

Published in cooperation with NATO Scientific Affairs Division

Proceedings of the NATO Advanced Study Institute on
Oxygen Ion and Mixed Conductors and their Technological Applications
Erice, Sicily, Italy
15–30 July 1997

A C.I.P. Catalogue record for this book is available from the Library of Congress.

ISBN 0-7923-6253-5

Published by Kluwer Academic Publishers,
P.O. Box 17, 3300 AA Dordrecht, The Netherlands.

Sold and distributed in North, Central and South America
by Kluwer Academic Publishers,
101 Philip Drive, Norwell, MA 02061, U.S.A.

In all other countries, sold and distributed
by Kluwer Academic Publishers,
P.O. Box 322, 3300 AH Dordrecht, The Netherlands.

Printed on acid-free paper

Printed in the Netherlands.

TABLE OF CONTENTS

PREFACE

Progress in the development of oxygen ion and mixed conductors is responsible for innovations in the fields of gas sensors, fuel cells, oxygen permeation membranes, oxygen pumps and electrolyzers. Commercialization has been impeded by materials stability and compatibility issues, high costs of fabrication and inadequate understanding of the interfacial phenomena controlling the operation of these devices. In this text we assemble a unique group of experts whose articles straddle, for the first time, all the key topical areas ranging from fundamentals relating to (a) defects, electrochemical, and interfacial processes, (b) catalysis, electrocatalysis and gas reforming, to design and fabrication including (c) advanced electroceramic processing methods, (d) materials selection and optimization, (e) and applications including scale up, commercialization and competitive technologies.

This material was first presented at a NATO Advanced Study Institute held in Erice, Sicily, Italy during the period July 15-30, 1997. All the participants benefited from the integrated and synthetic approach taken to the subject matter with liberal use of examples and case studies. Many opportunities were made available for critical discussions of the key concepts and issues both within the formal sessions as well as in the cafes and restaurants which populate Erice. I join the co-organizers of the Advanced Study Institute, Professors J. Schoonman, I. Riess and M. Balkanski, in thanking NATO for providing support for the ASI. Thanks are also due to Dr. Gabrielli and the Ettore Majorana Centre for exceptional organizational skills and local support of the activities and to the Ettore Majorana Centre and the National Science Foundation for support of participants' boarding and travel for US students respectively.

This work will be of great interest to materials scientists, chemists, physicists and chemical and electrical engineers first entering as well as those active in this field.

Harry L. Tuller
Cambridge, MA, USA

SOLID STATE ELECTROCHEMICAL CELLS

I. RIESS
Physics Department, Technion-IIT
Haifa 32000, Israel

1. Introduction

Solid electrochemical cells are becoming of much interest for energy conversion and storage using all solid batteries,[1] and high temperature solid oxide fuel cells (SOFCs) [2,3] and for selective sensing of chemicals, mainly for environmental control.[4] They also provide a very accurate means for determining thermodynamic quantities, such as the free energy of formation and changes in the chemical potential of a component in a compound.[3] In particular, for oxides, the partial pressure of oxygen in equilibrium with the oxide can be measured as a function of temperature and composition using a solid state coulometric titration method.[5] Such a measurement allows the characterization of the point defects in the oxide[6-8] and the identification of minority oxide phases.[9]

A simple electrochemical cell consists of one solid that conducts ions, and two electrodes applied on it. The ion conducting solid may be a pure ionic conductor, i.e. a solid electrolyte (SE). It may, however, also be a mixed ionic electronic conductor (MIEC) conducting both ions and electronic charge carriers (electrons and/or holes). Since a SE is a MIEC in the limit of negligible electronic conductivity, we shall refer to the ion conductor as a MIEC. A schematic of an electrochemical cell is presented in Fig.1.

2. Electrodes

The electrodes in solid electrochemical cells have to provide or remove both electronic charge and material. An example is a silver metal electrode on the silver ion conductor AgI. A different possibility is a combination of an electronic conductor or MIEC which provides the electric charge and the gas phase which provides the material, as is the case in SOFCs. In special cases material transport is blocked on purpose, as in the Hebb-Wagner polarization method for determining the partial electronic conductivity of a MIEC.[3,10,11] Then the electrode provides only electric charge. Electrodes are discussed in more detail in other chapters of this proceeding. For the present chapter it is sufficient to say that the electrodes provide electric charge, that a voltage can be measured with a voltmeter connected to the two electrodes and that each electrode (gas included, if necessary) provides material and fixes the chemical potential of one chemical component in the MIEC near the contact with the electrode. In particular we shall assume that the electrode fixes the chemical potential, μ, of the component, the

H.L. Tuller et al. (eds.), Oxygen Ion and Mixed Conductors and Their Technological Applications, 1–20.

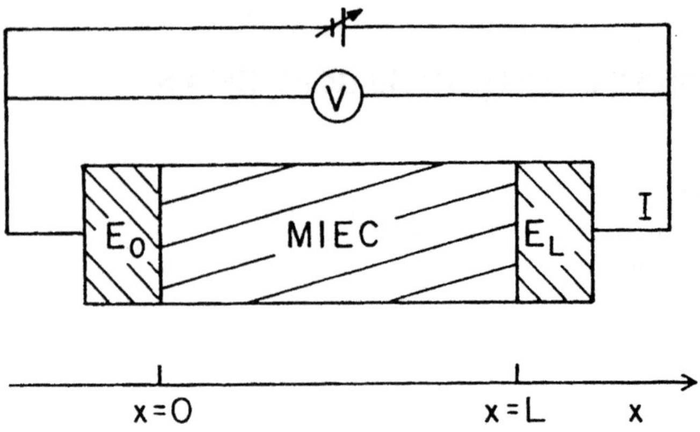

Figure 1. Schematic of a simple solid electrochemical cell. E_0, E_L - electrodes, MIEC - electrolyte.

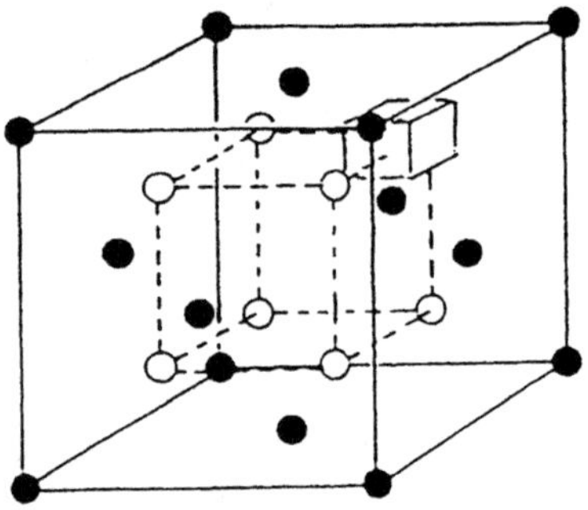

Figure 2. CaO stabilized ZrO_2 f.c.c. unit cell. ● - Zr or Ca ion, O - oxygen ion, small empty cube - oxygen vacancy.

ions of which are mobile. Thus for the SOFC it fixes the oxygen partial pressure, $\mu(O_2)$ (or equivalently, $P(O_2)$). In the cell $Ag|AgI|Ag_2S$, the electrode Ag fixes $\mu(Ag)$.

The electrodes are assumed in this chapter to be reversible. Non reversibility of electrodes is discussed in other chapters with special emphasis on oxygen cathode electrodes in SOFCs. Reversibility of an electrode has to be defined with respect to electronic charge transfer and with respect to material transfer. If the transfer of the corresponding particles requires only a negligible driving force, then the electrode is defined as reversible with respect to that transfer. Normally one is interested in electrodes reversible with respect to both electrical charge transfer and material transfer. This is the case in SOFCs. However, as mentioned before, for certain purposes, as in the Hebb-Wagner method, ion blocking electrodes are used. E.g. in the cell (-)Ag|AgI|C(+) the graphite electrode is blocking the silver ion transfer to the ion conductor AgI, on purpose, and one is looking for a reversible transfer of electrons only at the graphite electrode.

For electrodes reversible with respect to the electronic current, the voltage, V, measured on the electrodes equals the voltage across the MIEC. This voltage equals, up to a proportionality constant, the difference in the electrochemical potential, $\tilde{\mu}_e$, of the electrons in the electrodes and in the surface region of the MIEC, just under the electrodes,[3,12]

$$-qV = \Delta\tilde{\mu}_e \tag{1}$$

where q is the elementary charge.

For electrodes reversible with respect to material transport the chemical potential at the electrode equals that in the MIEC just under the electrode. This holds even under transfer of material across the electrode/MIEC interface, as the driving chemical potential difference needed is assumed to be negligible. Let the mobile specie be e.g. (in the Kröger-Vink notation [13]) $M_i^{z\bullet}$ and the chemical potential of the corresponding component M: μ_M. The electrodes (gas included, if necessary) fix the difference $\Delta\mu_M$ on the cell. It is convenient to express this difference in units of voltage rather than energy, by defining the parameter,

$$V_{th} = -\frac{\Delta\mu_M}{zq} \tag{2}$$

which is called the (theoretical) Nernst voltage.

In a battery or SOFC based on a SE, the open circuit voltage, denoted as the cell EMF, will be shown to equal the Nernst voltage. When the electronic current cannot be neglected this is not true anymore and $EMF<V_{th}$.

When the electrodes are not reversible, then one has to differentiate between $\Delta\mu_M$ and $\Delta\tilde{\mu}_e$ on the MIEC and $\Delta\mu_M$ and $\Delta\tilde{\mu}_e$ on the electrodes. The discussion below applies also to this case with $\Delta\mu_M$ and $\Delta\tilde{\mu}_e$ taken to be those on the MIEC.

3. Ionic Conductivity in Solids

Ionic conduction in solids is provided by the presence of point defects. A perfect crystal cannot conduct ions. Significant ionic conductivity, σ_i, can be found in solids. It can reach 1S/cm, the value observed in the best liquid electrolytes. In many solids high ionic conductivity is reached only at elevated temperatures. However, ionic conductors with high σ_i at room temperature exist for Ag^+ and Cu^+ ions.[3,14]

Ionic conduction prevails if:[15]

I) There is a vacant site for the ion to jump into.

II) The potential barrier for the jump of the ion is not too high.

III) The vacant sites form a continuous path.

Vacant sites that can accommodate ions are either vacancies or interstitial sites.[16] Vacancies can be introduced by three means:[17]

I) By change of stoichiometry. For example, reduction of CeO_2 to CeO_{2-x} introduces charged oxygen vacancies, $V_O^{\bullet\bullet}$ (and quasi free electrons, e').

II) By doping. An example is ZrO_2 doped substitutionally with CaO (stabilizing the CaF_2 f.c.c. structure of the oxide). This yields $V_O^{\bullet\bullet}$ (and Ca_{Zr}'') defects. This is shown schematically in Fig. 2.

III) By thermal excitation. An example is Ag_2S where Frenkel pairs[13] of silver are formed by the reaction,

$$Ag_{Ag}^{x} + V_i^{x} \Leftrightarrow Ag_i^{\bullet} + V_{Ag}' \tag{3}$$

A high density of point defects can be introduced by a sub-lattice quasi melting. This is the case in AgI at and above 149°C. As AgI is heated it reaches a phase transition at 149°C changing its lattice symmetry and forming a highly disordered silver sub-lattice with silver ions quasi flowing between the iodine ions.[18] In view of the high concentration of point defects it is difficult to speak of vacancies and interstitials in this case.

The quasi melting of the sub-lattice can also occur gradually without change in the symmetry of the other sub-lattice. This disordering of one sub-lattice occurs close, but definitely below the melting temperature of the solid. It is called a Faraday transition.[15,19,20]

There are solids that have empty sites that look like vacancies but are part of the ordered crystal and have formally to be considered as empty interstitial sites. An example is δ-Bi_2O_3. [21] In each unit cell there are four Bi^{3+} ions and six O^{2-} ions. The cations form an f.c.c. sub-lattice The anions occupy six out of eight near equivalent sites. This structure can be derived from the CaF_2 structure of YSZ shown in Fig. 2 by removing two of the eight oxygen ions, along the main diagonal, in the unit cell. The two unoccupied sites look like vacancies in the oxygen sub lattice but are strictly speaking interstitial sites of the crystal. This leaves plenty of empty sites for nearby oxygen to jump into and propagate via this route. Experiments show that this phase has a very high ionic conductivity of ~1S/cm at 800°C.[21]

The mobile ions need not be the only point defects. Electrons and holes can also be present due to:[17]

I) Deviation from stoichiometry.

II) Doping.

III) Thermal excitation.

However, in many of the ionic solids the electron energy band gap is large. It is then quite likely that donor and acceptor levels formed by deviation from stoichiometry, doping or thermal excitation are deep levels and electrons and holes cannot be excited into the corresponding conducting band. If charged, the ionic defects are then balanced by other ionic point defects. Examples are, for doping of ZrO_2: $CaO \leftrightarrow Ca_{Zr}^{\backslash\backslash} + V_O^{\bullet\bullet} + O_O^x$ and for thermal excitation: $Ag_{Ag}^x + V_i^x \leftrightarrow Ag_i^{\bullet} + V_{Ag}^{\backslash}$. These result in SEs. However, the mobility of ions is small relative to that of electrons and holes and it may happen that solids with an electron/hole concentration low with respect to the concentration of ions, are MIECs. An example is Ag_2S at T>177°C which is a MIEC with a high ion conductivity but an even higher and dominant electron conductivity, though the concentration of the electrons is more than two orders of magnitude lower than that of the Ag^+ ions.[3]

Solids can be found that have a more complex collection of point defects. In the most general case a MIEC can have more than one type of mobile ion, mobile ions with fixed and variable charge, immobile donors or acceptors with fixed and/or variable charge, electrons and holes. We shall limit our discussion from here on to three types of solids,

I)MIECs with a single type of mobile, positive ionic defects, e.g. $M_i^{z\bullet}$ or $V_O^{\bullet\bullet}$, a small concentration of electrons, $e^{\backslash}$, and charge compensating, immobile and fixed charge acceptors. As an example we consider Gd_2O_3 doped CeO_2 with $V_O^{\bullet\bullet}$ as the mobile ionic specie and $Gd_{Ce}^{\backslash}$ as the fixed charge, immobile acceptor. The concentration of electrons, n, (generated under reducing conditions) is small and that of holes, p, is negligible,

$$p << n << 2[V_O^{\bullet\bullet}] \approx [Gd_{Ce}^{\backslash}] \tag{4}$$

where [ ] denotes concentration.

II) MIECs with a single type of mobile, positive ionic defects, $M_i^{z\bullet}$ or $V_O^{\bullet\bullet}$, a small concentration of electrons under reducing conditions, a small concentration of holes under oxidizing conditions and charge compensating immobile fixed charge acceptor. An example is CaO doped ZrO_2 with $V_O^{\bullet\bullet}$ as the mobile ionic specie, $Ca_{Zr}^{\backslash\backslash}$ as the fixed charge, immobile acceptor, and,

$$p, n << 2[V_O^{\bullet\bullet}] \approx 2[Ca_{Zr}^{\backslash\backslash}] \tag{5}$$

III) MIECs with a single type of mobile, positive ionic defects, e.g. $M_i^{z\bullet}$ or $V_O^{\bullet\bullet}$, being charge compensated by quasi free electrons. An example is reduced CeO_{2-x} with $V_O^{\bullet\bullet}$ as the mobile ionic defect. The local neutrality equation is,

$$n = 2[V_O^{\bullet\bullet}] \tag{6}$$

4. Defect Distributions and I-V Relations in Cells Based on MIECs of the Three Types.

4.1 CURRENT DENSITY EQUATIONS AND I-V RELATIONS COMMON TO THE THREE MIEC TYPES

The I-V relations as well as the point defect distribution under current and applied chemical potential gradient, are calculated from the current density (j) equations of the mobile defects and the neutrality equation. For the ions $M_i^{z\bullet}$ with z=2,

$$j_i = -\frac{\sigma_i}{2q}\nabla\tilde{\mu}_i \tag{7}$$

for the electrons,

$$j_e = \frac{\sigma_e}{q}\nabla\tilde{\mu}_e \tag{8}$$

and for the holes,

$$j_h = -\frac{\sigma_h}{q}\nabla\tilde{\mu}_h \tag{9}$$

where each of the electrochemical potentials $\tilde{\mu}$ can be expressed in terms of a corresponding chemical potential, μ, and the internal (Galvani) electrical potential, φ,

$$\tilde{\mu} = \mu + zq\varphi \tag{10}$$

where zq is the charge of the defect. σ denotes conductivity,

$$\sigma = zqv[\,] \tag{11}$$

where v is the mobility and [ ] the concentration of the corresponding defect.

The creation annihilation reaction of electrons-hole pairs,

$$e' + h^\bullet \Leftrightarrow 0 \tag{12}$$

is assumed to be in equilibrium. Then,

$$\nabla\tilde{\mu}_e = -\nabla\tilde{\mu}_h \tag{13}$$

and Eq.(9) can be combined with Eq. (8) to yield,

$$j_{el} = j_e + j_h = \frac{\sigma_e + \sigma_h}{q}\nabla\tilde{\mu}_e \tag{14}$$

where j_{el} is the electronic (electron and hole) current density.

We shall now assume a linear cell in the x direction so that, $\nabla \rightarrow \partial / \partial x$. The length of the cell is L, i.e. 0<x<L.

In the steady state the ionic current, $I_i=Sj_i$ (S cross sectional area of the cell) is uniform as no material is introduced or removed from within the MIEC. Similarly the total electrical current, $I=I_i+I_{el}$, is uniform. From these two facts we conclude that the partial electronic current, $I_{el}=Sj_{el}$, is also uniform. Eqs. (7) and (14) can then be integrated to yield the general I-V relations for the three types of MIECs,[22,23]

$$I_i = \frac{V_{th} - V}{R_i} , \qquad R_i = \frac{1}{S}\int_0^L \frac{dx}{\sigma_i} \tag{15}$$

and,

$$I_{el} = -\frac{V}{R_{el}} , \qquad R_{el} = \frac{1}{S}\int_0^L \frac{dx}{\sigma_e + \sigma_h} \tag{16}$$

R_i and R_{el} are positive parameters that represent the resistance to ionic and electronic current respectively. These are not necessarily fixed parameters as will become apparent from the detailed discussion for the three MIEC types below. They may depend on the values of μ_M at the two electrodes as they govern the stoichiometry and defect concentration at the MIEC boundaries. R_i and R_{el} may also depend on the applied voltage.

An important conclusion of Eqs. (15) and (16) is that I_{el} vanishes when the applied voltage vanishes while I_i vanishes when the difference between the applied voltage and the applied chemical potential difference (expressed in terms of V_{th}), vanishes. Therefore I_i and I_{el} cannot vanish simultaneously, except when $V_{th}=0$. This result is the basis for a method for determining the ionic partial conductivity in a MIEC.[24] By shortcircuiting the cell one sets V=0. Then the current through the MIEC is purely ionic. Measuring the current through the cell with a low impedance ammeter and knowing the driving force V_{th} allows one to determine R_i from Eq. (15) with V=0. R_i may depends on V, therefore the method allows one to determine the value of R_i only for the specific condition, V=0.

The cell EMF is measured under open circuit conditions, i.e. for $I_{total}=I_i + I_{el} = 0$. Since $I_i > 0$ and $I_{el} < 0$ Eqs. (15) and (16) yield that EMF < V_{th}.[22,23]

4.2 DEFECT DISTRIBUTION FOR TYPE I MIECs

In type I (and II) MIECs, Eqs. (4) and (5), the concentration of mobile ions is large. For the materials of interest, e.g. YSZ, the concentration of ionic defects is so large that it can be assumed to be, approximately, uniform even under an oxygen potential gradient that prevails in a fuel cell (where $P(O_2)$ may vary over 20 orders of magnitude and $\Delta\mu(O_2)$ is large, of the order of 4eV). Thus

$$\nabla\mu_i = 0, \qquad \nabla\sigma_i = 0 \tag{17}$$

Combing Eqs. (7), (10) and (17) one finds,

$$j_i = -\sigma_i \nabla\varphi \tag{18}$$

σ_i is uniform as given in Eq. (17). j_i is uniform in the steady state. Then $\nabla\varphi$ is uniform as well. One can integrate Eq.(18) for the steady state. Comparing the result with Eq. (15) yields,

$$\nabla\varphi = -\frac{V_{th} - V}{L} \tag{19}$$

and,

$$R_i = \frac{L}{S\sigma_i} \tag{20}$$

The electron distribution is not uniform. Under steady state j_{el} is uniform and the electron distribution can be calculated by integrating Eq. (8). Given the low concentration of the electrons, one may express their chemical potential as,

$$\mu_e = \mu_e^0 + kT\ln\{n\} \tag{21}$$

where μ_e^0 is the standard value, k Boltzmann constant and T the temperature. The result of the integration is,[22,23]

$$n(x) = n_0\left[1 - \left(1 - \exp\{-\beta q V_{th}\}\right)\frac{1 - \exp\{-\beta q(V_{th} - V)x/L\}}{1 - \exp\{-\beta q(V_{th} - V)\}}\right] \tag{22}$$

where n_0 is the boundary value of n at x=0. The sign convention is that V and V_{th} are positive when the voltage at the right electrode is higher than at the left electrode. When $P(O_2)$ is lower at the left electrode (at x=0) then $n_0 > n_L$ and $V_{th} > 0$.

n(x) is shown in Fig. 3. The dependence on V is evident. The oxygen partial pressures at the electrodes fix n_0 and n_L but the distribution of electrons inside the MIEC, between these two boundary values, depends on V. This has an important implication for SOFCs. n(x) given here represents the electron distribution in cells based on doped CeO_2. It is seen that the MIEC is significantly reduced on the fuel (low $P(O_2)$) side. On the air side (high $P(O_2)$) n is small and the MIEC there is an insulator with respect to electronic current. Holes are assumed to be a minority throughout the whole cell, $p(x) \ll n(x)$, in this type of MIEC. The extent of reduction within the MIEC depends on the working conditions, i.e. on V/V_{th}. At maximum power, $V/V_{th} \sim 0.5$ and the degree of reduction is rather limited. A quantitative analysis for doped CeO_2 shows that indeed reduction is of little importance for $V/V_{th} < 0.6$.[23] On the other hand, for larger V, reduction cannot be neglected. In particular, under open circuit conditions, fuel is consumed while no power is generated. The dependence of n(x) on V/V_{th} was the basis[22,25] for the suggestion to consider doped CeO_2 for SOFCs though it exhibits significant electronic conduction under reducing conditions at $T \sim 800°C$.[26-28]

4.3 DEFECT DISTRIBUTION FOR TYPE II MIECs

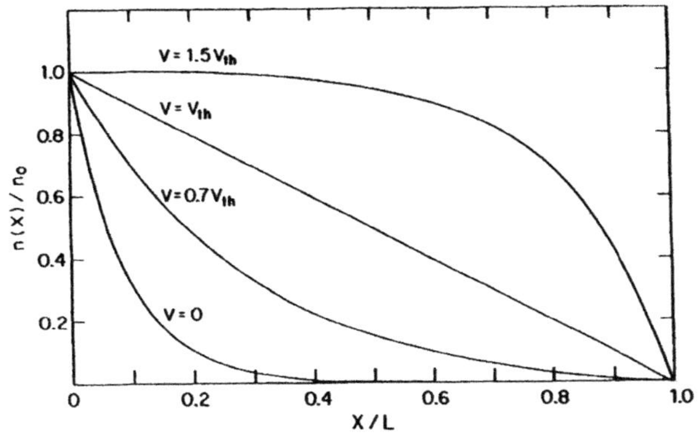

Figure 3. Electron distribution, n(x), in type I MIECs under current, for different values of V. [V_{th}=1Volt, T=1000K].

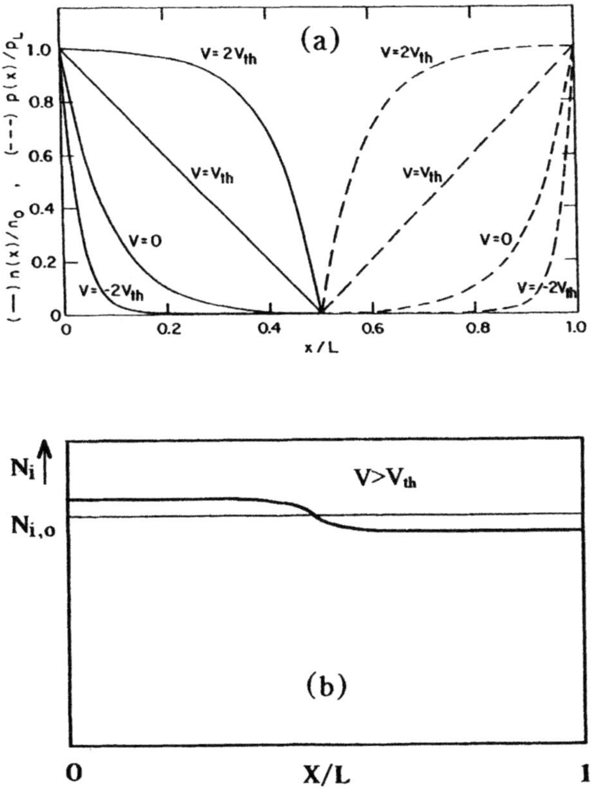

Figure 4. a - Distribution of electrons, n(x), and holes, p(x), in type II MIECs under current, for different values of V. [V_{th}=1Volt, T=1000K, n_0=p_L, ν_e=ν_h].
b - Schematic of the distribution of the mobile ionic defects in a type II MIEC under current.

Type II MIECs are similar to type I MIECs except that the electronic charge carriers are not dominated by either electrons or holes throughout the whole solid. Instead in one side of the MIEC, subject to, reducing conditions, n>>p while in the other side, subject to oxidizing conditions, p>>n. As a result Eqs. (17) - (21) hold also here. However the electronic current density equation that should be integrated is Eq. (14) which depends on the conductivity of both the electrons and holes rather than Eq. (8) which depends only on the conductivity of electrons. The integration of Eq. (14) under steady state condition i.e. for uniform j_{el} is given in Ref. [29]. Fig. 4a presents the calculated distribution n(x) and p(x) for a symmetric choice of parameters $\nu_e=\nu_h$ and $n_0=p_L$. Now an n region and a p region can be identified. The width of each region depends on the applied voltage. The dimension of the change between the n and p region can be made narrower or wider depending on the voltage used. n and p also represent the deviation between the concentration of mobile ions and the immobile ions as can be seen from Eq. (5). For the example of CSZ,

$$n + 2[Ca_{Zr}^{\prime\prime}] = 2[V_O^{\bullet\bullet}] + p \quad (23a)$$

and in the n region (n>>p),

$$n = 2[V_O^{\bullet\bullet}] - 2[Ca_{Zr}^{\prime\prime}] > 0 \quad (23b)$$

while in the p region (p>>n),

$$p = 2[Ca_{Zr}^{\prime\prime}] - 2[V_O^{\bullet\bullet}] > 0 \quad (23c)$$

The concentration of Ca impurities is fixed and uniform. The concentration of $V_O^{\bullet\bullet}$ is not strictly uniform as can be inferred from Fig. 4a and Eqs. (23b) and (23c). Eq. (17) holds only approximately. The distribution of the mobile ions is shown, schematically, in Fig. 4b.

For SOFCs applications considerable electronic conduction in the electrolyte is harmful. However one can infer from Fig. 4a, that for small V/V_{th}, the region with low n and p widens. Therefore the electronic leak current need not pose a problem for low V/V_{th} for type II MIECs, even if n and p at the MIEC boundaries are not negligible. This suggests that type II MIECs can be considered for SOFCs. Furthermore, the presence of electronic conductivity near the MIEC boundary may be of advantage by reducing the electrode impedance.

It is suggested that common pn junctions can be formed from type II MIECs. The n and p distribution (e.g. Fig. 4a) and the corresponding changes in the ionic defect (e.g. Fig. 4b) are generated at elevated T where ionic defects are mobile. One may consider quenching the profiles generated to obtain pn junction, the width of which are controlled by V, at elevated T. The interesting point is that while at elevated T the MIEC is neutral, after quenching local neutrality is expected to be lost and a normal pn junction to be formed[29-31]. One word of caution is however required. The pn junction will form only if after cooling the electrons and holes are not trapped and their concentrations do not become negligible. That it is possible to observe electronic conduction in a reduced and quenched material has been demonstrated for the n type MIEC, CeO_{2-x}[32]. In addition, as these MIECs are usually ionic compounds the electron and hole mobility might be rather low compared to common semiconductors.

4.4 DEFECT DISTRIBUTION FOR TYPE III MIECs

For type III MIECs with electrons and mobile ionic defects the neutrality equation (6) yields the relation between the defects concentrations. The current density equations to be solved are Eqs. (7) and (8). For low electron and ion concentrations the relations between the chemical potential and the concentration is logarithmic, as in Eq. (21), for both the electrons and the ions. Substituting Eq. (21) in Eqs. (7) and (8) and integrating yields the defect distribution, for z=1,[23]

$$[M_i^{\bullet}] = n(x) = n_0 + n_0\left(1 - \exp\{-\tfrac{1}{2}\beta q V_{th}\}\right)\frac{x}{L} \qquad (24)$$

This distribution is independent of V and depends only on the boundary values of μ_M via n_0 and n_L. The MIEC becomes a linearly graded n type semiconductor. Local neutrality is only approximately observed i.e., $n \sim [M_i^{\bullet}]$.[23,30] This type of MIEC cannot be used as the electrolyte in SOFCs. The reason is that the electronic conductivity is expected to be dominant throughout the whole MIEC as (cf. Eq. (11)) $n \sim 2[V_O^{\bullet\bullet}]$ but $\nu_e >> \nu_i$. It might be considered for electrodes of SOFCs and for the purpose of quenched pn junctions, though the flexibility of shaping the junction by V does not exist here.

4.5 I-V RELATIONS FOR CELLS BASED ON TYPE I MIECs AND REVERSIBLE ELECTRODES

The ionic current is given in Eq. (15). As the conductivity is approximately uniform (Eq.(17)) R_i is given by a constant as shown in Eq. (20),

$$I_i = \frac{V_{th} - V}{R_i}, \qquad R_i = \frac{L}{S\sigma_i} \qquad (25)$$

Eq. (25) shows that the ionic current I_i depends linearly on V and V_{th}.

The dependence of the electron current I_e on V and V_{th} is more complex. To calculate it, one uses Eqs. (10) and (21), differentiates n(x) of Eq. (22) at x=0 and insert, the result as well as Eq. (19) into Eq. (8). This yields,[22,23]

$$I_e = -\frac{V_{th} - V}{R_e^0}\,\frac{\exp\{-\beta q(V_{th} - V)\} - \exp\{-\beta q V_{th}\}}{1 - \exp\{-\beta q(V_{th} - V)\}}, \qquad R_e^0 = \frac{L}{Sq\nu_e n_0} \qquad (26)$$

I_e does not depend linearly on either V or V_{th}. I_e vanishes for V=0 in agreement with Eq.(16). A comparison with Eq. (16) shows that R_{el} is a function of V, V_{th} and n_0. The dependence on V_{th} and n_0 reflects a dependence on $P(O_2)$ at the cathode and the anode. The later two values of $P(O_2)$ fix the non-stoichiometry in the MIEC under the electrodes hence, n_0 and n_L. V_{th} is related to n_0 and n_L by,

$$V_{th} = \frac{1}{\beta q} \ln \left\{ \frac{n_0}{n_L} \right\} \tag{27}$$

The linear dependence of I_i on V and the non linear dependence of $-I_e$ on V are both shown in Fig. 5 for $0<V<V_{th}$. Under open circuit (oc) conditions, $I_i = - I_e$. As V decreases below V_{oc} (or EMF), $|I_e/I_i|$ decreases. Therefore it is possible to operate a cell based on such a MIEC with little loss due to electronic current leak, when V is small. This result is also the basis for determining the ionic partial conductivity in a MIEC, by shortcircuiting, as mentioned before.[25]

The asymptotic dependence of I_e on V is calculated from Eq. (26). For a large applied voltage in the "forwards direction" (same polarity as V_{th}), the MIEC is highly conductive as most of it is n type, as is shown in Fig. 3 and,

$$I_e \approx \frac{V_{th} - V}{R_e^0}, \qquad V>>V_{th} \tag{28a}$$

For a large reverse applied voltage, most of the sample is approximately intrinsic (cf. Fig.3) and therefore a poor electronic conductor and,

$$I_e \approx \frac{V_{th} - V}{R_e^L}, \qquad R_e^L = \frac{L}{Sqv_e n_L}, \qquad -V>>V_{th} \tag{28b}$$

In both Eqs. (28a) and (28b), I_e is linear in V and proportional to the difference V_{th} - V. This, however, holds only for large values of |V|.

4.6 I-V RELATIONS FOR CELLS BASED ON TYPE II MIECs AND REVERSIBLE ELECTRODES

The I_i - V_{th}, V relations are the same as for cells based on type I MIECs as σ_i is uniform in both cases (Eq. (17)) and Eq. (25) holds also here. The electronic current depends in a complex way on V_{th}, n_0 and V and as given in Ref. (29). We quote here only the asymptotic relations for large $|V/V_{th}|$ and for a symmetrical case, n_L=p_0 and $v_e = v_h$. For a large applied voltage in the forward direction, the MIEC is highly conductive as most of it is either n or p type, as is shown in Fig. 4a and,

$$I_{el} \approx \frac{V_{th} - V}{R_{el}^0}, \qquad R_{el}^0 = \frac{L}{Sqv_e n_0}, \qquad V>>V_{th} \tag{29a}$$

For a large reverse applied voltage most of the sample is approximately intrinsic (cf. Fig. 4a) and therefore a poor electronic conductor and,

$$I_{el} \approx \frac{V_{th} - V}{R_{el}^{mid}}, \qquad R_{el}^{mid} = \frac{L}{Sqv_e 2n_{int}}, \qquad -V>>V_{th} \tag{29b}$$

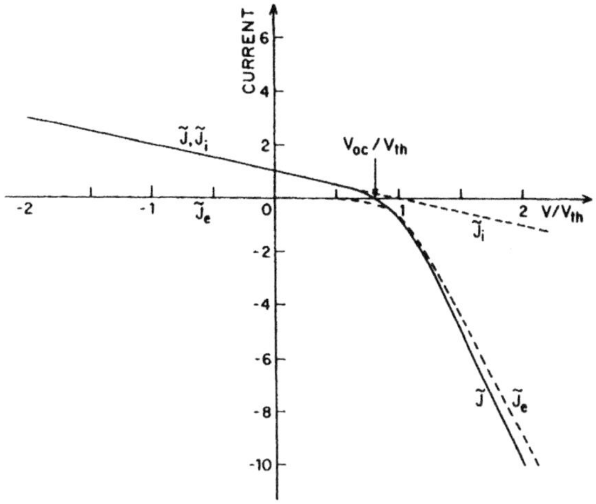

Figure 5. $\tilde{J}_e$ - Electronic current, I_{el}, in relative units, vs. V, $\tilde{J}_i$ - ionic current, I_i, in relative units, vs. V, $\tilde{J} = \tilde{J}_e + \tilde{J}_i$, in electrochemical cells based on type I MIECs having two reversible electrodes. [V_{th}=1Volt, T=1000K].

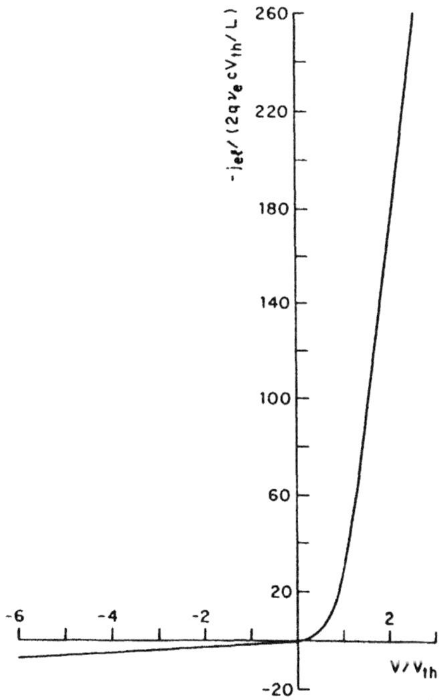

Figure 6. j_{el} - Electronic current, I_{el}, vs. V for electrochemical cells based on type II MIECs having two reversible electrodes. [V_{th}=1Volt, T=1000K, n_0=p_L, ν_e=ν_h].

where n_{int} is the concentration of electrons (and holes) in the intrinsic region. In both Eqs.(29a) and (29b) I_{el} is linear in V and proportional to the difference V_{th} - V. This is shown in Fig. 6. This however holds only for large values of |V|. As V vanishes I is proportional to V and I vanishes at V = 0, as required by Eq. (16).
The I-V relations are asymmetrical. However they are not the asymmetrical I-V relations of a common pn junction. The later exhibits exponential asymptotic I-V relations not linear ones. The reason for the difference is that the local neutrality can be maintained here by the motion of the ions together with the electrons and holes. This situation changes on quenching and a normal pn junction may be recovered under favorable conditions as mentioned before.

4.7 I-V RELATIONS FOR CELLS BASED ON TYPE III MIECs AND REVERSIBLE ELECTRODES

The I-V relations for type III MIECs are linear[23]. Inserting $n = [M_i^{\bullet}]$ of Eq. (24), (with z = 1, for simplicity) into Eqs. (7) and (8) allows one to calculate $\partial\varphi / \partial x$ and then the I-V relations. The result is,

$$I_e = -\frac{2kT}{qR_e^0}\left(1-\exp\{-\tfrac{1}{2}\beta qV_{th}\}\right)\frac{V}{V_{th}}, \qquad R_e^0 = \frac{L}{Sqv_e n_0} \tag{30a}$$

and

$$I_i = \frac{2kT}{qR_i^0}\left(1-\exp\{-\tfrac{1}{2}\beta qV_{th}\}\right)\left(1-\frac{V}{V_{th}}\right), \quad R_i^0 = \frac{L}{Sqv_i n_0} \tag{30b}$$

I_e is linear in V and I_i is linear in $(1-V/V_{th})$ while V_{th} is fixed by the reversible electrodes.

4.8 I-V RELATIONS FOR CELLS BASED ON TYPE I, II, and III MIECs UNDER ION BLOCKING CONDITIONS.

The I_e-V relations for MIEC type I and II are linear for large values of V ($|V| >> V_{th}$). For MIEC type III they are linear for all values of V. This, however, is true under the boundary conditions that the chemical potential at the boundaries of the MIEC are fixed. This means that n_0 and n_L (as well as V_{th}, p_0, and p_L) are fixed.

When one or both electrodes are perfectly blocking the transfer of ions, the ionic current vanishes, in the steady state. This holds for the cells based on either of the three types of MIECs. Then from Eq. (15),

$$V = V_{th}, \; I_i = 0 \tag{31}$$

i.e. the chemical potential at the MIEC boundary or boundaries changes until V_{th} reaches the value of the applied voltage, V. This is possible because at the ion blocking electrode the MIEC does not communicate with the outside material reservoir via material exchange. Therefore the local chemical potential in the MIEC, under the blocking electrode, is not fixed by that reservoir. Instead it is fixed internally by the applied voltage. To determine the value of the chemical potential two cases have to be considered.

I) One electrode say at x=L is the ion blocking one and the second electrode at x=0 is reversible with respect to ion transport (and electron transport). The chemical potential under the ion blocking electrode, $\mu(M)_L$, is fixed by the chemical potential at the reversible electrode, $\mu(M)_0$, and the applied voltage. To see this we use Eqs. (2) and (31),

$$\mu(M)_L = \mu(M)_0 - zqV \tag{32}$$

Inserting Eq. (31) into Eq. (26) yields, for type I MIECs,

$$I_e = -\frac{kT}{qR_e^0}\left(1 - \exp\{-\beta qV\}\right), \qquad I_i=0 \tag{33}$$

Under these conditions the I_e - V relations are exponential as n_0 and R_e^0 are constants fixed by the reversible electrode at x=0. These exponential relations are formally equal to those applicable to pn junction[33]. Yet one has to bear in mind that the situation here is different as local neutrality was assumed to hold. These relations do not originate from a change in a electrical potential barrier that is due to a space charge but originate from an exponential dependence of the electron concentration n(x) on voltage.

For type II MIECs the I_e - V relations are,[11,23]

$$I_{el} = -\frac{kT}{qL}\left(\sigma_e^0(1 - \exp\{-\beta qV) + \sigma_h^0(\exp\{\beta qV - 1\}\right), \qquad I_i=0 \tag{34}$$

This double exponential dependence on V is not comparable to the I - V relations of a common pn junction.

For type III MIECs using Eq. (31) in Eq. (30a) yields,

$$I_e = -\frac{2kT}{qR_e^0}\left(1 - \exp\{-\tfrac{1}{2}\beta qV\}\right), \qquad I_i=0 \tag{35}$$

The exponential dependence on V in Eq. (35) has a different coefficient, by a factor 0.5, as compared to Eq. (33).

II) When both electrodes are blocking ion transport, then the total amount of ions in the MIEC is constant. This is in contradiction to the case when an electrode is reversible and material can be exchanged with the surroundings allowing for small changes in stoichiometry and in the total amount of mobile ions. For two ion blocking electrodes,

$$\int_0^L [M_i^{z\bullet}]dx = \text{Const.} \tag{36}$$

Under the applied voltage, the concentrations of the mobile ions and electrons (and holes) in the MIEC under both electrodes do not stay constant as there is no reversible electrode that can impose a constant composition. The boundary values depend on V in different forms for the different MIECs types.

For type I MIECs, the neutrality equation is well represented by the example of Eq.(23b) (or (23c)), $n = 2[V_O^{\bullet\bullet}] - 2[Ca_{Zr}^{\backslash\backslash}] > 0$. The distribution of the acceptors is fixed. Then in view of Eq. (36),

$$\frac{1}{L}\int_0^L n(x)dx = C_0 \tag{37}$$

where C_0 is a constant, the average electron concentration, which is fixed by the preparation procedure of the MIEC, i.e. depends on its history. n(x) is given by Eq. (22) with the constraint of Eq. (31),

$$n(x) = n_0 - n_0(1 - \exp\{-\beta qV\})\frac{x}{L} \tag{38}$$

The, now unknown parameter n_0 is determined by integrating n(x) of Eq. (38) according to Eq. (37),

$$n_0 = 2C_0 \frac{1}{1 + \exp\{-\beta qV\}} \tag{39}$$

which yields the dependence of n_0 on V. Inserting n_0 into Eq. (33) (via R_e^0 defined in Eq. (26)) yields,

$$I_e = -\frac{2C_0 kTS\nu_e}{L}\tanh\{\tfrac{1}{2}\beta qV\}, \qquad I_i=0 \tag{40}$$

which exhibits an hyperbolic-tangent dependence on $\beta qV/2$.

Eq. (37) holds for type III MIECs as well. n(x) is linear in x (Eq. (24)) (see also ref. 34). Integration yields n_0 in terms of C_0. Inserting this into Eq. (35) yields,

$$I_e = -\frac{4C_0 kTS\nu_e}{L^2}\tanh\{\tfrac{1}{4}\beta qV\}, \qquad I_i=0 \tag{41}$$

which also exhibits an hyperbolic-tangent dependence on $\beta qV/4$.

5. Local neutrality

The calculations discussed above assume that local neutrality prevails. The calculations do not require strict local neutrality but only an approximate one in the sense demonstrated in the case of CaO doped CeO_{2-x} by,

$$\frac{\left|2[C_{Ce}^{\backslash\backslash}] + n - 2[V_O^{\bullet\bullet}]\right|}{2[V_O^{\bullet\bullet}]} << 1 \tag{42}$$

The condition of quasi local neutrality holds as long as the concentration of mobile ions is large enough to fix a short screening length.[30]

$$\lambda_D << \lambda_G, \qquad \lambda_D = \sqrt{\frac{\varepsilon kT}{q^2 N}} \tag{43}$$

where λ_D is a screening length determined by the majority mobile charged defect concentration, ε is the dielectric constant, N a typical concentration of the majority mobile charged defects and λ_G is shown in Fig. 7 to represent the range over which the change in the defect distribution takes place. When Eq. (43) is fulfilled then gradients in defect distribution can be screened by moving compensating charges to avoid the generation of significant space charge.

One immediate conclusion from Eq. (43) is that at the interface of two different phases, a space charge must, in general, exist. The reason is that changes in charged defect concentrations are expected to be abrupt, discontinuous, at the interface, with $\lambda_G = 0$. Then the relation $\lambda_D << \lambda_G$ cannot be fulfilled, i.e. no screening can be obtained from the mobile charges that can eliminate the space charge.

6. Power output and efficiency of a SOFC based on a MIEC

The power output of a SOFC based on a MIEC has been argued to be reduced significantly for high values of V. Calculations done for SOFCs based on 10 mol% Gd_2O_3 doped CeO_2 , at 1000K[22,25] are presented in Fig. 8. The power exhibit a maximum and vanishes at the open circuit voltage as well as at V=0. For $V<0.6V_{th}$ the power is very close (with 2% or less) of the power one would obtain if the electronic conductivity of the MIEC would be negligible. The efficiency shown in Fig. 9 reveals also that the deviation form the efficiency due to the electronic leak is insignificant for $V<0.6V_{th}$.

7. Summary

The analyse of the I-V relations as well as the point defect distribution are done by evaluating the current density equations of the mobile species. We have considered in particular three types of MIECs which represents a large group of MIECs. They contain quasi free electrons and/or holes and one type of mobile ionic defect with fixed charge. In two types of MIECs immobile defects with fixed charge were also allowed. The defect distribution and not only the I - V relations, can be calculated from the current density equations because the latter depend on the defect concentrations.

The solution of the current density equations to obtain both the I - V relations and the defect distributions, depend on the boundaries conditions. For reversible electrodes the concentration of electrons and holes at the boundary are fixed by the reaction of the MIEC with the electrode material (gas included, if necessary). When an electrode blocks the material transfer, then the concentration at the MIEC boundary is fixed by the applied voltage. One has to distinguish between two cases. First, one electrode is reversible and the second blocks ions. Then the MIEC composition under the blocking

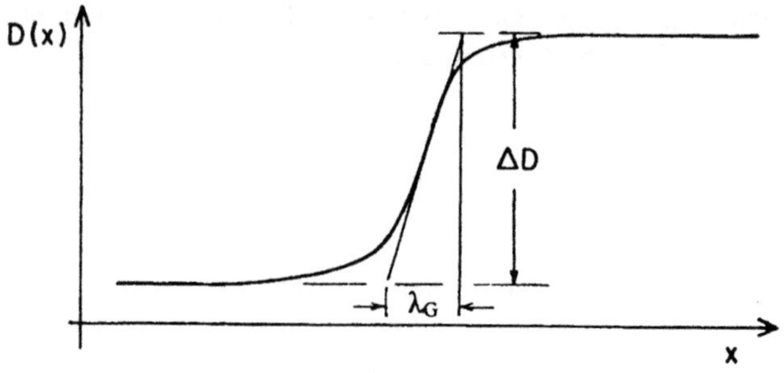

Figure 7. Graphic definition of λ_G as the range over which the non uniform distribution of charged defects, D(x), varies.

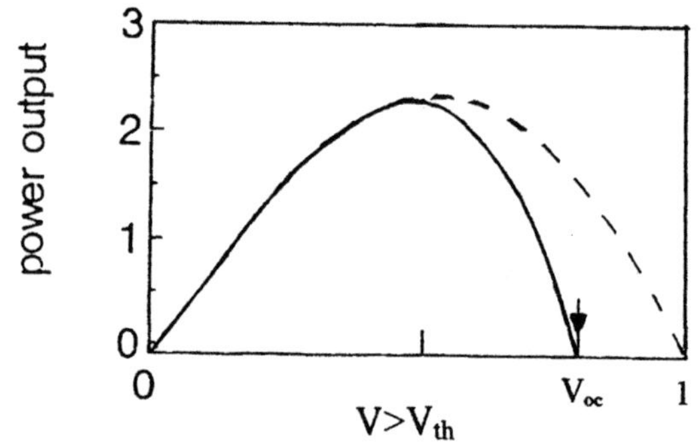

Figure 8. Calculated power output of a SOFC (relative units). ——— calculated for SOFC based on the MIEC CeO_2 doped with 10 mol% Gd_2O_3, having two reversible electrodes. [V_{th}=1Volt, T=1000K]. - - - - - calculated for a SOFC based on a SE.

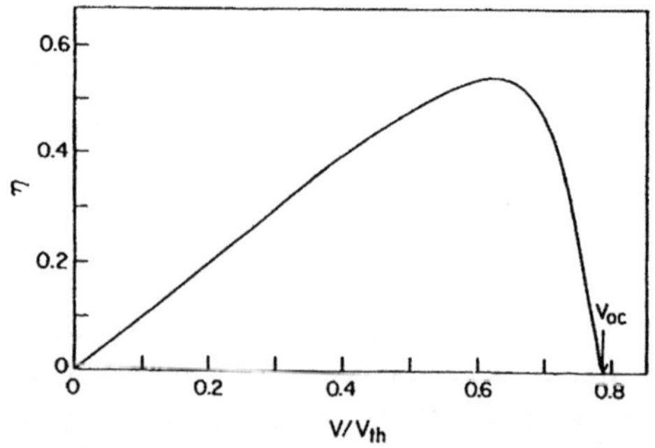

Figure 9. Calculated efficiency with respect to changes in the Gibbs free energy, $\eta(\Delta G)$, for the cell referred to in Fig. 8.

electrode is determined by the reaction of the MIEC with the other reversible electrode (gas included, if necessary) and the applied voltage. In the second case both electrodes block material transfer. Then the defect distributions in the MIEC are determined by the history of the sample and the applied voltage.

The solution of the current density equations assumed that the steady state condition prevails. Under this condition the total electric current is uniform and so is the material flux through the MIEC. For the three MIEC types considered, this yields that the ionic current, I_i, and the total electronic current, I_{el}, are uniform.

The analysis was done under a further assumption, of quasi local neutrality. This can be justified for high enough concentrations of mobile ionic defects.

The mobility of ionic defects allows local changes in stoichiometry. For a non ion blocking electrode, the mobility allows also material exchange with the electrode and thus overall stoichiometry changes. The excess or missing ionic defects can be considered as mobile donors or acceptors. These donors and acceptors may be present in addition to immobile ones introduced by doping.

Acknowledgment

This research was supported by the Basic Research Foundation Administered by the Israel Academy of Sciences and Humanities.

References:

1. Julien, C. and Nazri, A.-G. (1994) Solid state batteries: materials design and optimization, Kluwer Acad. Publ. Comp.
2. Minh, N.Q. and Takahashi T. (1995) Science and technology of ceramic fuel cells, Elsevier.
3. Rickert, H. (1982) Electrochemistry of solids, Springer Verlag.
4. Goto, K.S. (1988) Solid state electrochemistry and its application to sensors and electronic devices, Elsevier.
5. Porat, O. and Riess, I. (1994) Double electrochemical cell for controlling oxygen concentrations in oxides, *J. Electrochem. Soc.* **14**, 1533-1538.
6. Riess, I., Porat, O. and Tuller, H.L. (1993) Investigation of the dominant point defects in tetragonal $YBa_2Cu_3O_x$ at elevated temperatures, *J. Supercond.* **6**, 313-316.
7. Porat, O. and Riess, I. (1994) Defect chemistry of $Cu_{2-y}O$ at elevated temperatures, part I: non-stoichiometry, phase width ad dominant point defects, *Solid State Ionics* **74**, 229-238.
8. Porat, O. and Riess, I. (1994) Defect chemistry of $Cu_{2-y}O$ at elevated temperatures, part II: electrical conductivity, thermoelectric power and charged point defects, *Solid State Ionics* **81**, 29-41.
9. Porat, O. and Riess, I. (1994) Identification of CuO as a minority phase in YBCO: x-ray diffraction vs. EMF and coulometric titration measurements, *Mater. Sci. & Eng.* **B22**, 310-312.
10. Hebb, M.H. (1957) Electrical conductivity of silver sulfide, *J. Chem. Phys.* **20**, 185-190.
11. Wagner, C. (1957) Galvanic cells with solid electrolytes involving ionic and electronic conduction, *Proc. Int. Comm. Electrochem. Thermodyn. Kinetics (CITCE)* 7, 361-377. For limitation of the method see: Riess, I. (1996) *Solid State Ionics* **91**, 221-232.
12. Riess, I. (1997) What does a voltmeter measure?, *Solid State Ionics* **95**, 327-328.
13. Kröger, F.A. (1974) The chemistry of imperfect crystals, North Holland Publ. Comp., Vol. II.
14. Kudo, T. (1997) Survey of types of solid electrolytes, in: P.J. Gelling and H.J.M. Bouwmeester (eds.) *CRC handbook of solid state electrochemistry*, CRC Press, Chap. 6, pp. 195-221.
15. Riess, I. (1989) Crystalline anionic fast conductors, in: H.L. Tuller and M. Balkanski (eds.) *Science and technology of fast ion conductors*, Plenum Press, pp. 23-50.
16. Manning, J.R. (1968) Diffusion kinetics for atoms in crystals, D. van Nostrand, pp. 2-9.
17. Riess, I. and Tannhauser, D.S. (1989) Mixed ionic electronic conductors, in: T. Takahashi (ed.) *High conductivity solid ionic conductors, recent trends and applications*, World Scientific, pp. 478-512.

18. Hayes, W. and Hutchings, M.T. (1989) Ionic disorder in crystals at high temperatures with emphasis on fluorites, in: A.M. Stoneham (ed.) *Ionic solids at high temperatures,* World Scientific, Chap. 4 pp 247-362.
19. Sato, H. (1977) Some theoretical aspects of solid electrolytes, in: S. Geller (ed.) *Solid electrolytes,* Springer Verlag, Chap. 2 pp. 3-6.
20. Schröter, W. and Nölting, J. (1980) Specific heats of crystals with the fluorite structure, *J. de Physique* **41**, C6-20.
21. Shuk, P., Wiemhöfer, H.-D., Guth, U., Göpel, W. and Greenblatt, M. (1996) Oxide ion conducting solid electrolytes based on Bi_2O_3, *Solid State Ionics* **89**, 179-196.
22. Riess, I. (1981) Theoretical treatment of the transport equations for electrons and ions in a mixed conductor, *J. Electrochem. Soc.* **128**, 2077-2081. [Please note the following typographical errors: 1) ν_i is missing in the right hand side of Eq. (14), 2) replace (-) by (+) in the denominator of Eq. (28), 3) sixth line beyond Eq. (33), change the inequality sign to read $r \geq 0.9$, 4) V is missing in Eq. (40) that should read, $P = (S\sigma_i/L)V(V_{th}-V)r = \ldots$].
23. Riess, I. (1986) Current-voltage relations and charge distribution in mixed ionic electronic solid conductors, *J. Phys. Chem. Solids* **47**, 129-138.
24. Riess, I. (1991) Measurement of electronic and ionic partial conductivities in mixed conductors, without the use of blocking electrodes, *Solid State Ionics* **44**, 207-214.
25. Riess, I. (1992) The possible use of mixed ionic electronic conductors instead of electrolytes in fuel cells, *Solid State Ionics* **52**, 127-134.
26. Choudhury, N.S. and Patterson, J.W. (1971) Performance characteristics of solid electrolytes under steady state conditions, *J. Electrochem. Soc.* **118**,1398-1403.
27. Ross, Jr., P.N. and Benjamin, T.G. (1977) Thermal efficiency of solid electrolyte fuel cells with mixed conduction, *J. Power Sources* **1**, 311-321.
28. Tannhauser, D.S. (1978) The theoretical energy conversion efficiency of a high temperature fuel cell based on a mixed conductor, *J. Electrochem. Soc.* **125**, 1277-1282.
29. Riess, I. (1987) Voltage-controlled structures of certain p-n and p-i-n junctions, *Phys. Rev. B* **35**, 5740-5743.
30. Riess, I. (1994) Conditions for neglecting the space charge effects on distribution of point defects and I-V relations, *Solid State Ionics* **69**, 43-52.
31. Riess, I. (1995) pn junctions in mixed ionic-electronic conductors induced by chemical and electrical potential gradients, *Solid State Ionics* **75**, 59-66.
32. Ban, Y. and Nowick, A.S. (1972) Defects and mass transport in reduced CeO_2 single crystals, *NBS special publ.* **No. 364**, 353-365.
33. Ashcroft, N.W. and Mermin, N.D. (1976) Solid state physics, Holt-Saunders Int. Ed., pp. 592-600.
34. Allen, L.H. and Buhks, E. (1984) Copper electromigration in polycrystalline copper sulfide, *J. Appl. Phys.* **56**, 327-335.

ELECTRODE PROCESSES IN SOLID OXIDE FUEL CELLS

I. RIESS
Physics Department, Technion-IIT
Haifa 32000, Israel

1. Introduction

The electrodes in SOFCs, as usually in electrochemical cells, act as source or sink of electrical charge and allow for material to be transferred between the gas phase and the solid electrolyte. The gases of interest are oxygen coming from ambient air, fuel, in particular hydrogen (H_2) and methane (CH_4) and the corresponding exhaust gases, water vapor (H_2O) and carbon di-oxide (CO_2).

A fuel cell is shown schematically in Fig. 1. The electrolyte is an oxide conducting oxygen ions. The most common oxide solid electrolytes (SEs) are Y_2O_3 stabilized ZrO_2 and doped ceria (e.g. Sm_2O_3 doped CeO_2 (SCO) and Gd_2O_3 doped CeO_2 (GCO)). The cathodes are now-a-days porous layers of electronically conducting oxides such as $La_{0.84}Sr_{0.16}MnO_3$ (LSM) applied on YSZ or $La_{1-x}Sr_xCoO_3$ (LSC) applied on the ceria based electrolytes SCO and GCO. On top of these oxides a current collector is applied which can be of the same material but being thicker, to facilitate a low impedance path for the electrons, and exhibits much larger pores to facilitate a low impedance path for the oxygen gas. Different materials are applied to YSZ and SCO or GCO electrolytes because of the necessary compatibility of the electrode material with the electrolyte, concerning thermal expansion and chemical stability.

There is another difference between LSC and LSM which is of key importance: LSC at ~800°C conducts both electronic charge carriers and oxygen ions and is therefore a mixed ionic electronic conductor (MIEC) while the ionic conductivity of LSM is rather small. Mixed conductivity is important as it allows the whole electrode surface to be active. The reason is that oxygen as well as electrons or holes can diffuse through the MIEC. The oxygen can then enter the SE all over the common interface. In only electronically conducting electrode materials, such as LSM, or inert metals, such as Pt, oxygen cannot diffuse through the electrode material. It can diffuse only in the pores. In the latter case the effective area for oxygen ions incorporation into the SE is therefore considerably limited, and might require additional diffusion along interfaces, increasing the electrode impedance. Silver, in contradiction to Pt, allows for oxygen diffusion through the metal and can serve instead of a MIEC cathode.

H.L. Tuller et al. (eds.), Oxygen Ion and Mixed Conductors and Their Technological Applications, 21–56.

For anodes no suitable MIEC material has been found for zirconia and ceria based SOFCs. One can, however, use a quasi mixed conducting anode by preparing a cermet made of a metal and the electrolyte material. For that purpose a mixture of fine powder of app. 50%, in volume, of Ni and the electrolyte are normally used.

The performance of SOFCs, is characterized by the efficiency and power output per unit area. (Of cause life time, price and other parameters are also of interest). The power and current outputs have been low in the past and have been improved dramatically in recent years. These parameters are limited by the resistance of the cell, contributed by the SE bulk and the electrode impedance. These impedances have been reduced by using thin electrolytes and more suitable electrode materials and better electrode morphology. The efficiency at maximum power, on the other hand, has always been rather high, not too far from the theoretical one. This could be reached because the efficiency is not reduced by a large bulk resistance of the SE, only by the electrode impedance. It increases as the ratio between bulk to electrode impedance increases.

A recent report[1] describes a cell composed of Ni+YSZ/YSZ/LSM which yields, at 800°C, an exceptionally high power density of $2W/cm^2$ at app. 4.5 A/cm^2 with an efficiency of app. 40% as compared to the theoretical efficiency of 50%. This reflects a low impedance of the solid electrolyte (SE) YSZ as well as a low electrode impedance.

It is usually observed [see e.g. Ref. 2] that the current (I) vs. electrode over potential (η) relations have, for sufficiently high η, either an exponential dependence, (see curve CT in Fig. 2), or exhibit a limiting current (see curve D in Fig. 2). In other cases an exponential dependence is observed at comparatively low currents which then turns into a current limiting dependence, as shown by curve CT+D in Fig. 2. This is significant as it suggests that probably a single elementary step (or a group of consecutive elementary steps) that yields an exponential relation, or a single elementary step that yields a current limitation, dominate these relations, at least for a given current. If this is so it simplifies considerably the derivation of the electrode characteristics from the measured I-η relations.

It has been demonstrated that a SOFC can also be operated by applying a mixture of the fuel and oxygen to both electrodes.[3] This forms, so-called, mixed gas fuel cells (MGFCs). We emphasize that the whole cell is exposed to the same mixture of fuel and oxygen prepared outside the SOFC (with the same mixture being introduced into the cathode as well as the anode side). It is observed that voltage and current are generated in a particular direction. The symmetry is broken by the use of different electrode materials with different catalytic properties. Thus when discussing electrode properties for MGFCs one has to consider also the selectivity of the electrodes for the oxidation of the fuel and reduction of oxygen.[4] This topic will not be discussed further here.

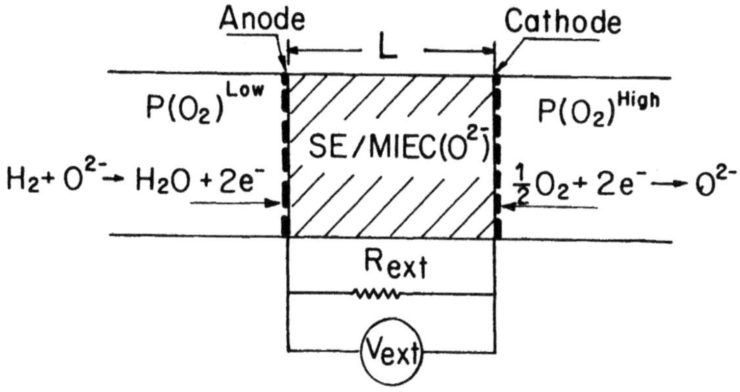

Fig. 1 Schematics of a SOFC.

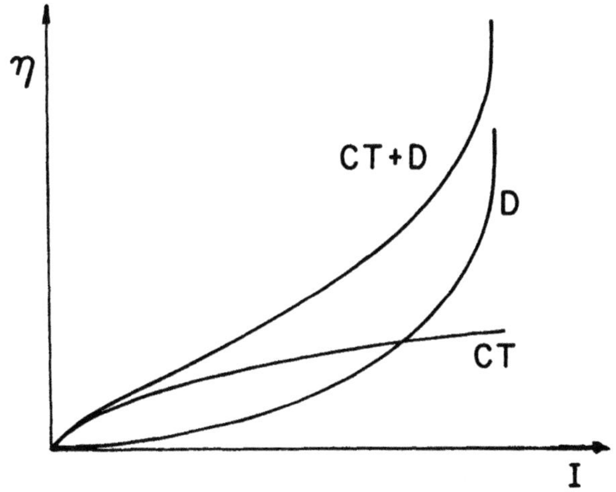

Fig. 2 I - η relations for the cathode. a - charge transfer type, b - diffusion limited type, c - combination of a and b.

2. Elementary Steps in the Electrode Reaction

2.1 CATHODE

The possible elementary steps of the cathode reaction are presented schematically in Fig. 3.[5] The series of the dominant elementary steps occurring in a cathode reaction depends on the properties of the electrode material. Decisive questions are:

1) Is it a MIEC or a pure electronic conductor?
2) Does it accelerate oxygen adsorption?
3) Does it enhance charge transfer rate of one or two electrons to oxygen adsorbed on it?
4) What is the rate of diffusion of adsorbed oxygen or oxygen ions along the electrode material?
5) For MIECs: What is the rate of transfer of oxygen ions from the surface into the MIEC, the rate of diffusion of oxygen ions via the MIEC and the rate of transfer of oxygen ions from the MIEC into the SE?
6) What is the reaction rate of oxygen reduced and introduced directly into the SE (with the electrode supplying only electrons to the oxygen via the SE)? This rate can be fast when the SE is not a pure ionic conductor but a MIEC. In this case the direct oxygen reduction on the SE may compete with that on the electrode material. The SE can be made a MIEC at the surface by ion implantation[6] or chemical diffusion of suitable dopants.[7] Ceria based SEs turn into MIECs near the anode side under the reducing conditions there.
7) What is the role of the pore? The gas phase in the pores is part of the electrode and the diffusion rate through the gas in the pores is of importance.

Fig. 3 is aimed at demonstrating the possible electrode processes at the cathode. It does not exhibit the realistic complex electrode morphology. The latter is shown schematically in Fig. 4 with larger grain and pores on the outer side becoming smaller towards the SE. This structure enables reduction in the impedance of molecular oxygen diffusion in the pores. The processes in Fig.3 are:

I) Oxygen molecules diffuse all the way through the pores to the SE, are reduced there and enter the SE. The last process takes place at the so-called three phase boundary (tpb) region where oxygen from the gas phase, the solid electrolyte and the electrode are close together. Electrons are supplied from the metal electrode via the SE. Due to the high resistance of the SE to electronic current the distance of the reaction zone from the SE-metal-gas three phase boundary (tpb), is small. This means that the active reaction area is small and the electrode impedance is expected, therefore, to be high. This mechanism has been suggested for Pt and Au electrodes on Er_2O_3 doped Bi_2O_3 [8] and for Au electrodes on Sm_2O_3 doped CeO_2 [2].

I') Oxygen is adsorbed on the metal. The molecule dissociates and diffuses as an adatom on the metal to the SE/metal interface. There, electrons are transferred to the adatoms and the oxygen ions enter the SE. This mechanism has been suggested for Ag

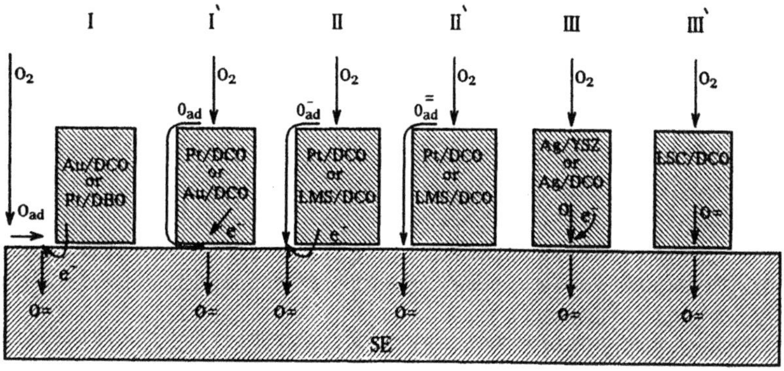

Fig. 3 Possible processes at the cathode of a SOFC.

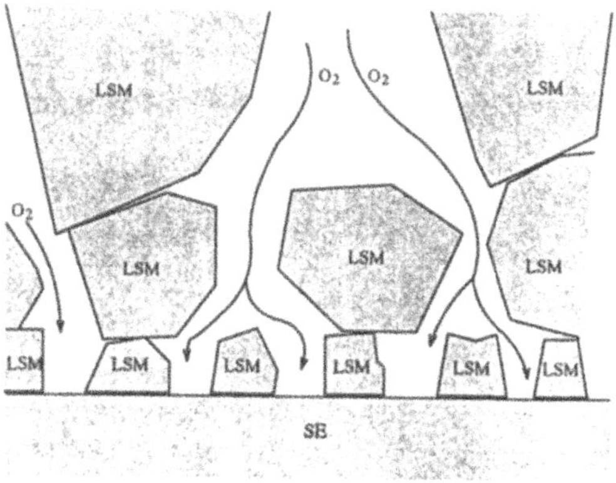

Fig. 4 The porous non uniform morphology of the cathode.

electrodes on CaO and Y_2O_3 doped CeO_2 [9] and for Pt electrodes on Gd_2O_3 doped CeO_2.[10]

II) Oxygen is adsorbed on the metal. The molecule dissociates into two adatoms. One electron is transferred to each adatom to form singly ionized oxygen ions. The ions diffuse to the SE where they obtain a second electron before they enter the SE. This mechanism has been suggested for Pt[2,11,12] and LSM[2] electrodes on doped CeO_2.

II') Oxygen is adsorbed and dissociated on the metal. Two electrons are transferred to form doubly ionized adsorbed oxygen ions. The ions diffuse along the metal to the SE and enter the SE. This mechanism has been suggested as an alternative one for Pt[11,12] and LSM [2] electrodes on doped CeO_2.

III) This mechanism can be found only in those metal electrodes that allow diffusion of oxygen through them. Oxygen is adsorbed, enters the metal and diffuses through it as an ion (accompanied by an electron/hole current) or as a neutral species. At the interface with the SE an oxygen ion, O^{2-}, is transferred from the metal to the SE. This mechanism was suggested for Ag electrodes on YSZ[13] and on doped Ceria.[2]

III') This mechanism can, in non-metallic electrodes, be found for MIECs. Oxygen is adsorbed and enters the MIEC as an ion. The ion diffuses through the MIEC and enters the SE. This mechanism was suggested for LSC electrodes on doped ceria[2] and for LSM electrodes on YSZ at T=960°C where LSM becomes a MIEC.[14]

It happens that different interpretations for seemingly similar electrode/SE pairs are suggested by different authors. This is, probably, due to the fact that the electrodes are not really identical but the data do not reveal this. The diffusion path in the gas phase or on the electrode material can be different due to a difference in the morphology. The charge transfer reaction of electrons depends on the work function of the electrode and a small amount of impurities on the metal might therefore have a dramatic effect on the transfer rate.[15]

In cases III) and III') the electrode can be made in the form of a thin dense layer. The oxygen diffuses through this layer. This increases the active area of the electrode. The layer should be made thin to reduce it serial resistance and be backed by a coarse current collector.

Some of the processes shown in Fig. 3 can occur in parallel. By singling out a particular series of steps one points at the fastest process believed to dominate. This choice depends on the operation condition of the SOFC. It can change with temperature as the temperature dependence of the rate of various elementary steps, is different. Furthermore, the composition (stoichiometry) of materials also depends on temperature. In particular, LSC is a MIEC at 800°C under air, exhibiting an ionic transference number $t_i \sim 0.5$, while at 600°C it is practically an electronic conductor, $t_i \ll 1$.[16] The preferred path can depend also on the current drawn. The reason is that the impedances of the various elementary steps are, in general, not ohmic. The impedances exhibit, then, a dependence on the current density. This dependence will, generally, be different for the different steps and the rate of competing parallel processes may change with the current, i.e. with the load on the fuel cell.

2.2. ANODES

The elementary steps in the anode process are less understood than in the cathode. For the cermet anode used, the relevant elementary steps can take place on the surface of either the SE or the Ni grains, or less likely in the gas phase.[5] One specific mechanism that was suggested is:[17]

$$H_2(g) \rightarrow 2H^+_{ad,Ni} + 2e^-_{Ni} \tag{1a}$$

$$H^+_{ad,Ni} \rightarrow H^+_{ad,SE} \tag{1b}$$

$$2H^+_{ad,SE} + 2O^{2-}_{ad,Se} \rightarrow 2OH^-_{ad,SE} \tag{1c}$$

$$2OH^-_{ad,SE} \rightarrow H_2O(g) + O^{2-}_{ad,SE} \tag{1d}$$

where protons and oxygen ions adsorbed, react on the SE, forming intermediate OH^- species, in the process of reacting to H_2O.

In addition there may be a diffusion problem for the fuel gas diffusing towards the anode and the exhaust gas diffusing away from the anode. Since the volume of the pores is fixed and small, the inward-diffusion of the fuel is coupled to the outward-diffusion of the exhaust gas.

3. SOFC electrodes I-V Relations

3.1 GENERAL

The current through the SOFC is determined by the oxygen chemical potential difference, $\Delta\mu(O_2)$, applied on the cell and the resistance of the load, R_{ext} (see Fig. 1). The oxygen chemical potential at the cathode is that of oxygen in air. The chemical potential of oxygen at the anode is that determined by the H_2/H_2O ratio (assuming H_2 to be the fuel). The difference in the chemical potentials can be expressed in terms of the theoretical Nernst voltage of the cell:

$$4qV_{th} = \Delta\mu(O_2)_{ext} \tag{2}$$

Rather than using the load R_{ext} explicitly in the analysis we consider the voltage on the load given by the product of load and the current, $V_{ext} = IR_{ext}$, where V_{ext} is the voltage on the load as well as on the SOFC

$$-qV_{ext} = \Delta\tilde{\mu}_{e,ext} \tag{3}$$

and I is the electrical current through the load as well as through the SOFC. $\tilde{\mu}_e$ denotes electrochemical potential of electrons and q the elementary charge.

In the SOFC the electrical current I is carried by oxygen ions and by electronic charge carriers (electrons or holes). Oxygen is transported in the cathode as a neutral molecule or an atom and as an oxygen ion. In the SE it is transported as an oxygen ion and in the anode as an ion or an atom and is finally reacted to form H_2O. The driving force for the electronic current, I_{el}, in each elementary step, whether in the electrode or in the SE (where I_{el} is rather small), is a difference in the electrochemical potentials of the electron, which can be expressed as a voltage drop:[18]

$$-q\Delta V = \Delta\tilde{\mu}_e \tag{4}$$

The driving force for the flow of neutral species is a difference in the chemical potential of oxygen: $\Delta\mu(O_2)$. The driving force for the ionic current is a difference in the electrochemical potentials of the corresponding ions ($\Delta\tilde{\mu}(O^-)$ or $\Delta\tilde{\mu}(O^{2-})$).

The reactions taking place at the electrode need not be in quasi equilibrium. For example, an adsorbed O_2 molecule that dissociates,

$$O_{2,ad} \Leftrightarrow 2O_{ad} \tag{5a}$$

need not be in equilibrium with the two adatoms formed, and the chemical potentials are not equal,

$$\frac{1}{2}\mu(O_{2,ad}) \neq \mu(O_{ad}) \tag{5b}$$

However, one can define a (virtual) chemical potential, $\mu(O_{ad})_{virt}$ of O_{ad} which would prevail if the reaction were in quasi-equilibrium under the given $\mu(O_{2,ad})$,

$$\mu(O_{ad})_{virt} = \frac{1}{2}\mu(O_{2,ad}) \tag{5c}$$

The reaction is then driven by the difference, $\mu(O_{ad})_{virt}$ - $\mu(O_{ad})$.

Similarly, the reaction,

$$O_{ad} + e^- \Leftrightarrow O^-_{ad} \tag{6a}$$

need not be in equilibrium, and

$$\mu(O_{ad}) + \tilde{\mu}_e \neq \tilde{\mu}(O^-_{ad}) \tag{6b}$$

Let us assume that $\mu(O_{ad})$ and $\tilde{\mu}(O_{ad}^-)$ are given. Then one can define a virtual electrochemical potential for electrons in equilibrium with the adatoms and adsorbed ions,

$$\mu(O_{ad}) + \tilde{\mu}_{e,virt} = \tilde{\mu}(O_{ad}^-) \quad (6c)$$

Comparing Eqs. (6b) and (6c) it is seen that the reaction, in this case, is driven by a difference $\tilde{\mu}_e - \tilde{\mu}_{e,virt}$. Alternatively, one can define a virtual electrochemical potential of the ion, defined as:

$$\mu(O_{ad}) + \tilde{\mu}_e = \tilde{\mu}(O_{ad}^-)_{virt} \quad (6d)$$

The driving force for the reaction Eq.(6a) can, under appropriate conditions, be driven by a difference $\tilde{\mu}(O_{ad}^-) - \tilde{\mu}(O_{ad}^-)_{virt}$.

We have argued that the cell is operated under the driving forces, $\Delta\mu(O_2)_{ext}$, $\Delta\tilde{\mu}_{e,ext}$ and $\Delta\tilde{\mu}(O^-)$ or $\Delta\tilde{\mu}(O^{2-})$. However ionic driving forces can be determined by the former ones if local equilibrium prevails.. This is expected at least at two points, at the relevant "beginning" and the "end" of the electrochemical cell. Then

$$\Delta\tilde{\mu}(O^-) = \frac{1}{2}\Delta\mu(O_2)_{ext} + \Delta\tilde{\mu}_{e,ext}, \quad \text{or} \quad \Delta\tilde{\mu}(O^{2-}) = \frac{1}{2}\Delta\mu(O_2)_{ext} + 2\Delta\tilde{\mu}_{e,ext} \quad (7)$$

We shall assume that it is sufficient to specify two of the driving forces only, and choose normally $\Delta\mu(O_2)_{ext}$ and $\Delta\tilde{\mu}_{e,ext}$.

The processes in the SOFC can be divided into elementary steps connected in series and in parallel. An example is presented in Fig. 5a. Oxygen is diffusing in the pore and enters the MIEC electrode at three sites, as oxygen ions. The ion diffuses inside the MIEC and then enters the SE. Obviously these are three processes that occur in parallel. In Fig. 5b one path is via the MIEC and the parallel one is directly from the gas phase into the SE. On the other hand Fig. 5c shows one path assumed to be of relative lowest impedance and therefore , dominant. The assumption that such a path exists and that the parallel ones can be neglected, is supported by the fact that, experimentally, one find, usually, one rate determining step in the electrode at any given current. Therefore as long as these steps are included in the series of elementary steps the error, introduced by neglecting parallel paths, is probably not large.

We shall therefore consider, for simplicity, only elementary steps of mass transport that are connected in series. We shall also assume that the elementary steps including electron/hole current are connected in series. Exchange of charge between the two series of steps is allowed, so that neutral oxygen can turn into ions. The potential drops corresponding to the series add up to the ones applied on the SOFC:

$$\sum_i \Delta\mu(O_2)_i = \Delta\mu(O_2)_{ext} \tag{8}$$

and

$$\sum_i \Delta\tilde{\mu}_{e,i} = \Delta\tilde{\mu}_{e,ext} \tag{9}$$

It has been shown that the currents in the SOFCs are not of a single type and so are also the driving forces. Therefore we use the term I-V relations loosely, to present all necessary relations between the ionic, electronic and neutral currents or fluxes and the two types of driving forces.

3.2 THE CATHODE I-V RELATIONS IN TERMS ADOPTED FROM CLASSICAL LIQUID STATE ELECTROCHEMISTRY

3.2.1 *Rate Equation*

The reaction at the electrode is treated as a chemical reaction.[19] The rate for that reaction is written accordingly as a product of the concentrations of the species involved and the relevant forward and reverse rate constants. Let us discuss the rate equation using specific examples. For instance, let us consider the incorporation of neutral oxygen into an oxide SE to form an interstitial neutral oxygen,

$$O_{ad} + V_i^x \Rightarrow O_i^x + V_s^x \tag{10a}$$

(using the Kröger-Vink notation of defects). This reaction can be considered to be an elementary one. The corresponding rate equation is then,

$$r = k_+(T)[O_{ad}][V_i^x] - k_-(T)[O_i^x][V_s^x] \tag{10b}$$

where [] denotes concentration and k the rate constant. The concentration of vacant interstitial sites is large and practically not altered by the occupation by a low concentration of interstitial oxygen. It can therefore be included in the forward rate constant, k_+. When the coverage is low the vacant surface site concentration, $[V_s^x]$ is a constant and can be included in k_-. Eq. (10b) reduces then to a first order rate equation,

$$r = k_+(T)[O_{ad}] - k_-(T)[O_i^x] \tag{10c}$$

3.2.2 *Energy Potential Barrier*

The rate constants in Eq. (10c) are temperature dependent since the incoming and out-going oxygen has to pass an energy barrier. This potential barrier originates from the fact that the adsorbed oxygen on the surface has to "squeeze" in between the atoms on the outer atomic layer of the oxide before reaching the vacant site inside the SE. This results

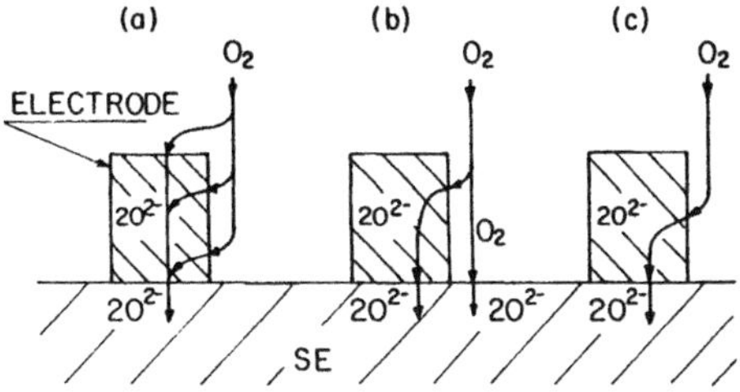

Fig. 5 Parallel reactions in the cathode: a - three paths for oxygen via a MIEC electrode, b - parallel paths, one via a MIEC electrode and one directly gas to SE. c - single path of least impedance.

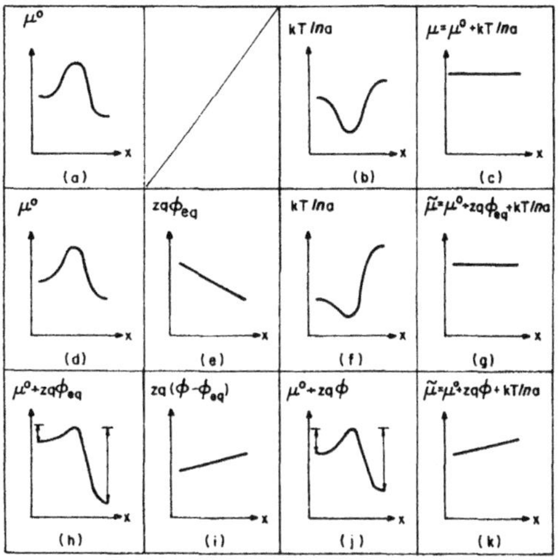

Fig. 6 Local potentials: a-c for neutral species under equilibrium , d-g for charged species under equilibrium, h-k for charged species with $\Delta\varphi$ being changed. h represents the sum of d and e. h-j exhibit the symmetric change in the two potential barriers under $\varphi-\varphi_{eq} \neq 0$. In k kTln(a) is from f and $\mu^0+zq\varphi$ from j.

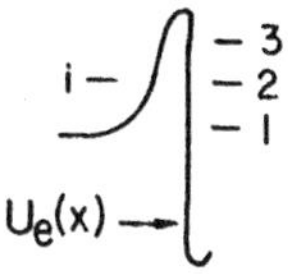

Fig. 7 Quantum tunneling of electrons through a energy barrier: i to 2 - between degenerate states, i to 1 and i to 3 phonon assisted between non degenerate states.

in an elastic deformation. Similarly the interstitial oxygen has to "squeeze" in between the atoms of the outer atomic layer on going out of the oxide. The two barrier heights are, usually, not the same. One usually considers not only the energy variation but also the variation in the entropy and local volume per particle, at constant T and P and therefore the chemical potential.

Under equilibrium the chemical potential of the neutral species is uniform. The chemical potential is usually defined as an average quantity over a volume large with respect to atomic dimensions. However, one can extend the definition to local, atomic scale, values. One can examine a standard chemical potential μ^0, the chemical potential without the configurational entropy.[20] It forms a potential barrier which together with a kTln{a} term yields, under equilibrium, a uniform chemical potential, μ. Then, the activity, a, and concentration is comparably low (Fig. 6b) where the standard chemical potential μ_0 (Fig. 6a) is high, so that $\mu=\mu_0 + kT\ln\{a\}$ is uniform, as shown in Fig.6c.

3.2.3 *Concentration Dependence*

The use of concentrations in the rate equations (see e.g. Eqs. 10b) and (10c)) is based on the argument that the probability of a particle to take part in a reaction is proportional to the concentration of that particle. Further more, in a second order (elementary) reaction, where two particle have to combine, the rate is proportional to the probability that the two particles collide, i.e. meet at the same site, and this is proportional to the product of their concentrations. For elementary steps the power of the concentration in the rate equation is equal to the stoichiometric coefficient of the corresponding substance in the chemical reaction equation.[19] These arguments hold in the dilute case, as long as the particles are statistically randomly distributed, as in an ideal gas, making occasionally collisions. For high concentrations this is not true. One can replace the concentrations by activities in the rate equations. This can be justified for small deviations from equilibrium. Doing this for instance for Eq(10c) yields,

$$r = k_+(T)\, a(O_{ad}) - k_-(T)\, a(O_i^x) \tag{11a}$$

This can be expressed alternatively in terms of concentrations and activity coefficient f as,

$$r = k_+(T)\,[O_{ad}]f_{ad} - k_-(T)\,[O_i^x]f_i \tag{11b}$$

3.2.4 *Relation to Voltage*

The reaction (10a) is driven by a difference in the chemical potential of the oxygen. There is no direct connection to the voltage of the cell. However, the I-V relations of the cell depend on the Nernst voltage on the SE, and on R_{ext}. The drop in the oxygen chemical potential required to drive the electrode reaction reduces the net oxygen chemical potential on the SE. As a result the Nernst voltage on the SE is smaller than the one calculated from the applied gases compositions. This affects the cell voltage, for a fixed load resistance R_{ext}.

The reaction described by Eq.(10a) can take place also in the cathode when the cell is placed in a uniform atmosphere and the current is driven by an applied voltage. The voltage cannot affect directly the neutral species but it can affect charged species and change their concentration. If these charged species interact to form neutral ones, as for example in Eq. (6a), then the voltage can generate a difference in the chemical potential of the neutral species. This argument can be also reversed, in the sense that a difference in the chemical potential of neutral species can reflect a voltage difference.

The energy potential barriers, for the neutral species, are not affected by the applied voltage. The reaction is driven only by a change in concentration.

3.2.5 *Transfer of Charged Species*

The charge transfer reaction is one in which electrical charge is transferred across a boundary in the electrode . In classical, aqueous electrochemistry, charge transfer is considered mainly as an electronation or de-electronation process.[21-24] Thus, for instance, a metal ion, M^+, accepts an electron at the cathode and turns into a neutral metal atom. The rate constant reflects the electron transfer rate. M^+ and M enter the rate equation via their concentrations. The transfer of ions across the interface has received only limited attention.[24]

We follow ref. [25] and argue that in the electrode reaction in SOFCs, an ion transfer step, and not an electron transfer step, is a rate determining charge transfer step. We therefore consider below charge transfer steps across interfaces, of the ions.

Let us consider now the incorporation of oxygen ions into the SE, e.g. by the reaction,

$$O_{ad}^{\backslash\backslash} + V_i^x \Rightarrow O_i^{\backslash\backslash} + V_s \tag{12}$$

where V_s is a vacant surface site and V_i a vacant interstitial site. Considering this reaction as an elementary one, the corresponding rate equation is,

$$r = k_+(T)\theta - k_-(T)[O_i^{\backslash\backslash}](1-\theta) \tag{13}$$

where θ is the coverage of the surface and (1-θ) is the concentration of the vacant surface sites assuming that the main adsorbed species is $O_{ad}^{\backslash\backslash}$.

The temperature dependence of the reactions constants k_+ and k_- reflects the energy potential barrier that an ion has to cross. This barrier consists, essentially, of two contributions. One, as for the neutral oxygen (see Fig. 6d) , is due to an elastic deformation of the lattice when an ion sqeeze through the outer atomic layer. The second contribution is due to a gradient in the electrical potential φ(x) which acts on the charge of the ion. A simple dependence of φ on x is a linear one, as for a two plate capacitor (Figs. 6e and 6i). This holds when the charges that generate the electric field form two planar layers. The potential barriers that result from μ^0 (Fig. 6d) and $Zq\varphi_{eq}$ (where Zq is the ionic charge) (Fig. 6e) under equilibrium, are given in Fig. 6h. Under equilibrium $\tilde{\mu}$ (Fig. 6g) is uniform. When an electric potential is applied it may modify the second contribution to the barrier ($\Delta\varphi$) (Fig. 6i) hence the potential barries (cf. Figs 6h and 6j) and therefore also the rates constants. Under current $\tilde{\mu}$ is not uniform (Fig. 6k).

3.2.6 *Voltage Over Potential at the Interface*

Let us examine the dependence of the rate constants on the voltage drop (over potential) η on the interface. This voltage drop reflects a difference in the electrochemical potential of the electrons, $\delta\tilde{\mu}_e$. It can be generated by a change $\delta(\Delta\varphi)$ in the electrical potential difference $\Delta\varphi$ across the interface , or a change $\delta(\Delta\mu_e)$ in the difference in the chemical potential of the electrons $\Delta\mu_e$ across the interface, or a combination of the two changes. For an electrode consisting of a metal in contact with a high concentration (liquid) electrolyte it is assumed that the applied voltage drop on the interface results in a change of the electrical potential barrier $\delta(\Delta\varphi)$ only. The justification for this is that in the metal the electron activity is constant, independent of the applied voltage. In electrolytes with high concentration of two types of charged species that can exchange electrons, in equilibrium, the activity of the electrons is also fixed. Then:

$$-q\eta = \delta\tilde{\mu}_e = -q\delta(\Delta\varphi), \quad \text{as } \delta\Delta\mu_e = 0 \tag{14}$$

To maintain a constant concentration of ions the liquid may need vigorous stirring.

Let us come back now to the ionic current across the interface. If the concentrations of ions on the two sides of the interface are not altered, then the driving force for the ion current is only the change $\delta(\Delta\varphi)$ which, by Eq(14), is related to the over-potential.

3.2.7 *Symmetry Factor*

The change $\delta(\Delta\varphi)$ is in part ($1-\alpha$) due to a lowering of the potential of the forward reaction potential barrier and in part (α) due to the increase of the reverse reaction potential barrier (see Figs. 6h-6j). The factor α is called the symmetry factor (or transfer coefficient). It reflects the asymmetry of the shape of the potential barrier and how the shape is distorted under an applied electric field. When the reaction is rate limited by electron transfer the rate - over-potential relation is the Butler-Volmer equation,

$$r = r_0\left(\exp\{\alpha\beta q\eta\} - \exp\{-(1-\alpha)\beta q\eta\}\right) \tag{15a}$$

where $\beta=1/k_BT$, and k_B is the Boltzmann constant.

3.2.8 *Multi-step Reactions*

A multi-step reaction is composed of a series of elementary steps. The r-η relations for the multi-step reaction can be evaluated if the charge species involved in all the elementary steps cross the same interface, and thus are subject to the same $\Delta\varphi$. It is further assumed that one of the steps is rate determining and the other are quasi-reversible. For a mutiple electron transfer the rate r can be expressed in terms of the over-potential η (Eq. 14)) and the concentrations of species before and after the mutli-step reaction,[26]

$$r = r_0\left(\exp\{\alpha_+\beta q\eta\} - \exp\{-\alpha_-\beta q\eta\}\right) \tag{15b}$$

where r_0 is the exchange current which depends on the concentration of the species involved in the reaction and α_+ and α_- are constants, called transfer coefficients.

3.2.9 *Galvani Potential*

Calculating the drop at each interface as well as the overall change on the cell, in terms of the (not measureable) internal (Galvani) electrical potential raises a problem. One has to relate this change to a measureable voltage i.e. to a difference in $\tilde{\mu}_e$. This requires the knowledge of the corresponding electron chemical potential differences. This is most conventiently done by using metallic electrodes of equal composition. For them the difference in the electron chemical potential is zero. Yet the procedure is, in our opinion, inconvenient and leads to confusion in the literature where galvani potentials and real voltage differences are sometimes mentioned together as if they were the same quantity. We prefer to calculate and follow the differences in the electrochemical potential of the electrons ($\delta\tilde{\mu}_e$) at each interface. This is directly related to the overpotential at that interface.

3.3 FURTHER CONSIDERATION OF THE METHOD FOR CALCULATING ELECTRODE REACTION RATES FOR SOFCS

3.3.1 *Electron Transfer*

Electron transfer occurs in changing the charge of ionic species. In the cathode this occurs in forming oxygen ions. Usually the electrons are transferred from the electrode material directly to the adsorbed oxygen atom. However, as discussed before (Sec. 2.1, cases I and II), the transfer via the SE can also take place. In that case one has also to consider the transfer of electrons across the metal/SE interface. This is significant for SOFCs based on electrolytes which are MIECs, since the electronic current through the MIEC cannot be neglected. It has to be considered also for the contact of the current collector and a MIEC electrode.

Let us first consider the transfer of electrons from a metal electrode into the SE and compare the exchange rate and expected impedances with those of the ion transfer. The impedance to electron transfer can be, in many cases, relatively low. First the introduction of the electron into the lattice does not require a large elastic deformation. Thus one contribution to the potential barrier can be neglected. The electron can follow a different path than the ion while crossing the outer layer of the SE. This may exhibit a potential barrier or a potential well (which for rapidly relaxing electron acts as a trap). However a significant reduction in the impedance of the barrier or trap height is due to the ability of the electrons to tunnel through potential barriers.[21] For a barrier of a height of the order of 1eV and a width of an atomic layer the transfer rate is fast, provided the electron can tunnel to a state of equal energy. (See in Fig. 7 the transfer from state i to state 2. U_e is the potential acting on the electron). If the state on the other side has a different energy, a phonon has to be involved. This lowers the probability of occurrence and the corresponding transfer rate. In Fig. 7 the i to 3 transfer requires the absorption of a phonon while the i to 1 transfer, the emission of a phonon. However even so the electron need not go over the whole potential barrier as does an ion. In conclusion we can assume that the impedance of the transfer of an electron into a SE can be neglected as compared to the impedance for the transfer of an oxygen ion. This suggests that the electrode charge transfer impedance is often due to ionic transfer. The ionic impedance results in a drop of the electrochemical potential of the ions, which however is not directly measured by a voltmeter. A voltage that would be measured using a voltmeter across the interface, would vanish, i.e. no over potential would be measured, as $\delta\tilde{\mu}_e = 0$. We shall show (in Sec 6) that the way "over-potentials" are measured with respect to properly chosen reference electrodes, the drop in the ion electrochemical potential on an interface can be determined.

The transfer of an electron to an oxygen adatom on a metal or on an MIEC may also be expected to be fast. The (phonon assisted) tunneling process should facilitate a quick chemical bonding in the chemisorbed layer. If so, the fact that some metals, e.g. Au, exhibit a high electrode impedance, may be attributed to a low equilibrium concentration of chemisorbed species that can participate in the current, rather than a low transfer rate of the electrons to adatoms. Alternatively, the reason can be a kinetic one, namely, that the adsorption process requires first adsorption of atoms and this may be slow.

3.3.2 *Details of the Reaction*

In order to write the rate equation of a reaction one has to consider carefully the details of that reaction.[27] For instance, introducing an oxygen atom adsorbed onto the SE into the SE may proceed according to the reaction,

$$O_{ad}^{x} + V_{O}^{\bullet\bullet} + 2e^{\backslash} \Leftrightarrow O_{O}^{x} + V_{s}^{x} \tag{16a}$$

It is obviously meant that the adatom is on the SE, while the oxygen vacancy comes from the SE. The electron can stem from the SE or from the electrode. The concentration of the electrons in the two phases is different and so are the potential barriers associated with a transfer of electrons from either phase. It is therefore important to specify the origin of the electron. If the electron is delivered from the SE Eq.(16a) becomes:

$$O_{ad}^{x} + V_{O}^{\bullet\bullet} + 2e_{SE}^{\backslash} \Leftrightarrow O_{O}^{x} + V_{s}^{x} \tag{16b}$$

3.3.3 *Using Structure Elements*

A reaction can be described in two levels, either on the component level or the defect level. For example the adsorption of an oxygen molecule can be written as,

$$O_2(g) \Leftrightarrow O_{2,ad} , \qquad \text{level 1} \qquad (17a)$$

but it can also be described using structure elements as,

$$O_2(g) + V_s^x \Leftrightarrow O_{2,ad}^x , \qquad \text{level 2} \qquad (17b)$$

Under equilibrium there is no difference. One can write the corresponding mass action law as,

$$K_{\theta,1}\, P(O_2) = [O_{2,ad}], \qquad \text{level 1} \qquad (18a)$$

where $[O_{2,ad}]$ is the adsorbed oxygen concentration, $K_{\theta,1}$ is the equilibrium reaction constant, or as,

$$K_{\theta,2}(T)\, P(O_2)(1-\theta) = \theta , \qquad \text{level 2} \qquad (18b)$$

where $K_{\theta,2}(T)$ is the equilibrium reaction constant and θ is the fraction of surface site occupied by oxygen molecules. θ plays the role of an activity of the structure element $O_{2,ad}^x$, while $\theta/(1-\theta)$ is the activity of the building unit or the component. (In the rate equation one has to consider structure elements not building units). Comparing Eqs. (18a) and (18b) yields,

$$[O_{2,ad}] = \frac{K_{\theta,1}\theta}{K_{\theta,2}(T)(1-\theta)} \equiv \frac{1}{K_\theta}\frac{\theta}{1-\theta} \qquad (19)$$

and the two equations for the mass action laws (18a) and (18b), are equivalent. The situation is different when the rate equation for the reaction of adsorption is considered. This yields,

$$r = k_+ P(O_2) - k_-[O_{2,ad}] = r = k_+ P(O_2) - \frac{k_-\theta}{K_\theta(1-\theta)}, \quad \frac{k_+}{k_-} = K_{\theta,1}(T), \quad \text{level 1} \qquad (20a)$$

and

$$r = k_+' P(O_2)(1-\theta) - k_-'\theta , \quad \frac{k_+'}{k_-'} = K_{\theta,2}(T), \qquad \text{level 2} \qquad (20b)$$

respectively. In both equation (20a) and (20b) the rate of the reaction is proportional to the difference in $K_{\theta,1}P(O_2)$ and $[O_{2,ad}]$ (cf. Eqs. (18a), (18b) and (19)). However the dependence on θ is different. It turns out that Eq. (20b) yields results in agreement with experiments. We shall therefore always use the level 2, defect level, with structure elements, to describe the reaction, when a difference exits between the two rate equations.

3.3.4 *Dependence of the Rate Constants k_+ and k_- on Concentration*

The rate constant need not be independent of the concentration. For example, in reaction (10a) one can envision that the energy barrier for the incorporation of an oxygen atom, which is a member of a close cluster of oxygen adatoms, on a uniform surface, may be different from that for a single adatom. The reason is that the nearby adatoms interact with the atoms on the outer-most atomic layer of the oxides thereby changing the energy barrier for the atom that enters the oxide. A similar effect should be on the barrier of the reverse reaction. This dependence is expected when the adatom concentration, $\theta=[O_{ad}]$, is not very small so that the probability for a statistical occurrence of a cluster is not negligible, usually for $\theta>0.01$. The change of the rate constant with θ is week also for high values of the coverage, $\theta\sim1$, as then most of the adatoms are members of clusters and the change in coverage matters little. We, here, shall assume that the rate constants are independent of concentration.

3.4 EVALUATION OF THE I-V RELATIONS FOR CERTAIN ELECTRODE CHARGE TRANSFER STEPS

3.4.1 *Case I*

Wang and Nowick[28] have considered the electrode reaction:

$$O_{ad}^{x} + 2e_{M}^{\backslash} + V_{O}^{\bullet\bullet} \Leftrightarrow O_{O}^{x} + V_{s}^{x} \tag{21}$$

In this reaction an adsorbed atom combines with two electrons from the electron conductor, M, and a vacancy from the SE to yield an oxygen ion in the SE at the normal ion site. Let us first, for simplicity, treat this reaction assuming that it is an elementary step and the rate equation is written accordingly, with the coefficients of the reaction as the exponents of the corresponding activities. We shall latter reconsider it as a multi-step reaction. The reaction is driven by differences in the electrochemical potentials of electrons and of ions. It is assumed that these differences originate from a deviation, $\delta(\Delta\varphi)=\Delta\varphi-\Delta\varphi_{eq}$, of $\Delta\varphi$ from its equilibrium value $\Delta\varphi_{eq}$. $\Delta\varphi$ is the Galvani potential difference that an electron senses in each electronation and de-electronation process. It is assumed that there is no change in the concentrations of electrons and ionic defects when the current is switched on. Then,

$$r = k_{+0}\exp\{-\alpha\beta 2q\delta(\Delta\varphi)\}\,\theta\, n_M{}^2\,[V_O^{\bullet\bullet}] - k_{-0}\exp\{(1-\alpha)\beta 2q\delta(\Delta\varphi)\}(1-\theta) \tag{22a}$$

where the dependence of the $k_{+,0}$ and $k_{-,0}$ on $\Delta\varphi_{eq}$ is,

$$k_{+,0} = k_{+,0,0}\exp\{-\alpha\beta q\Delta\varphi_{eq}\} \quad \text{and} \quad k_{-,0} = k_{-,0,0}\exp\{(1-\alpha)\beta q\Delta\varphi_{eq}\} \tag{22b}$$

The coverage θ is assumed to be dominated by the neutral adatoms and n_M is the concentration of the electrons in the electrode material, M. n_M in a metal electrode is constant. The energy barrier for the forward reaction is assumed to be reduced by a fraction α of $2q\delta(\Delta\varphi)$ while for the reverse reaction the barrier is increased by a fraction $(1-\alpha)$ of $2q\delta(\Delta\varphi)$.

Under equilibrium $\delta(\Delta\varphi)=0$, and the rate r vanishes. Hence,

$$k_{+0}\,\theta\, n_M{}^2\,[V_O^{\bullet\bullet}] = k_{-0}(1-\theta) \equiv r_0 \tag{23a}$$

Hence

$$\frac{k_{+0}}{k_{-0}} n_M{}^2\,[V_O^{\bullet\bullet}] = \frac{1-\theta}{\theta} \tag{23b}$$

Eq. (23b) and an equation analogous to Eq. (18b) but for adatoms rather than for adsorbed molecules,

$$K_\theta\, P(O_2)^{1/2}(1-\theta) = \theta \qquad \text{or} \qquad \theta = \frac{K_\theta P(O_2)^{1/2}}{1 + K_\theta P(O_2)^{1/2}} \tag{24a}$$

yield,

$$\frac{k_{+0}}{k_{-0}} n_M{}^2\,[V_O^{\bullet\bullet}] = \frac{1}{K_\theta\, P(O_2)^{1/2}} \tag{24b}$$

n_M in a metal electrode and $[V_O^{\bullet\bullet}]$ in the SE of the type considered are independent of $P(O_2)$. Therefore k_{+0}/k_{-0} must depend on $P(O_2)$. The $P(O_2)$ dependence of Eq. (24b) can be satisfied with:

$$k_{+0} \propto P(O_2)^{-\gamma/2} \quad \text{and} \quad k_{-0} \propto P(O_2)^{(1-\gamma)/2}. \tag{24c}$$

(The $P(O_2)$ dependence of k_{+0} and k_{-0} originates from the dependence of the concentration of the electrons in the SE on $P(O_2)$. Under equilibrium the electrochemical potential of the electrons is the same in the two phases, SE and metal electrode for any given $P(O_2)$. The electron chemical potentials in the SE depends on $P(O_2)$. Then the Galvani potential difference between the metal and the SE, $\Delta\varphi_{eq}$ shown in Fig. 6e, must depend on $P(O_2)$. This in turn affects the rate constants as they depend on the Galvani potential, see Eq. 22b). Using Eq. (22) in Eq (24b) yields,

$$\frac{k_{+0}}{k_{-0}} = \frac{k_{+0,0}}{k_{-0,0}} e^{-\beta q \Delta \varphi_{eq}} \propto \frac{1}{P(O_2)^{1/2}} \tag{25}$$

This choice is in agreement with Eq. (24).

Eq. (23a) can be rewritten as,

$$r_0 \equiv \sqrt{k_{+0}\, n_M{}^2 [V_O^{\bullet\bullet}] k_{-0}\, \theta(1-\theta)} \tag{26}$$

Using Eqs. (24a) (24b) and (25) in Eq. (26) yields,

$$r_0 \propto P(O_2)^{(1-2\alpha)/4} \frac{P(O_2)^{1/4}}{1+P(O_2)^{1/2}} \tag{27a}$$

The experimental results for the dependence of r_0 on $P(O_2)$ are found to be consistent with α=0.5.[28-30] Using Eq. (24c) in Eq. (26) would yield

$$r_0 \propto P(O_2)^{(1-2\gamma)/4} \frac{P(O_2)^{1/4}}{1+P(O_2)^{1/2}} \tag{27b}$$

The experimental results for the dependence of r_0 on $P(O_2)$ require: γ=0.5.

The next step is to relate the change in the Galvani potential difference $\delta(\Delta\varphi)$ under current to a measurable over potential, η. This is done by assuming that the chemical potential of the electrons in the SE (as well as in the metal electrode) does not change with current. Then,

$$\delta(\Delta\varphi) = \frac{\delta(\Delta\tilde{\mu}_e)}{-q} = \eta, \qquad \delta(\Delta\mu_e) = 0 \tag{28}$$

Combining Eqs. (22), (23a) and (28) yields the rate equation,

$$r = r_0(\exp\{-\alpha\beta 2q\eta\} - \exp\{(1-\alpha)\beta 2q\eta\} \tag{29}$$

This is a Butler-Volmer rate equation relating the rate of the reaction, or the current to the over-potential. The exchange rate r_0 depends on $P(O_2)$ as given in Eqs. (27a) and (27b).

It is doubtful whether it is justified to consider a complex reaction that involves the transfer of two electrons to an adatom and the reaction of these species with a charged vacancy, as an elementary step.[28] Different series of elementary steps have been suggested to represent the multi-step one.[29] The transfer of an electron from the metal electrode to the adsorbed oxygen atom or ion layer is considered. The rate reaction for a multi-step reaction (Eq. (15b)) is then used. The difference is in the transfer coefficients, α_+, and α_-.

There is a serious question that can be raised concerning the derivation leading to Eq.(29). Two interfaces can be identified to be involved in the electrode reaction considered. One is between the metal M and the adsorbed oxygen and the second one between the adsorbed oxygen and the SE. In deriving Eq. (29) the processes in the second interface are in quasi equilibrium. The over-potential originates from the electrons transfers at the first interface.

We have argued before that the driving force for electron transfer should be small compared to that needed to the transfer of ions. This is due to the possibility of the electrons to perform tunneling or phonon assisted tunneling, while the ions have to go over the full potential barrier. This is in contradiction to the approach in deriving Eq. (29) where the current is driven by a difference in the electron electrochemical potential. (If one were able to connect a voltmeter to the two sides of the interface one would measure a voltage η). We consider a different approach to the cathode over potential where the rate determining charge transfer step is not the electron transfer..

3.4.2 *Case II*

Two reactions have been considered, where the charge transfer rate determining step at the cathode of an SOFCs is the ion transfer one: a) the one given in Eq.(12), $O_{ad}^{\backslash\backslash} + V_i^x \rightarrow O_i^{\backslash\backslash} + V_s$,[25] and b) the reaction, $O_{ad}^{\backslash} + V_O^{\bullet\bullet} \rightarrow O_O^x + h^\bullet + V_s^x$.[30] For both reaction the rate equation is not Eq.(29), where the rate is driven by a difference in the electrochemical potential of the electrons. Instead, r depends exponentially on $\Delta\mu(O_2)$. Both reactions take place across the SE surface.

We concentrate here on Eq.(12) as the rate determining, elementary step. The step that follows this reaction is the relaxation of the interstitial oxygen ion into a vacant oxygen site,

$$O_i^{\backslash\backslash} + V_O^{\bullet\bullet} \Leftrightarrow V_i^x + O_O^x \tag{30}$$

Eq. (30) is assumed to be in quasi equilibrium. For the rate determining (elementary) step, Eq. (12), the rate equation is,

$$r = k_{+0}(T)\exp\{\alpha 2q\Delta\varphi\}\theta \ - \ k_{-0}(T)\exp\{-(1-\alpha)2q\Delta\varphi\}[O_i^{\backslash\backslash}](1-\theta) \tag{31}$$

where $\Delta\varphi$, here is the Galvani potential difference across the SE/adsorbed oxygen interface, and θ is the coverage of the SE surface with oxygen ions, $O_{ad}^{\backslash\backslash}$.

The concentration $[O_i^{\backslash\backslash}]$ is a constant, as can be concluded from reaction (30), which is in equilibrium and from the fact that the concentration of oxygen vacancies are fixed by the high doping level in a SE such as YSZ. The coverage θ is independent of the current and is fixed by the local $P(O_2)$ as can be inferred from the (multi step) reaction,

$$\frac{1}{2}O_2 + V_s^x + 2e_M^{\backslash} \Leftrightarrow O_{ad}^{\backslash\backslash} \tag{32}$$

considered to be in quasi equilibrium. Eq. (31) can then be written,

$$r = r_0\left(\exp\{\alpha\beta 2q\delta(\Delta\varphi)\} - \exp\{-(1-\alpha)\beta 2q\delta(\Delta\varphi)\}\right) \tag{33}$$

where the r_0 dependence on $P(O_2)$ is given in Eqs.(27a) and (27b). As the concentrations $[O_i^{\backslash\backslash}]$ and θ are independent of the current so are the corresponding chemical potentials. Therefore the change $\delta(\Delta\mu(O_i^{\backslash\backslash}))$, under current, vanishes. One can therefore substitute for $\delta(\Delta\varphi)$, in Eq.(33), the change in the electrochemical potential of the ions,

$$\delta(\tilde{\mu}(O_i^{\backslash\backslash}) = -2q\delta(\Delta\varphi), \qquad \delta(\Delta\mu(O_i^{\backslash\backslash}) = 0 \tag{34}$$

Furthermore, in view of the claim that the driving force for electron transfer across the interface is negligible compared to the one required to drive the ions: $\delta\tilde{\mu}_e = 0$ and,

$$\frac{1}{2}\delta\mu(O_2) = \delta\tilde{\mu}(O_i^{\backslash\backslash}), \qquad \delta\tilde{\mu}_e = 0 \tag{35}$$

One can express the atomic chemical potential of oxygen in terms of the chemical potential of molecular oxygen if it were in equilibrium with the local material. As a consequence one can define also the local $P(O_2)$ as the one that would exist in that virtual gas. In terms of this chemical potential, one can combine Eqs. (34) and (35) to yield,

$$\frac{1}{2}\delta\mu(O_2) = -2q\delta(\Delta\varphi) \tag{36}$$

Inserting Eq. (35) into Eq. (33) yields,

$$r = r_0\left(\exp\{-\frac{1}{2}\alpha\beta\delta\mu(O_2)\} - \exp\{\frac{1}{2}(1-\alpha)\beta\delta\mu(O_2)\}\right) \tag{37}$$

The rate equation (37) relates the rate to a chemical over-potential of the oxygen. No voltage drop is expected over the electrode. I.e. if it were possible to insert an inert metallic probe just under the electrode and measure the voltage difference between that probe and the electrode then the voltage measured would be zero. The difference in the oxygen chemical potential $\delta\mu(O_2)$ needed to drive the reaction has an effect on the I-V relations as it affects the net Nernst voltage on the SE. The latter is smaller than $V_{th,app} = \Delta\mu(O_2)_{ext}/4q$, determined by the gases applied to the cell. It is smaller by the difference

$\delta\mu(O_2)/4q$. A smaller V_{th} reduces the power output of the cell and the current that is obtained for a given load resistance.

In the Section 6 it is shown that the difference $\delta\mu(O_2)$, i.e. the loss in V_{th}, can be measured, under certain experimental conditions, as a voltage drop between the electrode and an auxiliary reference electrode.

3.5 DIFFUSION STEPS

3.5.1 *Case III*

The rate equation for diffusion is,

$$r \approx D\frac{\Delta c}{L} \tag{38}$$

where D is the diffusion coefficient, c the concentration and L the diffusion path length. The diffusion rate equation can also be cast in an exponential form. Let c be the concentration of oxygen say in the gas phase, $c=P(O_2)$, then

$$r = \frac{D}{L}\left(\exp\{\beta[\mu(O_2)^H - \mu^0]\} - \exp\{\beta[\mu(O_2)^L - \mu^0]\}\right) \tag{39}$$

where H and L denote the high and low concetration values along the diffusion path, respectively.

When the high pressure side is constant, Eq.(39) is better presented as,

$$r = r_0\left(1 - \exp\{-\beta[\mu(O_2)^H - \mu(O_2)^L]\}\right) \tag{40a}$$

where r_0 is a constant,

$$r_0 = \frac{D}{L}\exp\{\beta[\mu(O_2)^H - \mu^0]\} = \frac{D}{L}P(O_2)^H \tag{40b}$$

Eq. (40a) exhibits a limiting rate, r_0, reached when the exponential term in Eq.(40a) vanishes. However, not all diffusion steps exhibit a limiting rate. If the low pressure would be fixed then it is advantagous to rewrite Eq. (39) in the form,

$$r = r_0\left(\exp\{\beta[\mu(O_2)^H - \mu(O_2)^L]\} - 1\right) \tag{41a}$$

where r_0 is a constant,

$$r_0 = \frac{D}{L}\exp\{\beta[\mu(O_2)^L - \mu^0]\} = \frac{D}{L}P(O_2)^L \tag{41b}$$

Eq. (41a) exibits no limiting current. This is the case, for instance, at the anode of an electrochemical pump for oxygen. However this is not the case for the anode of a SOFC with respect to the out-diffusion of water vapor produced there. The low water vapor pressure, $P(H_2O)^L$, in the exhaust gas can be fixed, while $P(H_2O)^H$ in the anode is higher and may vary with the current. However, the out-diffusion of H_2O in the pores of the anode is coupled to the in-diffusion of H_2 (under constant total pressure). As a result H_2O diffuses with a Darken type effective diffusion coefficient which equals that for H_2 (this diffusion coefficient depends on the concentration and thus is not uniform).[33] As H_2 is diffusion limited, so is H_2O.

3.5.2 *Case IV*

Diffusion of adsorbed atoms on the surface of the electrode yields an equation similar to Eq. (39) except that the atoms diffuse under the gradient of the atom concetration hence,

$$r = \frac{D}{L}\left(\exp\{\frac{1}{2}\beta[\mu(O_2)^H - \mu^0]\} - \exp\{\frac{1}{2}\beta[\mu(O_2)^L - \mu^0]\}\right) \tag{42}$$

The same rate equation is expected to hold for neutral species diffusion within a metal such as Ag.

3.5.3 *Case V*

Diffusion of oxygen ions adsorbed on an electrode metal: The ionic current on the surface is forced by the gradient of the electrochemical potential of the ions. However, the electrical potential along the surface of the metal is expected to be uniform. Then the gradient is due to a gradient in the chemical potential of the ions, and hence in their concentration. Then for singly charged ions,

$$r = \frac{D}{L}\left(\exp\{\beta[\mu(O_{ad}^{\backslash\backslash})^H - \mu(O_{ad}^{\backslash\backslash})^0]\} - \exp\{\beta[\mu(O_{ad}^{\backslash\backslash})^L - \mu(O_{ad}^{\backslash\backslash})^0]\}\right) \tag{43}$$

It is assumed that the charge transfer reaction Eq.(32), resulting in $O_{ad}^{\backslash\backslash}$ on the metal, is in equilibrium. The activity of the electrons n_M in the metal is uniform. Then a gradient in $\mu(O_{ad}^{\backslash\backslash})$ equals a gradient in $\frac{1}{2}\mu(O_2)$. The diffusion equation (43) can therefore be written in terms of the different chemical potentials of the oxygen,

$$r = r_0\left(\exp\{\frac{1}{2}\beta[\mu(O_2)^H - \mu^0]\} - \exp\{\frac{1}{2}\beta[\mu(O_2)^L - \mu^0]\}\right) \tag{44a}$$

where,

$$r = \frac{D}{L} n_M{}^2 \exp\{\frac{1}{2}\mu^0 - 2\mu_e^0 - \mu(O_{ad}'')^0\} \tag{44b}$$

Eq. (44a) is expected to hold also for ions diffusing within a MIEC electrode, having a high concentration of electrons or holes.

4. I-V Relations in SEs/MIECs

The I-V relations in the SE/MIEC, serving as the electrolyte in a SOFC, are governed by the current density equations for the ions, j_i, and electrons, j_e. The two SEs of interest are, Y_2O_3 stabilized ZrO_2 (YSZ) and doped ceria (DCO). For YSZ the electronic current can be neglected. For DCO the electronic current cannot be neglected under low oxygen partial pressures for the temperature range of interest T~800°C. The electronic current in DCO is contributed mainly by the electrons. The relevant current density equations are:

$$j_i = \frac{\sigma_i}{2q} \nabla \tilde{\mu}_i \tag{45}$$

and,

$$j_e = \frac{\sigma_e}{q} \nabla \tilde{\mu}_e \tag{46}$$

where subscript i denotes the doubly charged oxygen ions and σ is the conductivity. The ionic conductivity is uniform for the two SEs under consideration and Eq.(45) can be integrated to yield a linear I-V,V_{th} relation,

$$I_i = \frac{V_{th} - V}{R_i} \tag{47a}$$

where

$$I_i = j_i S \tag{47b}$$

represents the ionic current through the SE (S is the cross sectional area), and

$$R_i = L / S\sigma_i \tag{47c}$$

represents the resistance of the SE to ionic current (L is its length, see Fig. 1),

$$V = -\Delta\tilde{\mu}_e / q \tag{47d}$$

represents the voltage across the SE, and

$$V_{th}=\Delta\mu(O_2)/4q \tag{47e}$$

is the Nernst theoretical voltage across the SE defined by the difference in the chemical potential of the oxygen at the two edges of the SE, inside, a few atomic layers deep, just beyond the surface region (which is expected to contain a space charge). The electron conductivity is not uniform. The integration of Eq. (46) yields a non linear relation between the electronic current I_e and V, V_{th} ,[32,33]

$$I_e = -S\sigma_e(0)\frac{V_{th}-V}{L}\frac{\exp\{-\beta q(V_{th}-V)\}-\exp\{-\beta q V_{th}\}}{1-\exp\{-\beta q(V_{th}-V)\}} \tag{48a}$$

where $\sigma_e(0)$ is the electron conductivity in the SE near the fuel side. I_e depends not only on the differences in the potentials via V_{th} and V but also on the absolute values of $\mu(O_2)$ near the edges of the SE, as $\sigma_e(0)$ is a function of $\mu(O_2)$ on the fuel side

$$\sigma_e(0) = \sigma_e^0 P(O_2)^{\frac{1}{4}} \tag{48b}$$

where σ_e^0 is a constant.

Eq.(46) can be formally integrated to yield a quasi linear I-V relation,

$$I_e = -\frac{V}{R_e}, \qquad R_e = \frac{1}{S}\int_0^L \frac{dx}{\sigma_e} \tag{48c}$$

where R_e is the resistance of the SE to electron current. However, comparing Eqs. (48a) and (48c) shows that R_e is not a constant but a function of the potentials on the SE.

5. I-V Relations for SOFCs Taking Electrodes and Electrolyte Impedance into Consideration

The discussion in Sec. 3.1 shows that the rate of transport of material and charge in each step is determined by the values of two potentials, the chemical potential of oxygen and the electrochemical potential of electrons, before and after the step. Sec. 4 shows that on the SE the partial currents are also determined by the values of these two potentials on the SE.

The rate of mass transport in the electrode elementary steps is directly related to the ionic current, I_i, in the SE, as it is assumed that every oxygen molecule that is transported through the cathode is, eventually, converted into two doubly charged ions in the SE.

The electrical current, I, that enters through the current collector into the electrode is equal the total current through the SE.

$$I = I_i + I_e \tag{49}$$

The oxygen chemical potential drops on all steps (including the SE) add up to the difference in the values $\mu(O_2)_{cathode}$ and $\mu(O_2)_{anode}$ applied to the SOFC, which determines $V_{th,ext}$. V_{th}, the Nernst voltage on the SE and $V_{th,ext}$ are related by,

$$V_{th,ext} = V_{th} + \delta V_{th,C} + \delta V_{th,A} \tag{50}$$

where, $\delta V_{th,C}$ and $\delta V_{th,A}$ are the corresponding drops in V_{th} on the cathode and anode, respectively.

The drops in the electrochemical potential of the electrons on all steps (including the SE) add up to the voltage V_{ext} generated on the load. The voltage V on the SE is related to V_{ext} and to the total current, I, through the cell,

$$I = \frac{V_{ext}}{R_{ext}}, \qquad V_{ext} = V + \delta V_C + \delta V_A \tag{51}$$

where R_{ext} is the resistance of the external load connected the SOFC, V is the voltage on the SE, δV_C and δV_A are the corresponding drops on the cathode and anode, respectively. We have argued that δV_C and δV_A can be neglected.

Experimentally (see Sec. 6) one observes for the cathode a charge transfer over-potential at low current densities and a diffusion limited current at high current densities.[2,25] The $P(O_2)$ dependence of the diffusion-limited current shows that the diffusing species is a single oxygen atom or ion. We consider three model processes for the cathode with two rate determining steps (rds), one for ionic charge transfer and one for oxygen atomic/ionic diffusion:

I) Diffusion of adsorbed oxygen atoms, or ions on the electrode material (Eqs.(42), (43) or (44a)) preceded by quasi reversible (fast) steps so that $P(O_2)^H$ is the externally applied oxygen partial pressure. The rate represents the ionic current and the equation can be written in general as,

$$I_i = I_0^D\left(1 - \exp\{-2\beta\delta V_{th}^D\}\right) \tag{52}$$

where δV_{th}^D is the drop in $\mu(O_2)/4q$ in this step and the coefficient I_0^D depends on T and is proportional to $[P(O_2)^H]^{1\backslash 2}$. This first rds in the series, being dominating the cathode impedance at high currents, is followed by a second rds (not necessarily the immediate next step) which is a charge transfer (CT) by oxygen ions into the SE (Eq. (37)). The CT impedance is dominant at low currents. The ionic current equation for the CT step is,

$$I_i = I_0^{CT}(\exp\{(1-\alpha)\beta 2q\delta V_{th}^{CT}\} - \exp\{-\alpha\beta 2q\delta V_{th}^{CT}\} \tag{53}$$

where δV_{th}^{CT} and I_0^{CT} have the corresponding meanings for the second rds.

II) Alternatively, we consider a series of steps (applicable only to MIEC cathodes): the first in the series, rate determining at high currents, is a diffusion of oxygen ions through a MIEC electrode (Eq.(44a) with $P(O_2)^H$ fixed) given by an equation of the form of Eq. (52). The second step, rate determining at low currents, is transfer of oxygen ions across the MIEC/SE interface, in the form of Eq.(53).

III) To obtain a limiting current, the pressure, $P(O_2)^H$, on the high pressure side must be constant. Therefore models I) and II) considered above had the diffusion step first. Yet, one can allow an ion charge transfer step to precede the diffusion one within a MIEC. Under high currents the CT over-potential increases only slowly with current, due to the exponential r-δV_{th} relations (Eq. (53)). Thus $P(O_2)^H$ on the high pressure side of the diffusion step that follows the CT step, can be considered to reach, under high current densities, a constant value, to a good approximation. One then expects to see a quasi limited current (somewhat "rounded") at high current densities. Therefore the first step, rate determining at low currents, can be that of oxygen ion transfer into the MIEC (Eq. (53)) and the second, rate determining at high currents, is diffusion of the ion through the MIEC (Eq. (52)).

As will be shown below (Sec. 6) there are experimental ways to measure separately the cathode (and anode) over-potential. We can therefore assume, for simplicity, that the anode is reversible and concentrate on the cathode over-potential. We consider the I-V relations for models I and II for which the same mathematical treatment is valid.

One approach to solving the I-V relations is starting with the I_i, I_e - V, V_{th} relations for the SE (Eqs. (47a) and (48a)). The values of V and V_{th} to be used are those that prevail on the SE. These values equal to the external values minus the losses on the electrodes. For the models considered, the electron electrochemical loss on the electrode (in Eq. (51)) can be neglected hence: $V = V_{ext}$. The value of V_{th} is, using Eq. (50),

$$V_{th} = V_{th,ext} - \delta V_{th}^{D} - \delta V_{th}^{CT} \tag{54}$$

To write two equations for the currents (I_i and I_e) we first invert Eqs. (52) and (53) and express the potential drops in terms of the current. Inverting Eq. (52) for the diffusion, yields,

$$\delta V_{th}^{D} = -\frac{k_B T}{2q} \ln\left(1 - \frac{I_i}{I_0^D}\right) \tag{55}$$

The CT equation (53) cannot be inverted for an arbitrary value of α. However a very good approximation can be substituted for one branch of the relations. For $I_i > 0$ and $\delta V_{th}^{CT} > 0$ Eq. (53) can be approximated by,

$$I_i = 2I_0^{CT} \sinh\{(1-\alpha)2\beta\delta V_0^{CT}\} \tag{56a}$$

Eq. (56a) can be inverted,

$$\delta V_{th}^{CT} = \frac{k_B T}{(1-\alpha)2q} \operatorname{arcsin h}\left(\frac{I_i}{2I_0^{CT}}\right) \tag{56b}$$

For example, for YSZ I_e can be neglected, and $I = I_i$. Inserting Eqs. (54), (55) and (56b) with $V=V_{ext}$ into Eqs. (47a) yields,

$$I = \frac{V_{th,ext} + \frac{k_B T}{2q} \ln\left(1 - \frac{I}{I_0^D}\right) - \frac{k_B T}{(1-\alpha)2q} \operatorname{arcsin h}\left(\frac{I}{2I_0^{CT}}\right) - V_{ext}}{R_i} \tag{57}$$

where $V_{th,ext}$ is known from the oxygen partial pressures at the two electrodes and V_{ext}, the cell voltage under current, can be measured. Instead of measuring V_{ext} one can use Eq. (51) to express V_{ext} in terms of I and R_{ext}. Then Eq. (57) becomes an implicit equation for I in terms of a set of parameters, including R_{ext}.

The procedure of fitting the theory to measurements outlined above is not practical in view of the many parameters involved, all of which have to be adjusted. It is therefore better to measure the cathode (and anode) over-potential and subtract the measured value from the external potential. This reduces the number of unknown parameters in fitting the I-V relations for the bulk SE. However this approach requires a model in order to decide if the over-potential represents a drop in V_{th} or in V or a combination of both. Assuming model I or II to hold then subtracting the measured over-potential yields V_{th} on the SE. This allows the evaluation of the I-V relations on the SE and the separation of the ionic and electronic currents I_i and I_e in the SE. It then allows discussing separately the cathode over-potential and fitting the parameters of the cathode I-V relations.

The discussion above assumed two types of elementary rds. The measurements cannot always distinguish between consecutive steps. Let us consider, for instance, two diffusion steps in the electrode (the first starting at the external fix, high, oxygen partial pressure) and the current in the SE in series. This will exhibit a limiting current as for a single diffusion step. Therefore, the fitting cannot resolve the data into two consecutive diffusion steps. (The limiting current measured will reflect the diffusion coefficient of both steps and the limiting current will be smaller than that for the first single step).

6. Experimentally Separating the Electrode Over Potential from the Bulk Impedance

The electrode over potential can be measured separately using a four electrode arrangement, as shown in Figs. (8a) and in (8b) for the cathode, as an example. A voltage, η_C, is measured between the working cathode electrode, C, and the corresponding reference electrode, refC. The question is: under what conditions can this voltage, η_C, be related to the potential loss $\delta V_{th,C}$ that is the chemical over potential, one is really looking for.

The measurement of η_C is done under three experimental conditions:

1) The working electrode and the corresponding reference electrode are both exposed to the same gas. Thus for the cathode side the gas is air.

2) The electrode and reference electrode are made of the same material and prepared in the same manner. This is of importance when the measurements are done on electrolytes that are MIECs. The reason is that in the case of a MIEC, a current is leaking through all four electrodes under open circuit conditions. Therefore an over-potential is generated on all electrodes under open circuit conditions. Yet one can match the over-potential on a working electrode and on the corresponding reference electrode so that the measured η_C (and η_A) is negligible. To achieve this the nature (i.e. impedance) of the electrode and corresponding reference electrode should be equal.

3) The time evolution of η_C is chosen as follows: First η_C is measured till a steady state is reached. Then the current is abruptly stopped and η_C (denoted as η_C^+) is measured as a function of time. Here we are interested in particular in the initial value of η_C^+ just after current interruption. It is essential that the switching time is shorter than the relaxation time of the electrode polarization. The switching time needed is typically of the order of 1μSec.

4) The size (area) of the reference electrode should be small with respect to the size of the working electrodes. This assures that the electrical potential distribution is not disturbed by large, equipotential reference probes. For MIEC electrolytes it assures also that the current through the reference electrode is negligible.

5) The distance, Δy, between the reference electrodes and the working electrodes should be large compared to the diameter of the working electrode. The current density in a working electrode is not uniform, with higher values existing near the edge. Therefore $\delta V_{th,C}$ is not uniform. η_C is more affected by the current density closer to refC, unless this distance is large compared to the working electrode diameter. In the latter case η_C reflects the average $\delta V_{th,C}$. When the distance is small, any shift in the position of one working electrode with respect to the another one changes the current density near the edge and affects considerably η_C.[34]

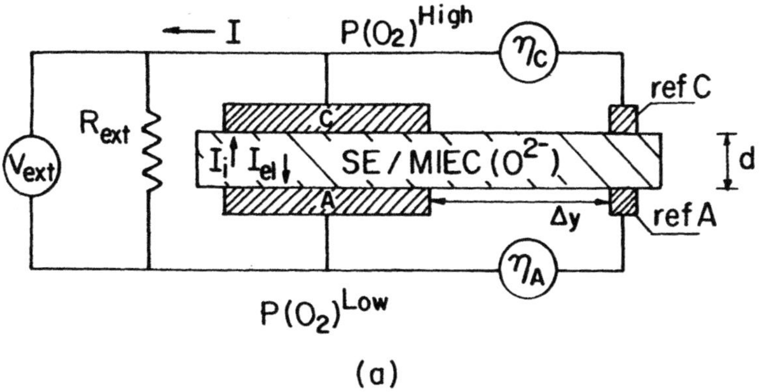

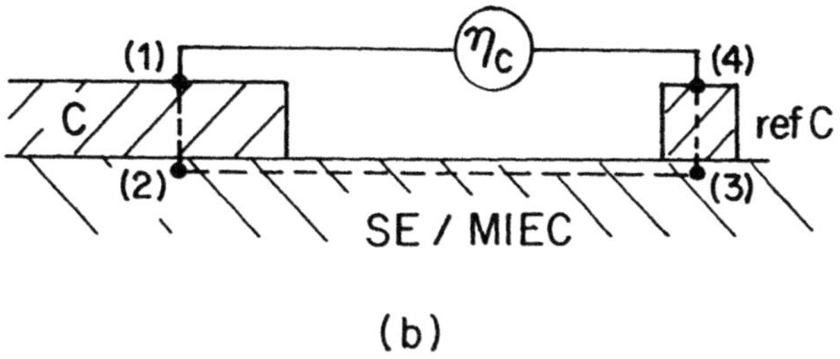

Fig. 8 Four electrode arrangement for electrode over potential easurements in SOFCs. a - overview, b - cathode side. ---- integration path used in the analysis.

We show in Appendix A that

$$\eta_C^+ = \pm\delta V_{th,C}, \qquad \text{for SEs} \qquad (58a)$$

for an electrolyte which is a pure ionic conductor. When the electrolyte is a MIEC then the two parameters are, to a good approximation equal up to a constant,

$$\eta_C^+ \approx \pm\delta V_{th,C} + \text{Const.}, \qquad \text{for MIECs} \qquad (58b)$$

7. Summary

The process at the electrodes in SOFCs and the corresponding rate equations have been discussed with special emphasis on the reactions at the cathode. It was argued that the electrode impedance is dominated by ionic transfer at low current density and by diffusion limitation at high current density. Electron transfer impedance may be negligible because of quantum effects.

The I-V relations of a SOFC were discussed, taking both the electrolyte and electrode impedance into consideration. Based on the model calculation it was shown how to interpret results of 4-electrode measurements, done under fuel operation conditions combined with current interruption. This allows to determine experimentally the electrode over-potential. The latter is interpreted as a difference in the chemical potential of the oxygen over the electrode.

Acknowledgment

This research was supported by the Fund for the Promotion of Research at the Technion. The author acknowledges many helpful discussions with Prof. J. Maier.

Appendix A

We show now that:

$$\eta_C^+ = \pm\delta V_{th,C}, \qquad \text{for SEs} \qquad (A\text{-}1a)$$

for an electrolyte which is a pure ionic conductor. When the electrolyte is a MIEC then the two parameters are, to a good approximation equal up to a constant,

$$\eta_C^+ \approx \pm\delta V_{th,C} + \text{Const.}, \qquad \text{for MIECs} \qquad (A\text{-}1b)$$

To show this let us consider Fig. (8b). η_C can be related to the difference in the electrochemical potential of the electrons between points (1) and (4),

$$-q\eta_C = \tilde{\mu}_e(4) - \tilde{\mu}_e(1) \equiv \Delta\tilde{\mu}_{e,41} \qquad \text{(A-2)}$$

We consider the path (1)-(2)-(3)-(4) from the cathode to refC. The difference in Eq.(A-2) can be divided into three contributions,

$$\Delta\tilde{\mu}_{e,41} = \Delta\tilde{\mu}_{e,43} + \Delta\tilde{\mu}_{e,32} + \Delta\tilde{\mu}_{e,21} \qquad \text{(A-3a)}$$

Points (2) and (3) are inside the SE, close to the surface, in a depth of a few atomic layers, just under the charge transfer region of the (working and reference) electrode. As claimed before, $\Delta\tilde{\mu}_{e,21}$ and $\Delta\tilde{\mu}_{e,43}$ can be neglected,

$$\Delta\tilde{\mu}_{e,21} = 0, \quad \Delta\tilde{\mu}_{e,43} = 0 \qquad \text{(A-3b)}$$

hence,

$$\Delta\tilde{\mu}_{e,41} = \Delta\tilde{\mu}_{e,32} \qquad \text{(A-3c)}$$

Assuming local equilibrium at (2) and (3) (as well as (1) and (4)) for the reaction $0.5O_2 + 2e^- = O^{2-}$,

$$\Delta\tilde{\mu}_{e,32} = -\frac{1}{4}\Delta\mu(O_2)_{32} + \frac{1}{2}\Delta\tilde{\mu}(O^{2-})_{32} \qquad \text{(A-4)}$$

Making use of the fact that both electrodes C and refC are exposed to the same gas,

$$\Delta\mu(O_2)_{41} = 0 \qquad \text{(A-5a)}$$

hence,

$$\Delta\mu(O_2)_{43} + \Delta\mu(O_2)_{32} + \Delta\mu(O_2)_{21} = 0 \qquad \text{(A-5b)}$$

$\Delta\mu(O_2)_{21}$ is the chemical over potential, in the chemical potential of oxygen, at C.

$$4q\delta V_{th,C} = \Delta\mu(O_2)_{21} \qquad \text{(A-5c)}$$

$\Delta\mu(O_2)_{43}$ is the chemical over potential at refC.

$$4q\delta V_{th,refC} = \Delta\mu(O_2)_{43} \qquad \text{(A-5d)}$$

$\delta V_{th,refC}$ vanishes when the electrolyte is a pure ionic conductor, since the current via refC vanishes. For a MIEC electrolyte there is a constant leak current via refC and $\delta V_{th,refC}$ does not vanish. However it is a constant. Hence,

$$\Delta\mu(O_2)_{43} = C_1 \tag{A-5e}$$

where C_1 is a constant. Combing Eqs. (A-5b), (A-5c) and (A-5e) yields,

$$\Delta\mu(O_2)_{32} = -4q\delta V_{th,C} - C_1 \tag{A-6}$$

The difference $\Delta\tilde{\mu}(O^{2-})_{32}$ is the driving force required to move the oxygen ions within the bulk of the SE/MIEC between points (2) and (3), (as can be deduced from Eq.(45)).
Hence,

$$\Delta\tilde{\mu}(O^{2-})_{32} = 2\gamma_C q I_i R_i \tag{A-7a}$$

where I_i is the ionic (O^{2-}) current in the bulk SE/MIEC, R_i is the resistance of the bulk and $0<\gamma_C<1$ is the fraction of the I_iR_i drop between points (2) and (3). (For $\Delta y >> d$ in Fig. 8a: $\gamma_C \sim 0.5$). The value of $\Delta\tilde{\mu}(O^{2-})_{32}$ after current interruption depends on the nature of the electrolyte. For a pure ionic conductor $I_t = I_i$ and under open circuit condition I_t and I_i vanish. Then $\Delta\tilde{\mu}(O^{2-})_{32}$ vanishes. For a MIEC $I_t \neq I_i$. Under open circuit conditions, I_t vanishes but $I_i^+ \neq 0$ and it relaxes to a steady state value. The steady state leak current I_i^+ is independent of the operation conditions before current interruption. The initial current I_i^+, however, depends on the operation conditions. Yet we argue that to a first approximation I_i^+ can be approximated by a constant. The changes in the over potential in $\Delta\mu(O_2)$, under different currents in the SOFC, are only a fraction of the order of 10% of the total driving force $\Delta\mu(O_2)$ value, thus modifying V_{th} only to that extent. To a first approximation the driving force and the corresponding leak current can, therefore, be assumed to be constant, independent of the current before interruption. Then,

$$\Delta\tilde{\mu}(O^{2-})_{32} = 2\gamma_C q I_i^+ R_i \approx C_2, \quad \text{after current interruption} \tag{A-7b}$$

where C_2 is a constant.

Combining Eqs. (A-2), (A-3c), (A-4), (A-5b) and (A-6) yields,

$$-q\eta_C = q\delta V_{th,C} + C_1/4 + C_2/2 \tag{A-8}$$

which is equivalent to Eq. (A-1b) and (58b). For a pure ionic conductor, C_1 and C_2 vanish.

References:

1) De Sauza, S., Visco S.J. and De Jonghe L.C. (1997) Reduced-temperature solid oxide fuel cell based on YSZ thin-film electrolyte, *J. Electrochem. Soc.* **144,** L35-37.
2) Gödickemeier, M., Sasaki, K., Gauckler L.J. and Riess, I. (1997) Electrochemical characteristics of cathodes in solid oxide fuel cells based on ceria electrolytes, *J. Electrochem. Soc.* **144,** 1635-1646.
3) Asano, K., Ihibino T. and Iwahara, H. (1995) A novel solid oxide fuel cell system using the partial oxidation of methane, *J. Electrochem.* Soc. **142,** 3241-3245.
4) Riess, I., van der Put P.J. and Schoonman, J. (1995) Solid oxide fuel cells operating on mixtures of fuel and air, *Solid State Ionics* **82,** 1-4.
5) Riess I. and Schoonman, J. (1997) Electrodics, in: *CRC Handbook of Solid State Electrochemistry*, Gellings P.J. and Boumeester H.J.M. eds., CRC Press, Chap. 8, pp. 269-294.
6) Van Hassel B.A. and Burggraaf, A.J. (1989) Structural, electrical and catalytic properties of ion-implanted oxides, *Appl. Phys. A* **49,** 33-40.
7) Widmer, S., Ravindranathan Thampi K. and McEvoy, A.J. (1994) Impedance spectroscopy of an electrocatalytic cathode domain in a SOFC, *Solid State Ionics* **73,** 165-168.
8) Verkerk, M.J., Hammink M.W.J. and Burggraaf, A.J. (1983) Oxygen transfe on subtituted ZrO2, Bi_2O_3, and CeO_2 electrolytes with platinum electrodes, I. electrode resistance by dc polarization, *J. Electrochem. Soc.* **130,** 70-78.
9) Wang D.Y. and Nowick, A.S. (1981) Diffusion controlled polarization of Pt, Ag and Au electrodes with doped ceria electrolytes, *J. Electrochem. Soc.* **128,** 55-63.
10) Braunshtein, D., Tannhauser D.S. and Riess, I. (1981) Diffusion limited charge transport at platinum electrodes on doped ceria, *J. Electrochem. Soc.* **128,** 82-89.
11) Wang D.Y. and Nowick, A.S. (1979) Cathodic and anodic polarization at platinum electrodes with doped ceria as electrolyte, I. steady state overpotential, *J. Electrochem. Soc.* **126,** 1155-1165.
12) Wang D.Y. and Nowick, A.S. (1979) Cathodic and anodic polarization at platinum electrodes with doped ceria as electrolyte, II. transient overpotential and ac impedance, *J. Electrochem. Soc.* **126,** 1166-1172.
13) Van Herle, J., McEvoy A.J. and Ravindranathan, T.K. (1994) Oxygen reduction at porous and dense cathode fuel cells, *Electrochim. Acta* **39,** 1673.
14) Hammouche, A., Siebert, E., Hammou, A., Kleitz M. and Caneiro, A. (1991) Electrocatalytic properties and nonstoichiometry of the high temperature air electrode $La_{1-x}Sr_xMnO_3$, *J. Electrochem. Soc.* **138,** 1212-1216.
15) Olmer L.J. and Isaacs, H.S. (1982) *J. Electrochem. Soc.* **129,** 345.
16) Gödickemeier, M. (1995) Mixed ionic electronic conductors for solid oxide fuel cells, PhD Thesis, ETH Zürich.
17) Møgensen M. and Lindegaard, T. (1993) in: *Proc. 3rd Intl. Symp. SOFC*, Singhal S.C. and Iwahara H. eds., *The Electrochemical Soc.* Vol **93-4,** pp. 484-493.
18) Riess, I. (1997) What does a voltmeter measure?, *Solid State Ionics* **95,** 327-328.
19) Castellan, G.W. (1983) *Physical Chemistry*, 3rd ed., The Benjamin/Cummings Publ. Comp. Inc., Chap. 32.
20) Maier, J. (1993) Defect chemistry: transport and reaction in the solid state; part II: kinetics, *Angew. Chem. Int. Ed. Engl.* **32,** 528-542.
21) Bockris J.O'M. and Reddy, A.K.N. (1970) *Modern Electrochemistry*, Vol 2, a Plenum/Rosetta edition, pp. 847-888, 918-929, 947-981.
22) Bard A.J. and Faulkner, L.R. (1980) *Electrochemical methods, fundamentals and applications*, John-Wiley & Sons, pp. 92-100.
23) Oldham K.B. and Myland, J.C.(1994) *Fundamentals of Electrochemical Science, Academic Press Inc.* pp. 167-172.
24) Gerischer, H. (1997) Principles of electrochemistry, in: *CRC Handbook of Solid State*

Electrochemistry, Gellings P.J. and Boumeester H.J.M. eds., CRC Press, Chap. 2, pp. 9-73.
25) Riess, I., Gödickemeier M. and Gauckler, L.J.,(1996) Characterization of solid oxide fuel cells based on solid electrolytes or mixed ionic electronic conductors, *Solid State Ionics* **90,** 91-104.
26) Bockris J.O'M. and Reddy, A.K.N. (1970) Modern Electrochemistry, Vol 2, a Plenum/Rosetta edition, pp. 997-1002.
27) Jamnik J. and Maier, J. (1997) Charge transfer and chemical diffusion involving boundaries, *Solid State Ionics* **94,** 189-198.
28) Inoue, T., Seki, N., Eguchi K. and Arai, H. (1983) *J. Electrochem. Soc.* **130,** 70.
29) van Heuveln F.H. and Boumeester H.J.M. (1997) Electrode properties of Sr-doped $LaMnO_3$ yttria-stabilized zirconia, II. electrode kinetics, *J. Electrochem. Soc.* **144,** 134-140.
30) Adler, S.B., Lane J.A. and Steele, B.C.H. (1996) Electrode kinetics of porous mixed-conducting oxygen electrodes, *J. Electrochem. Soc.* **143,** 35543564.
31) Manning, J.R. (1968) *Diffusion Kinetics for Atoms in Crystals*, D. Van Nostrand Comp. Inc., pp. 193-194.
32) Riess, I. (1981) Theoretical treatment of the transport equations for electrons and ions in a mixed conductor, *J. Electrochem. Soc.* **128,** 2077-2081.
33) Riess, I. (1992) The possible use of mixed ionic electronic conductors instead of electrolytes in fuel cells, *Solid State Ionics* **52,** 127-134.
34) Møgensen, M. Private communication.

DEFECTS AND TRANSPORT: IMPLICATIONS FOR SOLID OXIDE ELECTROLYTES AND MIXED CONDUCTORS

H.L. Tuller
Crystal Physics and Electroceramics Laboratory
Department of Materials Science and Engineering
Massachusetts Institute of Technology
Cambridge, MA 02139, U.S.A.

I. Introduction

Solid state electrochemical systems are increasingly being considered in the areas of sensing and combustion control, energy conversion and storage, and chemical processing. Examples include zirconia-based auto exhaust sensors and solid oxide fuel cells, mixed ionic-electronic conducting (MIEC) oxides as oxygen separation membranes and Li-ion battery electrodes. In all these applications, the key elements of these cells, the solid electrolyte or semipermeable membrane, the cathode, the anode and the interconnect material must satisfy certain specified criteria. These include (1) controlled levels of ionic and electronic conductivity, (2) phase, morphological and dimensional stability and (3) thermal, mechanical and chemical compatibility. It may come as a surprise to some, that defects and their transport properties dominate or substantially influence nearly all of these factors. In this chapter, we review defect formation and transport processes, thereby providing a basis for treating often complex interactions of defects in terms of defect chemical models. In our later chapter, we apply these concepts toward the design and optimization of materials for solid state electrochemical devices.

II. Defect Processes

The simplest lattice defects include vacancies and interstitials. Such defects are essential for providing diffusion pathways for ions. The self diffusivity of a given ion depends on the density of corresponding defects. Further, since ions are charged, they also contribute to ionic conductivity. In solid electrolytes, this form of conduction predominates over electronic conduction.

As is customary in describing semiconductors and insulators, electrons and holes refer to populations of electronic states in the conduction and valence bands which are occupied and empty, respectively. The density and mobility of these carriers determine the electronic conductivity of these materials. Likewise, the thermal and optical generation of electron-hole pairs determine the intrinsic temperature dependence of the electronic conductivity and the wavelength dependence of photoconductivity and optical absorbtivity.

Impurities, with net electrical charge relative to the ideal lattice, result in the formation of electronic and/or ionic defects with opposite charge to maintain overall charge neutrality. Both impurities and lattice defects often lead to the creation of energy levels in the band gap of the host crystal.

1.1 DEFECT EQUILIBRIA

1.1.1 Intrinsic Defects

Defects cannot be treated in isolation. Thermal disorder in a stoichiometric compound is accommodated by the generation of pairs of defects to maintain stoichiometry. This may be achieved, for example, in a binary compound by formation of (1) Schottky disorder: cation-anion vacancy pairs, (2) Frenkel disorder: cationic or anionic interstitial-vacancy pairs, (3) Anti-site disorder: cation-anion exchange and (4) electron-hole pairs. By "subtracting" the ideal from the actual lattice, Fig. 1, defect visualization becomes easier. It also emphasizes that these defects contain the net excess charges which control the electrical properties of the solid.

H.L. Tuller et al. (eds.), Oxygen Ion and Mixed Conductors and Their Technological Applications, 57–74.

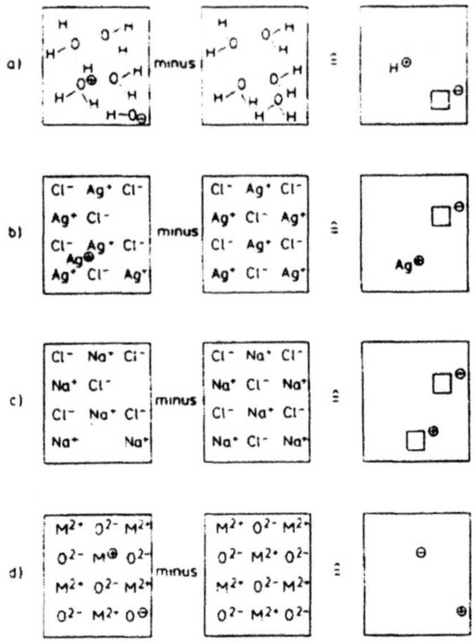

Figure 1. By "subtracting" the ideal structure (middle column) from the actual structure (left-hand column), the defects remain as the "particles" that determine the behavior.[1]

One derives expressions for thermally induced disorder by considering the free energy G of an elemental defective crystal [2].

$$G = G_0 + G_1 + G_2 \tag{1}$$

where G_0 is the free energy associated with the perfect crystal, G_1=ng is the work done in creating n defects and G_2=-TS_{config} is the configurational entropy. At equilibrium

$$\frac{\partial G}{\partial n} = 0 = g - T\left(\frac{\partial S_{config}}{\partial n}\right) \tag{2}$$

with

$$S_{config} = k \ln\{N!/[(N-n)!n!]\} \tag{3}$$

in which N is the total number of lattice sites of a given type and n the subset of defective sites. Applying the Sterling approximation, one obtains for the second term on the right side of Eq. 2

$$T\frac{\partial S}{\partial n} \approx -kT\ln(n/N) \tag{4}$$

which when substituted into Eq. 2 and solving for n results in

$$n/N = \exp(-g/kT) \tag{5}$$

This treatment can be readily extended to a compound by defining g_i = work to form a defect pair and noting that

$$T\frac{\partial S}{\partial n} \approx -kT\frac{n}{N}\frac{n}{N'} \tag{6}$$

where N and N′ are the number of lattice sites on the two respective lattice
s and n the number of defect pairs. Note, for Schottky disorder, N and N′ correspond to the cation and anion lattices while in Frenkel disorder they refer to say the cation sublattice and the cation interstitial site lattice, respectively. Again solving the appropriate form of Eq. 2 for n, one obtains

$$\frac{n}{N}\frac{n}{N'} = \exp(-g_i/kT) \tag{7}$$

or

$$n = (NN')^{1/2}\exp(-g_i/2kT) \tag{8}$$

Some key observations to be made include (1) that at any temperature above absolute zero, disorder is thermodynamically stable due to a lowering of the free energy by the entropic term, (2) the smaller the defect formation enthalpy, the higher the equilibrium defect density and (3) the higher the temperature, the higher the equilibrium defect density. Indeed, as evident from Eq. 8, defect disorder increases exponentially with temperature. Table 1 lists the defect formation enthalpies for a number of ionic compounds [3].

Table 1. The formation enthalpy of Schottky and Frenkel defects in some selected compounds [3].

	Compound	$\Delta H (10^{-19} J)$	$\Delta H (eV)^*$
Schottky defects	MgO	10.57	6.60
	CaO	9.77	6.10
	LiF	3.75	2.34
	LiCl	3.40	2.12
	LiBr	2.88	1.80
	LiI	2.08	1.30
	NaCl	3.69	2.30
	KCl	3.62	2.26
Frenkel defects	UO_2	5.45	3.40
	ZrO_2	6.57	4.10
	CaF_2	4.49	2.80
	SrF_2	1.12	0.70
	AgCl	2.56	1.60
	AgBr	1.92	1.20
	β-AgI	1.12	0.70

*The literature often quotes values in eV, so these are included for comparison: 1 eV = 1.60219×10^{-19} J.

The defect formation enthalpies reflect the bond strengths of the host lattices. A number of investigators have observed linear relationships between, e.g., Schottky pair formation enthalpies and atomization energies of halides or electron-hole formation energies (band gap) and melting temperatures [4-6]. It is therefore not surprising that intrinsic defect disorder is difficult to observe in solids with high melting temperature, T_M, except for temperatures approaching T_M.

1.2 DEFECT NOTATION

Three features of each defect must be specified, i.e. its (1) nature (2) site and (3) relative charge. These are included in the notation D_y^x in which:

D: refers either to an element (e.g., Na, Ti, O, F) or a vacancy, V

y: refers to atomic sites in the structure (e.g., in NaCl, y might be Na, Cl, or "i", an interstitial site)

x: represents the net effective charge with x-neutral, '-negative, •-positive

Examples of common defects and their host lattices include:

Vacancies (Ta_2O_5)	$V_{Ta}^{5'}, V_0^{\bullet\bullet}$
Interstitials (ZnO)	$Zn_i^{\bullet\bullet}$
Substitutional ions (ZrO_2)	$Y_{Zr}^{'}$
Electrons and holes	$e', h^{\bullet}$
Defect Associates (ZrO_2)	$\left(Y_{Zr}' - V_0^{\bullet\bullet}\right)^{\bullet}$ dimer
	$\left(Y_{Zr}' - V_0^{\bullet\bullet} - Y_{Zr}'\right)^{x}$ trimer

Aside from the examples of thermal disorder discussed above, defects also form in response to changes in the surrounding gaseous environment and to the dissolution of impurities. One commonly treats all such processes in the form of quasi-chemical reactions which must satisfy:

(1) Mass balance

$$\text{e.g., } O_0 \rightarrow 1/2\,O_2 + V_o^x \tag{9}$$

Oxygen vacancies which form in this reduction reaction have no mass.

(2) Charge Balance

$$\text{e.g., } M_M \rightarrow M_i^{\bullet} + V_M^{'} \tag{10}$$

$$\text{e.g., } M_i^{\bullet} \rightarrow M_i^{\bullet\bullet} + e' \tag{11}$$

The net charge must be retained following the reaction.

(3) Site Balance

$$\text{e.g., } O_2(g) \rightarrow 2O_0 + 2V_M \tag{12}$$

The ratio of lattice sites fixed by the crystalline structure, in this case MO, must be retained. A complete set of reactions will include intrinsic ionic and electronic disorder, ionization, interaction with the ambient and defect association reactions.

Table 2 summarizes reactions relevant to an oxide whose intrinsic ionic disorder is controlled by Frenkel equilibria on the oxygen sublattice [Eq. 14]. Also included are vacancy and interstitial ionization (Eqs. 15, 16), intrinsic electron-hole generation (Eq. 18), gaseous reduction (Eq. 13), donor and acceptor ionization (Eq. 17, 19) and dopant-lattice defect association (Eq. 20, 21) as well as the corresponding mass action equations.

Table 2
Defect reactions and corresponding mass action relations

Defect Reaction	Mass Action Relations	
$O_0 \Leftrightarrow \frac{1}{2}O_2 + V_0^{\cdot\cdot} + 2e'$	$K_R(T) = Po_2^{1/2}\left[V_0^{\bullet\bullet}\right]n^2$	(13)
$O_0 \Leftrightarrow V_0^{\cdot\cdot} + O_i''$	$K_F(T) = \left[V_0^{\bullet\bullet}\right]\left[O_i''\right]$	(14)
$V_0^{\cdot} \Leftrightarrow V_0^{\cdot\cdot} + e'$	$K_{V2}(T)\left[V_0^{\bullet}\right] = \left[V_0^{\bullet\bullet}\right]n$	(15)
$O_i' \Leftrightarrow O_i'' + h^{\cdot}$	$K_{I2}(T)\left[O_i'\right] = \left[O_i''\right]p$	(16)
$D_M^x \Leftrightarrow D_M^{\cdot} + e'$	$K_{D1}(T)\left[D_M^x\right] = \left[D_M^{\bullet}\right]n$	(17)
$nil \Leftrightarrow e' + h^{\cdot}$	$K_e = n \cdot p$	(18)
$A_M^x \Leftrightarrow A_M' + h^{\cdot}$	$K_{A1}(T)\left[A_M^x\right] = \left[A_M'\right]p$	(19)
$\left(A_M' - V_u^{\cdot\cdot}\right) \Leftrightarrow A_M' + V_0^{\cdot\cdot}$	$K_{Assoc}^A(T)\left[\left(A_M' - V_0^{\cdot\cdot}\right)\right] = \left[A_M'\right]\left[V_0^{\bullet\bullet}\right]$	(20)
$\left(D_M^{\cdot} - O_i''\right)' \Leftrightarrow D_M^{\cdot} + O_i''$	$K_{Assoc}^D(T)\left[\left(D_M^{\bullet} - O_i''\right)'\right] = \left[D_M^{\cdot}\right]\left[O_i''\right]$	(21)

Mass conservation equations provide a means for accounting for dopants in their various charge states. Thus

$$\left[D_M^{x}\right] + \left[D_M^{\bullet}\right] = \left[D_M\right]_{total} \tag{22}$$

$$\left[A_M^{x}\right] + \left[A_M'\right] = \left[A_M\right]_{total} \tag{23}$$

where $\left[D_M\right]_{total}$ and $\left[A_M\right]_{total}$ correspond to the total level of soluble donor and acceptor impurities added to the materials system.

Charge neutrality must also be conserved as specified in the following equation:

$$2\left[O_i''\right] + \left[O_i'\right] + n + \left[A'\right] + \left[\left(D^{\cdot} - O_i''\right)'\right] =$$

$$2\left[V_o^{\cdot\cdot}\right] + \left[V_o^{\cdot}\right] + p + \left[D^{\cdot}\right] + \left[\left(A' \quad V_o^{\cdot\cdot}\right)^{\cdot}\right] \tag{24}$$

Equations 13-24 can then be solved simultaneously in terms of the equilibrium constants and Po_2.

1.3 INTRINSIC DISORDER AND NONSTOICHIOMETRY

It is interesting to examine the relationship between the extent of nonstoichiometry exhibited by a solid and its level of intrinsic disorder. Consider the compound MY_{1-x} with x defined by

$$x=[V_o]-[O_i] \tag{25}$$

Assuming un-ionized defects, we have the reduction and Frenkel reactions given by

$$O_o \rightarrow 1/2 O_2 + V_o \qquad [V_o] = K_R Po_2^{-1/2} \tag{26}$$

$$O_o \rightarrow V_o + O_i \qquad [V_o][O_i] = K_F = \delta^2 \tag{27}$$

solving for $[V_o]$ and $[O_i]$ and substituting into Eq. 25 gives

$$x = K_R Po_2^{-1/2} - \delta^2 Po_2^{1/2}/K_R \tag{28}$$

Rearranging in terms of $Po_2(0)$, the oxygen partial pressure corresponding to stoichiometry or x=0, one obtains

$$\{Po_2(x)/Po_2(O)\}^{1/2} = \left[x + (x^2 + 4\delta^2)^{1/2}\right]/2\delta \tag{29}$$

This relation demonstrates that solids with higher intrinsic disorder exhibit correspondingly greater deviations from stoichiometry at a given Po_2.

1.4 THE DEFECT DIAGRAM

A reasonably complete description of the defect equilibria in a solid often requires the solution of a large number of simultaneous equations (viz. Eq's. 13-24). The solutions often lead to very high order equations which can only be solved numerically. However, under specified experimental conditions, one can generally ignore all but several dominant defects and obtain an approximate solution very close to the precise solution. We illustrate this approach by considering a solid, as in Table 2, but simplifying the analysis by assuming (1) all defects are fully ionized under conditions of interest and (2) that the total concentrations of acceptor and donor impurities are sufficiently small to be ignored relative to the lattice defects.

Thus, we need only consider Eq's 13, 14 and 18 listed in Table 2. Furthermore, we can simplify the electroneutrality relation, Eq. 24 to read

$$2[V_0^{\cdot\cdot}] + p = 2[O_i''] + n \tag{30}$$

This provides us with four simultaneous equations with which to solve for the four unknowns in Eq. 30 in terms of K_F, K_e, K_R and Po_2.

Because Eq. 30 is in the form of a summation, this still leads to solutions with higher order equations. Further approximations, called Brouwer approximations, are commonly applied to further simplify matters. By considering different Po_2 regimes (e.g., high Po_2 - oxidizing, low Po_2 - reducing), certain defect species become dominant while others become small in concentration by comparison. The simplest condition is achieved when only one type of defect on either side of the electroneutrality (EN) relation is dominant as assumed in the Brouwer approximation. This is illustrated by application to Eq. 30.

Under given conditions, the electroneutrality relation may be approximated by a single pair of defects, as summarized in Table 3. Substituting these simplified electroneutrality equations into Eq's 13, 14 and 18 results in the solutions listed in Table 3.

Table 3

Defect concentrations for oxide characterized by Frenkel disorder on the oxygen sublattice. Solutions provided for Brouwer regimes. See text for details.

	Limiting neutrality equation	Po_2 range	n	p	$[V_o^{\bullet\bullet}]$	$[O_i'']$
(31)	$n = 2[V_o^{\bullet\bullet}]$	Low	$(2K_R)^{1/3} Po_2^{-1/6}$	$K_e(2K_R)^{-1/3} Po_2^{+1/6}$	$\frac{(2K_R)^{1/3}}{2} Po_2^{-1/6}$	$2K_F(2K_R)^{-1/3} Po_2^{+1/6}$
(32)	$n = p = K_e^{1/2}$	Intermediate	$K_e^{1/2}$	$K_e^{1/2}$	$(K_F/K_e)Po_2^{-1/2}$	$(K_F K_e/K_R)Po_2^{+1/2}$
(33)	$[O_i''] = [V_o^{\bullet\bullet}] = K_F^{1/2}$	Intermediate	$(K_R/K_F^{1/2})^{1/2} Po_2^{-1/4}$	$K_e(K_F^{1/2}/K_R)^{1/2} Po_2^{+1/4}$	$K_F^{1/2}$	$K_F^{1/2}$
(34)	$p = 2[O_i'']$	High	$(K_R K_e^2/2K_F)^{1/3} Po_2^{-1/6}$	$(2K_F K_e/K_R)^{1/3} Po_2^{+1/6}$	$(2K_F^2 K_R/K_e)^{1/3} Po_2^{-1/6}$	$\frac{(2K_F K_e/K_R)^{1/3}}{2} Po_2^{+1/6}$

Note that there are two simplified electroneutrality equations which are possible at intermediate Po_2. Both relate to intrinsic thermal disorder at or near stoichiometry. The dominant disorder will simply depend on whether K_e, the constant defining electron-hole generation is smaller or greater than K_F, the constant defining ionic Frenkel disorder.

Schematic defect diagrams, sometimes called Kröger-Vink (KV) diagrams, showing log defect density versus log Po_2, are shown below for both possible cases [7]. Fig. 2 illustrates the case for which $K_e > K_F$ while Fig. 3 illustrates the case for which $K_F > K_e$.

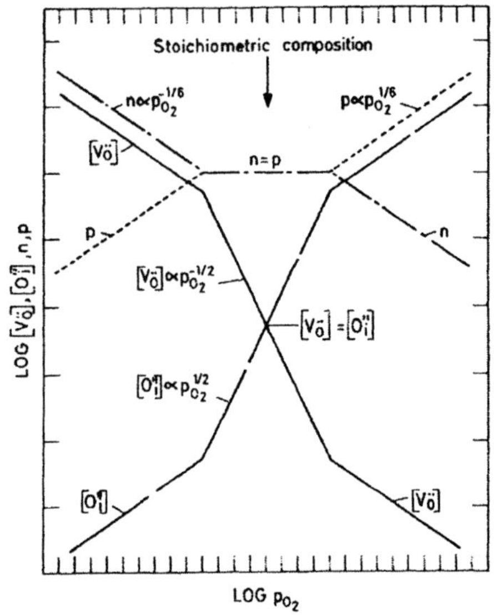

Figure 2. Schematic presentation of concentration of oxygen point defects and electronic defects as a function of oxygen pressure in an oxide which depending on the partial pressure of oxygen may have an excess or deficit of oxygen. Intrinsic electronic equilibrium is assumed to pre-dominate at stoichiometric composition [7].

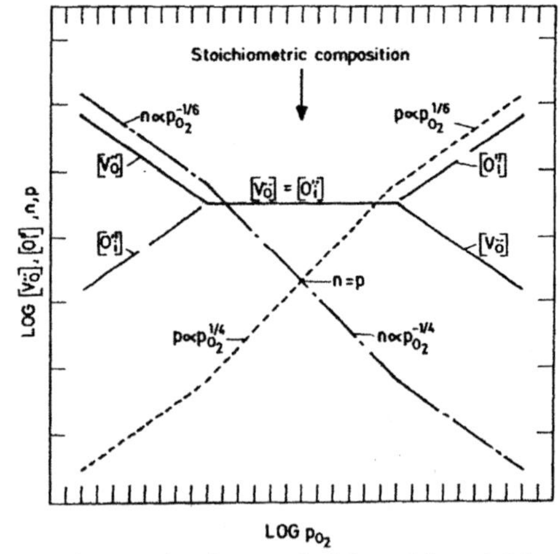

Figure 3. Schematic presentation of concentration of oxygen point defects and electronic defects as a function of oxygen. Oxygen Frenkel defects are assumed to predominate at stoichiometric composition [7].

An experimental illustration of the case illustrated in Fig. 3 for intermediate Po_2 is shown in Fig. 4 in which one observes the superposition of the intrinsic ionic conductivity which is Po_2-independent and the n- and p-type electronic contributions with $Po_2^{-1/4}$ and $Po_2^{+1/4}$ dependencies respectively [8].

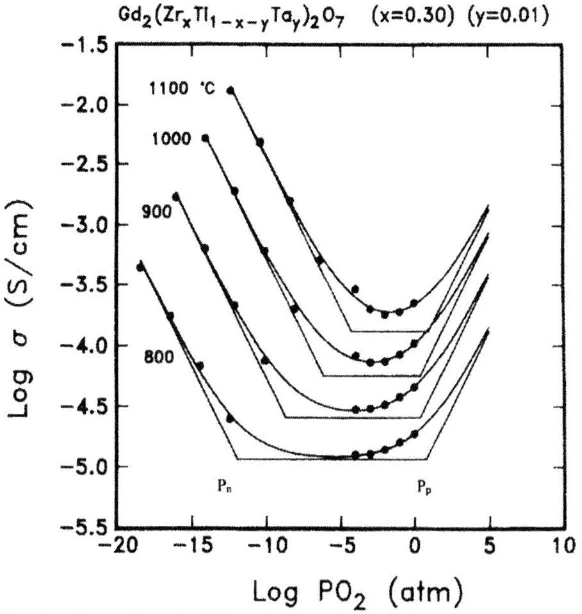

Figure 4. The Po_2 dependence of the electrical conductivity of $Gd_2(Zr_{0.3}Ti_{0.69}Ta_{0.01})_2O_7$. The solid curves through the data points represent a least square fit of the sum of Eqs. 35-37 to the data. The straight line segments represent the n, ionic and p-type contributions as described in the text. From [8].

For purposes of application of such materials as solid electrolytes, it is important, not only to specify the magnitude of ionic conductivity but also the conditions of temperature and Po_2 over which the material remains predominantly ionic. These bounds are called the "electrolytic domain boundaries" and specify at a given temperature the Po_2's at which the n-type or p-type conductivity becomes equal to the ionic conductivity. These points are designated as P_n and P_p, respectively in Fig. 4 at 800°C.

Expressions for the domain boundaries can be obtained by first writing down general expressions for the partial conductivities (see Table 3, solutions corresponding to Eq. 33):

$$\sigma_i = \sigma_i^\circ \exp - E_i/kT \tag{35}$$

$$\sigma_p = \sigma_p^\circ Po_2^{+1/4} \exp(-E_p/kT) \tag{36}$$

$$\sigma_n = \sigma_n^\circ Po_2^{-1/4} \exp(-E_n/kT) \tag{37}$$

And then equating σ_i and σ_n or σ_i and σ_p to solve for P_n and P_p, respectively [9]. These are given by

$$\ln P_n = \frac{-4(E_n - E_i)}{k}\frac{1}{T} + 4\ln\left(\frac{\sigma_i^0}{\sigma_i^0}\right) \tag{38}$$

$$\ln P_p = \frac{4(E_i - E_p)}{k}\frac{1}{T} + 4\ln\left(\frac{\sigma_i^0}{\sigma_p^0}\right) \tag{39}$$

Note that since the mobilities of vacancies in such oxides have been found to be much greater than those of interstitials [10], we ignore the latter's contributions. The domain boundaries for stabilized zirconia are shown plotted in Fig. 5 [11]. Note, as commonly observed, the electrolytic domain shrinks with increasing temperature due to the fact that E_n and E_p are typically greater than E_i. The same trend may be observed in Fig. 4.

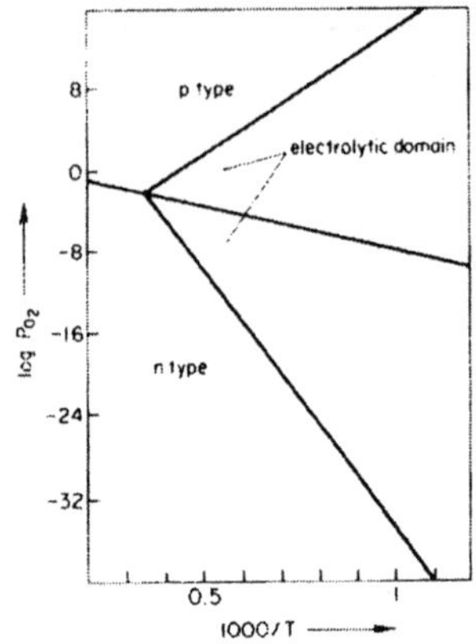

Figure 5. Domain boundaries of stabilized zirconia as projected on the log Po_2-1000/T plane [11].

On the other hand, the electrolytic domain is expected to expand with increasing levels of ionic conductivity. This is illustrated in Fig. 6, in which P_n is observed to shift by over 10 orders of magnitude to lower Po_2 as x in $Gd_2(Zr_x Ti_{1-x})_2O_7$ (GZT) is increased from 0.30 to 0.60 [12]. This has been shown to result in a significant increase in disorder on the oxygen sublattice of this pyrochlore [12].

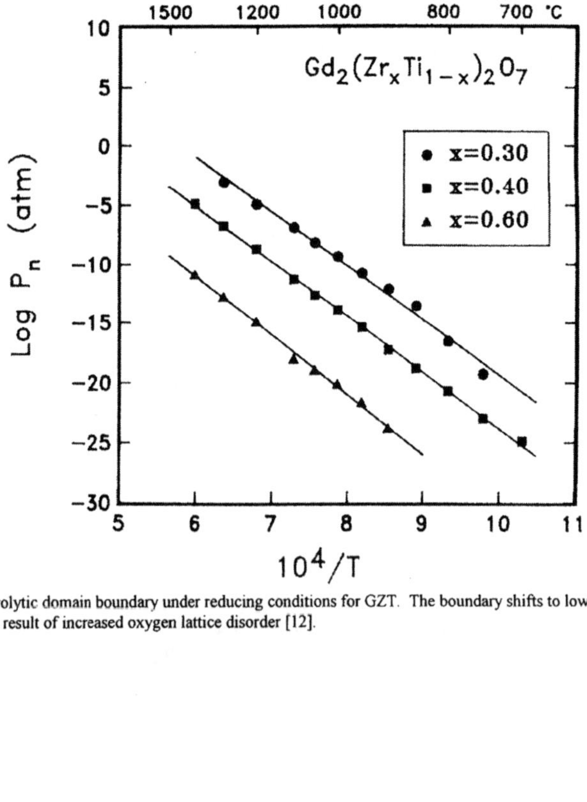

Figure 6. Electrolytic domain boundary under reducing conditions for GZT. The boundary shifts to lower Po_2 with increasing x as a result of increased oxygen lattice disorder [12].

1.5 DOPED SYSTEMS

While some intrinsically disordered compounds exhibit reasonably high levels of ionic conductivity, e.g. $Gd_2Zr_2O_7$, most materials systems require dopants to enhance their ionic and/or electronic conductivities. Dopants were explicitly incorporated in Eqs. 17, 19-24. In the following we discuss means for extracting thermodynamic and kinetic parameters from experimental results obtained from such systems. Furthermore, we evaluate their impact on both semiconducting and electrolytic regimes. In the latter case, we discuss how dopant-defect association may impact ionic conduction.

1.5.1 *Defect Equilibria for Doped Systems*

To illustrate the application of the Brouwer approximation to doped systems, we extend the above treatment to include an acceptor impurity such as Ca^{2+} substituted on a Gd^{3+} site in e.g. $Gd_2Ti_2O_7$. We need only modify Eq. 25 to include Ca'_{Gd} such that the electroneutrality equation becomes

$$2[V_o^{\cdot\cdot}] + p = 2\left[O_i''\right] + n + \left[Ca_{Gd}'\right] \tag{40}$$

As before at very low Po_2, Eq. 35 is approximated by (see also Eq. 31)

$$2[V_o^{\cdot\cdot}] \approx n \tag{41}$$

and at very high Po_2, by (see also Eq. 34)

$$p \approx 2\left[O_i''\right] \tag{42}$$

At intermediate Po_2, where deviations from stoichiometry are limited, impurities are expected to dominate. Eq. 40 requires that $\left[Ca_{Gd}'\right]$ be compensated by the positive defect species $[V_0^{\cdot\cdot}]$ and/or p. At lower intermediate Po_2, where oxygen vacancies predominate,

$$2[V_0^{\cdot\cdot}] \approx \left[Ca_{Gd}'\right] \tag{43}$$

while at higher intermediate Po_2, the Brouwer approximation will be satisfied by

$$p = [Ca_{Gd}^{\cdot}] \tag{44}$$

One may readily solve for the remaining deficit species by substituting Eqs. (43) or (44) into Eqs. 13, 14 and 18. In the vacancy controlled regime, it is straightforward to show that

$$[O_i^{\cdot\cdot}] = K_F \Big/ \left[Ca_{Gd}'\right] \tag{45}$$

$$n = \left(2K_R/[Ca'_{Gd}]\right)^{1/2} Po_2^{-1/4} \tag{46}$$

$$p = K_e\left([Ca'_{Gd}]/2K_R\right)^{1/2} Po_2^{+1/4} \tag{47}$$

While in the hole controlled regime, the defects satisfy the following relationships:

$$n = K_e \Big/ \left[Ca_{Gd}' \right] \tag{48}$$

$$\left[V_0^{\bullet\bullet}\right] = K_R \left(\left[Ca_{Gd}'\right] / K_e\right)^2 Po_2^{-1/2} \tag{49}$$

$$\left[O_i''\right] = (K_F / K_R)(K_e / \left[Ca_{Gd}'\right])^2 Po_2^{+1/2} \tag{50}$$

The appropriate defect diagram is illustrated in Fig. 7. Expressions for the boundaries separating the different regimes may be determined by e.g. equating Eqs. 43 and 49 for $V_0^{\cdot\cdot}$ or Eqs. 44 and 47 for p and solving for Po_2. See Ref. 9 for details.

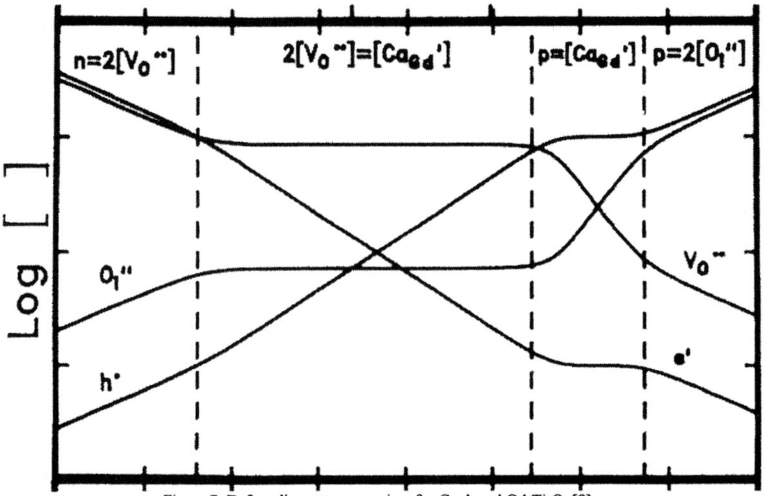

Figure 7. Defect diagram appropriate for Ca doped $Gd_2Ti_2O_7$ [8].

As is obvious from Fig. 7, only the regime for which the electroneutrality equation includes no electronic carriers can we expect to support electrolytic behavior. Given that the mobilities of electrons and holes are typically orders of magnitude greater than that of ions, even in this regime, predominant ionic behavior is possible only towards the center of this regime where the electronic carriers are at a minimum, i.e. n=p.

Since the intrinsic electron density n_i is a constant

$$n = p = n_i = K_e^{1/2} \tag{51}$$

the larger the acceptor dopant density, the higher the ionic conductivity and the wider the ionic domain. This is illustrated in a series of plots of experimental data obtained for Ca doped $Gd_2Ti_2O_7$ in Figs. 8 and 9.

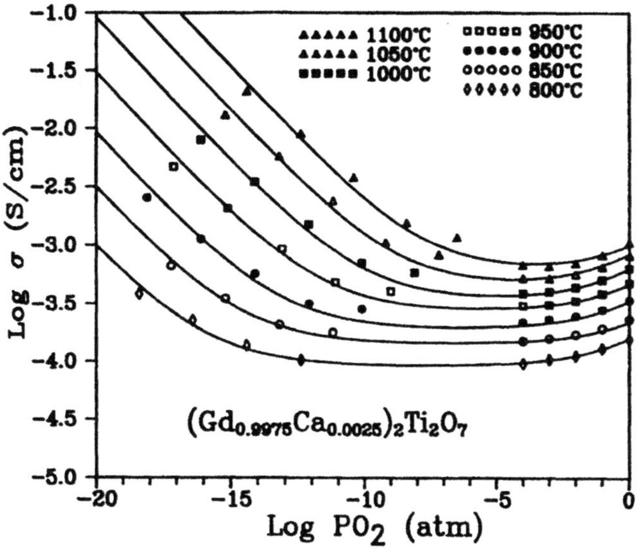

Figure 8. log σ versus log Po_2 for $(Gd_{0.9975}Ca_{0.0025})_2Ti_2O_7$ [13]

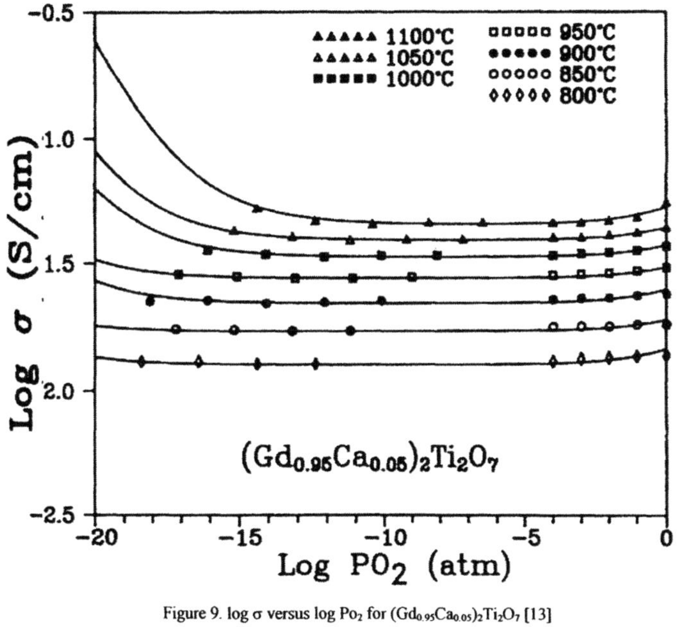

Figure 9. log σ versus log Po_2 for $(Gd_{0.95}Ca_{0.05})_2Ti_2O_7$ [13]

Note that the dopant level is as high as 5 mol % in the latter figure. This represents a fairly high concentration and is reflected in the complex dependence of the activation energy and pre-exponential term in the ionic conductivity shown plotted as a function of dopant level in Fig. 10 [13]. Possible explanations for these trends are discussed in a subsequent section.

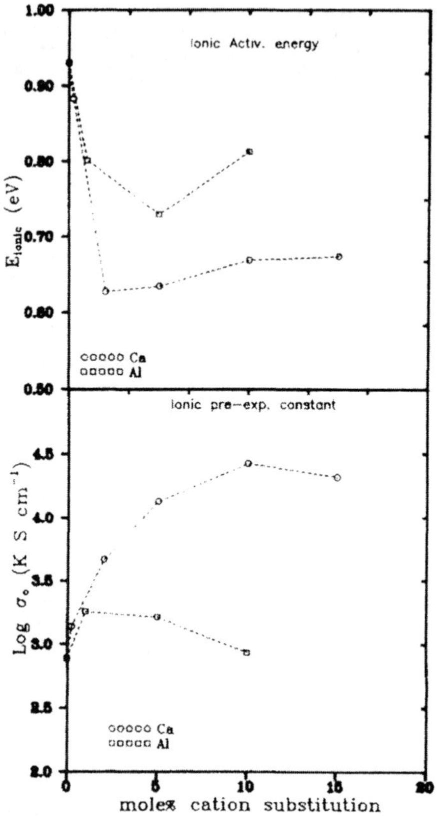

Figure 10. The ionic conductivity of $Gd_2Ti_2O_7$ at 1000°C as a function of acceptor doping on the "A" (Ca'_{Gd}) and "B" (Al'_{Ti}) cation sublattices [13].

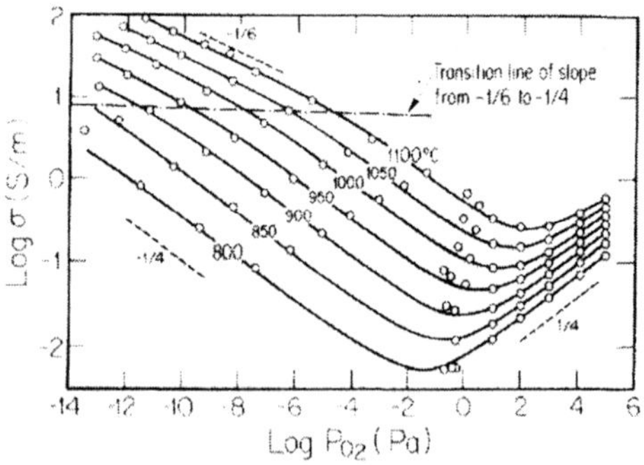

Figure 11. Measured and calculated conductivities of $Ba_{0.03}Sr_{0.97}TiO_3$. Dots are experimental data; lines are calculated values [14].

We complete this section by showing results for another acceptor doped titanate, $Ba_{0.03}Sr_{0.97}TiO_3$ [14] in Fig. 11. Here no ionic plateau is observed even though we are largely operating in the Brouwer regime given by $n = 2[V_0^{\cdot\cdot}]$. $SrTiO_3$ has both a considerably smaller band gap and acceptor solubility than $Gd_2Ti_2O_7$ combining to give $\sigma_{electronic} >> \sigma_{ionic}$. This work illustrates how key thermodynamic and kinetic parameters may be extracted by analysis of such data in terms of the above defect model. The results are summarized in Table 4.

Table 4 Constants Derived from Experimental Data for Single-Crystalline $Ba_{0.03}Sr_{0.97}TiO_3$ [14]

Constant	Calculated values
E_g^0	314 ± 5 kJ/mol
β	0.0482 kJ/(mol·K)
ΔH_R	500 ± 17 kJ/mol
ΔH_{Ox}	125 ± 2 kJ/mol
μ_e	2.3×10^{-5} m^2/(V·s) ($T = 950$°C)
μ_h	10^{-5} m^2/(V·s) ($T = 950$°C)
K'_R	$(1.52 \pm 0.17) \times 10^{87}$ Pa$^{1/2}$·m^{-9}
K'_{Ox}	$(2.45 \pm 0.18) \times 10^{25}$ Pa$^{-1/2}$·m^{-3}
[I']	$(1.85 \pm 0.19) \times 10^{24}$ m^{-3}
	$= (1.15 \pm 0.12) \times 10^{-4}$/Ti $= 115 \pm 12$ ppm of Ti
K'_i	$(7.67 \pm 0.57) \times 10^{54}$ m^{-6}

1.6 ANALYTICAL TREATMENT OF DEFECT EQUILIBRIA

As we attempt to improve the performance of solid state electrochemical devices, we turn increasingly to more complex systems. These systems often include both host and dopant elements with multiple oxidation states which markedly complicate the defect structure. Under these circumstances, we have found application of the Brouwer approximation often inadequate in treating experimental data which often fall within transition regions between the Brouwer regimes.

Progress has recently been made in solving a series of simultaneous defect equations analytically by solving for Po_2 in terms of the defect concentrations rather than the inverse. In many cases, this leads to a quadradic expression which leads to analytical solutions. This approach is described by Porat and Tuller [15] and applied to the case of donor doped $Gd_2Ti_2O_7$. The defect diagram shown in Fig. 12 illustrates the type of detailed information which now becomes accessible. Note that there are three different donor-controlled regimes.

1.7 DEFECT INTERACTIONS

Materials of interest to the field of solid state electrochemistry are generally required to exhibit high ionic conductivities, typically above 10^{-2}S/cm. Given the low mobilities relative to that of semiconductors, this requires ionic defect densities above 1%, a level at which defect interactions are expected to become significant.

There are a number of implications for our treatment of defect chemistry. First, defect activities rather than concentrations should be used in the mass action relations. Second, the equilibrium constants characterizing the defect generation reactions are no longer independent of composition[9]. Third, in dealing with electron-hole generation, Fermi rather than Boltzmann statistics need be applied when the Fermi energy approaches within 3kT of the conduction or valence band edge [16].

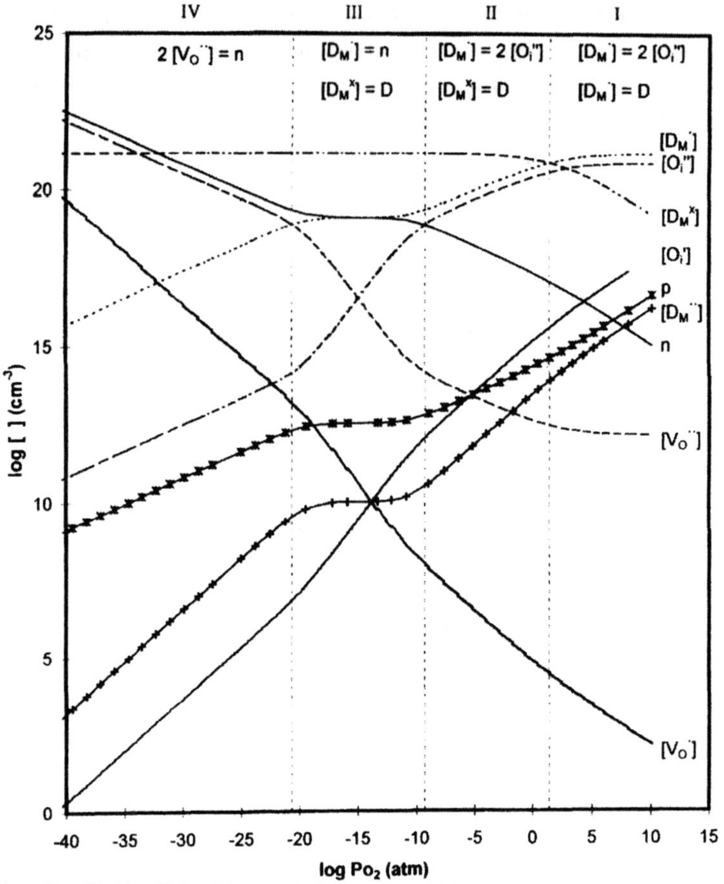

Figure 12. Model prediction of the atmospheric dependence of defect concentrations in donor doped ($Gd_2Ti_2O_7$). Roman numerals represent the different approximate electroneutrality conditions. Vertical dashed lines represent region boundaries [15].

Fortunately, as we shall see, the major features of the defect equilibria derived for dilute solutions remain intact. While the ionic defect activities may not be known precisely, they normally remain constant and leave the general form of Eqs. 13-22 unchanged. On the other hand, the minority carriers, e.g., electrons in a solid electrolyte, remain negligibly small and so may continue to be characterized by their concentrations within the mass action relations.

As the defect interactions become stronger, pairwise interactions give way to long range order and ultimately to intermediate ordered compounds or entirely new phases. Examples include the interstitial vacancy clusters which form in rocksalt structured $Fe_{1-x}O$ and resemble building blocks in the spinel structure [17]. Of more direct relevance to this work are the so-called Bevan clusters, which form in highly reduced fluorite structure materials, such as CeO_{2-x}. Here, as shown in Fig. 13 oxygen vacancies at the corner of cubes align as strings along the (111) direction with the reduced cation, e.g., Ce^{3+}, at the body center. Both pairwise and longer range interactions tend to lock potentially mobile defects into deeper potential wells. Consequently, an optimum defect density exists which provides high enough levels of defects to achieve useful magnitudes of ionic conductivity but low enough to minimize trapping effects. In a later section we discuss the relationship between ionic conductivity and degree of disorder in some pyrochlore oxides which illustrates some of these issues.

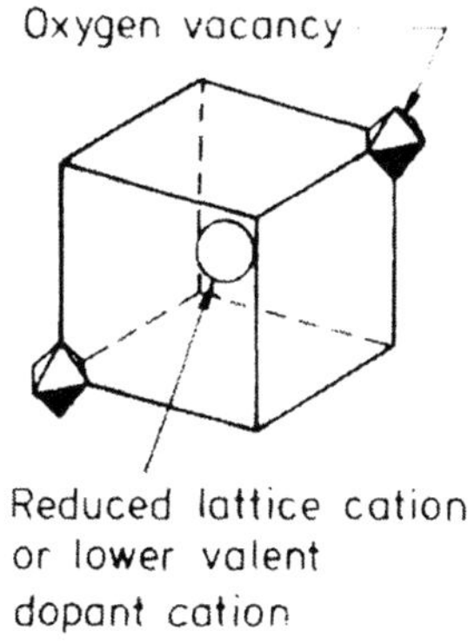

Figure 13. <111> Vacancy dimer in fluorite oxides [17].

III. Summary

Solid state electrochemical devices such as sensors, fuel cells, permeation membranes, batteries, etc. rely on materials which support high level of ionic conductivity (solid electrolytes) or mixed conductivity (electrodes or permeation membranes). In order to understand the source of these charge carriers and their dependence on composition, temperature and potential, it is essential to be able to model their defect equilibria. In this chapter we have reviewed the means for assembling the key defect reactions and solving them initially in a piece-wise fashion but later analytically. The subsequent construction of a defect diagram provides useful insight into the expected influence of dopant, and atmosphere on the transport properties.

Special emphasis was placed on means for defining and optimizing the electrolytic domain by impurity doping or inducing disorder in the crystalline lattice. Specific illustrations were drawn from our own work with the pyrochlore oxides. In our later chapter, we apply these concepts further towards the design and optimization of materials for use in, for example, solid oxide fuel cells.

Acknowledgements

The author gratefully acknowledges the work of his previous students on the pyrochlore oxides: P. Moon, S. Kramer, T.H. Yu and M. Spears. The author's research in this area has been supported by the Department of Energy under grant number DE-FG02-86ER45261 and the National Science Foundation under grant number DMR-9701699.

References

1. Maier, J. (1993) Defect chemistry: composition, transport, and reactions in the solid state; part I: thermodynamics, *Angew. Chem. Int. Edit.* **32,** 313-335.
2. Hayes, W. and Stoneham, A.M. (1985) *Defect and Defect Processes in Nonmetallic Solids*, John Wiley, New York, NY.
3. Smart, L. and Moore, F. (1992) *Solid State Chemistry*, Chapman and Hall, London.
4. Kubashewski, O., Evans, E.L. and Alcock, C.B. (1967) *Metallurgical Thermochemistry*, Pergamon Press, London.
5. Vigh, A.K. (1968) Comments on the relation between band gap energy in semiconductors and heats of formation, *J. Phys. Chem. Sol.* **29**, 2233-2236.
6. Vigh, A.K. (1970) A thermochemical approach to the bandgaps of semiconducting and insulating materials, *J. Mat. Sc.* **5**, 379-382.
7. Kofstad, P. (1972) *Nonstoichiometry, Diffusion and Electrical Conductivity in Binary Metal Oxides*, Wiley-Interscience, New York.
8. Moon, P.K. (1988) Electrical conductivity and structural disorder in $Gd_2Ti_2O_7$ – $Gd_2Zr_2O_7$ and $Y_2Ti_2O_7$ – $Y_2Zr_2O_7$ solid solution, Ph.D. Thesis, Dept. of Materials Sc. And Eng., MIT, Cambridge, MA.
9. Tuller, H.L. (1981) Mixed conduction in nonstoichiometric oxides, in O.T. Sorensen (ed.), *Nonstoichiometric Oxides*, Academic Press, New York, pp. 271-335.
10. Moon, P.K. and Tuller, H.L. (1988) Ionic conduction in the $Gd_2Ti_2O_7$-$Gd_2Zr_2O_7$ system, *Solid State Ionics* **28-30**, 470-474.
11. Heyne, L. (1977) Electrochemistry of mixed ionic-electronic conductors, in S. Geller (ed.), *Solid Electrolytes*, Springer Verlag, Berlin and New York, pp. 169-221.
12. Tuller, H.L. and Moon, P.K. (1988) Fast ion conductors-future trends, *Mat. Sc. and Eng.* **B1**, 171-191.
13. Kramer, S.A. and Tuller, H.L. (1995) A novel titanate-based oxygen ion conductor: $Gd_2Ti_2O_7$, *Solid State Ionics* **82,** 15-23.
14. Choi, G.M. and Tuller, H.L. (1988) Defect structure and electrical properties of single crystalline $Ba_{0.03}Sr_{0.97}TiO_3$, *J. Am. Ceram. Soc.* **17**, 201-205.
15. Porat, O. and Tuller, H.L. (1997) Simplified analytical treatment of defect equilibria: application to oxides with multi-valent dopants, *J. Electroceramics* **1**, 41-49.
16. Pierret, R.F. (1996) *Semiconductor Device Fundamentals*, Addison-Wesley, Reading, MA.
17. Catlow, C.R.A. (1981) Defect clustering in nonstoichiometric oxides, in O.T. Sorensen (ed.), *Nonstoichiometric Oxides*, Academic Press, New York, pp. 61-99.

INTERFACES

JOACHIM MAIER

Max-Planck-Institut für Festkörperforschung

Heisenbergstraße 1, 70569 Stuttgart, Germany

1. General Remarks

The term "Interface" denotes strictly speaking the two-dimensional transition region between two three-dimensional regions which are homogeneous in equilibrium. The transition region is, however, not necessarily restricted to a single discontinuity plane and may form an extended region itself. Even if the structural adjustment is restricted to a single plane or a few planes (in the following termed core region), there is — also in equilibrium — usually a zone which is characterised by point defect (i. e. charge) redistribution. In non-equilibrium also one-dimensional defects (dislocations) can be induced by the proper boundary (which itself may be conceived in some cases as a dislocation array or even as a point defect aggregation). Depending on the fact whether or not the two connected regions are different phases (of different or equal composition) the interfaces are named phase boundary or grain boundary. Whilst the latter is a non-equilibrium boundary (in all cases of interest), the first one may also — depending on the situation — be present in thermodynamic equilibrium. Another distinction refers to the crystallographic junction of the two lattices (Fig. 1). If in the ideal case

H.L. Tuller et al. (eds.), Oxygen Ion and Mixed Conductors and Their Technological Applications, 75–121.

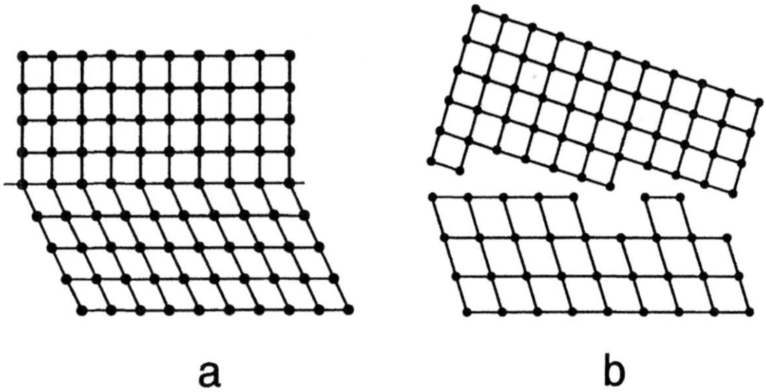

Figure 1. Coherent (l.h.s.) and incoherent (r.h.s.) solid/solid interface [1].

the proper plane of discontinuity is element of both lattices, the interface is a coherent interface. The opposite is usually called an incoherent interface. In this case often amorphous "inter-phases" are forming the region of structural adjustment. The majority of interfaces are more or less of intermediate type ("semi-coherent"), and may be characterised by a certain fraction of coincidence points (atomic positions belonging two both lattices). These considerations have to be taken "cum grano salis" since slight changes in bond lengths and strengths are almost ubiquitous.

2. Interface Thermodynamics

Without restriction of generality we consider the contact of two phases α and β. The extended nature of the interface (of area Σ) is taken into account by the definition (Gibbs) of excess quantities $G^{\Sigma}, S^{\Sigma}, n^{\Sigma}$. They refer to the integral of changes introduced by the contact compared to the homogeneous bulk phases. The differential of the excess free energy follows as $(\underline{\mu} \equiv (\mu_1, \mu_2, \ldots), \underline{n} \equiv (n_1, n_2, \ldots))$

$$dG^{\Sigma} = d\left(G_{\text{total}} - G_{\alpha} - G_{\beta}\right) = -S^{\Sigma}dT + \underline{\mu}d\underline{n}^{\Sigma} + \gamma d\Sigma. \tag{1}$$

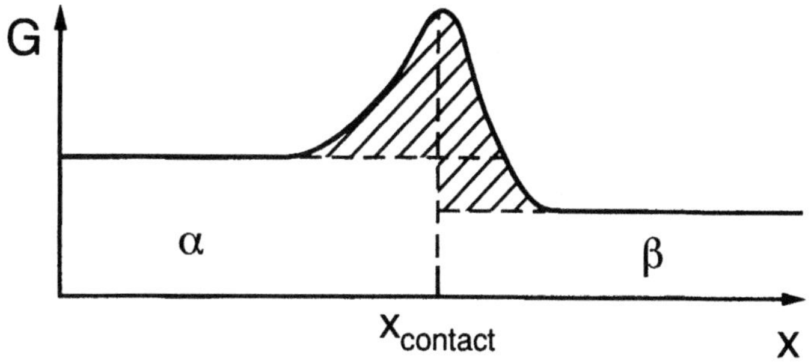

Figure 2. The shaded area corresponds to G^{Σ} as defined in the text.

G_{total} is the integral over the entire system, G_{α} (or G_{β}) is the integral over the α-phase (or β-phase) up to the contact (s. Fig. 2):

$$dG_{total} = Vdp - SdT + \underline{\mu}d\underline{n} + \gamma d\Sigma$$
$$dG_{\alpha,\beta} = V_{\alpha,\beta}dp - S_{\alpha,\beta}dT + \underline{\mu}d\underline{n}_{\alpha,\beta}. \tag{2}$$

The central quantity is the surface tension which can be defined as the increase of the Gibbs energy when at constant pressure, temperature and composition the interfacial area is increased

$$\gamma = \left.\frac{\partial G}{\partial \Sigma}\right)_{p,T,\underline{n}} = \left.\frac{\partial G^{\Sigma}}{\partial \Sigma}\right)_{T,\underline{n}^{\Sigma}} \tag{3}$$

or as increase of G^{Σ} with Σ at constant T and excess composition. The term γ has to be distinguished from the so-called surface stress. While the first corresponds to the work of creating a unit area of new surface, the latter refers to the work involved in deforming a surface. The difference between surface stress and surface tension is given by the dependence of γ on the deformation (see e.g. Ref. [2]).

As can be seen from Eq. (1), for the interface γ plays a role analogous to the pressure in the mechanical volume work term $\delta w = -pdV$. The generalisation is given by

$$\delta w = -\int_V dV \Sigma_{ik} p_{ik} \delta e_{ik} \tag{4}$$

(p_{ik}: components of the pressure tensor, e_{ik}: components of the deformation tensor (see e. g. Ref. [3])). For a plane interface (with the normal vector parallel to $\underline{x}$) it holds that

$$\delta w = -p_N dV + d\Sigma \int_{-\infty}^{+\infty} (p_N - p_T)\, dx, \tag{5}$$

p_N being the normal component which is spatially invariant (unless the interface is curved) and p_T being the x-dependent tangential component. It follows that

$$\gamma = \int_{-\infty}^{+\infty} (p_N - p_T)\, dx. \tag{6}$$

Since γ is positive in all cases of our interest, interfaces are only present in thermodynamic equilibrium if they are unavoidable, e. g. as surfaces (or generally phase boundaries) due to mass conservation. All grain boundaries considered in our context are non-equilibrium (frozen-in) phenomena (see also Table 1). In particular, the configurational entropy of two- (also one-) dimensional defects is (unlike point defects) extremely small and does not give rise to a realistic equilibrium concentration (see Table 1, in the cases of interest). To be exact it has to be mentioned that the existence of the interface edges requires the introduction of line tensions (cf. also point tensions for corners). The obvious first question in this context is: What is the equilibrium form of a crystal embedded in a fluid environment (e. g. crystal in air)? Obviously for an isotropic phase (i. e. γ = const) the minimum condition

$$\left(\delta \int \gamma d\Sigma\right) = 0 \tag{7}$$

is fulfilled (at given volume) for a sphere (smallest Σ), whereas for a given crystal γ depends on orientation and the problem is more complicated [3]. The solution is given by the so-called Wulff construction (Fig. 3), the information content of which is the following: One draws lines from a central

TABLE 1. Statistics and configurational entropy

Statistics and Configurational Entropy

Point Defects

N_0 point defects in a system of N atomic sites

$$S = k \ln \begin{pmatrix} N \\ N_0 \end{pmatrix} = \ln \frac{N!}{N_0!(N-N_0)!}$$

$$\mu_0^{id} = \mu_0^0 + RT \ln \frac{N_0}{N-N_0} \simeq \mu_0^0 + RT \ln \frac{N_0}{N}$$

Vacancies in an elemental crystal : $\mu = 0$

$$\boxed{\hat{x}_0 = \frac{\widehat{N}_0}{N} = \exp - \frac{\mu_0^0}{RT}}$$

Higher Dimensional Defects

analogously, but $\widehat{N} \ll 1$!

Example: N_2 boundaries, $3N^{1/3}$ possibilities

$$S = k \ln \begin{pmatrix} 3N^{1/3} \\ N_2 \end{pmatrix}$$

$$\mu_2^{id} = \mu_2^0 + RT \ln \frac{N_2}{3N^{1/3}}$$

$$\widehat{N}_2/N^{1/3} \simeq \exp - \frac{\mu_2^0}{RT}$$

$$\mu_2^0 = \alpha\mu_0^0 \qquad \left(1 \ll \alpha < N^{2/3}\right)$$

$$\boxed{\widehat{N}_2/\widehat{N}_0 = (\hat{x}_0)^{\alpha-1}/N^{2/3} \simeq \text{zero}}$$

$$\left(\text{dislocations}: \ \widehat{N}_1/\widehat{N}_0 = \hat{x}^{\alpha-1}/N^{1/3} \simeq \text{zero}\right)$$

typically $\widehat{N}_0 \simeq 10^{15}, \hat{x}_0 \simeq 10^{-6}, N \simeq 10^{21}$

$$\widehat{N}_2\,(\alpha = 1) = 10$$

$$\widehat{N}_2\,(\alpha = 2) = 10^{-5}(!)$$

$$\widehat{N}_2\,(\alpha = N^{2/3}) - \left(10^{-6}\right)^{10^{14}} \quad 10 \lll 1$$

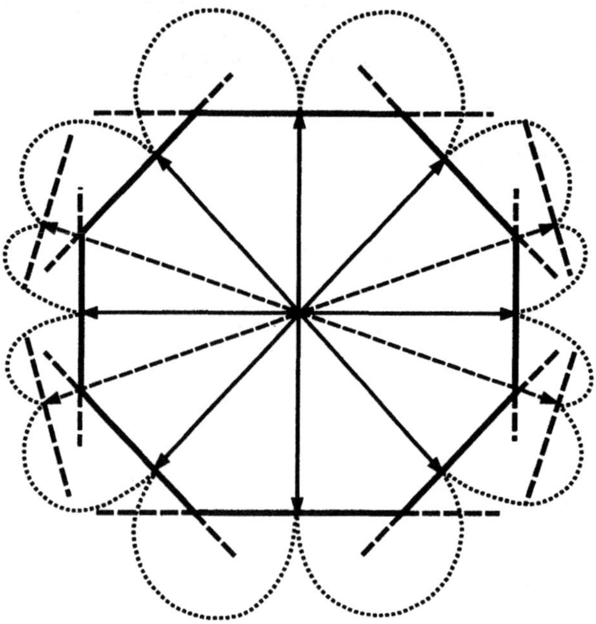

Figure 3. Wulff construction leads to the equilibrium shape of a crystal.

point, the length of which (h) is proportional to the surface tension of the corresponding orientation (γ/h is constant). The inner envelope then represents the equilibrium form. Unfortunately, this result is not of great significance, since large crystals usually do not reach morphological equilibrium due to kinetic requirements, and in the case of very small crystals, additional effects, line tensions as well as point tensions must not be neglected. In addition, the standard chemical potentials in the "bulk" may be changed (see below).

2.1. LOCAL EQUILIBRATION

Of practical importance are, however, local equilibration phenomena. Consider e. g. the phenomenon of faceting (Fig. 4). Let us assume that a certain boundary is formed at high temperatures and becomes morphologically unstable during cooling. A total rearrangement of the crystal may be impos-

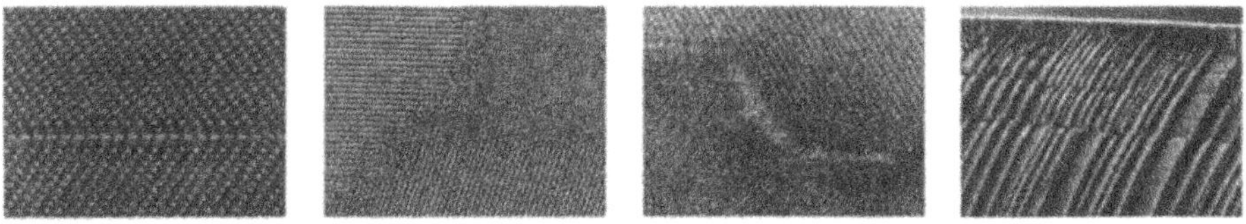

Figure 4. $SrTiO_3$ boundaries: Electron micrographs of a) $\Sigma 5$ grain boundary; b) triple grain junction (with amorphous phase); c) faceted grain boundary, d) crack surface [4, 5, 6]. (In the Σ-nomenclature, the number following Σ gives the inverse fraction of coincidence point.)

sible, nevertheless a local reconstruction under the formation of interfacial planes of lower energy even at the expense of a larger overall area may occur. In all these cases, local force considerations are the convenient way of considering the matter. If we neglect orientational dependencies, the contact equilibrium (Fig. 5) of three phases α, β, γ is given by [7, 8]

$$\gamma_{\alpha\beta}\underline{e}_{\alpha\beta} + \gamma_{\alpha\gamma}\underline{e}_{\alpha\gamma} + \gamma_{\beta\gamma}\underline{e}_{\beta\gamma} = \underline{0}. \tag{8}$$

(If the orientation dependencies have to be considered, a rotational term has to be added [9].) Eq. (8) regulates, e.g., the equilibrium morphology of inclusions in a ceramic microstructure. Other examples (see e.g. [7]) are the wetting angle Θ of a liquid phase on a substrate

$$\gamma_{\mathrm{sg}} = \gamma_{\mathrm{sl}} + \gamma_{\mathrm{lg}} \cos\Theta, \tag{9}$$

or the grooving angle ψ of a grain boundary at a surface

$$\gamma_{\mathrm{grain,grain}} = 2\gamma_{\mathrm{grain,air}} \cos\frac{\psi}{2}. \tag{10}$$

From Eq. (8) also follows the result that the ideal three grain contact (same γ-values) angle is 120°. The kinetics of morphology evolution can be extremely complicated and is usually very hard to predict. A few simple examples are considered at the end of the paper.

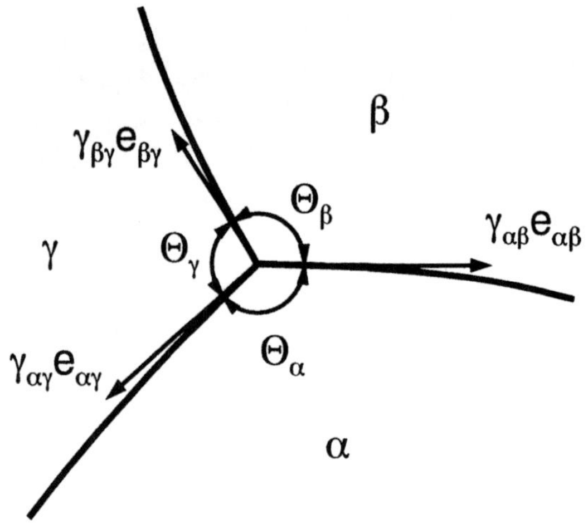

Figure 5. Force equilibrium at the three phase contact (neglect of dependence of γ orientation).

For the purpose of the workshop frozen-in interfaces are most important. They will be considered as metastable, timely invariant surface structure elements playing an important role as hosts, sources and sinks for charge carriers (point defects) and by setting boundary conditions for the carrier kinetics.

2.2. CHARGE CARRIER REDISTRIBUTION AT INTERFACES

Consider Fig. 6. The introduction of an interface (e. g. of a foil into water) means a symmetry break of the initial homogeneous situation (which implies structurisation). As a consequence, the particles dissolved in the homogeneous phase now become distributed differently. This is especially relevant to charge carriers. The affinity of the cation under consideration to get there may be higher relative to the one of the anion. As a consequence, the cation will be preferably adsorbed and the interfacial core will be positively charged. The effect is restricted in size by the appearance of an electric field. The field effect itself will be restricted in space, in that the excess of anions

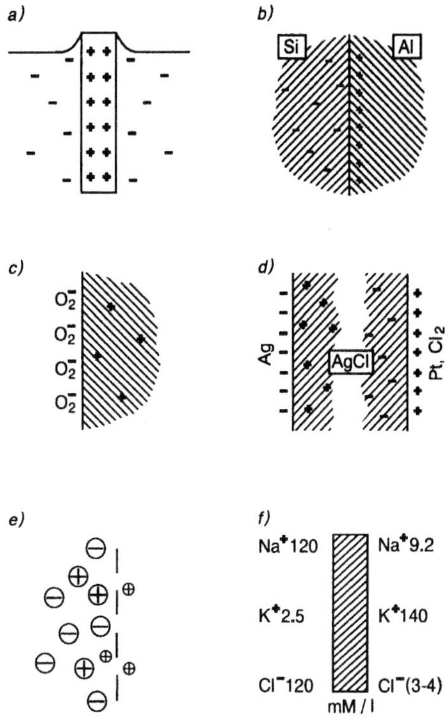

Figure 6. Examples of important interfaces (see text) [10].

in the neighbourhood will die out with increasing distance. The examples given in Fig. 6, metal-semiconductor interface, gas-semiconductor interface, electrode-electrolyte interfaces, carrier distribution over a frog muscle cell membrane, testify the extreme importance of the phenomenon of charge separation at interfaces for the field of physics (microelectronics), chemistry (electrochemistry, sensors, batteries) and biology (nerve propagation, electrophysiology). If the charged particles are considered as components, electrochemical potentials have to be used in the thermodynamic relations (cf. electrocapillarity) instead of chemical ones. The fact that the gradients of the thermodynamic potentials have to vanish in equilibrium, implies that changes of the chemical potentials occur at the expense of electrical fields.

In the following we will adopt the core-space charge model which is set out in more detail in Ref. [11, 12]. There we replace the boundary region by a zone of distinct structure (core) and neglect any structural changes outside. In other words, the standard chemical potential changes in a step function way over the core. The neglect of structural changes outside the core is of course an approximation (as can be seen from the band gap change). In this approximation the bending of μ means a bending of ln (activity) and, thus, everything is cooked down to concentration changes of the charge carriers. If the point defect concentration is low, it follows (from the uniformity of the electrochemical potential of k) that

$$\left(\frac{c_k}{c_{k\infty}}\right)^{1/z_k} \equiv \zeta_k^{1/z_k}(x) = \exp - \left(F\frac{\phi(x) - \phi_\infty}{RT}\right). \tag{11}$$

Eq. (11) predicts a definite change of each carrier in the space charge field solely determined by its charge. The calculation of the space charge field (which itself depends on the the c_k's) requires the solution of Poisson's equation which contains the full electrostatic information as long as time dependent magnetic fields are excluded (see Table 2):

$$\left(\partial^2/\partial x^2\right)\phi = -\frac{\Sigma_k z_k c_k}{\varepsilon}. \tag{12}$$

The condition under which electroneutrality holds are investigated in Refs. [13, 14]. How $c_k(k)$ and the conductance can be solved using ϑ, the "degree of influence", as a parameter and how conductivity effects can be calculated is shown extensively in Ref. [15]. Here we will consider only two simplified ("Debye case" and "Schottky case") examples depending on whether the space charge layer majority carrier (index 1) is mobile or not. In both cases we will assume that $|\vartheta| \longrightarrow 1$, i.e. strong interfacial effects. In the first case, $\Delta\phi'' \propto c_1(x) \propto \exp - (z_1 F \Delta\phi/RT)$ or $(\ln c_1)'' \propto c_1$. This is solved by ($\xi = x/\lambda$, λ: Debye length, $\zeta = c/c_\infty$, the index 0 refers to the first layer adjacent to the core)

$$\zeta_1 = \frac{\zeta_{10}}{\left(1 + \sqrt{\zeta_{10}}\,\xi/2\right)^2} \tag{13}$$

TABLE 2. Poisson-Equation

Poisson-Equation

a) Maxwell-Equation for the case of the electric field ($\underline{E}$) and the absence of time dependent magnetic fields.

$$\nabla x \underline{E} = \begin{pmatrix} \partial E_z/\partial y - \partial E_y/\partial z \\ \partial E_x/\partial z - \partial E_z/\partial x \\ \partial E_y/\partial x - \partial E_x/\partial y \end{pmatrix} = \underline{0}$$

Since $\partial E_i/\partial j = \partial E_j/\partial i$,

$$E_x dx + E_y dy + E_z dz = \underline{E} d\underline{r}$$

is a total differential which we call $-d\phi$

Hence

$$\underline{E} = -\nabla\phi.$$

b) Maxwell-Equation for the divergence of the dielectric displacement $\underline{D}$

$$\underline{\nabla}\,\underline{D} = \underline{\nabla}\,(\varepsilon\underline{E}) = \varrho.$$

c) Poisson's equation follows

$$\boxed{\nabla^2\phi = -\varrho/\varepsilon}$$

d) At boundaries: Gauss' law results

$$\Delta\nabla\phi = \Sigma/\varepsilon$$

$$\left(\int_V \nabla(\nabla\phi)dV = \int_O (\nabla\phi)d\underline{a} = \int_a (\nabla\phi)_{II}da - \int_a (\nabla\phi)_I da = -\int_V \varrho/\varepsilon dV \right.$$

$$\left. = -\int_a \Sigma/\varepsilon da\right)$$

Of great significance are conductance effects caused by charge carrier enhancements near the interface. The result for the excess conductance parallel to the interface is obtained by integration to be (u: mobility)

$$\Delta Y_1^{\|} = z_1 F u_1 (2\lambda)\sqrt{c_{10} c_\infty} = u_1 \sqrt{2\varepsilon RT c_0}. \tag{14}$$

The second limiting case which is especially met in doped systems in which the (native) counter carrier (index 2) is depleted. There $\phi'' = \text{const}$, thus ϕ is quadratic in x and as a consequence

$$\zeta = \exp - \left|\frac{z_2}{z_1}\right| \left(\frac{x - \lambda^*}{\lambda}\right)^2. \tag{15}$$

In such a Schottky barrier situation the depletion resistance follows as

$$\Delta Z^{\perp} = \frac{\lambda^*}{|z_2| F u_2 c_{20}^*}. \tag{16}$$

with c_{20}^* being given by $c_{20}\left(2 \ln \frac{c_\infty}{c_{20}}\right)$ [16]. In the extreme case of a rectangular profile ("depth" c_{20}, width λ^*) $c_{20}^* = c_{20}$. It is worth noting that the width of a Schottky barrier λ^* can — in contrast to the first case — be wider than the Debye length (such that even in YSZ in which λ is negligible ($\sim$ 1 Å), Schottky barriers may have an extension of $\sim$ 10Å (all defects are considered to be completely dissociated, space charge field is assumed to be $\sim$ 1V, T = 1000 K)).

Let us briefly consider some examples involving ionic and mixed conductors.

Let us first concentrate on the Debye case. Figure 7 gives different situations of interest, involving ionic conductors. In the first example the mobile cation of the conductor is adsorbed at the interface of the surface active but insulating second phase. This refers to the well-known effect that admixtures of Al_2O_3 and other insulators to ionic conductors can enhance the ionic conductivity by orders of magnitude. In many cases this can be

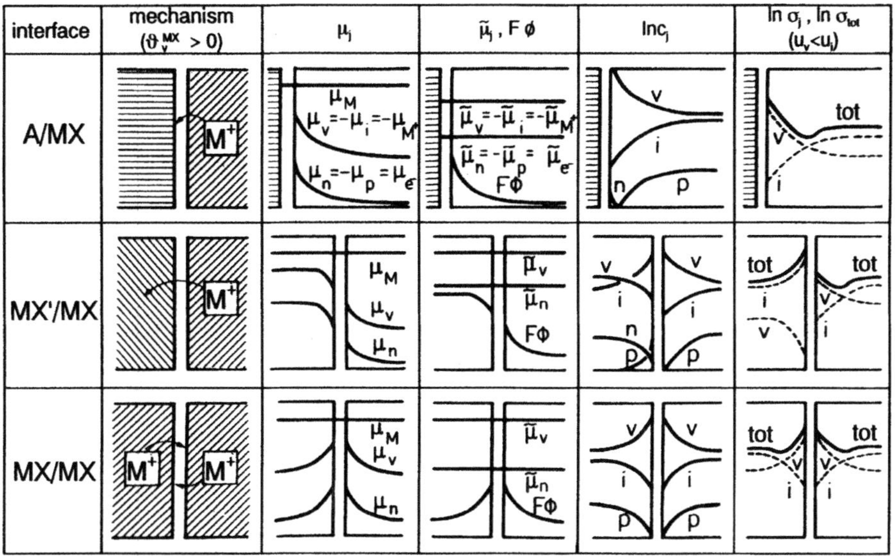

Figure 7. Solid/Solid interfaces involving ionic conductors [10].

fully understood by the above space charge effect and even far-reaching predictions can be made. The situation is — in many respects — analogous to the homogeneous doping and has been termed "heterogenous doping". Knowing the surface change (sign of interaction), one can predict the effect of each single defect type, as it is possible in conventional doping when the effective charge of the dissolved dopant is known (cf. "rules of homogeneous and heterogenous doping" [17]). Similar effects occur in fluorides (CaF_2, PbF_2). There the more acidic SiO_2 is more active than Al_2O_3 as a second phase, resulting in F^- adsorption [18]. (Finally, a charge segregation combined with space charge effects can also occur in grain boundaries of polycrystalline materials.)

While the picture seems to fully account for the conductivity enhancement in the systems AgCl-Al_2O_3 and AgBr-Al_2O_3 [11], in AgI-Al_2O_3 even surface phase transformations occur with great impact on the electrical properties [19]. The driving force stems from the surface-surface interaction. Note that there is also a relation between defect-defect interactions and phase transitions which may be relevant in this context by setting an upper limit. In the bulk of AgI experiment and theoretical considerations show that the excess term of the chemical defect potential scales with the mean distance between the defects and, thus, the cube root of the concentration [20]. Conductivity anomalies and even the phase transition temperature can be interpreted this way. Similar interactions may lead to surface phase transitions (consider e. g. an already partly disordered core which takes up additionally segregated ions).

In all cases it is for a proper analysis, indispensable to heal out one-dimensional defects (dislocation) formed during the preparation [21, 15, 22].

An equally important case is met in the miscibility gap of two ionic conductors, e. g. AgCl-AgI (first row in Fig. 7), there space charge effects can occur even if no change is stored at the interfacial core directly, since ions can redistribute over both space charge layers [23]. An extensive treatment is given in Ref. [24].

Let us concentrate for a moment on grain boundaries (second row in Fig. 7). Here a space charge effect is expected to occur since the core offers different standard formation energies. In the silver halides an accumulation of Ag^+ is very probable according to the temperature dependencies involved. The resulting profiles are shown in Fig. 8. In pure materials an inversion from interstitial conduction to vacancy conduction results. Such a profile itself can give rise to significant anisotropies, the difference of the parallel (along the interface) and perpendicular (across the interface) effects is additionally influenced by the core. In the silver halides the space charge layers lead to highly conducting parallel pathways short-circuiting the bulk paths whereas additional transfer resistances may appear due to the boundaries

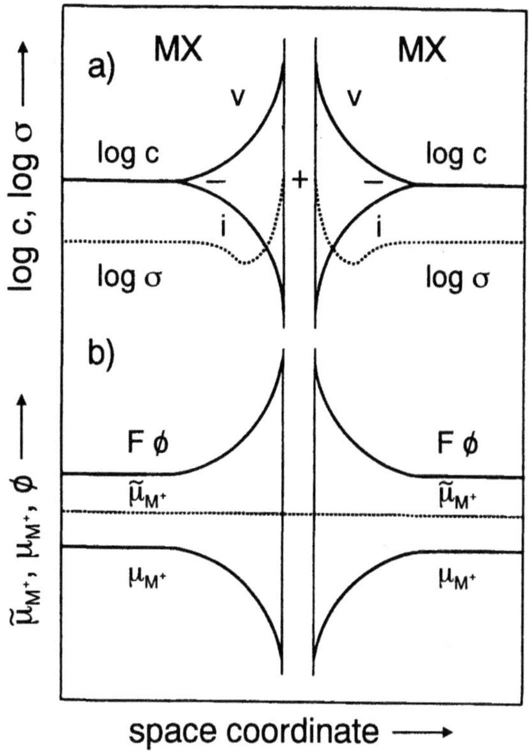

Figure 8. A nucleophilic effect (accumulation of cations) of a grain boundary: a) defect concentrations, b) electrochemical functions [25].

perpendicular to the bulk. As a consequence upon grain growth the low frequency semicircle representing the latter effect shrinks in the impedance plot while the high frequency one increases due to a loss of boundary effects [24]. It is worth mentioning that the (serial) grain boundary resistance is partly due to current constriction effects. This often neglected phenomenon arises from a lateral inhomogeneity, as e.g. realised by a porous boundary. The porosity is "ignored" at high frequencies but reduces the d.c. current through the boundaries. (The same happens in the composites. But note, that under the bottom line the resistance decrease by parallel pathways is much more significant for the overall d.c. current.) The superposition can be analysed in a brick layer model. Recently the influence of more realistic patterns has been investigated for the special case of (isotropic) low conducting boundaries by finite element computations [26]: The same numerical method can be applied to make use of the current constriction-effect for local conductivity measurements (favourably coupled with atomic force microscopy) [27]. One example to be mentioned in this context is the measurement of highly conducting surface layer on top of a AgCl single crystal if the crystal surface is polished. After the treatment the damaged region (a few μm thick) contains dislocations and grain boundaries. The first can be easily healed out by annealing [27]. Again, this points towards the necessity of careful pre-treating in order to obtain steady state values.

In nanocrystalline materials the fraction of interfaces is unusually large and, thus, core and space charge effects are of extreme importance. If the grain size (ℓ) under-steps the width of space charge layers, the effect qualitatively changes in that even in the center the grain will be charged [28]. The additional conductivity effect can be described by a nano-size factor

$$\text{n.s.f.} = \frac{4\lambda}{\ell}\left[\frac{c_{10} - c_1^*}{c_{10}}\right]^{1/2} . \tag{17}$$

(The center concentration c_1^* depends on ℓ and the boundary concentration (c_0) in a complicated but defined way.)

In nano-sized CaF_2 the enhanced conductivity has been explained by space charges [29]. On the other hand, in nano-crystalline CeO_2 the conductivity modification has been ascribed to structural changes in the core [30]. Note that structural changes in the core necessarily demand space charges. A decisive question is in how far the core space-charge model with unchanged structure outside the core applies or if already there the structure gradually changes. On one hand it is shown that the energy levels (e.g. band gap) change as a function of grain size, on the other hand cluster chemistry and physics evidences that in many cases the bulk structure is essentially

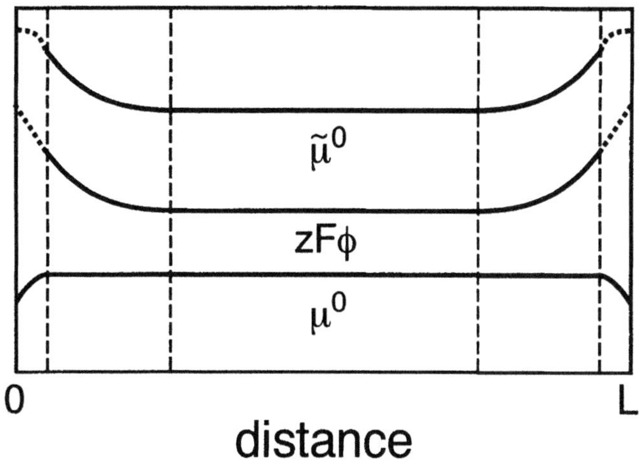

Figure 9. The profile in a thin film in which space charge and structural effects occur. A reduction of the thickness will lead to two mesoscopic effects.

reached already at small cluster sizes [31, 32]. If we also allow for a change of μ^0 (structural changes) outside the real core (where the counter charge is accumulated) say in a range of $0 < x < 2\ell$, while space charge effects decay within $0 < x < 4\lambda$, we expect two separate mesoscopic effects as long as $\ell \ll \lambda$ (Fig.9). Another important interface is the solid/gas contact. There the core is formed by particles adsorbed from the gas phase plus particles segregated from the solid bulk. Gas phases which are expected to interact with ions such as NH_3 (with Ag^+) or BF_3 (with F^-) show the expected depletion effects resulting in an enhancement of the respective vacancy conduction [15]. This can be used as a sensor effect for acid-base active gases (see "Solid State Gas Sensors").

As already mentioned, the concentration profiles can be described as a function of the bulk and boundary ($x = 0$) values. The dependence of the bulk value on the control parameters is quantitatively tractable for a simple defect chemical situation. This is more difficult for c_0; often even the direction of the effect (i. e. which defect type is enriched) is not so easy

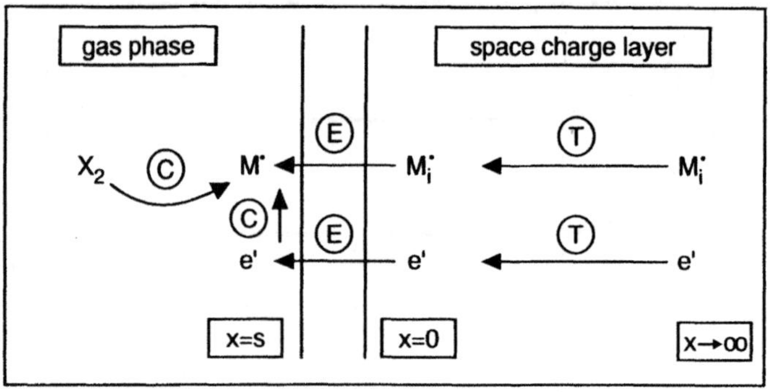

Figure 10. Boundary equilibration of a phase MX with X_2-gas.

to predict. Within the core space charge model, the most important bottlenecks for the analysis are the core defect chemistry as well as the details of the transition between core and space charge layer (see Fig. 10). The problem is treated in more detail in Refs. [11, 12]. The models of Kliewer and Köhler and Pöppel and Blakely (see Ref. [15]) represent very crude and partly inconsistent approximations. In order to obtain the dependence on the component activity (e.g. P_{X_2} or P_{O_2} in binary halides or oxides) a simplified procedure may be sufficient in some cases. In AgCl the overall excess conductance is proportional to the integral enrichment ($\propto \sqrt{c_0}$) and, thus, to the core charge (ρ_{co}). The change of the logarithm of c_0 and of $\Delta Y^{||}$ is then proportional to the relative change of ρ_{co} upon P_{Cl_2}. Since electronic effects will be quite restricted in AgCl, the quantities $\rho_{co}, \Delta Y^{||}$, and also the local point defect concentration of the majority carriers will not change significantly with P_{Cl_2}. Formulating a mass action law for the chlorine incorporation for a certain x-value, it is found that the minority carriers then vary with P_{Cl_2} as they do in the bulk $\left(P_{Cl_2}^{\pm 1/2}\right)$. Note that for the enhanced majority carriers the characteristic exponent of $\Delta Y^{||}$ turns out to be half of the bulk value (since $\Delta Y^{||} \propto \sqrt{c_0}$). For Schottky barriers some of these features are different [33]. Not only can the width be significantly larger than λ and will vary with the applied potential [34], but also the overall resistance $\Delta Z^{\perp}$ depends more sensitively on the component partial pressure. For the extreme case of an abrupt layer, the analysis shows that $\rho_{co}\,(\delta\rho_{co}/\delta \ln P)$ rather than $\rho_{co}^{-1}\,(\delta\rho_{co}/\delta \ln P)$ is decisive in this case [33]. Fe-doped $SrTiO_3$ grain boundaries reveal pronounced Schottky barriers. There the space charge potential is governed by frozen-in cation segregation [35, 36] and, thus, essentially constant. As Eq. 16 shows (and in contrast to $\Delta Y^{\parallel}$ in the Debye case) $\Delta Z^{\perp}$ is inversely proportional to the bulk concentration, and in this case characterised by the characteristic exponent of the bulk. Figure 11 confirms this picture: R_{gb}^{-1} varies with $P_{O_2}^{+1/4}$ (hole conduction) while the capacitance is a constant. Even though the relationships can be given to a much better approximation than above, the general case, however, which combines enhancement layers and Schottky barriers cannot be solved [37]. However, the space charge capacitance can be obtained under much more general conditions, since only the first integration is necessary. In Fig. 12 the capacitance change is shown for different potentials. Curves like this have not yet been observed for solid electrolytes probably due to the interference with adsorption capacitances or due to finite size effects. For intrinsic carriers

$$C_{sc}/a = \frac{\varepsilon}{\lambda} \cosh \frac{|z|\,F\,(\phi_0 - \phi_\infty)}{2RT} \tag{18}$$

is obtained. The difference $(\phi_0 - \phi_\infty)$ reflects the applied voltage (corrected with respect to point of zero charge). For large effects an exponential dependence follows, while close to the point of zero charge (small effects)

$$C_{sc}/a \simeq \varepsilon/\lambda \tag{19}$$

holds valid. A similar result is fulfilled for the Schottky case, in which the charge is proportional to doping concentration m and width λ^*. From there

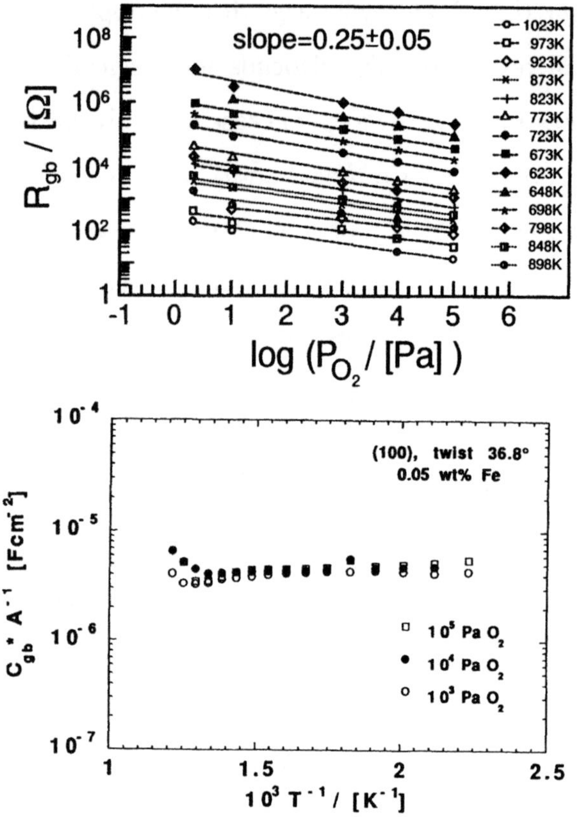

Figure 11. Partial pressure dependence of the Schottky barrier resistance in bicrystalline $SrTiO_3$ [33, 5]

we have [34]

$$C_{sc}/a = \frac{d\left(zFm\lambda^*\right)}{d\lambda^*}\frac{d\lambda^*}{d\left(\phi_0-\phi_\infty\right)} = \sqrt{\frac{|z|\,Fm\varepsilon}{2\left(\phi_0-\phi_\infty\right)}} = \frac{\varepsilon}{\lambda^*}. \qquad (20)$$

However, now λ^* is voltage dependent reflecting the fact that we face large effects here.

It has to be added that the total capacitance (without adsorption) includes a rigid capacitance value according to $\frac{\delta(\phi_s-\phi_\infty)}{\delta\rho_{co}} = \frac{\delta(\phi_0-\phi_\infty)}{\delta\rho_{co}} + \left(\frac{C_{sc}}{a}\right)^{-1}$. The last term refers to $\frac{\delta(\phi_s-\phi_0)}{\delta\rho_{co}}$, where s denotes the core layer. Since between x = s and 0 there is no charge density, ϕ is linear in x with the slope $\phi'|_{x=0} \propto \rho_{co}$. As a consequence $\left(\phi_s-\phi_0\right)/s \propto \rho_{co}$ and, thus $C_H/a = \varepsilon/s$.

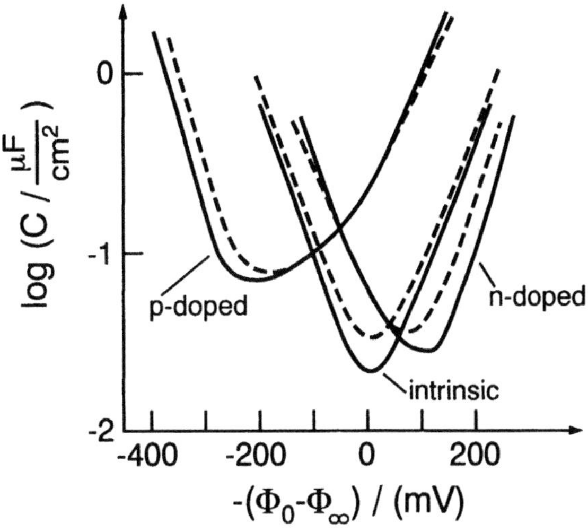

Figure 12. Double layer capacitance as a function of bias [37] (corrected with respect to flat band potential).

3. Kinetics

In this context the term kinetics means kinetics of compositional changes such as stoichiometry changes (e.g. oxygen incorporation in oxides) or compound formation (catalysis, corrosion) which usually imply ambipolar fluxes of at least two carriers. Even if we concentrate on the interior, external interfaces play an important role as boundary conditions, and grain boundaries act as pathways or hindrances for the chemical transport. In a chemical relaxation experiment we suddenly change the outer partial pressure and symmetrical diffusion profiles evolve as shown in Fig. 16. Figures 13, 14, 15 give an overview on the basic laws of (electro-)chemical kinetics, irreversible thermodynamics and their range of validity. For more details cf. [17].

If the surface reaction is very fast the boundary conditions are given by the new equilibrium concentration. The time dependence yields the chemical diffusion coefficient $\tilde{D}$. Besides this chemical relaxation we can also consider the transient or the steady state cases of chemical polarisation

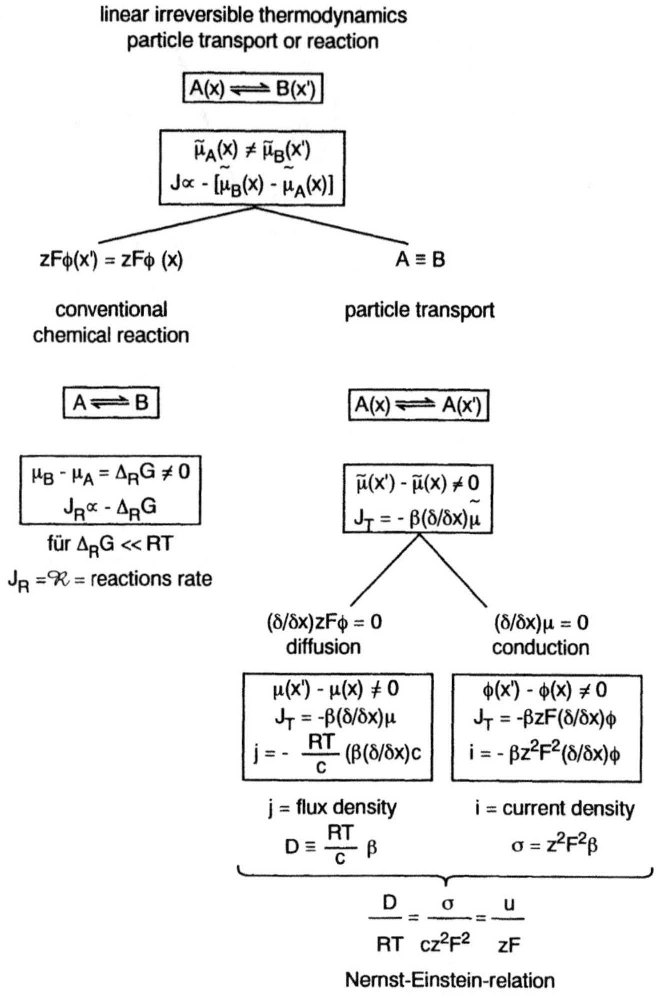

Figure 13. Electrochemical, chemical and diffusional process in terms of irreversible thermodynamics.

or depolarisation. As far as electrochemical measurements are concerned we can distinguish between electrochemical polarisation or depolarisation methods with (one or two) reversible, ionically blocking and electronically blocking electrodes. All combinations are possible. Generally from the transients $\tilde{D}$ can be evaluated, whilst the steady state, in which a linear profile establishes, yields partial conductivities [38]. These cells can be studied — in an electrical or chemical sense — by galvanostatic, potentiostatic, or

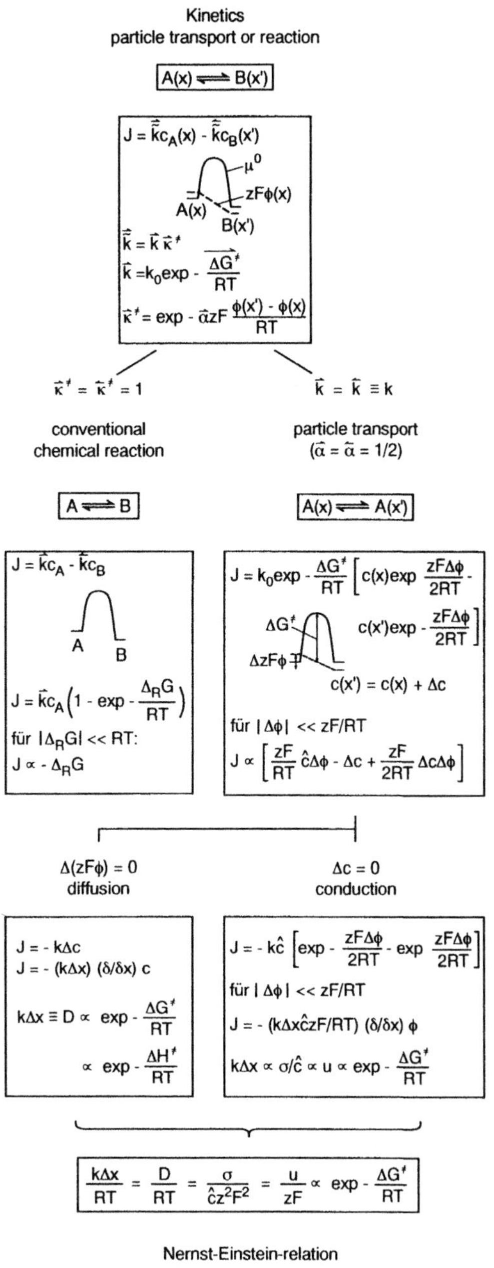

Figure 14. Electrochemical, chemical and diffusional process in terms of simple chemical kinetics (no interactions).

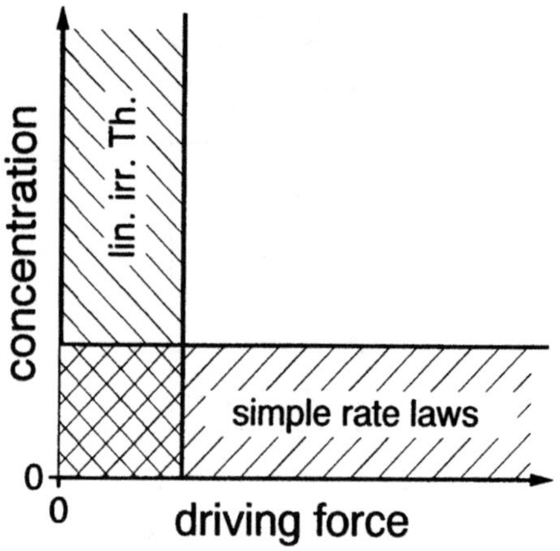

Figure 15. While linear irreversible thermodynamical laws are only valid close to equilibrium, the simple kinetic laws are also valid far from it but presuppose dilute cases. The range of validity of linear relationships usually decreases from Fick's law to Ohm's law and is extremely narrow for chemical reactions [17].

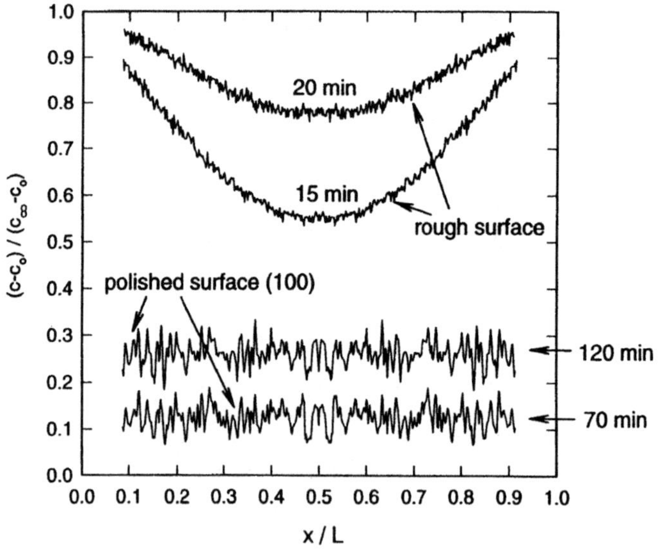

Figure 16. Concentration profiles during oxygen incorporation in $SrTiO_3$ (from optical absorption relaxation) [39, 4]. Top: diffusion controlled (rough surface), Bottom: surface reaction controlled (polished surface).

impedance measurements to mention the most common ones. The term galvanostatic, potentiometric and impedance can also be understood in chemical terms ($\Delta\mu_{O_2}$ as a driving force instead of voltage, chemical impedance etc.).

3.1. CHEMICAL DIFFUSION AND SURFACE REACTION

The chemical transport through boundaries is poorly understood if space charge layers are present (Fig. 17). While in the bulk $j_O = -\tilde{D}_O\,(\partial/\partial x)\,c_O$ holds, it does not apply for space charge effects since $(\partial/\partial x)\,c_O$ is not a good driving force ($\tilde{D} \propto \tilde{\sigma}/\tilde{c}$, $\tilde{\sigma}/\tilde{c}$ are essentially harmonic means of the individual σ, and c-values of the two species carrying the ambipolar flux). In equilibrium j_O vanishes, unlike $\partial c_O/\partial x$. The more general statement $j_O = -\tilde{\sigma}_O\,(\partial/\partial x)\,\mu_O$ is still valid for the steady state ($\tilde{\sigma}$: ambipolar conductivity). Let us consider a key example [40]. Let σ_+ and σ_- be the two decisive conductivities for the ambipolar process, $\left(\tilde{\sigma} = \frac{\sigma_+\sigma_-}{\sigma_+ + \sigma_-}\right)$ and let them be equal in the bulk. In the space charge regions the profiles bend upwards and downwards depending on charge and space charge potential. Since $\tilde{\sigma}$ is determined by the lowest value, $\tilde{\sigma}$ is in this case always decreased irrespective of the sign of the space charge potential. The effects

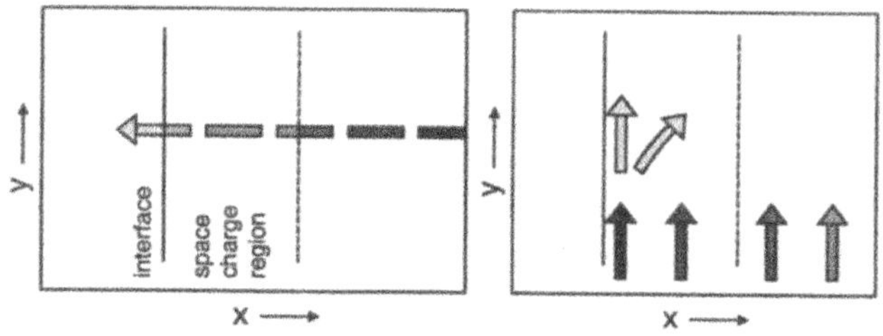

Figure 17. Chemical transport through and along interfaces (schematically) [33].

of equilibrium space charge layers on the chemical diffusion can be very pronounced. In particular worth mentioning is the jam of charge in front of the boundary. The observed hindrance of oxygen diffusion by Schottky barriers can be measured and calculated [4, 40]. In Refs.[40, 41] the dependence of the out-coming flux in a permeation experiment on a bicrystal for different space charge potentials (the boundary of which is oriented perpendicular to the flux) is discussed. Here we will only consider the case that the thin grain boundaries are insulating. Then $\tilde{\sigma}$ will be composed of the σ_+ and σ_- values of the grain boundary (space charge regions). Since most of the stoichiometry changes occur in the bulk, the effective concentration $\tilde{c}$, i. e. the thermodynamic factor $d\mu_O/dc_O$ (or the chemical storage capacity $dc_O/d\mu_O$) is, however, determined by the bulk value. Hence the measured, overall value is:

$$\tilde{D}_m \propto \frac{\ell}{d}\frac{\tilde{\sigma}_{\text{grain boundary}}}{\tilde{c}_{\text{bulk}}} \tag{21}$$

(ℓ: grain size, d: grain boundary width) [41].

Another boundary problem of practical significance is the preferential diffusion along boundaries. This is even complicated without space charge regions. At a first glance one might assume that the flux lines are parallel to the boundaries and the problem may be solved by simple integration similar as in the conductivity case. Here, however, we change composition, and once the compound flux is faster along the boundary and occurred in a parallel way only, immediately there would be a gradient of the chemical compound potential and, thus, a driving force perpendicular to the boundary. In essence the influx will broaden the deeper the boundary penetrates into the sample. Solutions to this, but without space charges, have been worked out by Fisher, Whipple, Preis and others [42]. The inclusion of space charges has still to be done. Only if $\tilde{D}$ is extremely fast along the boundaries compared to the bulk, the broadening can be neglected. On the time scale of grain diffusion, all the grain boundaries are immediately equilibrated and the diffusion problem is reduced to the in-diffusion into the grain interior (L: sample thickness)

$$\tilde{D}_m = \frac{L^2}{\ell^2}\tilde{D}_{bulk} \propto \frac{L^2}{\ell^2}\frac{\tilde{\sigma}_{bulk}}{\tilde{c}_{bulk}} \tag{22}$$

(time constant $\propto$ (thickness)$^2/\tilde{D}$).

Surface space charge layers or space charge layers at the contact to the electrodes are chiefly of impact on the chemical transport across boundaries. It is worth noting that surface space charge layers will even appear as a consequence of the chemical diffusion process if we start from the "flat-band" situation due to Gauss' law [40, 43].

Retardation of chemical diffusion by initial or induced space charge layers as well as by the surface reaction itself will show up in non-ideal diffusion profiles (Fig. 16) as indicated by the dotted line. The deviation in the surface concentration corresponds to the "jump of μ_O" over the boundary. In fact the flux can be taken as proportional to this value ($\tilde{k}'\Delta\mu_O = \tilde{k}\Delta c_O$) as long as we stay close to the equilibrium or deal with first order reactions.

If the surface reaction is rate limiting, the profiles will be horizontal (see Fig. 16). The surface process itself can be very complicated, it is composed of gas diffusion, adsorption, dissociation, ionisation, transfer through the interface (and space charge transport). Figure 16 compiles data obtained on the k rate constants ($\tilde{k}$) will be given in Ref. [44].

The elucidation of the elementary steps for gas-solid interaction is one of the most serious bottlenecks in solid state physical chemistry and is especially important for fuel cells and sensors where electrode kinetics limit low temperature application.

Much information was successfully extracted from tracer profiles. Rate constants (k_{Tr}) obtained in that way are not necessarily comparable to the above ones ($\tilde{k}$), since they may refer to different mechanisms. Moreover, the k-values are conceptionally different. The k_{Tr}-values certainly reflect better the kinetics of charge transfer reactions, while $\tilde{k}$-values are appropriate if the composition is changed. A treatment of the correlation of the empirical k-values obtained by tracer, electrical and chemical measurements between each other, with respect to the microscopic rate constants and with respect to the corresponding D-values will be given in Ref. [45]. Some of these problems will be treated in other contributions to this workshop. Here only a few related examples are to be considered.

3.2. ADSORPTION

As already mentioned we may distinguish between a pure chemical reaction ($A \rightleftharpoons B$), transport processes ($A(x) \rightleftharpoons A(x')$) and electrochemical reactions ($A(x) \rightleftharpoons B(x')$). In the latter case (as for transport) electrical fields are important if charged species are involved. As an example of a pure chemical reaction we will consider the adsorption process, while the electrochemical phase transfer reaction may serve as a prototype for an electrochemical step. Before doing so, let us make a few general remarks. If two reactions are in parallel, the fastest is rate limiting. If

$$\begin{aligned} A &\overset{\vec{k}_1}{\to} Z \\ A &\overset{\vec{k}_2}{\to} Z \end{aligned} \tag{23}$$

and $k_1 \gg k_2$ then

$$\mathcal{R} = \frac{d\,[Z]}{dt} = [A]\left(\vec{k}_1 + \vec{k}_2\right) \simeq k_1\,[A]\,. \tag{24}$$

For linear response the parallel switching is more generally expressed as

$$J = J_1 + J_2 = (\beta_1 + \beta_2)\,X \simeq \beta_1 X. \tag{25}$$

If they are in series

$$A \overset{\vec{k}_1}{\to} B \overset{\vec{k}_2}{\to} Z \tag{26}$$

the slowest one is determining. In the general linear formulation this is generally expressed by

$$X = X_1 + X_2 = \left(\beta_1^{-1} + \beta_2^{-1}\right) J \simeq \beta_2^{-1} J. \tag{27}$$

In this way the total reaction scheme can be condensed essentially to a rate limiting step. In the steady state of a sequence

$$A \rightleftharpoons B \rightleftharpoons \ldots K \overset{\overrightarrow{k}_{lim}}{\rightarrow} L \rightleftharpoons \ldots \rightleftharpoons Z, \tag{28}$$

the conversion from K to L may be rate limiting. For each partial step ($i \rightleftharpoons j$) the rate is given by

$$\overrightarrow{\mathcal{R}}_{ij} = \overrightarrow{k}_{ij}\,[i] - \overleftarrow{k}_{ij}\,[j] = \overrightarrow{k}_{ij}\,[i]\left(1 - \frac{[j]\,/\,[i]}{K_{ij}}\right), \tag{29}$$

if $i = K$ and $j = L$, the rate is by definition very low, viz. $\overrightarrow{\mathcal{R}} \simeq \overrightarrow{k}_{lim}\,[K]$, since $\overrightarrow{k}_{lim}$ is small. If all rates are similarly small — in the steady state they are all equal — for the other steps now the term $\left(1 - \frac{[j]/[i]}{K_{ij}}\right)$ is close to zero. Hence they are quasi in equilibrium. We finally obtain for the overall rate $\mathcal{R} = k_{eff}[A]$ (k_{eff} contains $\overrightarrow{k}_{lim}$ and equilibrium constants). The same applies of course for a coupling of chemical reaction and transport

$$A\,(x_s) \rightleftharpoons B\,(x_s) \rightleftharpoons B(x), \tag{30}$$

as in the case of oxygen incorporation. If diffusion is rate limiting, A and B at x_s are in chemical equilibrium, i. e. the surface immediately follows a P_{O_2}-change. If the surface reaction is rate limiting, the transport process is in equilibrium, i. e. uniform profiles are established (as discussed above).

The equality of the rates is restricted to the steady state, in the transient the situation may be more delicate, since the rates change as a function of time. We can, however, make the following statement: Those steps are in quasi-equilibrium, the rate of which ($\overrightarrow{\mathcal{R}}_{ij} - \overleftarrow{\mathcal{R}}_{ij}$) is depressed much below the level of the partial rates $(\overrightarrow{\mathcal{R}}_{ij} - \overleftarrow{\mathcal{R}}_{ji} \ll \overrightarrow{\mathcal{R}}_{ij}, \overleftarrow{\mathcal{R}}_{ij})$ by virtue of a serial coupling to slow steps. Close to equilibrium $\overrightarrow{\mathcal{R}}$ and $\overleftarrow{\mathcal{R}}$ reduce to the exchange rate which is inversely proportional to the respective reaction resistance (cf. linearisation of Eq. 41). In the sequence the overall driving force is roughly given by the driving force of the rate limiting process. The effective rate constant is proportional to the average over the forward and backward rate constants of the slowest process.

Let us first consider a simple sorption reaction (s) of the type

$$A(g) + V_A \rightleftharpoons A_A \tag{31}$$

with the rate

$$\mathcal{R}_{ad} = \overrightarrow{k}_S \, P_A \, (1 - \theta_A) - \overleftarrow{k}_S \, \theta_A . \tag{32}$$

This Langmuir-equation assumes that rate constants are independent of the coverage θ_A. In equilibrium

$$\frac{\theta_A}{P_A \, (1 - \theta_A)} = K_S = \frac{\overrightarrow{k}_S}{\overleftarrow{k}_S} . \tag{33}$$

The adsorption process is important, e.g. for electrochemical processes in fuel cells or sensors but also for heterogenous catalysis. There the adsorbed species undergoes a reaction and the product is desorbed. Heterogenous catalysis is already efficient by leading to a high effective concentration in the adsorption layer, so to speak, by condensing the reaction partner in a narrow (two-dimensional) reactor. In addition, the bond to the surface may be helpful in lowering the (effective) activation energies. Generally a catalyst (cat) is defined by $dK_{eff}/d\,[cat] = 0 \neq dk_{eff}/d\,[cat]$ (K_{eff}: effective mass action constant, k_{eff}: effective rate constant). In the ammonia synthesis e.g. the bond to the Fe-catalyst is a prerequisite of breaking the N_2-triple bond. Highest catalytic activity is often provided by a mean reactivity. If e.g. the adsorption strength is too low, the adsorption will not occur efficiently, if it

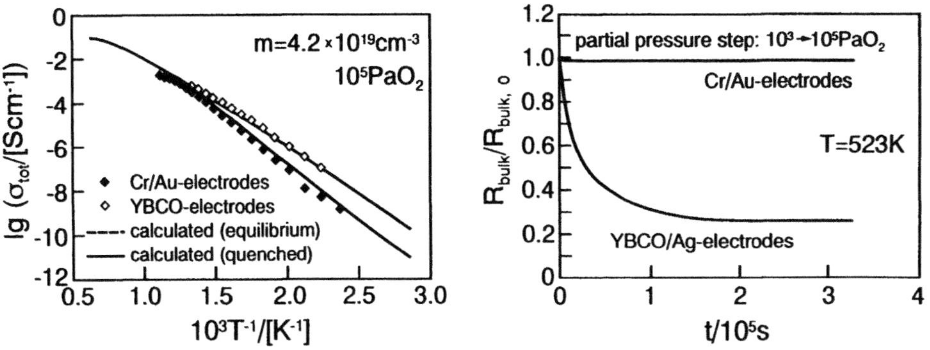

Figure 18. The use of $YBa_2Cu_3O_{7-\delta}$ electrodes enhances the surface reaction rate. Equilibriums profiles can be observed also at lower temperatures [4].

is too high, the further reaction steps may be difficult. Curves of the type catalytic activity versus reactivity exhibit then a maximum and are called volcano curves. Another general observation is that mixed conductors are often good catalysts. The supply of reactive centers is easy in those materials due to a high electronic and ionic defect density and due to a fast diffusion the desired species to the site under regard.

Figure 18 shows that coatings by $YBa_2Cu_3O_7$ on $SrTiO_3$ increase the rate of the surface reaction such that an overall equilibrium can be reached even at quite low temperatures (according to $\tilde{D}$) [4]. Often the importance of acid-base active centers (i.e. ionic point defects) is underestimated in heterogeneous catalysis. Their significance is well known in homogeneous catalysis. Simkovich and Wagner [46] have demonstrated the catalytic activities of silver vacancies in AgCl for the elimination of HCl from tertiary butyl/chloride. The effective rate constant increases linearly with the $CdCl_2$-impurity content (i.e. linear with $\left[V'_{Ag}\right]$). Also heterogeneous doping ($AgCl{:}Al_2O_3$) proved effective [47]. In these reactions, the process seems to be rate limiting and accelerated by basic centers (V'_{Ag}), also the counter species ($Cd^{\cdot}_{Ag}$, $Ag^{\cdot}_{ad}$) may be active as acid centers.

Here we follow the work by Wang and Nowick [48] on electrode kinetics of Pt, O_2/CeO_2 (doped). We do not want to discuss the relevance of the kinetic assumptions underlying this work [45], rather the example may serve as a master example in this context. The fast dissociative adsorption

$$\frac{1}{2}O_2(g) + V_{ad} \rightleftharpoons O_{ad} \tag{34}$$

is supposed to be followed by the rate determining transfer at the three phase boundary formulated as:

$$O_{ad} + 2e' + V_O^{\cdot\cdot} \underset{\overleftarrow{k}_T}{\overset{\overrightarrow{k}_T}{\rightleftharpoons}} O_O^{\times} + V_{ad}. \tag{35}$$

The diffusion of O_{ad} to the three phase boundary is assumed to be fast. Formulating the anodic and cathodic current densities according to the simple kinetic rules and by taking into account that the applied voltage enhances the activation energy in one and depresses the activation energy in the opposite direction, we obtain the corresponding Butler-Volmer-equation, which predicts a Tafel behaviour (exponential dependence of i on the overvoltage η) at high and an ohmic behaviour at low voltages. If we set the transfer coefficients to 1/2 we obtain the exchange current density (which is inversely proportional to the transfer resistance) as a function of T and P_{O_2} as (c_{eff}, k_{eff}: effective surface concentration, effective rate constant defined by the geometric means)

$$i_0 = zF\sqrt{\Theta(1-\Theta)}\sqrt{\overrightarrow{k}_T\overleftarrow{k}_T} \equiv zFc_{eff}k_{T_{eff}}. \tag{36}$$

Note the isomorphism when compared to the bulk conductivity. Since the adsorption enthalpy is smaller than zero we derive from Eq. (21) for high

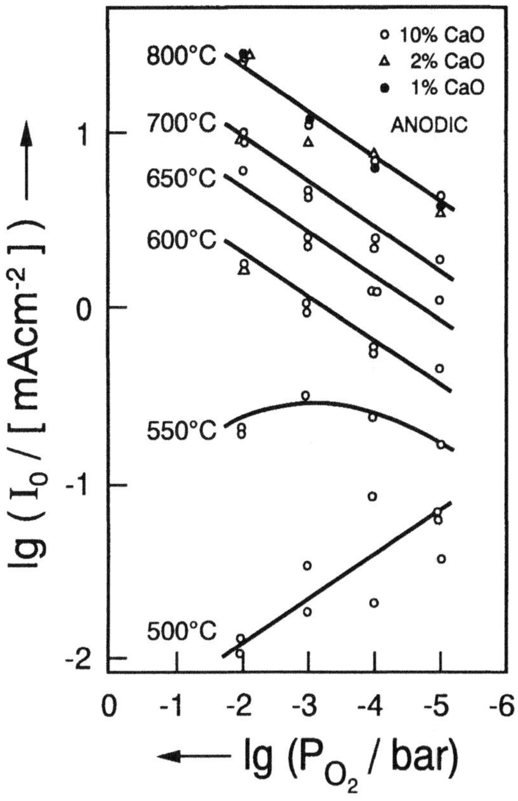

Figure 19. The exchange current density as a function of oxygen partial pressure for different temperatures confirming the electrode kinetical model given in the text ([48]).

temperatures and/or low P_{O_2}, that $K_s P_{O_2}^{1/2} \ll 1$, and it follows ($\Theta \ll 1$)

$$\Theta = c_{eff}^2 = K_s P_{O_2}^{1/2}. \tag{37}$$

The P_{O_2}- and T-dependencies of the exchange current density are then given by

$$\begin{aligned} \frac{\partial \ln i_0}{\partial \ln P} &= 1/4 \\ -R\frac{\partial \ln i_0}{\partial 1/T} &= \frac{1}{2}\Delta H_s^\circ + \frac{1}{2}\left(\Delta \overset{\rightharpoonup}{H}_T^{\neq} + \Delta \overset{\leftharpoonup}{H}_T^{\neq}\right). \end{aligned} \tag{38}$$

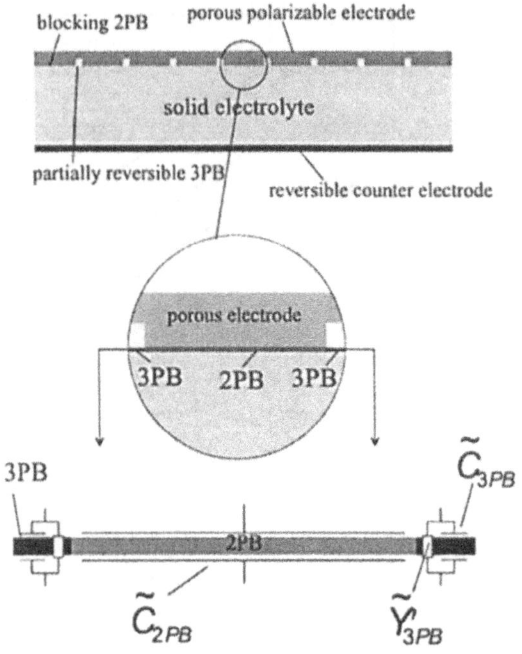

Figure 20. The presence of pores, and of three (3PB) and two phase boundaries (2PB) leads to serious current constriction phenomena treated in Ref. [49]. ($\tilde{C}$ and $\tilde{Y}'$ denote differential capacitances and conductances.)

In the other limit $K_sP^{1/2} \gg 1$ we obtain analogously

$$\frac{\partial \ln i_0}{\partial \ln P} = -1/4$$
$$-R\frac{\partial \ln i_0}{\partial 1/T} = -\frac{1}{2}\Delta H_s^\circ + \frac{1}{2}\left(\Delta \overset{\rightharpoonup\neq}{H_T} + \Delta \overset{\leftharpoonup\neq}{H_T}\right). \tag{39}$$

Thus, analogously to the bulk concentration, we find a power law

$$c_{eff} \propto P^{N'}\left(\Pi_r K_r^{\gamma_r'}\right) C^{M'}. \tag{40}$$

The temperature dependence of i_0 (cf. σ) also includes $\Delta H^{\neq}_{T_{eff}}$ (cf. mobility). In consistency with the assumed model* Fig. 19 confirms the above model.

*According to Ref. [48] a fast ionisation of O_{ad} to O'_{ad} followed by a slow incorporation (to form $O_O^{\cdot}$) and a quick full ionisation to O_O gives the same effective transfer coefficients. This is, however, only true if the O'-incorporation does not involve electrical fields which are changed by the applied overvoltage [45] which may generally be doubted.

If e. g. adsorption is rate limiting the P_{O_2}-dependence is different since P_{O_2} enters the expression for the exchange rate explicitly [45].

The information on the phase boundary quantities is often obtained from AC measurements. It has to be denoted in this context that the frequency behaviour can be strongly influenced by current constriction the ignoring of which can lead to serious misinterpretations. This essentially applies to fuel cell electrodes where three phase boundaries and two phase boundaries are involved (see Fig. 20) [49].

In Ref. [49] the superposition with a polarisation resistance has been discussed on a numerical basis.

3.4. SOLID STATE REACTION

Let us consider a true solid state reaction, namely the oxidation of a metal such as under the formation of a new phase, i. e. ZnO [50, 34, 51]. The reaction starts with nucleation and early growth phenomena (lateral mass transport, tunnelling, space charge phenomena). Usually at larger thicknesses the interface smoothens and at sufficiently large thicknesses the transport becomes controlled by chemical diffusion. In the diffusion controlled regime the flux and thus reaction rate is given by $\tilde{\sigma}\nabla\mu_O$ which is equal to $\frac{1}{L}\int\tilde{\sigma}d\mu_O$ due to flux constancy. Thus the growth rate measured by the thickness change $\dot{L}$ is inversely proportional to L and and the well-known parabolic rate law follows.

For ZnO, $\sigma_{eon} \gg \sigma_{ion}$, thus $\tilde{\sigma} \simeq \sigma_{ion}$ and the steep temperature dependence is explained by the activation enthalpy of σ_{ion}. Furthermore, the doping effect by Al^{3+} or Li^{+} in depressing or enhancing the corrosion rate is immediately derived from the defect chemistry. $Al_{Zn}^{\cdot}$ defects decrease σ_{ion} by decreasing $[V_O^{\cdot\cdot}]$ and $[Zn_i^{\cdot\cdot}]$ while the Li_{Zn}' defects increase σ_{ion}. The partial pressure dependence follows through integration.

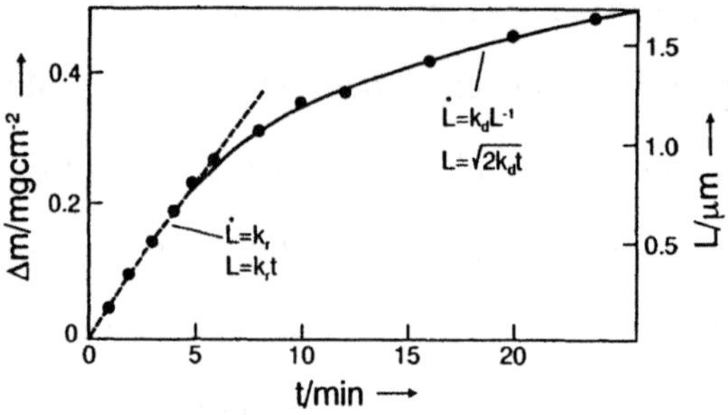

Figure 21. Change from reaction to diffusion control in a Spinel formation reaction [50].

Figure 21 shows a parabolic law for long times and a linear law for short times, the latter being indicative of a surface control [50]. In the steady state of the surface control $\dot{L}$ does not depend on t, neither explicitly or implicitly (If $\mathcal{R}$ is the rate of the surface reaction we may write $\dot{L} = f(L(t), \mathcal{R}(t))$, hence $\ddot{L} = \frac{\partial \dot{L}}{\partial L}\frac{dL}{dt} + \frac{\partial \dot{L}}{\partial \mathcal{R}}\frac{\partial \mathcal{R}}{dt} = 0$, since the first term vanishes through $\partial\dot{L}/\partial L = 0$ and the second because of the steady state condition $\partial\mathcal{R}/\partial t = 0$.) In the stage of early growth $\dot{L}$ depends also on L, and a more complicated law follows. Generally, the description of real solid state reactions is awfully complicated owing to the role of the space co-ordinate in general, and the interfacial problems in particular. Qualitatively speaking it is clear from the above considerations that good contact, high conductivities (high T, appropriate doping), high driving forces and in particular small diffusion lengths are necessary to provide fast reaction.

Walking around in the lecture hall or in admiring the natural beauty of the Erice area, we see essentially two categories of structures. Firstly, simple mostly planar geometries, particularly due to artificial architecture. There we rely on non-equilibrium processes such as cutting, the structure being maintained thanks to the extremely low diffusion constants. Secondly, complex interface morphologies, partly self-similar or self-affine, created under extreme conditions in the earth's history (e.g. rock-formation) and now

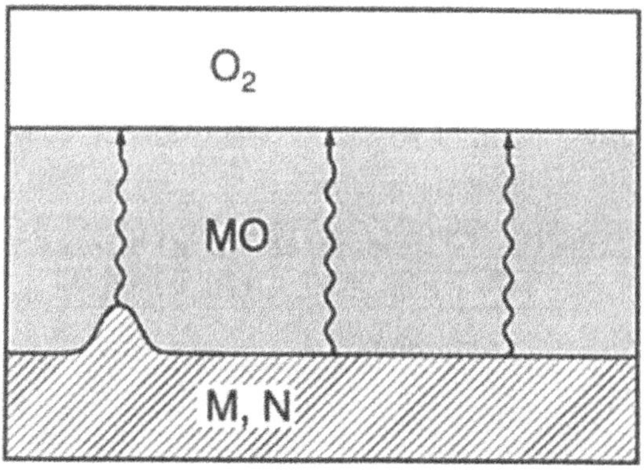

Figure 22. Morphological stability/instability of the interfaces in a corrosion experiment [51].

frozen-in, or formed under mild in-situ conditions such as natural growth but still far from equilibrium. Such typical non-linear processes will be considered below. At this stage let us consider Fig. 22. It highlights that an interface is not necessarily stable even close to equilibrium: If the diffusion of the metal M through the oxide is rate limiting for the corrosion of an M-N alloy (N: noble metal), any perturbation of the M, N/MO interface will be washed out (as shown in Fig. 22), but it will be augmented, possibly leading to dendrites, if the diffusion of M through the alloy is rate limiting [51]. In this case the noble metal will remain embedded in the oxide matrix. Local stability criteria like this may be extremely helpful even if they fail to predict the growth process [52]. (In the stable case the perturbation δL is positive, while the time change of it $(\partial/\partial t)(\delta L)$ is negative, such a function guaranteeing stability is called a Ljapunow-function.)

3.5. NON-LINEAR PROCESSES

Let us consider the reaction $A(x) \rightleftharpoons B(x')$ again, which may represent a chemical reaction ($x = x'$), a pure transport ($A = B$) or an electrochemical reaction. We follow Ref. [17]. The transport step itself may exhibit two extreme cases, (i) the transport of neutral species and (ii) a pure electrical transport if the relevant chemical potentials are essentially constant (e′ in a metal, Ag^+ in α-AgI etc.). Assuming dilute states, it can

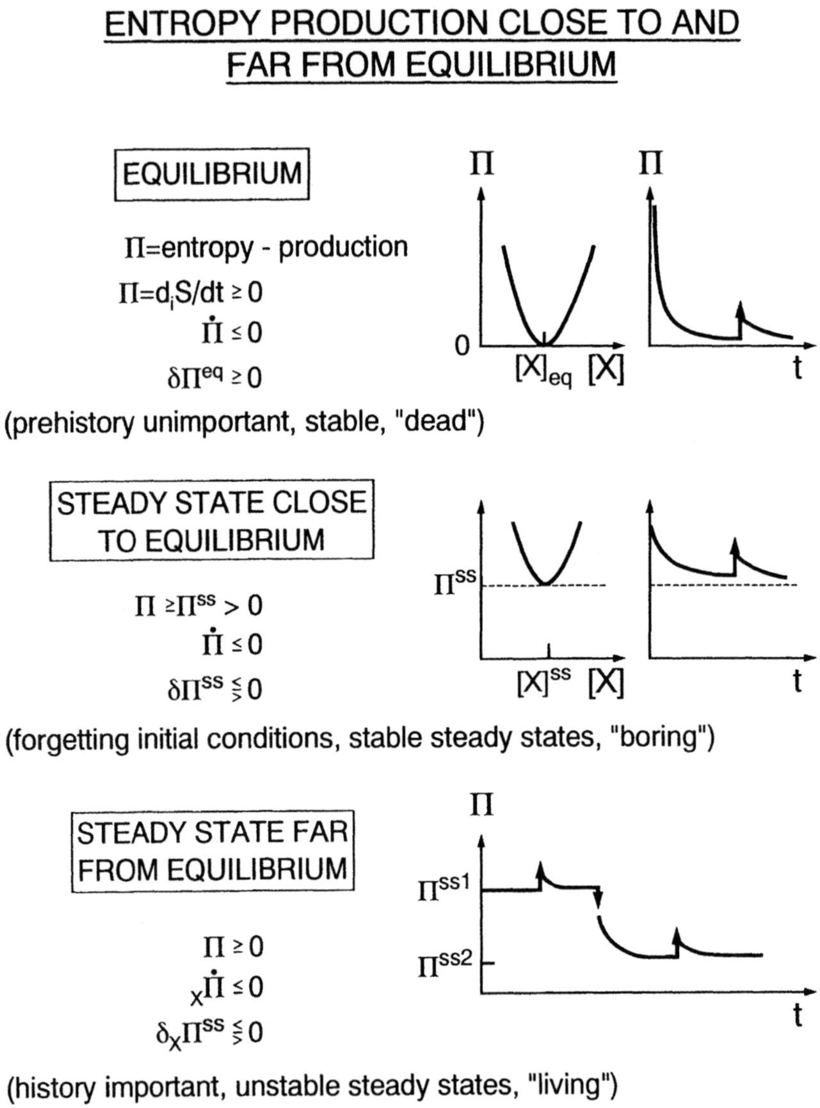

Figure 23. Entropy production close to and far from equilibrium.

be immediately seen by formulating rate equations in the pure transport case that a linear flux-driving force law (Fick's law) holds over a wide range ($J \propto k\,(c'(x') - c(x))$). This is not true if fields are involved. Here $J \propto c(k' - k)$. k' and k are different due to different fields, modifying the activation threshold, thus the field effect appears in the exponent. A linear law results only if $F\Delta\phi \ll RT$ (Taylor expansion). However, for not too thin samples this is usually fulfilled for the bulk transport (Ohm's law). For boundaries, however, a sinh-law or more generally a Butler-Volmer-law results. For chemical reactions, finally, linear force-flux relations are usually very poor approximations. This we recognise from Eq. (29):

$$J = \mathcal{R} = \vec{k}\,[A]\left(1 - \exp -\frac{\mathcal{A}}{RT}\right), \tag{41}$$

$\left(\text{where } \mathcal{A} = -\Delta_R G = \Delta_R G^0 + RT \ln \frac{[B]}{[A]} = RT \ln \frac{[B]/[A]}{K}\right)$.

In the majority of cases $|\Delta_R G|$-values are great compared to RT. Thus a proportionality between $\mathcal{R}$ and $\mathcal{A}$ resulting after expansion and requiring $|\mathcal{A}| \ll RT$, is mostly of academic interest. On the other hand it is this non-linearity of chemical reactions which is decisive for the vividness of our world. In equilibrium entropy production (Π) is zero. If we are off equilibrium $\Pi \equiv T\delta_i S/\delta t > 0$ according to the second law of thermodynamics. If we perturb the system the entropy production decreases while we reestablish equilibrium $\dot{\Pi} < 0$ (Fig. 23). The entropy production can be written for the case of interest as a product of fluxes and corresponding forces (JX, e.g. $\mathcal{RA}$). If we are fixing some of the external forces, equilibrium cannot be achieved, only a steady state occurs. In the linear regime the steady state corresponds to a minimum of entropy production (but not zero). Again this steady state is stable, since any perturbation corresponds to a higher Π-value ($\delta\Pi \gg 0$) and $\dot{\Pi} < 0$ [53].

Let us consider an important example: We will prove that the linear concentration profile in a steady state of a diffusion experiment corresponds to a minimum entropy production $\Pi \propto \int J\nabla c dV \propto \int (\nabla c)^2 dV$. We argue

Chemical kinetics far from equilibrium

autocatalytic reaction	A + X ⟶ 2X	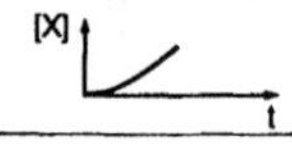	growth, positive feedback
autocatalytic + decay reaction no selection pressure	A + X $\overset{k_1}{\rightleftharpoons}$ 2X X $\xrightarrow{k_2}$ Z [A = const $W_c = k_1[A] - k_2$	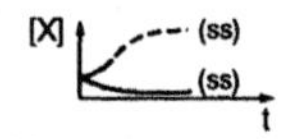 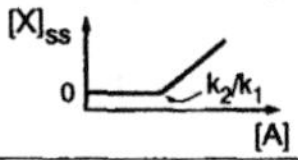	growth or death nonequilibrium phase transformation bifurcation
autocatalytic + decay + selection pressure (+ perturbation)	A + X ⟶ 2X X ⟶ Z [A] ≠ const $W_c = f(t)$	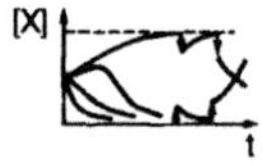	competition selection mutation
Lotka-Volterra reaction scheme	A + X ⟶ 2X X + Y ⟶ 2Y Y ⟶ Z	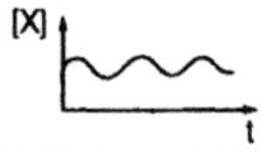	structurally unstable oscillation
Brusselator reaction scheme or catalytic CO oxidation	A ⟶ X 2X + Y ⟶ 3X B + X ⟶ Y + D X ⟶ Y ($CO + O \rightleftharpoons CO_L$; $Pt_{hex} \rightleftharpoons Pt_{sq}$)	 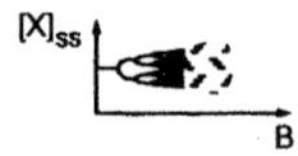	limit cycle oscillation symmetry breaking, deterministic chaos
above reaction scheme + diffusion	(z. B. X ⇄ X)	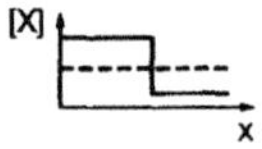	compartmentation, dissipative structure in space

Figure 24. Typical non-equilibrium phenomena far from equilibrium [17].

as follows: The function c(x) sufficing this condition will be the same if we change the integrand of the variational problem to $\nabla c + 1$ or even to $|\nabla c + 1| = \sqrt{(\nabla c^2 + 1)}$. While the first point is trivial, the second one can be justified by Euler's variational theorem. Since $\int \sqrt{1 + (dc/dx)^2}dx = \int \sqrt{(dx)^2 + (dc)^2} = \int(\text{length element})$, the variational problem is obviously analogous to finding the curve of minimum length between two terminals, which is the line in the x-c-plane.

All this does no longer apply to non-linear systems, there (if at all) only that part of Π which is caused by the forces, reaches a minimum. For those states, however, a stability is not guaranteed ($\delta_x \Pi \lesseqgtr 0$) (Fig. 23).

It can be shown that autocatalytic processes can rapidly destabilise such states and all the processes such as oscillations, non-equilibrium processes,

TABLE 3. Fractals

Fractals – Self Similarity – "Zoom Symmetry"

A) $l(\varepsilon_0) = 1\varepsilon_0$

$$l(\varepsilon_0/3) = 4\frac{\varepsilon_0}{3} = \frac{4}{3}\varepsilon_0$$

$$l(\varepsilon) = 4/3\, l(\varepsilon/3)$$

$$l(\varepsilon) \propto \varepsilon^{1-d} \text{ with } d = \log 4/\log 3 = 1.26$$

$$\left(\begin{array}{l} l(\varepsilon) = \alpha\varepsilon^{1-d} \\ l(\varepsilon/3) = \alpha(\varepsilon/3)^{1-d} = \alpha\varepsilon^{1-d}\left(\frac{1}{3}\right)^{1-d} = l(\varepsilon)3^{d-1} \\ l(\varepsilon/3) = 4/3 l(\varepsilon) \\ \hline 3^{d-1} = \frac{4}{3} \Longleftrightarrow 3^d = 4 \end{array}\right)$$

B)

smooth curve	M(bL)	=	bM(L)	$\Longleftrightarrow$	$M(L) \propto L$
smooth area	M(bL)	=	$b^2M(L)$	$\Longleftrightarrow$	$M(L) \propto L^2$
fractals	M(bL)	=	$b^dM(L)$	$\Longleftrightarrow$	$M(L) \propto L^d$

(L is projected length)

$$M(3L) = 4M(L) = 3^dM(L)$$

$$d = \log 4/\log 3 = 1.26$$

Scale invariance of the power laws

$$\left(b^d\right)^e = b^{de} = b^{d'}$$

Power of a power is still a power.

biological pattern formation, chaotic behaviour, dissipative structures making life possible are allowed to occur. In most cases interfaces play a key role. The reader is referred to Fig. 24 and also to the relevant literature (see [53], [17] and references cited therein). Many examples have been studied in

Figure 25. Fractal pattern obtained by diffusion limited aggregation after Ref. [56]

chemistry, biology or in semiconductors physics, but these considerations are also relevant to ionic conductors and mixed conductors [52, 54, 55].

A useful (also extreme) pendant to the also idealised linear geometry is fractal geometry which plays a key role in non-linear processes (Table 3). If we measure the length of a fractal interface with different scales, we will see that it increases with decreasing scale since more and more details are included. The number which measures how often the scale ε can be applied to the interfacial curve, is not proportional to ε but is a power law function of ε with the exponent $1 - \mathrm{d}$ being characteristic for the self-similarity of the structure; d is called the Hausdorff-dimension (see Table 3). Diffusion limited aggregation is a process that evidently leads to fractal structures. That this is a non-linear process follows from the complete neglect of the back-reaction. Obviously it holds for a given process that

$$\frac{\overrightharpoon{\mathcal{R}}}{\overleftharpoon{\mathcal{R}}} = \frac{\mathrm{K}}{[\mathrm{B}]\,/\,[\mathrm{A}]} = \exp\frac{\mathcal{A}}{\mathrm{RT}}. \tag{42}$$

$\overrightharpoon{\mathcal{R}} \gg \overleftharpoon{\mathcal{R}}$ implies $\mathcal{A} > \mathrm{RT}$ and vice versa [56]. Figure 25 gives a fractal obtained by diffusion limited aggregation [56]. The impedance of the tree-like metal (Fig. 26) synthesised by electrolysis, shows the impedance behaviour expected for a fractal electrode [58]. Percolation laws are characterised by

Figure 26. Computer generated fractal landscape after [57].

non-trivial problems. In the context of our interface considerations this has been evidenced for AgCl:α-AgI composites (see Refs. [59], [23b]). Many rate laws in chemical kinetics involve fractal dimensions due to self-similar interfaces. Finally the picture computer generated shown in Fig. 26, according to the rules of fractal geometry [57], reveals an astonishing similarity to Sicily's landscape and highlights the significance of non-linear processes for the description of our world.

Acknowledgement

Discussions with J. Claus, J. Fleig, J. Jamnik and I. Riess are appreciated. Special thanks to Dr. Fleig and Prof. Riess for reading the manuscript.

References

1. Bohm, J. (1995) *Realstruktur von Kristallen*, Schweizerbartsche Verlagsgesellschaft, Stuttgart.
2. Blakely, J.M. (1973) *Introduction to the Properties of Crystal Surfaces*, Pergamon Press, Oxford.
3. Rusanov, A.I. (1978) *Phasengleichgewichte und Grenzflächenerscheinungen*, Akademie Verlag, Berlin.

4. Denk, I., Noll, F., and Maier, J. (1997) In-situ profiles of oxygen diffusion in $SrTiO_3$: Bulk behaviour and boundary effects, *J. Am. Ceram. Soc.* **80**, 279–285.
5. Denk, I., Claus, J., and Maier, J. (1997) Electrochemical investigations of $SrTiO_3$ boundaries, *J. Electrochem. Soc.* **144**, 3526–3536.
6. Pan, X., Gu, H., Stemmer, S., and Rühle, M. (1996) Grain boundary structure and composition in strontium titanate, *Mater. Sci. Forum* **207-209**, 421–424.
7. Lupis, C.H.P. (1983) *Chemical Thermodynamics of Materials*, North Holland, New York.
8. Adamson, A.W. (1982) *Physical Chemistry of Surfaces*, Wiley, New York.
9. Herring, C. (1951) Some theorems on the free energies of crystal surfaces, *Phys. Rev.* **82**, 87–93.
10. Jamnik, J. and Maier J. (1997) Charge transport and chemical diffusion involving boundaries, *Solid State Ionics* **94**, 189–198.
11. Maier, J. (1987) Defect chemistry and conductivity effects in heterogeneous solid electrolytes, *J. Electrochem. Soc.* **134**, 1524–1535.
12. Jamnik, J., Maier, J., Pejovnik, S. (1995) Interfaces in solid ionic conductors: Equilibrium and small signal picture, *Solid State Ionics* **75**, 51–58.
13. Riess, I. (1994) Conditions for neglecting space charge effects on distributions of point defects and *I-V* relations, *Solid State Ionics* **69**, 43–52.
14. Riess, I. (1995) *pn* junctions in mixed ionic-electronic conductors induced by chemical and electrical potential gradients, *Solid State Ionics* **75**, 59–66.
15. Maier, J. (1995) Ionic conduction in space charge regions, *Prog. in Solid St. Chem.* **23**, 171–263.
16. Fleig, J., unpublished results; Henisch, H. K. (1984)*Semiconductor Contacts*, Clarendon Press, Oxford.
17. Maier, J. (1993) Defect chemistry: Composition, transport, and reactions in the solid state. Part I: Thermodynamics, *Angew. Chem. Int. Ed. Engl.* **32**, 313–335; Maier, J. (1993) Defect chemistry: Composition, transport, and reactions in the solid state. Part II: Kinetics *Angew. Chem. Int. Ed. Engl.* **32**, 528–542.
18. Hariharan, K., Maier, J. *Solid State Ionics*, in preparation.
19. Lee, J.-S. and Maier, J. (1997) Transport and phase transition characteristics in $AgI:Al_2O_3$ composite electrolytes, in T. A. Ramanarayanan, W.L. Worrell, H. L. Tuller, M. Mogensen, and A. C. Khandkar (eds.), *Proc. Third Int. Symp. Ionic and Mixed Conducting Ceramics*, Electrochem. Soc., **Vol. PV 97-24**, Pennington/NJ, pp. 751-763; Uvarov, N.F., Khairetdinov, E.F., Bratel, N.B. (1993) Composite solid electrolytes in the AgI-Al_2O_3 System, *Russ. J. Electrochem.* **29**, 1231–1235.
20. Hainovsky, N. and Maier, J. Simple phenomenological approach to premelting and

sublattice melting in Frenkel disordered ionic crystals, (1995) *Phys. Rev. B* **51**, 15789–15797; Zimmer, F., Ballone, P., Maier, J., Parrinello, M. (1997) Defect-defect interactions in ionic conductors: A classical MD and MC study, *Ber. Bunsenges. Phys. Chem.* **101**, 1333–1338.

21. Maier, J. (1986) On the conductivity of polycrystalline materials, *Ber. Bunsenges. Phys. Chem.* **90**, 26–33.
22. Dudney, N.J. (1989) Composite electrolytes, *Ann. Rev. Mat. Sci.* **19**, 103–120.
23. a) Shahi, K. , Wagner, J.B. (1981) Ionic conductivity and thermoelectric power of pure and Al_2O_3-dispersed AgI, *J. Electrochem.* **128**, 6–13; b) Lauer, U. and Maier, J. (1992) Electrochemical analysis of anomalous conductivity effects in the AgCl-AgI two phase system, *Ber. Bunsenges. Phys. Chem.* **96**, 111–119.
24. Maier, J. (1985) Space charge regions in solid two phase systems and their conduction contribution. II: Contact equilibrium at the interphase of two ionic conductors and the related conductivity effect *Ber. Bunsenges. Phys. Chem.* **89**, 355–362.
25. Maier, J. (1989) Heterogeneous solid electrolytes, in S. Chandra, A. Laskar (eds.), *Recent Trends in Superionic Solids and Solid Electrolytes*, Academic Press, New York, pp. 137–184.
26. Fleig, J. and Maier, J. (1997) The influence of the microstructure on the grain boundary impedance: A finite element study, in T. A. Ramanarayanan, W.L. Worrell, H. L. Tuller, M. Mogensen, and A. C. Khandkar, *Proc. Third Int. Symp. Ionic and Mixed Conducting Ceramics*, Electrochem. Soc., **Vol. PV 97-24**, Pennington/NJ, 1997, pp. 166–177 ; (1998) A finite element study on the grain boundary impedance of different microstructures, *J. Electrochem. Soc.* **145**, 2081–2089.
27. Fleig, J., Noll, F., and Maier, J. (1996) Surface conductivity measurements on AgCl single crystals using microelectrodes, *Ber. Bunsenges. Phys. Chem.* **100**, 607–615.
28. Maier, J. (1987) Space charge regions in solid two phase systems and their conduction contribution. III: Defect chemistry and ionic conductivity in thin films, *Solid State Ionics* **23**, 59—67; Maier, J., Prill, S., and Reichert, B. (1988) Space charge effects in polycrystalline, micropolycrystalline and thin film samples: Application to AgCl and AgBr, *Solid State Ionics* **28-30**, 1465–1469.
29. Denk, I., Münch, W., and Maier, J. (1995) Partial conductivities in $SrTiO_3$: Bulk polarization experiments, oxygen concentration cell measurements, and defect-chemical modeling, *J. Am. Ceram. Soc.* **78**, 3265–3272.
30. Chiang, Y.-M., Lavik, E.B., Kosacki, I., Tuller, H.L., and Ying, J.Y. (1997) Nonstoichiometry and electrical conductivity of nanocrystalline CeO_{2-x}, *J. Electroceramics* **1**, 7–14.
31. Martin, T.P. (1984) The structure of elemental and molecular clusters, in P. Grosse

(ed.), *Festkörperprobleme (Advances in Solid State Physics)*, Vieweg, Braunschweig, **Vol. XXIV**, 1–24.

32. Fritsche, H.G. (1989) Particle size effects of ion crystallites, *phys. stat. sol. (b)* **154**, 603–608.
33. Maier, J. (1996) Interface effects in mixed conductors, in F.W. Poulsen, N. Bonanos, S. Linderoth, M. Mogensen, B. Zachau-Christiansen (eds.), *High Temperature Electrochemistry: Ceramics and Metals*, Risø National Laboratory, Roskilde, pp. 67–76.
34. Sze, S.M. (1981) *Physics of Semiconductor Devices*, Wiley-Interscience, New York.
35. Waser, R., Vollmann, M. (1994) Grain boundary defect chemistry of acceptor-doped titanates: Space charge layer width, *J. Am. Ceram. Soc.* **77**, 235–243.
36. Pike, G.E., Seager, C.H. (1983) Electronic properties of silicon grain boundaries, *Adv. Ceram.* **1**, 53–66.
37. Bohnenkamp, K., Engell, H.-J.(1957) Messungen der Impedanz der Phasengrenze Germanium-Elektrolyt, *Zeitschr. Elektrochemie* **61**, 1184–1196.
38. Hariharan, K. and Maier, J. (1995) Enhancement of the fluoride vacancy conduction in PbF_2: SiO_2 and PbF_2: Al_2O_3 composites, *J. Electrochem. Soc.* **142**, 3469–3473.
39. Denk, I., Traub, U., Noll, F., and Maier, J. (1995) In-situ optical investigation of oxygen diffusion profiles in $SrTiO_3$, *Ber. Bunsenges. Phys. Chem.* **99**, 798–801.
40. Jamnik, J. and Maier, J. (1996) Chemical diffusion through interfaces, in F.W. Poulsen, N. Bonanos, S. Linderoth, M. Mogensen, B. Zachau-Christiansen (eds.), *High Temperature Electrochemistry: Ceramics and Metals*, Risø National Laboratory, Roskilde, pp. 287–293.
41. Jamnik, J. and Maier, J. (1997) Chemical diffusion through grain boundaries in mixed conductors, in T. A. Ramanarayanan, W.L. Worrell, H. L. Tuller, M. Mogensen, and A. C. Khandkar (eds.), *Proc. Third Int. Symp. Ionic and Mixed Conducting Ceramics*, Electrochem. Soc., **Vol. PV 97-24**, Pennington/NJ, pp. 379–389.
42. Preis, W., Sitte, W. (1996) Grain boundary diffusion through thin films. Application to permeable surfaces, *J. Appl. Phys.* **79**, 2986–2994, and literature cited therein.
43. Jamnik, J. and Maier, J. (1997) Transport across boundary layers in ionic crystals. Part I: General formalism and conception, *Ber. Bunsenges. Phys. Chem.* **101**, 23–40.
44. Leonhardt, M., Claus, J., Maier, J., in preparation.
45. Maier, J., (1998) On the Correlation of Macroscopic and Microscopic Rate Constants in Solid State Chemistry, *Solid State Ionics* **112** 197–228.
46. Simkovich, G., Wagner, C. (1962) The role of ionic point defects in the catalytic activity of ionic crystals *J. Catalysis* **1**, 521–525.

47. Murugaraj, P. and Maier, J. (1989) Heterogeneous catalysis with composite electrolytes, *Solid State Ionics* **32/33**, 993–999; Murugaraj, P. and Maier, J. (1990) The effect of heterogeneous doping on heterogeneous catalysis: Dehydrohalogenation of tertiary butyl chloride, *Solid State Ionics* **40/41**, 1017–1020.
48. Wang, D.Y., Nowick, A.S. (1979) Cathodic and anodic polarization phenomena at platinum electrodes with doped CeO_2 as electrolyte, *J. Electrochem. Soc.* **126**, 1155–1165; Wang, D.Y., Nowick, A.S. (1979) Cathodic and anodic polarization phenomena at platinum electrodes with doped CeO_2 as electrolyte. II. Transient overpotential and a-c impedance, *J. Electrochem. Soc.* **126**, 1166–1172.
49. Fleig, J. and Maier, J. (1997) The influence of current constriction on the impedance of polarisable electrodes: Application to fuel cell electrodes, *J. Electrochem.* **144**, L302–L305.
50. Schmalzried, H. (1981) *Solid State Reaction*, Verlag Chemie, Weinheim.
51. Wagner, C. (1956) Oxidation of alloys involving noble metals, *J. Electrochem. Soc.* **103**, 571–580.
52. Schmalzried, H. (1996) *Chemical Kinetics*, VCH, Weinheim.
53. Prigogine, I., Glansdorff, P. (1971) *Structure, Stability and Fluctuations*, Wiley, London.
54. Janek, J., Majoni, S. (1995) Investigation of charge transport across the Ag|AgI-interface: (I) Occurrence of periodic phenomena during anodic dissolution of silver, *Ber. Bunsenges. Phys. Chem.* **99**, 14–20.
55. Schimschal-Thölke, S., Schmalzried, H., and Martin, M. (1995) Instability of moving interfaces between ionic crystals KCl/AgCl, *Ber. Bunsenges. Phys. Chem.* **99**, 1–6, Schimschal-Thölke, S., Schmalzried, H., and Martin, M. (1995) Stability of diffusion profiles in quasi-binary solid solutions (Ag, Na)Cl, *Ber. Bunsenges. Phys. Chem.* **99**, 7–13.
56. Nittmann, J., Stanley, H.E. (1986) Tip splitting without interfacial tension and dendritic growth patterns arising from molecular anisotropy, *Nature* **321**, 663–668.
57. Mandelbrot, B. (1982) *The Fractal Geometry of Nature*, Freeman, San Francisco.
58. Chassaing, E., Sapoval, B. (1994) Electrochemical impedance of blocking quasi-fractal 3-d electrodes, *J. Electrochem. Soc.* **141**, 2711–2715; Daccord, G., Lenormand, R. (1987) Fractal patterns from chemical dissolution, *Nature* **325**, 41–43.
59. Bunde, A., Havlin, S. (1995) Percolations I, II, in A. Bunde S. Havlin (eds.), *Fractals and Disordered Systems*, Springer Verlag, Berlin pp. 59–175.

ELECTROCATALYSIS, CATALYSIS AND ELECTROCHEMICAL PROMOTION IN SOLID ELECTROLYTE CELLS

C.G. VAYENAS and S.I. BEBELIS
Department of Chemical Engineering
University of Patras, Patras, GR-26500, Greece

1. Introduction

Solid electrolyte galvanic cells have been studied intensively during the last thirty years both for sensor applications [1] and for their potential use as power producing devices [2]. Other emerging applications include their use for chemical cogeneration, i.e., the simultaneous production of power and chemicals [3,4] and for enhancing and controlling the catalytic properties of metal and metal oxide catalysts via the effect of non-Faradaic electrochemical modification of catalytic activity (NEMCA) or electrochemical promotion [5,6]. All these applications depend crucially on electrocatalytic (net charge-transfer) reactions occurring primarily at the solid electrolyte-electrode-gas three-phase-boundaries (tpb) and on catalytic (no net charge transfer) reactions occuring primarily on the gas-exposed electrode surface. Improving our current understanding of electrocatalysis, catalysis and electrochemical promotion in solid electrolyte galvanic cells is of considerable theoretical and practical importance. This paper attempts to provide an introduction to these areas.

2. General considerations

The operating principle of a current state-of-the-art power-producing solid electrolyte fuel cell is shown schematically in Fig. 1. The positive electrode (cathode) acts as an electrocatalyst, i.e., it promotes the electrocatalytic (net charge transfer) reduction of $O_2(g)$ to O^{2-}:

$$O_2(g) + 4e^- \rightarrow 2O^{2-} \tag{1}$$

H.L. Tuller et al. (eds.), Oxygen Ion and Mixed Conductors and Their Technological Applications, 123–164.

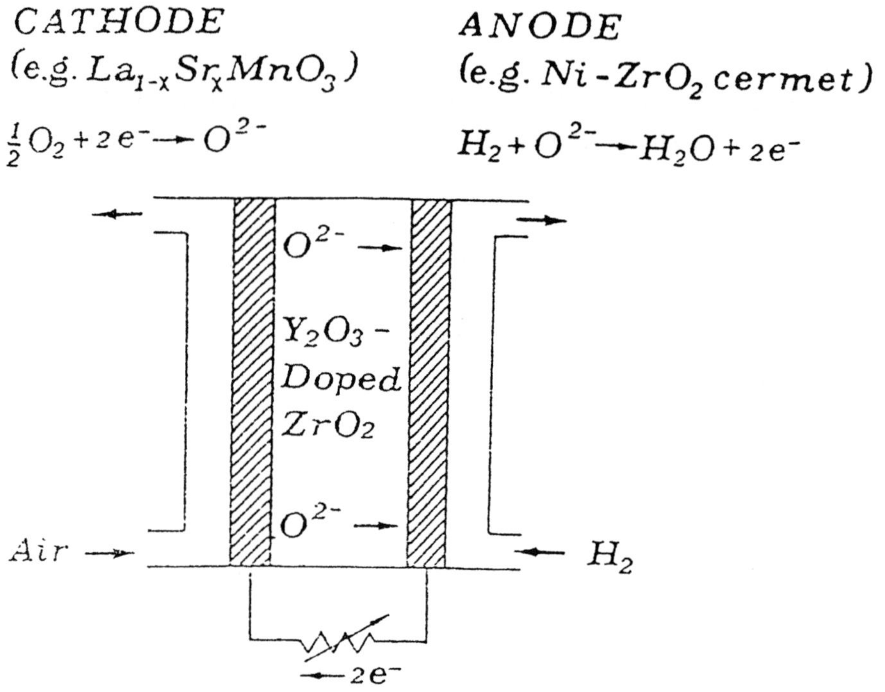

FIGURE 1: Operating principle of a solid oxide fuel cell. Reprinted with permission from ACS.

or, in Kröger-Vink notation:

$$O_2(g) + 2V_O^{\bullet\bullet} + 4e' \rightarrow 2O_O \quad (2)$$

where $V_O^{\bullet\bullet}$ stands for an O^{2-} vacancy in the yttria-stabilized-zirconia (YSZ) lattice and O_O denotes an oxygen anion O^{2-} in the lattice.

Although several metals, such as Pt and Ag, can also act as electrocatalysts for reaction (1) the most commonly used cathodic electrocatalysts in solid oxide fuel cells (SOFC) are perovskites such as $La_{1-x}Sr_xMnO_{3-\delta}$ (LSM). When a mixed ionic-electronic conductor is used as a cathode then the electrocatalytic sites are not restricted to the geometric three-phase-boundaries (tpb) but can also exist on the gas-exposed electrode surface as well, forming an electrocatalytically active zone which can, in principle, extend over the entire gas exposed electrode surface. The ionic conductivity of LSM is, however, low and thus LSM is a predominantly tpb system under most operating conditions.

The negative electrode (anode) acts as an electrocatalyst for the reaction of O^{2-} with the fuel, e.g. H_2:

$$2H_2 + 2O^{2-} \rightarrow 2H_2O + 4e^- \tag{3}$$

State-of-the-art solid oxide fuel cells (SOFC) utilize Ni-YSZ cermet as the anodic electrocatalyst for two reasons: First because Ni is an active electrocatalyst for reaction (3) and the porous cermet structure consisting of Ni and YSZ particles, typically 1 μm in size, has a large tpb length and the electrocatalytic reaction (3) takes place at the tpb. Consequently the use of Ni in a Ni-YSZ cermet to carry out electrocatalytic reactions is analogous to the use of Ni or other metals in a highly dispersed form on SiO_2 or Al_2O_3 (supported catalysts) to carry out catalytic reactions [7]. The second reason for using a Ni-YSZ cermet as the anode is that Ni is also an active catalyst for the steam-reforming reaction:

$$CH_4 + H_2O \rightarrow CO + 3H_2 \tag{4}$$

and also for the water-gas-shift reaction:

$$CO + H_2O \rightarrow CO_2 + H_2 \tag{5}$$

Consequently this permits efficient SOFC operation with CH_4, and also natural gas, as the fuel. Direct electrocatalytic CH_4 oxidation at the tpb is too slow for practical SOFC units with any known electrocatalyst, but, due to the catalytic properties of the Ni surface, CH_4 is converted to CO_2 and H_2 via reactions (4) and (5) and the latter is then easily oxidized electrocatalytically at the tpb.

Fuel cells of the type shown in Fig.1 convert H_2 to H_2O or CH_4 to CO_2 and H_2O and produce electrical power with no intermediate combustion cycle. Thus their thermodynamic efficiency compares favorably with thermal power generation which is limited by Carnot-type constraints. One important advantage of solid electrolyte fuel cells is that due to their high operating temperatures (typically 800 to 1000°C) they offer the possibility of "internal reforming" via reactions (4) and (5) which allows for the use of fuels such as CH_4 or natural gas without a separate external reformer.

3. Thermodynamic considerations

The maximum work obtainable from an isothermal continuous process is the negative of the Gibbs free energy change between the product and reactant streams $-\Delta G$. When an oxidation reaction, e.g.

$$H_2 + 1/2\, O_2 \rightarrow H_2O \tag{6}$$

$$CH_4 + 2O_2 \rightarrow CO_2 + 2H_2O \qquad (7)$$

is carried out in a fuel cell the useful electrical work produced tends to the upper limit of $-\Delta G$ when reversible operation is approached, i.e., when the operating cell voltage E approaches the open-circuit reversible voltage E_{rev}:

$$E_{rev} = -\,\Delta G/nF \qquad (8)$$

where n is the number of electrons involved in the oxidation reaction, e.g. 2 for reaction (6) and 8 for reaction (7).

The fuel cell efficiency ε is defined by:

$$\varepsilon = \frac{W}{-\Delta H^o} = \frac{E}{E_{TH}} = \frac{\Delta G}{\Delta H^o} \cdot \frac{E}{E_{rev}} \qquad (9)$$

where E_{TH} is the "thermoneutral" voltage defined as $E_{TH}=(-\Delta H^o)/nF$. According to Eq. (9) ε can exceed unity for reactions with a positive entropy change ΔS^o such as the ammonia oxidation to NO [4] or the partial oxidation of methane to synthesis gas [5,6]. Under such operating conditions ($\varepsilon>1$) a fuel cell absorbs heat from the environment instead of producing heat.

When the external resistive load of the fuel cell is finite and electrical power is produced, then in general the operating voltage E drops below the reversible voltage E_{rev} and the difference η:

$$\eta = E_{rev} - E \qquad (10)$$

is termed the cell overpotential (Fig. 2).

One of the key problems in electrochemical power producing devices, which hampered commercialization for years, is a thorough understanding and minimization of cell overpotential or "polarization".

4. Types of overpotential

The cell overpotential η can be considered as the sum of three major components termed ohmic overpotential η_{ohm}, concentration overpotential η_{conc} and activation overpotential η_{act}:

$$\eta = \eta_{ohm} + \eta_{conc} + \eta_{act} \qquad (11)$$

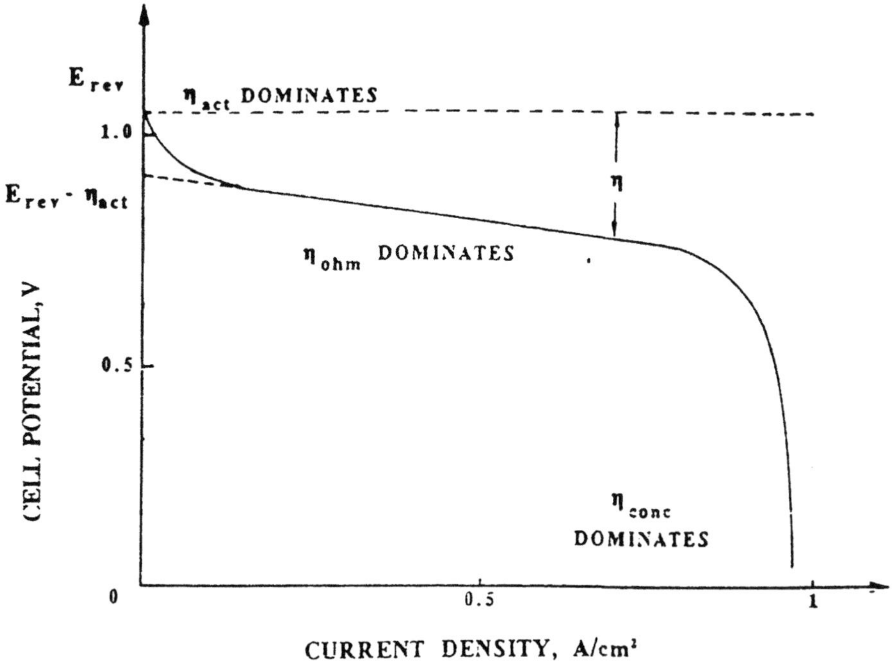

FIGURE 2: Typical current-voltage curve of a fuel cell obtained by varying the external load. Reprinted with permission from Elsevier Science Ltd.

a. The ohmic overpotential can be measured via AC impedance spectroscopy or via the current interruption technique in conjunction with a recording oscilloscope and is proportional to the cell current I. It is due to the ohmic resistance R_{el} of the electrodes, of the solid electrolyte R_i and the electrode-electrolyte contact resistance R_c:

$$\eta_{ohm} = I(R_{el} + R_i + R_c) \quad (12)$$

By introducing the current density i (A/cm^2) and the area-specific resistances, R'_{el}, R'_i, R'_c ($\Omega \cdot cm^2$) one can rewrite Eq. (12) as:

$$\eta_{ohm} = i(R'_{el} + R'_i + R'_c) \quad (12a)$$

To minimize the ohmic overpotential, one has to use thin, highly conductive solid electrolytes and highly conductive electrodes with good adherence to the solid electrolyte to minimize the contact resistance.

b. The cell concentration overpotential η_{conc} is the sum of the concentration overpotentials at the anode and at the cathode and is caused by slow mass transfer between the gas phase and the three-phase-boundaries. For an SOFC operating on H_2 fuel it can be expressed as:

$$\eta_{conc} = \frac{-RT}{4F}\left[\ln\left(1 - \frac{i}{i_{L,c}}\right) + 2\ln\left(1 - \frac{i}{i_{L,a}}\right)\right] \tag{13}$$

where $i_{L,a}$ and $i_{L,c}$ are the anodic and cathodic limiting current densities. When mass transfer in the gas phase is rate limiting, then $i_{L,a}$ and $i_{L,c}$ can be computed for many geometries of the anodic and cathodic compartments via standard analytical or empirical mass transfer correlations [8,9]. When surface diffusion on the electrode surface to the tpb is rate limiting then $i_{L,a}$ and $i_{L,c}$ must be measured experimentally. Concentration overpotential can be minimized by appropriate design of the anodic and cathodic compartments and by the use of porous electrodes. In this way limiting current density values $i_{L,a}$ and $i_{L,c}$ of at least 2-3 A/cm^2 can be achieved so that in view of Eq. (13) η_{conc} can be rather small (<100 mV) even for current densities up to 0.7 A/cm^2, which is a typical current density obtained with state-of-the-art SOFC units. In general due to the high values of gaseous and surface diffusivities at the operating temperature of SOFCs relative to liquid phase diffusivities, the concentration overpotential is a less serious problem in SOFCs than in low temperature aqueous electrolyte fuel cells.

c. The cell activation overpotential η_{act} is the sum of the activation overpotentials of the anode and cathode and is caused by slow electrocatalytic (charge transfer) reactions at both electrodes. The anodic and cathodic activation overpotentials, $\eta_{act,a}$ and $\eta_{act,c}$ respectively, can be related to the current density i via the Butler-Volmer equation:

$$i = i_{0,a}\left[\exp\left(\frac{\alpha_{a,a}F\eta_{act,a}}{RT}\right) - \exp\left(\frac{-\alpha_{c,a}F\eta_{act,a}}{RT}\right)\right] \tag{14a}$$

$$i = i_{0,c}\left[\exp\left(\frac{\alpha_{a,c}F\eta_{act,c}}{RT}\right) - \exp\left(\frac{-\alpha_{c,c}F\eta_{act,c}}{RT}\right)\right] \tag{14b}$$

where $i_{0,a}$, $i_{0,c}$ are the exchange current densities of the anode and cathode, respectively, $\alpha_{a,a}$ and $\alpha_{c,a}$ are the anodic (i>0) and cathodic (i<0) charge transfer coefficients of the anode and $\alpha_{a,c}$ and $\alpha_{c,c}$ are the anodic and cathodic charge transfer coefficients of the cathode. The anodic and cathodic charge transfer coefficients α_a and α_c depend on the electrocatalytic reaction mechanism and typically take values between zero and two. A zero value (a rather rare case) implies that the rate limiting step is catalytic or in general involves no net charge-transfer in which case the term chemical overpotential can also be used.

The exchange current densities $i_{0,a}$ and $i_{0,c}$ of the anode and cathode, respectively, are of crucial importance as they determine the magnitude of $\eta_{act,a}$ and $\eta_{act,c}$ via Eqs. (14). A good electrocatalyst is characterized by a high exchange current density i_0 value.

Two points need to be emphasized about the famous Butler-Volmer equation:

$$i = i_0\left[\exp\left(\frac{\alpha_a F\eta}{RT}\right) - \exp\left(\frac{-\alpha_c F\eta}{RT}\right)\right] \tag{14c}$$

and for its limiting high-field or Tafel approximation [8] to which it reduces for $|\eta|>200mV$:

$$\eta = \frac{RT}{\alpha_a F}\ln(i/i_0) \quad ; \quad i>0 \tag{15a}$$

$$\eta = -\frac{RT}{\alpha_c F}\ln(-i/i_0) \quad ; \quad i<0 \tag{15b}$$

Firstly that it is a kinetic expression, a rate law, such as e.g. the Langmuir-Hinshelwood-Hougen-Watson rate expressions in heterogeneous catalysis [5-7], and as such has no universal applicability. It is derived on the basis of mass action kinetics [8] and reduces to the fundamental thermodynamic Nernst equation for i=0, thus η=0. Nevertheless experimental deviations can be expected as with any other, even most successful, rate expression.

Secondly that the parameters i_0, α_a and α_c, which are usually treated as constants, are in reality dependent on the coverage of species adsorbed at the tpb. Consequently if the coverage of these species changes significantly with η, then i_0, but also α_a and α_c can vary significantly.

Figure 2 shows a typical current-voltage curve of a fuel cell obtained by varying the external resistive load which shows that, as expected from Eqs (12) to (14), the concentration overpotential becomes important at very high current densities, while the activation and ohmic overpotentials dominate at low and intermediate current densities where the cell power density P=Ei is maximized. State-of-the-art SOFC units can produce power densities of the order of 0.6 W/cm^2 and, to a good approximation, E decreases linearly with i with a slope, denoted by R'_{cell}, as low as 0.5 Ω cm^2. This area-specific resistance provides a good measure of the practical usefulness of a SOFC unit. Obtaining such small R'_{cell} values (<0.5 Ω cm^2) which are necessary for technological SOFC power producing applications, depends crucially on minimizing the cell ohmic and anodic and cathodic activation overpotentials.

5. Electrocatalytic activity

The magnitude of the activation overpotential for an electrocatalytic reaction occurring at the electrode-solid electrolyte tpb or interface depends crucially on the exchange current density i_0 (Eqs. 14). In order to measure the overpotential of a working electrode (W), anode or cathode, and thus extract i_0, α_a and α_c from the Butler-Volmer equation, it is necessary to utilize a reference (R) electrode [8]. When a current I flows between the working electrode and a counter electrode then the potential of the working electrode V_{WR} with respect to the reference electrode deviates from its open-circuit (I=0) value V^o_{WR} and the working electrode overpotential η_{WR} is defined from:

$$\eta_{WR} = V_{WR} - V^o_{WR} \tag{16}$$

The overpotentials η_W and η_R of the working and reference electrodes (the latter vanishes for an ideal reference electrode) are defined as the deviations of the inner (or Galvani) potentials [8] of these electrodes from their open-circuit values:

$$\eta_W = \varphi_W - \varphi^o_W \quad ; \quad \eta_R = \varphi_R - \varphi^o_R \tag{17}$$

One can write Eq. (16) in the form:

$$\eta_{WR} = \eta_W + \eta_R + \eta_{ohmic,WR} \tag{18}$$

Ideally no current flows through the reference electrode, therefore in principle η_R=0 and $\eta_{ohmic,WR}$=0. In practice the first assumption is usually adequate for reasonably non-polarizable reference electrodes, since the parasitic (uncompensated) current flowing via the reference electrode is usually very small. The ohmic drop, however, between the working and reference electrodes, i.e. $\eta_{ohmic,WR}$, may in general not be negligible and must be determined in each case using e.g. the current interruption technique in conjunction with a recording oscilloscope [5,6]. The ohmic component decays to zero within typically less than 1 μs and the remaining part of η_{WR} is η_W. As in aqueous electrochemistry, the reference electrode must be placed as near to the catalyst as possible to minimize $\eta_{ohm,WR}$.

It must be emphasized that although overpotentials are usually associated in electrochemistry with electrode-electrolyte interfaces, in reality they refer to, and are measured as, deviations of the potentials of the electrodes only. Thus the concept of overpotential must be associated with an electrode and not with an electrode-electrolyte interface, although the nature of the interface will, in general, dictate the magnitude of the overpotential.

The usual procedure for measuring the exchange current density i_0 is then to measure η as a function of I and to plot ln I vs η_W (Tafel plot). Such plots are shown in Fig. 3 for Pt and Ag electrodes deposited on YSZ. From the slopes of the linear part of these plots ($|\eta|$>200mV in which case Eqs. 15 are valid) one obtains the transfer coefficients α_a and α_c. By extrapolating the linear part of the plot to η=0 one obtains i_0. One can then plot i vs η and use the "low field" approximation of the Butler-Volmer equation which is valid for $|\eta|$<10mV, i.e.

$$i/i_0 = (\alpha_a + \alpha_c) F\eta/RT \tag{19}$$

in order to check the accuracy of the extracted i_0, α_a and α_c values.

The exchange current density i_0 is a measure of the electrocatalytic activity of the tpb or, more generally, the electrode-solid electrolyte interface for a given electrocatalytic reaction. It expresses the (equal and opposite) rates of the forward (anodic) and backward (cathodic) electrocatalytic reaction rates when no net current crosses the electrode-solid electrolyte interface or, equivalently, the tpb. It has recently been shown that quite often, the exchange current I_0 is proportional to the length l of the tpb [10] as one would intuitively expect. The tpb length l can be estimated via cyclic voltammetry [10] and electron microscopy. In principle different electrocatalysts should be compared via the parameter I_0/l. In practice this is seldom done due to the difficulty in measuring l accurately and the comparison is usually made via the exchange current density $i_0=I_0/A$, where A is the superficial surface area of the electrode-electrolyte interface.

The exchange current density i_0 is usually strongly temperature dependent. It increases with temperature with an activation energy which is typically 140-180 kJ/mol for Pt/YSZ and 80-100 kJ/mol for Ag/YSZ [5,6]. The i_0 dependence on gaseous composition is usually complex. Figure 4 shows the measured i_0 dependence on P_{O_2} and T for Pt/YSZ films [5,6] These results can be described adequately on the basis of Langmuir-type dissociative oxygen chemisorption at the tpb, i.e:

$$\theta_O = K_O P_{O_2}^{1/2} / (1+K_O P_{O_2}^{1/2}) \tag{20}$$

where θ_O is the coverage of atomic oxygen at the tpb and K_O is the adsorption equilibrium constant of oxygen on Pt. It can be shown [11] that :

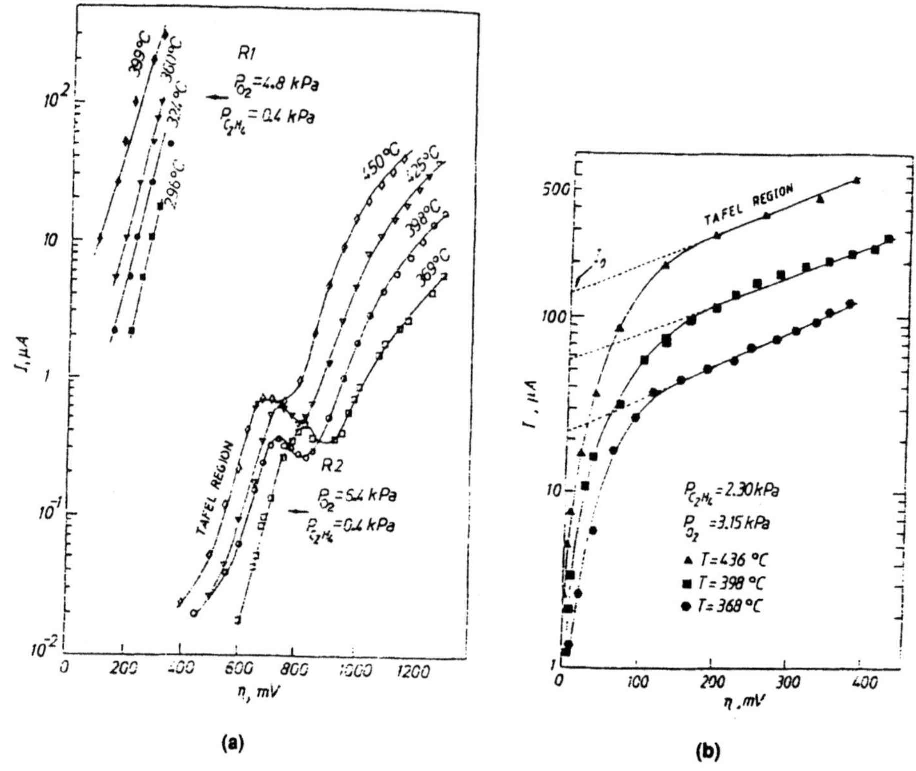

FIGURE 3: Typical Tafel plots for Pt catalyst-YSZ (a) and Ag catalyst-YSZ (b) interfaces. Reprinted with permission from Elsevier Science Publishers B.V. [5].

$$i_0 = c[\theta_O(1-\theta_O)]^{1/2} \tag{21a}$$

where c is a constant, or, equivalently:

$$i_0 = cK_O^{1/2} P_{O_2}^{1/4} / (1+K_O P_{O_2}^{1/2}) \tag{21b}$$

which explains nicely the observed i_0 maxima and the fact that i_0 is proportional to $P_{O_2}^{1/4}$ for low P_{O_2} and $P_{O_2}^{-1/4}$ for high P_{O_2} (Fig.4). According to this successful model, the i_0 maxima correspond to θ_O=1/2.

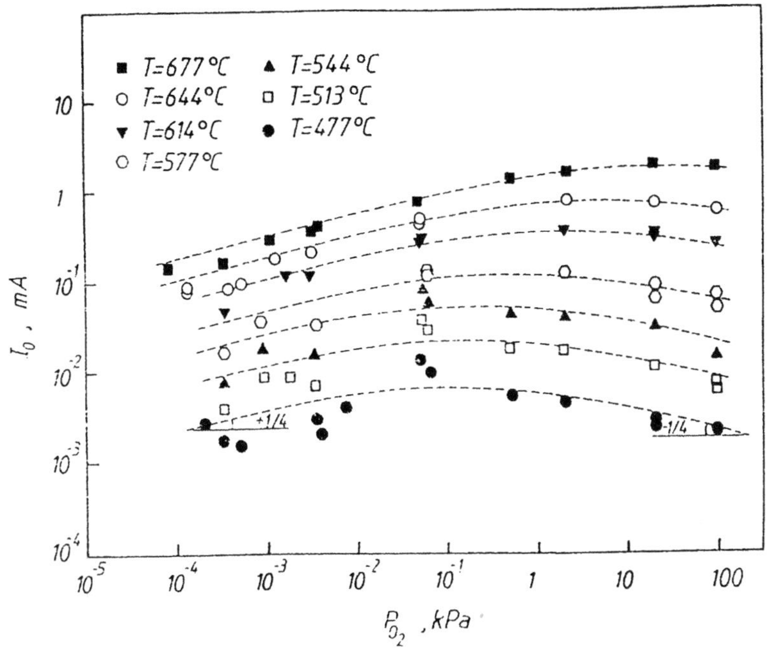

FIGURE 4: Effect of temperature and P_{O_2} on the I_0 of a Pt catalyst film (A=2 cm^2) on YSZ [5,6].

Extraction of I_0 and of the transfer coefficients α_a and α_c from steady-state current-overpotential data is frequently complicated by significant changes in surface coverage of oxygen at the tpb or by surface oxide formation. It has been shown recently [12] that several of these problems can be overcome by the use of cyclic voltammetry in conjuction with the equations:

$$\frac{dE_{p,a}}{d\ln\upsilon} = \frac{RT}{\alpha_a F} \tag{22a}$$

$$\frac{dE_{p,c}}{d\ln\upsilon} = -\frac{RT}{\alpha_c F} \tag{22b}$$

where υ is the sweep rate (V/s) and $E_{p,a}$, $E_{p,c}$ are the potentials corresponding to the anodic and cathodic oxygen peaks respectively. In this way it has been shown that for porous Pt/YSZ electrodes exposed to oxygen $\alpha_a=\alpha_c=1$.

The transfer coefficients α_a and α_c convey useful mechanistic information and can be used for mechanism discrimination. For an one-electron transfer process (n=1), e.g.:

$$H^+ + e^- \rightarrow H(a) \tag{23}$$

the Butler-Volmer equation is:

$$i/i_0 = \exp(\beta F\eta/RT) - \exp[-(1-\beta)F\eta/RT] \tag{24}$$

where β is the symmetry factor $(0<\beta<1)$ which is frequently found to equal 0.5 [8,13]

For multi-electron transfer processes ($n \geq 2$, e.g. oxygen reduction) the Butler-Volmer equation is:

$$i/i_0 = \exp(\alpha_a F\eta/RT) - \exp(-\alpha_c F\eta/RT) \tag{25}$$

It can be shown [8,13] that α_a and α_c are related to the symmetry factor β of the rate-determining-step (rds) via:

$$\alpha_a = \frac{n - \gamma}{\nu} - r\beta \tag{26a}$$

$$\alpha_c = \frac{\gamma}{\nu} + r\beta \tag{26b}$$

where:

n= total number of electrons transfered in the overall electrocatalytic reaction.

ν= stoichiometric number, i.e. number of times the rds occurs for one act of the overall reaction.

γ= number of electrons transferred before the rds in the forward reaction.

r= number of electrons transferred in the rds (0 or 1).

It follows from Eqs. (26) that

$$\alpha_a + \alpha_c = n/\nu \tag{27}$$

The above equations can be quite useful for discriminating between various mechanisms for a given overall electrocatalytic reaction. They give rise to several well-known Tafel slopes, the most common of which (2RT/F, RT/F, 2RT/3F, RT/2F and RT/4F) have been observed. However, the derivation of Eqs. (26) is based on three important assumptions:

I. A single clearly defined rds

II. Low coverages (θ<<1) which can be described by a linear isotherm.

III. A single reaction pathway.

6. Electrocatalytic operation of solid electrolyte cells

6.1 ELECTROCATALYSIS FOR THE PRODUCTION OF CHEMICALS

In recent years it has been shown that solid electrolyte fuel cells with appropriate electrocatalytic anodes can be used not only for power generation via oxidation of H_2 and CH_4 but also for "chemical cogeneration", i.e., for the simultaneous production of power and useful chemicals. This mode of operation, first demonstrated for the case of NH_3 oxidation to NO [3] combines the concepts of a fuel cell and of a chemical reactor (Fig. 5). The economics of chemical cogeneration have been modelled and discussed recently [4]. The economics appear promising for a few highly exothermic reactions which can be carried out at temperatures above 700°C, such as the oxidation of H_2S to sulfur and SO_2 and of NH_3 to NO [3,4]. It is likely that if SOFC units operating on H_2 or natural gas become commercially available, then they could be used with appropriate anodic electrocatalysts by some chemical industries.

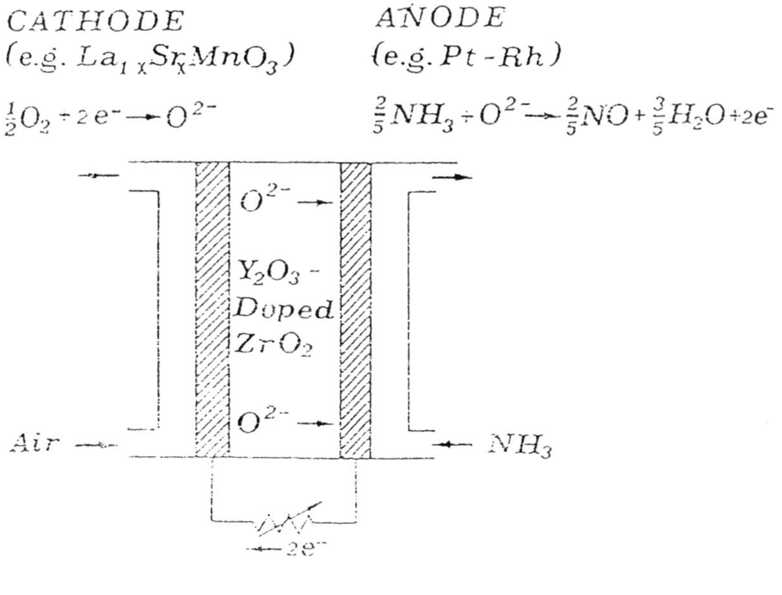

FIGURE 5: Operating principle of a chemical cogenerator. Reprinted with permission from ACS [4].

Table 1 lists the anodic reactions which have been studied so far in small cogenerative SOFC units. One simple and interesting rule which has emerged from these studies is that the selection of the anodic electrocatalyst for a selective electrocatalytic oxidation can be based on the heterogeneous catalytic literature for the corresponding selective catalytic oxidation. Thus the selectivity of Pt and Pt-Rh alloy electrocatalysts for the anodic NH_3 oxidation to NO:

$$2NH_3 + 5O^{2-} \rightarrow 2NO + 3H_2O + 10e^- \qquad (28)$$

turns out to be comparable (>95%) with the selectivity of Pt and Pt-Rh alloy catalysts for the corresponding commercial catalytic oxidation where oxygen is co-fed with NH_3 in the gas phase. The same applies for Ag which turns out to be equally selective as an electrocatalyst for the anodic partial oxidation of methanol to formaldehyde:

$$CH_3OH + O^{2-} \rightarrow H_2CO + H_2O + 2e^- \qquad (29)$$

as it is a catalyst for the corresponding heterogeneous catalytic reaction:

$$CH_3OH + 1/2O_2 \rightarrow H_2CO + H_2O \qquad (30)$$

TABLE 1. Electrocatalytic reactions investigated in doped ZrO_2 solid electrolyte fuel cells for chemical cogeneration refs. [4] - [6].

	Electrocatalyst
$2NH_3+5O^{2-} \rightarrow 2NO+3H_2O+10e^-$	Pt, Pt-Rh
$CH_3OH+O^{2-} \rightarrow H_2CO+H_2O+2e^-$	Ag
$C_6H_5\text{-}CH_2CH_3 + O^{2-} \rightarrow C_6H_5\text{-}CH{=}CH_2 + H_2O + 2e^-$	Pt, Fe_2O_3
$H_2S + 3O^{2-} \rightarrow SO_2+H_2O+6e^-$	Pt
$C_3H_6+ O^{2-} \rightarrow C_3$ dimers $+ 2e^-$	Bi_2O_3-La_2O_3
$CH_4+NH_3+3O^{2-} \rightarrow HCN+3H_2O+6e^-$	Pt, Pt-Rh
$2CH_4 + 2O^{2-} \rightarrow C_2H_4+2H_2O+4e^-$	Ag,Ag-Sm_2O_3

Another reaction which is of considerable interest from the view point of chemical cogeneration is the oxidative coupling of methane (OCM) to C_2 hydrocarbons ethane and ethylene [14]:

$$2CH_4 \xrightarrow{O^{2-}} C_2H_6 \xrightarrow{O^{2-}} C_2H_4 \xrightarrow{O^{2-}} 2CO_2 \tag{31}$$

There have been several OCM studies utilizing SOFC reactors. In many of these studies electrical power was supplied to the cell to increase cell current. This, however, tends in general to decrease the selectivity and yield of C_2 hydrocarbons. Very recently it was found that it is possible to obtain C_2 yields up to 88% by means of a novel gas-recycle SOFC reactor- separator [14]. The C_2H_4 yield is up to 85%. These systems are of considerable technological interest.

6.2 ELECTROCHEMICAL REACTOR ANALYSIS AND DESIGN

Solid oxide fuel cells can be viewed as special type of a chemical reactor which generate electrical power in addition to heat. The first rigorous engineering modelling of SOFC units was carried out in 1983 [15]. Well-mixed (CSTR-type) anodic and cathodic compartments were assumed, and isothermality within the SOFC structure for the overall reaction:

$$A + 1/2O_2 \rightarrow B \tag{32}$$

e.g., H_2 oxidation. It was found that the following dimensionless mass, energy and electron balances govern the SOFC behaviour:

Fuel (A) and oxygen mass balances:

$$X_A = A_1 \cdot \xi \tag{33a}$$

$$X_{O_2} = A_1 \cdot A_2 \cdot \xi \tag{33b}$$

Energy balance:

$$A_1A_3(1-A_4\xi)\xi = (1+A_6-A_1A_2A_7\xi)(\Theta-1) - A_5(\Theta_1^o-1) - A_6(\Theta_2^o-1) + A_8(\Theta-\Theta_c) \tag{33c}$$

Electron balance:

$$\left\{A_4(1+A_9)+\exp\left[A_{12}\left(\frac{1}{\Theta}-1\right)\right]\right\}\cdot\xi = A_{10}+$$

$$+A_{11}\Theta\left\{\ln\frac{P_o y^o_{O_2}(1-X_{O_2})(1-X_A)^2}{(1-y^o_{O_2}X_{O_2})X_A^2} + 2\ln(1-A_{13}\xi) +\right.$$

$$\left.+\ln(1-A_{14}\xi)-A_{15}\left[\ln(A_{16}\xi)+A_{17}\ln(A_{18}\xi)\right]\right\} \qquad (33d)$$

where the dimensionless current ξ and temperature Θ are defined by:

$$\xi = I\rho_o\delta/A_{el}E_{th} \qquad (34a)$$

$$\Theta = T/T_o \qquad (34b)$$

All symbols appearing in Eqs 33 and 34 are defined in reference [15]. The dimensionless parameters A_1-A_4 and A_6-A_9 play a key role in SOFC performance. Their physical significance and importance for scale-up is discussed in reference [15]. The left and right-hand side of the dimensionless energy balance (33c) can be viewed as a heat generation and heat removal term respectively. The dimensionless electron balance (33d) can be viewed as a dimensionless kinetic expression where the right-hand side represents the driving force for current flow, thus for reaction. Numerical solution of Eqs (33) with realistic parameter values has shown that, similar to chemical reactors, SOFC units can exhibit steady-state multiplicity as well as ignition and extinction phenomena over a wide range of design and operating parameters.

The above model is valid both for power producing and chemical producing SOFC units. It is a lumped-parameter CSTR-type model, i.e., it assumes uniform gas composition and temperature within the SOFC structure. After the first flat plate or cross-flow monolithic SOFC units were fabricated and tested [16] a new two-dimensional models were developed to account for the variation in gas-phase composition and temperature within the SOFC structure [17].

7. Catalysis on the electrodes of solid electrolyte cells

7.1. POTENTIOMETRIC INVESTIGATIONS

The interesting role which solid electrolytes can play in the study of heterogeneous catalysis was first recognized by C. Wagner who proposed the use of solid electrolyte cells for the measurement of the activity of oxygen on metal and metal oxide catalysts

[18]. This technique, first used to study the mechanism of SO_2 oxidation on noble metals [19] was subsequently termed solid electrolyte potentiometry (SEP) [20]. It has been used in conjunction with kinetic measurements to study the mechanism of several catalytic reactions on metals and, more recently, metal oxides. It is particularly suitable for the study of oscillatory catalytic reactions.

Figure 6 shows a typical experimental arrangement for the use of SEP. The bottom porous metal electrode (e.g. Pt) of the oxygen-ion-conducting solid electrolyte cell is exposed to ambient air and serves as a reference electrode (R). The top electrode, which we denote by W (for "working") acts both as an electrode and as a catalyst for the catalytic reaction to be studied. The open-circuit emf V^o_{WR} of the solid electrolyte cell can be written as :

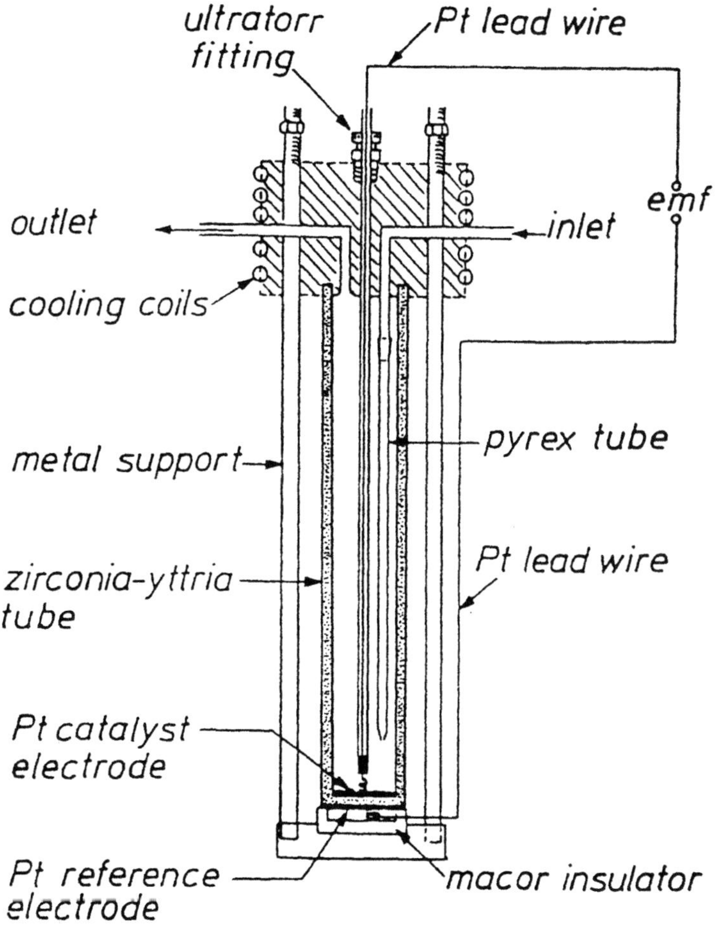

FIGURE 6: Schematic of a solid electrolyte reactor used for simultaneous kinetic and SEP studies.

$$V^o_{WR} = (1/4F)\left[\mu_{O_2,W} - \mu_{O_2,R}\right] \qquad (35)$$

where $\mu_{O_2,W}$ and $\mu_{O_2,R}$ are the chemical potentials of oxygen at the catalyst and reference electrode surfaces, respectively. Equation (35) is derived on the following four assumptions:

1. The solid electrolyte is a pure O^{2-} conductor
2. The catalyst and reference electrodes are made of the same bulk material
3. The dominant electrocatalytic reaction taking place at the metal-solid electrolyte-gas three-phase-boundaries (tpb) is :

$$O(a) + 2e^- \leftrightharpoons O^{2-} \qquad (36)$$

or

$$O_2(a) + 4e^- \leftrightharpoons 2O^{2-} \qquad (37)$$

where O(a) and O_2(a) denote oxygen dissociatively or molecularly adsorbed at the tpb. It is also assumed that equilibrium is established for reaction (36) or (37).

4. There are no concentration or chemical potential gradients within the porous catalyst film so that $\mu_{O_2,W}$ is the same at the tpb and over the entire catalyst surface. This implies that thin (e.g. 5-10μm) and sufficiently porous metal or metal oxide catalyst films must be used in order to ensure the absence of internal diffusional limitations.

Assumption 3 is certainly valid for the reference electrode, but may not be always valid for the catalyst electrode: If one of the reactants adsorbs strongly on the catalyst surface and has a high affinity for reaction with O^{2-} (e.g. H_2 or CO under fuel-rich conditions) then other electrocatalytic reactions, such as :

$$H_2 + O^{2-} \leftrightharpoons H_2O + 2e^- \qquad (38)$$

$$CO + O^{2-} \leftrightharpoons CO_2 + 2e^- \qquad (39)$$

may also take place at the tpb, leading to the establishment of mixed potentials. In this case V^o_{WR} provides only a qualitative measure of surface activities. In practice this means that one must be rather cautious when using Eq. (35) to treat very negative V^o_{WR} values, e.g. less than -400 mV, unless one proves the validity of assumption 3 via electrokinetic measurements by utilizing a three-electrode system, which is not an easy exercise. When assumption 3 is also satisfied, in addition to 1, 2 and 4, then the emf V^o_{WR} provides indeed an in situ quantitative measure of the chemical potential or activity of oxygen adsorbed on the catalyst under reaction conditions and this information can indeed be quite useful.

Before discussing how Eq. (35) has been used in the past to analyze SEP data, it is important to discuss first some recent findings. It was recently found both theoretically [5,6] and experimentally by means of a Kelvin probe [21], that the emf V^o_{WR} of solid electrolyte cells provides a direct measure of the difference in the work function $e\Phi$ of the gas-exposed surfaces of the working and reference electrodes:

$$eV^o_{WR} = e\Phi_W - e\Phi_R \tag{40}$$

Equation (40) shows that solid electrolyte cells are work function probes for their gas-exposed electrode surfaces. It also shows that SEP is essentially a work-function measuring technique and that several aspects of the SEP literature must be reexamined in the light of these findings. One can still use SEP to extract information about surface activities, provided the nature of the electrocatalytic reaction at the tpb is well known but, even when this is not the case, the cell emf still provides a direct measure of the work function difference between the two gas-exposed electrode surfaces. Equation (40) also holds under closed-circuit conditions, as is further discussed in section III.B.

Equation (35) has been used in the following way in SEP studies. First one notes that the chemical potential of oxygen adsorbed on the reference electrode is given by:

$$\mu_{O_2,R} = \mu^o_{O_2(g)} + RT \ln (0.21) \tag{41}$$

where $\mu^o_{O_2(g)}$ is the standard chemical potential of gaseous oxygen at the temperature T, and 0.21 (bar) corresponds to the oxygen activity in the reference gas, i.e., air. Defining the activity of atomic oxygen adsorbed on the catalyst a_O from:

$$\mu_{O_2,W} = \mu^o_{O_2(g)} + RT \ln a_O^2 \tag{42}$$

one obtains

$$V^o_{WR} = (RT/2F) \ln \left[a_O/(0.21)^{1/2} \right] \tag{43}$$

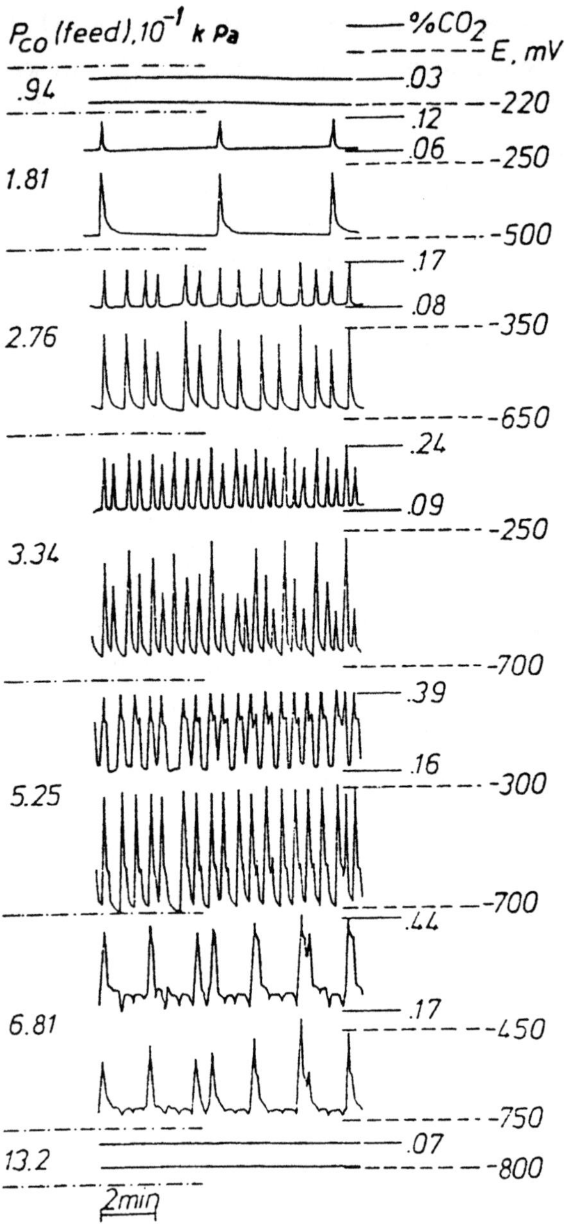

FIGURE 7: Effect of inlet CO partial pressure on reaction rate and on V^{o}_{WR} and eΦ oscillations during CO oxidation on Pt; P_{O_2}=5.4 kPa, T=337°C, total molar flowrate 2.7×10^{-4} mol/s. Reprinted with permission from Elsevier Science Publishers [5].

Consequently by measuring V^o_{WR} and T one computes a_O which expresses the square root of the partial pressure of gaseous oxygen that would be in thermodynamic equilibrium with oxygen adsorbed on the catalyst surface, if such an equilibrium were established. By comparing the potentiometrically obtained surface oxygen activity with the independently measured gas phase oxygen activity P_{O_2} it is possible to extract useful information about the rate determining step (rds) of the catalytic reaction. Thus if $a_O^2 = P_{O_2}$, then the oxygen adsorption step is in equilibrium and cannot be rate limiting. If, however $a_O^2 < P_{O_2}$, as is often found to be the case, then oxygen adsorption is rate limiting.

Despite its limitations, SEP is one of the very few techniques which can be used to extract in situ information about adsorbed species on catalyst surfaces without UHV requirements. It is particularly useful for the study of oscillatory reactions Fig. 7 [20].

7.2. ELECTROCHEMICAL PROMOTION

During the last few years a new application of solid electrolytes has emerged. It was found that the catalytic activity and selectivity of the gas-exposed electrode surface of metal electrodes in solid electrolyte cells is altered dramatically and reversibly upon polarizing the metal-solid electrolyte interface. The induced steady state change in catalytic rate can be up to 9000% higher than the normal (open-circuit) catalytic rate and up to $3x10^5$ higher than the steady state rate of ion supply. This new effect of non-Faradaic electrochemical modification of catalytic activity (NEMCA) or electrochemical promotion has been already demonstrated for more than 50 catalytic reactions on Pt, Pd, Rh, Ag, Au and Ni surfaces by using O^{2-}, F^-, Na^+ and H^+ conducting solid electrolytes. There are also recent demonstrations for aqueous electrolyte systems. In this section the common features of NEMCA studies are summarized and the origin of the effect is discussed in the light of recent in situ work function and XPS measurements which have shown that:

I. Solid electrolyte cells with metal electrodes are work function probes and work function controllers, via potential application, for their gas-exposed electrode surfaces.

II. NEMCA is due to an electrochemically driven and controlled backspillover of ions from the solid electrolyte onto the gas-exposed electrode surface. These backspillover ions establish an effective electrochemical double layer and act as promoters for catalytic reactions. This interfacing of electrochemistry and catalysis offers several exciting theoretical and technological possibilities. We note that in the catalytic literature the term spillover usually denotes migration of a species from a metal to a support, while the term backspillover denotes migration in the opposite direction and is thus more appropriate here.

The first reports on NEMCA appeared in the literature in 1988 [23]. Since then the NEMCA effect or electrochemical promotion has been described for more than fifty catalytic reactions and work prior to 1992 and prior to 1996 has been reviewed [5,6]. In addition to the group which first reported this novel effect [21-23] the groups of Sobyanin [24], Lambert [25,26], Stoukides [27], Haller [28,29] and Comninellis [30] have also contributed recently to the NEMCA literature. Recently the NEMCA effect was also demonstrated in an aqueous electrolyte systems [6]. The term "Electrochemical Promotion in Catalysis" has also been proposed by Pritchard to describe the NEMCA effect.

It has been found that the catalytic activity and selectivity of porous metal catalyst films deposited on solid electrolytes can be altered in a dramatic, reversible and, to some extent, predictable manner by carrying out the catalytic reaction in solid electrolyte cells of the type [22]:

gaseous reactants, metal catalyst I solid electrolyte I metal, O_2
(e.g. C_2H_4+O_2) (e.g. Pt) (e.g. YSZ) (e.g. Ag)

where the metal catalyst also serves as an electrode and by applying currents or potentials to the cell with a concomitant supply or removal of ions, e.g. O^{2-}, F^-, Na^+, H^+, to or from the catalyst surface.

The NEMCA effect was first demonstrated on Pt and Ag electrodes using 8mol% Y_2O_3-stabilized-ZrO_2 (YSZ), an O^{2-} conductor, as the solid electrolyte [23]. Electrochemical O^{2-} pumping to the catalyst-electrode was found to cause up to 60-fold (6000%) steady state reversible enhancement in the rate of C_2H_4 oxidation. Furthermore this steady state rate increase was found to be up to $3x10^5$ times higher than the steady state rate of supply of O^{2-} to the catalyst, i.e. the enhancement factor Λ [5,6] or apparent Faradaic efficiency for the process is $3x10^5$. More recently the rate of C_2H_4 oxidation on Rh electrodes supported on YSZ was found to reversibly increase by a factor of 90, again with Faradaic efficiency values of the order of 10^5 [31].

The effect of NEMCA or electrochemical promotion does not appear to be limited to any particular metal, solid electrolyte or group of catalytic reactions. Thus in addition to O^{2-} conducting solid electrolytes, the NEMCA effect has also been demonstrated using Na^+-conducting solid electrolytes such as β"-Al_2O_3, H^+-conducting solid electrolytes such as $CsHSO_4$ and F^- conducting solid electrolytes, such as CaF_2.

Also in addition to complete and partial oxidation reactions the NEMCA effect has also been demonstrated for dehydrogenation, hydrogenation and decomposition reactions [5,6]

7.2.1. *Experimental setup*

The basic experimental setup is shown schematically on Fig. 8a. The metal working catalyst electrode, usually in the form of a porous metal film 3-20 μm in thickness, is deposited on the surface of a ceramic solid electrolyte (e.g. Y_2O_3-stabilized-ZrO_2 (YSZ) an O^{2-} conductor, or β"-Al_2O_3, a Na^+ conductor). Catalyst, counter and reference electrode preparation and characterization details have been presented in detail elsewhere [5,6] together with the analytical system for on-line monitoring the rates of catalytic reactions by means of gas chromatography, mass spectrometry and IR-spectroscopy.

The superficial surface area of the metal working catalyst-electrode is typically $2cm^2$ and its true gas-exposed surface area is typically $5\text{-}10^3$ cm^2 as measured via surface titration of oxygen with CO or C_2H_4. The catalyst electrode is exposed to the reactive gas mixture (e.g. C_2H_4+O_2) in a continuous flow gradientless reactor (CSTR). Under open-circuit conditions (I=0) it acts as a regular catalyst for the catalytic reaction under study, e.g., C_2H_4 oxidation. The counter and reference electrodes are usually exposed to ambient air.

A galvanostat or potentiostat is used to apply constant currents between the catalyst and the counter electrode or constant potentials between the catalyst and reference electrodes. In this way ions (O^{2-} in the case of YSZ, Na^+ in the case of β"-Al_2O_3) are supplied from (or to) the solid electrolyte to (or from) the catalyst-electrode surface. The current is defined positive when anions are supplied to or cations removed from the catalyst electrode. There is convincing evidence that these ions (together with their compensating (screening) charge in the metal thus forming surface dipoles) migrate (backspillover) onto the gas-exposed catalyst electrode surface. Thus the solid electrolyte acts as an active catalyst support and establishes an effective electrochemical double layer on the gas-exposed, i.e. catalytically active, electrode surface.

The (average) work function of the gas-exposed catalyst-electrode surface can be measured in situ, i.e., during reaction at atmospheric pressure and temperatures up to 300°C, by means of a Kelvin probe (vibrating condenser method) using, e.g., a Besocke Delta-Phi Kelvin probe with a Au vibrating disc as described in detail elsewhere [5,6].

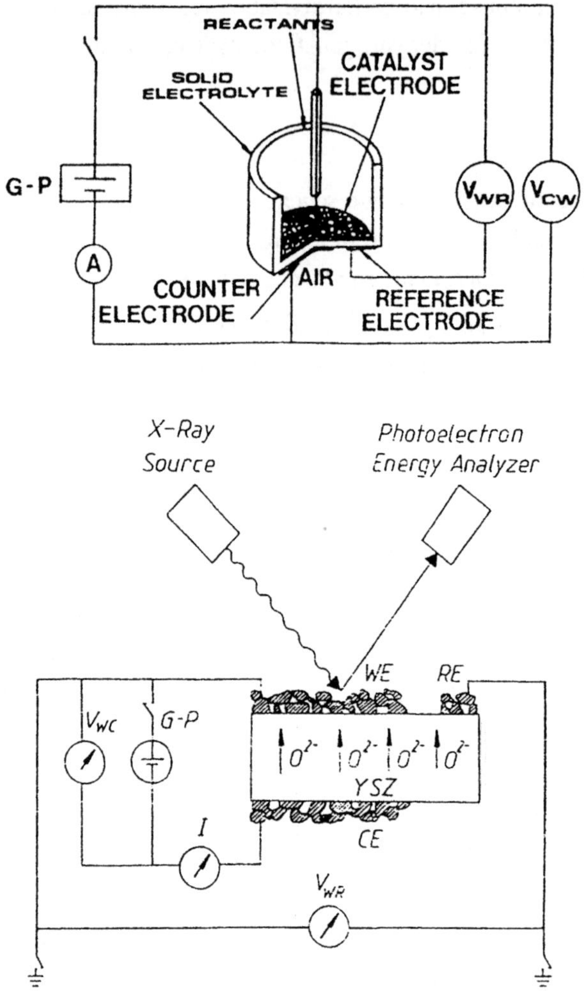

FIGURE 8: Schematic of the experimental setup for NEMCA studies (a), and for using XPS (b) G-P: Galvanostat-Potentiostat. Reprinted with permission from CRC Press [33].

X-ray photoelectron spectroscopy (XPS) is a useful tool for investigating metal electrode surfaces under conditions of electrochemical O^{2-} pumping [5,6]. The experimental setup is shown schematically in Fig. 8b. A 2-mm thick YSZ slab (10 mmx13mm) with a Pt catalyst electrode film, a Pt reference electrode and a Ag counter electrode was mounted on a resistively heated Mo holder in an ultra-high-vacuum (UHV) chamber (base pressure $5x10^{-10}$Torr) and the catalyst-electrode film (9mmx9mm) was examined at temperatures 25° to 525°C by XPS using a Leybold HS-12 analyzer operated at constant ΔE mode with 100-eV pass energy and a sampling area of 5 mm x 3mm. Electron binding energies were referenced to the metallic Pt $4f_{7/2}$ peak of the grounded catalyst electrode at 71.1 eV, which always remained unchanged with no trace of an oxidic component. Further experimental details are given elsewhere [32,33].

7.2.2. *Catalytic rate modification*

Figure 9 shows a typical NEMCA experiment carried out in the setup depicted in Fig. 8a.The catalytic reaction under study is the complete oxidation of C_2H_4 on Pt:

$$CH_2 = CH_2 + 3O_2 \rightarrow 2CO_2 + 2H_2O \tag{44}$$

The figure shows a typical galvanostatic transient, i.e., it depicts the transient effect of a constant applied current on the rate of C_2H_4 oxidation (expressed in g-atom O/s).

The Pt catalyst film with a surface area corresponding to $N=4.2\cdot10^{-9}$ g-atom Pt, as measured by surface titration techniques [5,6], is deposited on YSZ and is exposed to $P_{O_2}=4.6$ kPa, $P_{C_2H_4}=0.36$ kPa in the CSTR-type flow reactor depicted schematically in Fig. 10a. Initially ($t<0$) the circuit is open ($I=0$) and the open-circuit catalytic rate r_o is $1.5\cdot10^{-8}$ g-atom O/s. The corresponding turnover frequency (TOF), i.e., the number of oxygen atoms reacting per site per second, is 3.57 s^{-1}.

Then at $t=0$ a galvanostat is used to apply a constant current of +1μA between the catalyst and the counter electrode (Fig. 8a). Now oxygen ions O^{2-} are supplied to the catalyst-gas-solid electrolyte three-phase-boundaries (tpb) at a rate $G_O=I/2F=5.2\cdot10^{-12}$ g-atom O/s. The catalytic rate starts increasing (Fig. 9) and within 25 min gradually reaches a value $r=40.\cdot10^{-8}$ g-atom O/s, which is 26 times larger than r_o. The new TOF is 95.2 s^{-1}. The increase in catalytic rate $\Delta r=r-r_o=38.5\cdot10^{-8}$ g-atom O/s is 74,000 times larger than I/2F. This means that each O^{2-} supplied to the Pt catalyst causes at steady-state 74,000 additional chemisorbed oxygen atoms to react with C_2H_4 to form CO_2 and H_2O. This is why this novel effect has been termed Non-Faradaic Electrochemical Modification of Catalytic Activity (NEMCA).

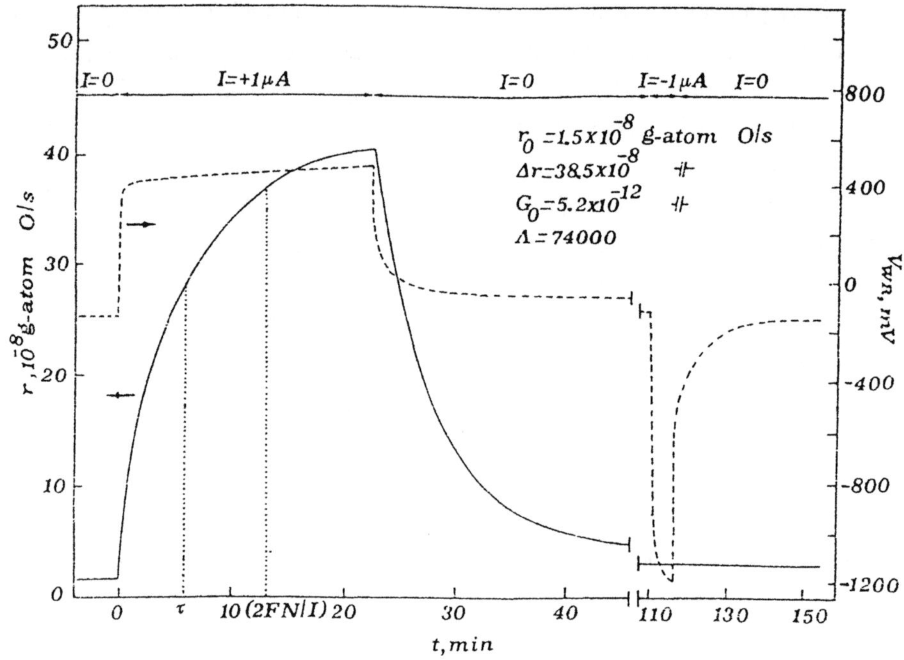

FIGURE 9: Rate and catalyst potential response to step changes in applied current during C_2H_4 oxidation on Pt; T=370°C, Po_2=4.6 kPa, Pc_2H_4=0.36 kPa. The steady-state rate increase Δr is 74,000 times higher than the steady-state rate of supply of O^{2-} to the catalyst (Λ=74,000). Reprinted with permission from Elsevier Science Publishers BV [5].

There is an important observation to be made regarding the time required for the rate to approach its steady-state value. Catalytic rate transients obtained during galvanostatic (i.e., constant current) operation are found in NEMCA studies to be usually, but not always, of the type:

$$\Delta r = \Delta r_{max}\left[1\text{-}\exp\left(\text{-}t/\tau\right)\right] \tag{45}$$

This is the response of a first order system with a characteristic time constant τ, and thus one can define the NEMCA time constant τ as the time required for Δr to reach 63% of its maximum, i.e., steady-state value. As shown in Fig. 9, τ is of the order of 2FN/I and this turns out to be a general observation in NEMCA studies utilizing doped ZrO_2, i.e.:

$$\tau \approx 2FN/I \qquad (46)$$

This observation shows that NEMCA is a catalytic effect, i.e., it takes place over the entire gas-exposed catalyst surface, and is not an electrocatalytic effect localized at the three-phase-boundaries (tpb) metal-solid electrolyte-gas. This is because 2FN/I is the time required to form a monolayer of an oxygen species on a surface with N sites when it is supplied at a rate I/2F. The fact that τ is found to be smaller than 2FN/I, but of the same order of magnitude, shows that only a fraction of the surface is occupied by oxygen backspillover species, as discussed in detail elsewhere [5,6]. It is worth noting that if NEMCA were restricted to the tpb, i.e., if the observed rate increase were due to an electrocatalytic reaction, then τ would be practically zero during galvanostatic transients.

Figure 9 further shows that NEMCA is reversible, i.e., upon current interruption the catalytic rate returns to its initial value within roughly 100 min. The rate relaxation curve upon current interruption conveys valuable information about the kinetics of reaction and desorption of the promoting oxygen species as discussed in detail elsewhere [5,6]. Negative current application has practically no effect on the rate of this particular reaction.

7.2.3. *Effect of gaseous composition*

Figure 10 shows the effect of O_2 to C_2H_4 ratio on the regular (open-circuit) steady-state rate of C_2H_4 oxidation on Pt and on the NEMCA-induced rate when the same catalyst film is maintained at a potential of +1V (V_{WR}=+1V) with respect to the reference Pt/air electrode (Fig. 8a). It can be seen that the effect is much more pronounced at high $P_{O_2}/P_{C_2H_4}$ ratios, i.e. for high oxygen coverages, where the NEMCA induced reaction rate or TOF values are a factor of 55 higher than the corresponding open-circuit rate. A quantitative description and explanation of the NEMCA behaviour of C_2H_4 oxidation on Pt can be found elsewhere.

Figure 11 shows the effect of C_2H_4 partial pressure at constant P_{O_2} on the rate of C_2H_4 oxidation on Rh at various imposed values of V_{WR}. Increasing V_{WR} causes up to 90-fold (9000%) rate enhancement relative to the open-circuit rate value. The dramatic rate enhancement with increasing V_{WR} depicted in Figures 10 and 11 is due to the weakening of the metal-covalently chemisorbed oxygen chemisorptive bond, cleavage of which is rate limiting, as discussed in detail elsewhere [6,31].

7.2.4. *Definitions and the role of the exchange current I_0*

Table 2 provides a list of the reactions which have been studied already and shown to exhibit NEMCA. In order to compare different catalytic reactions, it is useful to define two dimensionless parameters, i.e., the enhancement factor or Faradaic efficiency Λ and the rate enhancement ratio ρ. The former is defined as:

$$\Lambda = \Delta r/(I/2F) \tag{47}$$

where the change in catalytic rate Δr is expressed in terms of g-atom of O. More generally Λ is computed by expressing Δr in g-equivalent and dividing by I/F.

A catalytic reaction is said to exhibit NEMCA when $|\Lambda|>1$. When $\Lambda>1$ as, e.g. in the case of C_2H_4 oxidation on Pt, the reaction is said to exhibit positive or electrophobic NEMCA behaviour. When $\Lambda<-1$ then the reaction is said to exhibit electrophilic behaviour.

The rate enhancement ratio ρ is defined from :

$$\rho = r/r_o \tag{48}$$

In the C_2H_4 oxidation example presented on Figure 9 and discussed above the Λ and ρ values at steady-state are Λ=74000 and ρ=26.

As it turns out experimentally (Figure 12) and can be explained theoretically [5,6] one can estimate or predict the order of magnitude of the absolute value of the enhancement factor Λ for any given reaction, catalyst and catalyst-solid electrolyte interface from:

$$|\Lambda| \approx 2Fr_o/I_0 \tag{49}$$

where I_0 is the exchange current of the metal-solid electrolyte interface.

As noted previously, the parameter I_0 can be easily determined from standard ln I vs η (Tafel) plots [5,6,8]. The overpotential η of the catalyst-electrode is defined, according to section ID, from:

$$\eta = V_{WR} - V_{WR}^{o} \tag{50}$$

where V_{WR} is the catalyst (working electrode, W) potential with respect to a reference (R) electrode. The overpotential η is related to current I via the classical Butler-Volmer equation:

$$(I/I_0) = \exp(\alpha_a F\eta/RT) - \exp(-\alpha_c F\eta/RT) \tag{51}$$

where α_a and α_c are the anodic and cathodic transfer coefficients, respectively. Thus by measuring η as a function of I one can extract I_0, α_a and α_c. Physically I_0 expresses the (equal under open-circuit conditions) rates of the electronation and deelectronation reaction at the tpb, e.g.:

$$O^{2-} \leftrightharpoons O(a) + 2e^- \quad (52)$$

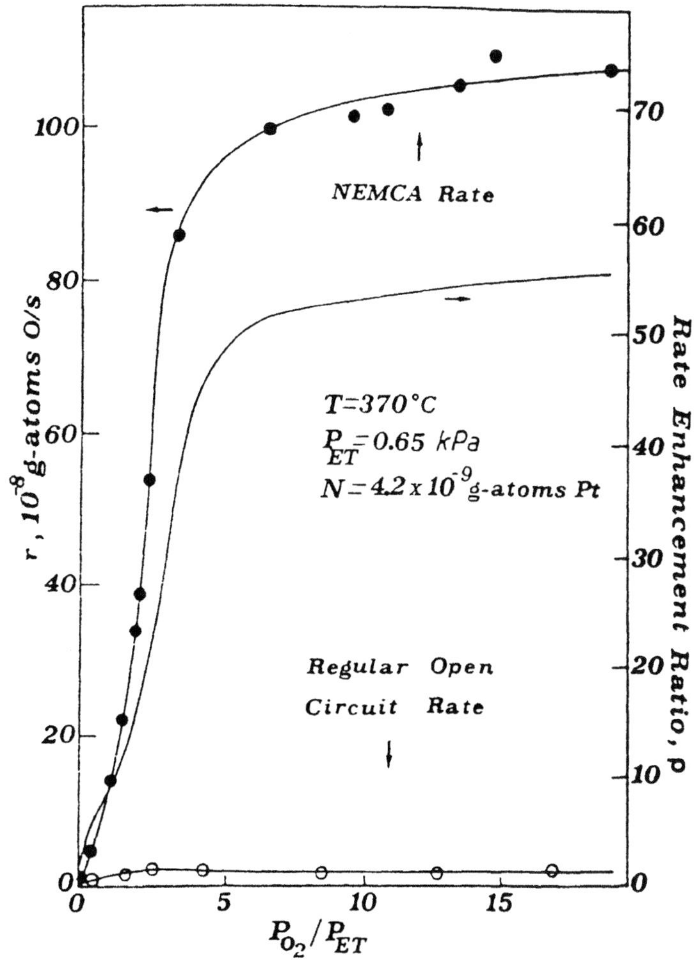

FIGURE 10: Effect of gaseous composition on the regular (open-circuit) steady-state rate of C_2H_4 oxidation on Pt and on NEMCA-induced catalytic rate when the catalyst film is maintained at V_{WR}=1V. Reprinted with permission from Elsevier Science Publishers BV [5].

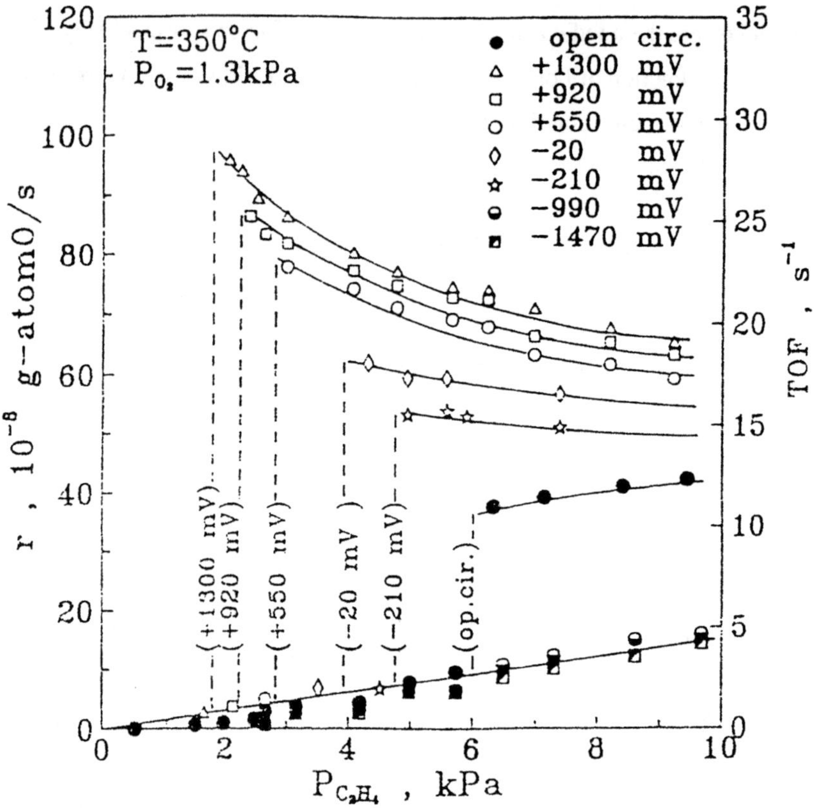

FIGURE 11: Effect of gaseous composition and catalyst potential V_{WR} on the steady-state rate of C_2H_4 oxidation on Rh. Reprinted with permission from Academic Press [31].

where O(a) stands for oxygen adsorbed on the metal catalyst in the vicinity of the tpb. Thus, the exchange current I_0 is a measure of the non-polarizability of the metal-solid electrolyte interface.

As shown in Figure 12, Eq. (49) shows good agreement with experiment for all catalytic reactions studied so far. The agreement extends over more than five orders of magnitude. Thus, contrary to fuel cell applications where non-polarizable, i.e., high I_0 electrode-electrolyte interfaces are desirable to minimize activation overpotential losses, the opposite holds for catalytic applications, i.e., I_0 must be low in order to obtain high Λ values, i.e., a strong Non-Faradaic rate enhancement. This is a very important point for Electrochemical Promotion studies and potential technological applications.

TABLE 2: Catalytic reactions investigated in electrochemical promotion studies (Adapted from Table 3 in ref. [6] which provides specific references to each reaction)

I. Electrophobic reactions $\partial r/\partial(e\Phi)>0$; $\partial r/\partial V_{WR}>0$; $\partial r/\partial I>0$; $\Lambda>0$

Reactants	Products	Catalyst	Electrolyte (Promoting Ion)	T(^oC)	Λ	ρ	P_i
C_2H_4, O_2	CO_2	Pt	YSZ (O^{2-})	260-450	$3\cdot10^5$	55	55
C_2H_4, O_2	CO_2	Rh	YSZ (O^{2-})	250-400	$5\cdot10^4$	90	90
C_2H_4, O_2	CO_2	IrO_2	YSZ (O^{2-})	350-400	200	6	5
C_2H_4, O_2	C_2H_4O,CO_2	Ag	YSZ (O^{2-})	320-470	300	30*	30
C_2H_6, O_2	CO_2	Pt	YSZ (O^{2-})	270-500	300	20	20
C_3H_6, O_2	C_3H_6O,CO_2	Ag	YSZ (O^{2-})	320-420	300	2*	1
CH_4, O_2	CO_2	Pt	YSZ (O^{2-})	600-750	5	70	70
CH_4, O_2	CO_2	Pt	YSZ (O^{2-})	590	50	3	3
CH_4, O_2	CO_2,C_2H_4,C_2H_6	Ag	YSZ (O^{2-})	650-850	5	30*	30
CO, O_2	CO_2	Pt	YSZ (O^{2-})	350-550	$2\cdot10^3$	3	2
CO, O_2	CO_2	Pd	YSZ (O^{2-})	400-550	10^3	2	1
CO, O_2	CO_2	Ag	YSZ (O^{2-})	350-450	20	5	4
CH_3OH,O_2	H_2CO,CO_2	Pt	YSZ (O^{2-})	300-500	10^4	4*	3
CO_2, H_2	CH_4, CO	Rh	YSZ (O^{2-})	300-450	200	3*	2
CO, H_2	C_xH_y,$C_xH_yO_z$	Pd	YSZ (O^{2-})	300-370	10	3*	2
CH_4, H_2O	CO,CO_2	Ni	YSZ (O^{2-})	600-900	12	2*	1
H_2S	S_x, H_2	Pd	YSZ (O^{2-})	600-750	-	11	10
C_2H_4, O_2	CO_2	Pt	β''-Al_2O_3(Na^+)	180-300	$5\cdot10^4$	0.25	-30
CO, O_2	CO_2	Pt	β''-Al_2O_3(Na^+)	300-450	10^5	0.3	-30
C_6H_6, H_2	C_6H_{12}	Pt	β''-Al_2O_3(Na^+)	100-150	-	~0	-10
CH_4	C_2H_6,C_2H_4	Ag	$SrCe_{0.95}Yb_{0.05}O_3$(H^+)	750	-	11*	10
C_2H_4, H_2	C_2H_6	Ni	$CsHSO_4$(H^+)	150-170	6-300	0.16-2	12
H_2, O_2	H_2O	Pt	Nafion (H^+)	25	20	6	5
H_2, O_2	H_2O	Pt	KOH-H_2O(OH^-)	25-50	20	6	5
CO, O_2	CO_2	Pt	CaF_2(F^-)	500-700	200	2.5	1.5
C_2H_4, O_2	CO_2	Pt	TiO_2(TiO^+_x, O^{2-})	450-600	$5\cdot10^3$	20	20

II. Electrophilic reactions $\partial r/\partial(e\Phi) < 0$; $\partial r/\partial V_{WR} < 0$; $\partial r/\partial I < 0$; $\Lambda < 0$

C_2H_6, O_2	CO_2	Pt	YSZ (O^{2-})	270-500	-100	7	-
C_3H_6, O_2	CO_2	Pt	YSZ (O^{2-})	350-480	$-3\cdot10^3$	6	-
CO, O_2	CO_2	Pt	YSZ (O^{2-})	300-550	-500	6	-
CO, O_2	CO_2	Au	YSZ (O^{2-})	400-600	-60	3	-
CO_2, H_2	CO	Pd	YSZ (O^{2-})	500-590	-50	10	-
CH_3OH, O_2	H_2CO, CO_2	Pt	YSZ (O^{2-})	300-550	-10^4	15*	-
CH_3OH	H_2CO, CO, CH_4	Ag	YSZ (O^{2-})	550-750	-25	6*	-
CH_3OH	H_2CO, CO, CH_4	Pt	YSZ (O^{2-})	400-500	-10	3*	-
CH_4, O_2	CO_2	Au	YSZ (O^{2-})	700-900	-3	3*	-
CH_4, O_2	C_2H_4, C_2H_6, CO_2	Ag	YSZ (O^{2-})	700-750	-1.2	8*	-
CO, O_2	CO_2	Pt	β''-$Al_2O_3(Na^+)$	300-450	-10^5	8	250
C_2H_4, NO	CO_2, N_2, N_2O	Pt	β''-$Al_2O_3(Na^+)$	400	-	∞	500
CO, NO	CO_2, N_2, N_2O	Pt	β''-$Al_2O_3(Na^+)$	320-400	-	13*	200

* Promotion-induced change in product selectivity

Although Λ is an important parameter for determining whether a reaction exhibits NEMCA, it is not a fundamental one. The reason is that for the same catalytic reaction on the same catalyst material one can obtain significantly different $|\Lambda|$ values by varying I_0 (Eq. 49). The parameter I_0 is proportional to the tpb length [10] and can thus be controlled during catalyst film preparation by varying the sintering temperature and in this way control the metal crystallite size and tpb length [10].

From a catalytic point of view a very important parameter which can be obtained via NEMCA is the promotion index P_i of the doping species defined from:

$$P_i = \frac{\Delta r/r_o}{\Delta\theta_i} \tag{53}$$

where θ_i is the coverage of the backspillover promoting species introduced on the catalytic surface (e.g. $O^{\delta-}$, $Na^{\delta+}$, F^- etc.). When $P_i>0$, then the backspillover species has a promoting effect on the catalytic reaction under study. When $P_i<0$, then the backspillover species has a poisoning effect on the catalytic reaction. When the backspillover species does not react appreciably with any of the reactants, as e.g. in the case of $Na^{\delta+}$, then $\Delta\theta_{Na}$ can be measured accurately via coulometry. Very recently a method based on the current interruption technique has been developed to also measure $\Delta\theta_i$ for the case of $O^{\delta-}$ and F^- backspillover species which react with the reactants of the catalytic reaction at a rate Λ times slower than the NEMCA induced catalytic rate. The method has been applied to only very few reactions yet and thus the P_i values listed in Table 2 for the case of ZrO_2-Y_2O_3 solid electrolytes as the ion donor are based on the approximation $\Delta\theta_i$=1 for the maximum measured ρ value for each reaction, in which case P_i equals (ρ-1).

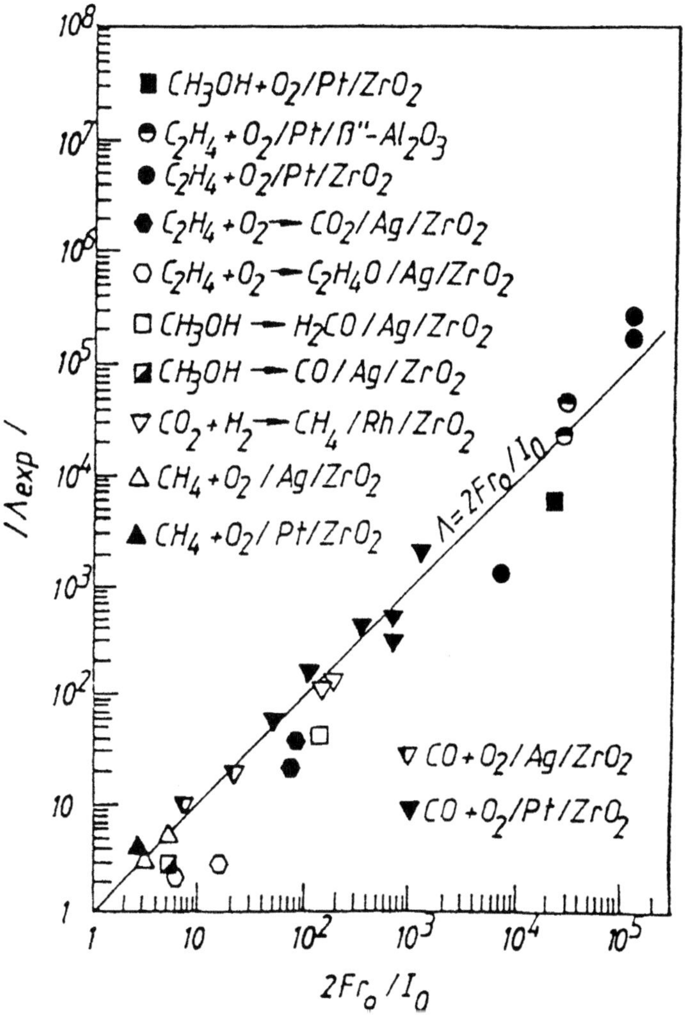

FIGURE 12: Comparison of predicted and measured enhancement factor Λ values for the first twelve catalytic reactions found to exhibit the NEMCA effect. Reprinted with permission from Elsevier Science Publishers B.V. [5]

7.2.5. *Selectivity modification*

One of the most promising applications of NEMCA is in product selectivity modification. An example is shown in Fig. 13 for the case of C_2H_4 oxidation on Ag [5,6]. The figure shows the effect of varying the catalyst potential V_{WR} on the selectivity to ethylene oxide (the other products being CO_2 and for V_{WR}<-0.4V some acetaldehyde) at various levels of addition of gas-phase chlorinated hydrocarbon "moderators". With no 1,2-$C_2H_4Cl_2$ present in the feed the selectivity to ethylene oxide is varied between 0 and 56% by varying V_{WR}. Combination of NEMCA and 1,2-$C_2H_4Cl_2$ addition gives selectivities well above 75% when YSZ is used as the solid electrolyte and up to 88% when using β"-Al_2O_3 as the ion donor. The beneficial effect of increasing eΦ and Cl coverage is due to the weakening of the binding strength of chemisorbed atomic oxygen which makes it more selective for epoxidation.

Another example of selectivity modification is shown in Figure 14 for the reduction of NO by CO on Pt/β"-Al_2O_3 [34]. Decreasing catalyst potential enhances significantly both the rate and the selectivity to N_2. This is due to sodium-enhanced NO dissociation [34].

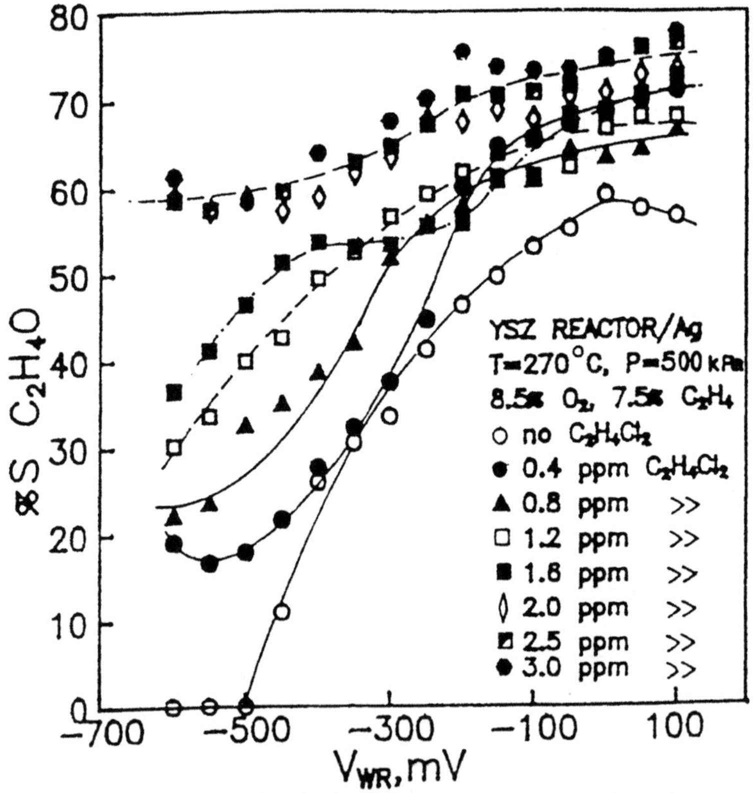

FIGURE 13: Effect of catalyst potential V_{WR} on the selectivity to ethylene oxide during C_2H_4 oxidation on Ag at various levels of gas-phase "moderator" $C_2H_4Cl_2$. Reprinted with permission from Elsevier Science Ltd. [5].

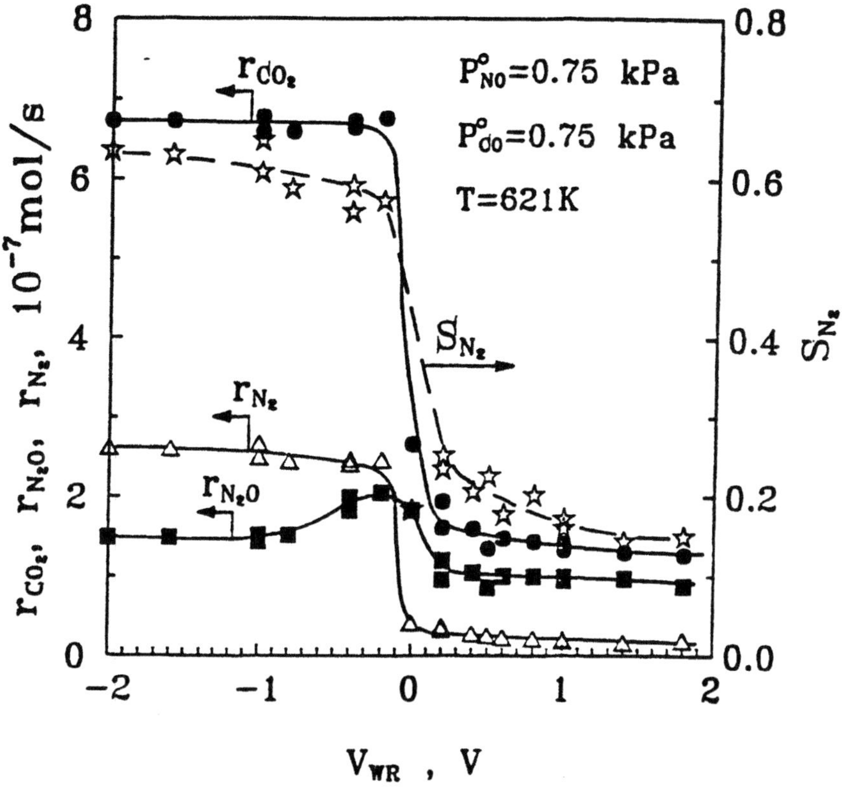

FIGURE 14: Effect of catalyst potential on the rates of formation of N_2, N_2O and CO_2 and on the selectivity to N_2 during NO reduction by CO on Pt/β ″-Al_2O_3. Reprinted with permission from Academic Press [34].

7.2.6.Work function measurements: An additional meaning of the EMF of solid electrolyte cells with metal electrodes

One of the key steps in understanding the origin of NEMCA was the realization that solid electrolyte cells can be used both to monitor and to control the work function of the gas-exposed surfaces of their electrodes. It was shown both theoretically and experimentally that :

$$eV^o_{WR} = e\Phi_W - e\Phi_R \tag{40}$$

and

$$e\Delta V_{WR} = \Delta (e\Phi_W) \tag{54}$$

where $e\Phi_W$ is the catalyst surface work function and $e\Phi_R$ is the work function of the reference electrode surface. The derivation of Eqs. (40) and (54) is based on the standard definition of the work function and on the fact that the average Volta electrode potential Ψ vanishes at the electrode-gas interface, since no net charge can be sustained there.

The validity of Eqs (40) and (54) was demonstrated by using a Kelvin probe to measure in situ $e\Phi$ on catalyst surfaces subject to electrochemical promotion [21].

Therefore by applying currents or potentials in NEMCA experiments and thus by varying V_{WR}, one also varies the average catalyst surface work function $e\Phi$ (Eq. 54). Positive currents increase $e\Phi$ and negative currents decrease it. Physically the variation in $e\Phi$ is primarily due to backspillover of ions to or from the catalyst surface.

7.2.7. *Dependence of catalytic rates and activation energies on eΦ*

In view of Eq. (54) it follows that NEMCA experiments permit to directly examine the effect of catalyst work function $e\Phi$ on catalytic rates. From a fundamental viewpoint the most interesting finding of all previous NEMCA studies is that over wide ranges of catalyst work function $e\Phi$ catalytic rates depend exponentially on $e\Phi$ and catalytic activation energies vary linearly with $e\Phi$.

A typical example is shown in Fig.15 for the catalytic oxidation of C_2H_4 and of CH_4 on Pt. Both reactions exhibit electrophobic behaviour which is due to the weakening of the Pt=O chemisorptive bond with increasing $e\Phi$. Chemisorbed atomic oxygen is an electron acceptor, thus increasing $e\Phi$ causes a weakening in the Pt=O bond, cleavage of which is involved in the rate-limiting-step of the catalytic oxidation, and thus a linear decrease in activation energy and an exponential increase in catalytic rate is observed. In general, increasing $e\Phi$ weakens the chemisorptive bond of electron acceptor adsorbates such as oxygen [5,6]. Figure 11 provides another example. The abrupt rate increases are due to reduction of surface Rh oxide. Increasing $e\Phi$ weakens the Rh-O bond and destabilizes the oxide, thus causing the observed dramatic rate enhancement.

7.2.8. *XPS Spectroscopic and voltammetric identification of backspillover ions as the cause of NEMCA*

A recent XPS study of Pt films interfaced with YSZ under conditions of electrochemical promotion has shown [32]:

I. Backspillover oxide ions (O1s at 528.8 eV) are generated on the gas-exposed electrode surface upon positive current application (peak δ in Figure 16 top).

II. Normally chemisorbed atomic oxygen (O1s at 530.2 eV) is also formed upon positive current application (peak γ in Figure 16 top). The coverages of the γ and δ states of oxygen are comparable and of the order of 0.5 each.

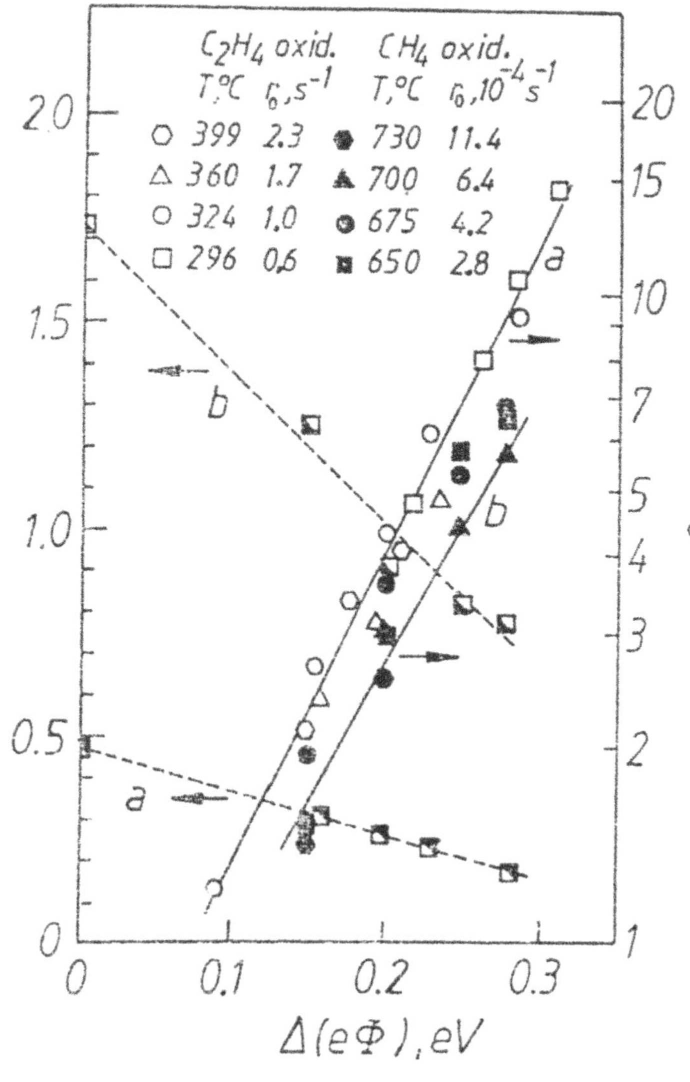

FIGURE 15: Effect of catalyst work function eΦ on the activation energy E and catalytic rate enhancement ratio r/r_o for C_2H_4 oxidation on Pt (a) and CH_4 oxidation on Pt (b). Reprinted with permission of Kluwer Academic Publishers [37].

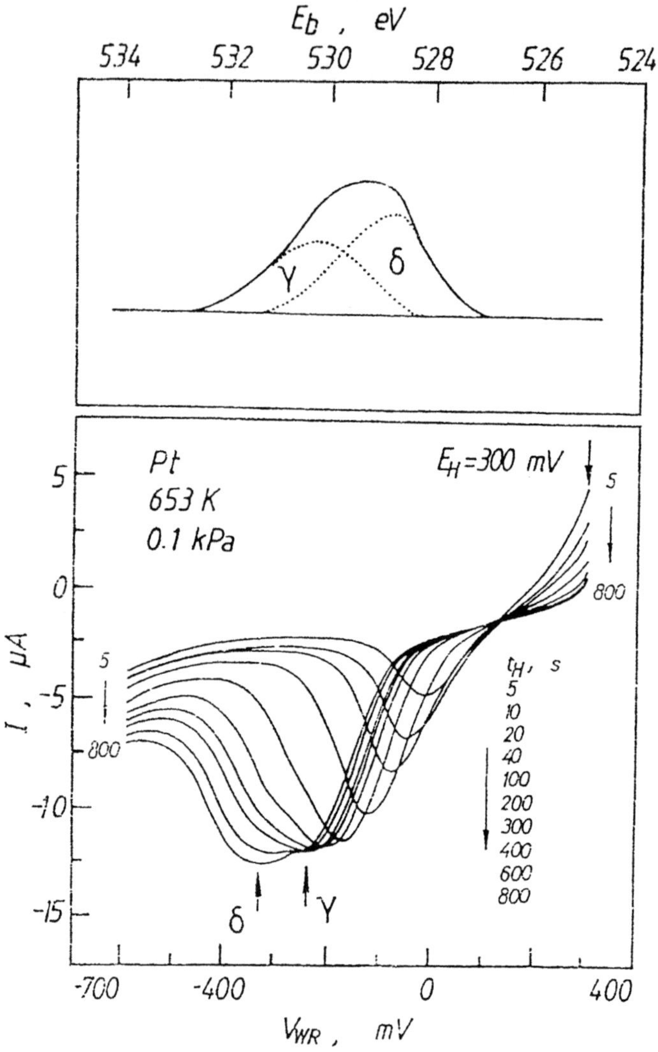

FIGURE 16: Top: O1s photoelectron spectrum of oxygen adsorbed on a Pt electrode supported on YSZ under UHV conditions after applying a constant overpotential ΔV_{WR}=1.2V corresponding to a steady state current I=40 μA for 15 min at 673K. The same O1s spectrum was maintained after turning off the potentiostat and rapidly cooling to 400 K. The γ-state is normally chemisorbed atomic oxygen (E_b=530.2 eV) and the δ-state is spillover oxidic oxygen (E_b=528.8 eV). Reprinted with permission from ACS [32]. Bottom: Linear potential sweep voltammogram obtained at T=653 K and P_{O_2}=0.1 kPa on a Pt electrode supported on YSZ showing the effect of holding time t_H at V_{WR} =300 mV on the reduction of the γ- and δ-states of adsorbed oxygen; sweep rate: 30 mV/s. Reprinted with permission from CRC Press [33].

III. Oxidic backspillover oxygen (δ-state) is less reactive than normally chemisorbed atomic oxygen (γ-state) with the reducing (H_2 and CO) ultra high vacuum background.

Similar conclusions can be reached from a recent XPS, UPS and EELS study of Ag/YSZ surfaces under NEMCA conditions [35].

These observations provide a straightforward explanation for the origin of NEMCA when using YSZ: Backspillover oxide ions (O^{2-} or O^{-}) generated at the tpb upon electrochemical O^{2-} pumping to the catalyst spread over the gas-exposed catalyst-electrode surface. They are accompanied by their compensating charge in the metal, thus forming backspillover dipoles. They thus establish an effective electrochemical double layer which increases the catalyst surface work function and affects the strength of chemisorptive bonds such as that of normally chemisorbed oxygen via through-the-metal or through-the-vacuum interactions. This change in chemisorptive bond strength causes the observed dramatic changes in catalytic rates.

The creation of two types of chemisorbed oxygen on Pt surfaces subject to NEMCA conditions has been recently confirmed by means of linear potential sweep voltammetry (Fig. 16 bottom [12]). The first oxygen reduction peak corresponds to normally chemisorbed oxygen (γ-state) and the second reduction peak, which appears only after prolonged application of a positive current [12] corresponds to the δ-state of oxygen, i.e. backspillover oxidic oxygen. Recent in situ XPS investigation of Pt films deposited on β''-Al_2O_3 has similarly shown that electrochemically controlled Na backspillover is the origin of NEMCA when using β''-Al_2O_3 as the solid electrolyte [36].

8. Conclusions

Electrocatalysis plays an important role in the efficient operation of solid electrolyte fuel cells used for power generation. Catalysis at the anode also has a significant role. In addition to power generation, solid electrolyte electrochemical reactors can also be used for chemical cogeneration, i.e. for the simultaneous production of electrical power and chemicals. If solid electrolyte fuel cells become commercially available, then chemical cogeneration can become an attractive option for the industrial production of some important chemicals.

Electrocatalysis in solid electrolyte cells can also be used to activate catalysis on the gas-exposed electrode surfaces. This novel application of solid electrolytes (NEMCA effect or Electrochemical promotion) is of significant importance both in electrochemistry and in heterogeneous catalysis. The solid electrolyte is used as a reversible promoter donor, via electrocatalysis, to precisely tune and control the catalytic activity and product selectivity of the electrodes. Aside from several potential technological applications, this new application of electrocatalysis allows for a systematic study of the action of promoters in heterogeneous catalysis.

Acknowledgement: Sincere thanks are expressed to the EEC TMR and JOULE programmes and to the Electric Power Research Institute (EPRI) for financial support over the years.

References

1. Gopel, W., Hesse, J., Zemel, J.N, (Eds), (1989,1991) *Sensors: A comprehensive Survey*, Vol 1,2,3, VCH, Manheim.
2. Singhal, S.C, Iwahara, H., (Eds), (1993) Proc. of the 3rd Int. Symposium on Solid Oxide Fuel Cells, Vol 93-4, The Electrochemical Society, Pennington, NJ.
3. Vayenas, C.G. and Farr, R.D., (1980) Cogeneration of Electric Energy and Nitric Oxide, *Science* **208**, 593-595.
4. Vayenas, C.G., Bebelis, S., Kyriazis, C.C. (1991) Solid Electrolytes and Catalysis. Part 1: Chemical Cogeneration, *Chemtech* **21**, 422-428.
5. Vayenas, C.G., Bebelis, S., Yentekakis, I.V. and Lintz, H.-G. (1992) Non-Faradaic Electrochemical Modification of Catalytic Activity: A Status Report, *Catalysis Today* **11**, 303-442.
7. Hegedus, L.L., Aris, R., Bell, A.T., Boudart, M., Chen, N.Y., Gates, B.C., Haag, W.O., Somorjai, G.A. and Wei, J. (1987) *Catalyst design: Progress and Perspectives*, John Wiley and Sons, New York.
8. Bockris, J.O'M and Reddy, A.K.N. (1970) *Modern Electrochemistry*, Vol. 2, Plenum Press, New York.
9. Newman, J.S. (1973) *Electrochemical Systems*, Prentice Hall Inc., 1973; p.3.
10. Vayenas, C.G., Ioannides, A. and Bebelis, S., (1991) Solid Elecrolyte Cyclic Voltammetry for in situ Investigation of Catalyst Surfaces, *J. Catalysis* **129**, 67-87.
11. Wang, D.Y. and Nowick, A.S. (1979) Cathodic and Anodic polarization phenomena at Pt electrodes with doped CeO_2 as electrolyte, *J. Electrochem. Soc.* **126**, 1155-1165.
12. Jiang, Y., Kaloyannis, A. and Vayenas, C.G. (1993) High Temperature cyclic voltammetry of Pt electrodes in solid electrolyte cells, *Electrochimica Acta* **38**, 2533-2539.

13. Bockris, J.O'M and Khan, S.U.M. (1993) *Surface Electrochemistry: a molecular level approach*, Plenum Press, New York.13.
14. Yiang, Y., Yentekakis, I.V. and Vayenas, C.G. (1994) Methane to ethylene with 85% yield in a gas-recycle electrocatalytic reactor-separator, *Science* **264**, 1583-1586.
15. Debenedetti, P.G. and Vayenas, C.G. (1983) Steady State Analysis of High Temperature Fuel Cells, *Chemical Engineering Sci.* **38**, 1817-1829.
16. Michaels, J.N., Vayenas, C.G. and Hegedus, L.L. (1986) A novel cross-flow solid electrolyte fuel cell, *J. Electrochem. Soc.* **133**, 522-527.
17. Vayenas, C.G., Debenedetti, P.G., Yentekakis, I.V. and Hegedus, L.L. (1985) Mathematical modeling of cross-flow solid-state electrochemical reactors, *Ind. Eng. Chem. Fundam.* **24**, 316-326.
18. Wagner, C. (1979) Adsorbed intermediates in heterogeneous catalysis, *Adv. in Catalysis* **21**, 323-361.
19. Vayenas, C.G. and Saltsburg, H.M. (1979) Chemistry at Catalyst Surfaces: The Oxidation of SO2 on noble metals, *J. Catalysis* **57**, 296-314.
20. Vayenas, C.G., Lee, B. and Michaels, J.N. (1980) Kinetics, Limit Cycles and Mechanism of Ethylene Oxidation on Pt, *J. Catalysis* **66**, 36-48.
21. Vayenas, C.G., Bebelis, S. and Ladas, S. (1990) The Dependence of Catalytic Activity on Catalyst Work Function, *Nature* **343**, 625-627.
22. Bebelis, S. and Vayenas, C.G. (1989) Non-faradaic Electrochemical Modification of Catalytic Activity 1. Ethyelene oxidation on Pt, *J. Catalysis* **118**, 125-146.
23. Vayenas, C.G., Bebelis, S. and Neophytides, S. (1988) Non-Faradaic Electrochemical Modification of Catalytic Activity, *J. Phys. Chem.* **92**, 5083-5085.
24. Politova, T.I., Sobyanin, V.A. and Belyaev, V.D. (1990) Ethylene hydrogenation in electrochemical cell with solid proton-conducting electrolyte, *React. Kinet. Catal. Lett.* **41**, 321-326.
25. Yentekakis, I.V., Moggridge, G., Vayenas, C.G. and Lambert, R.M. (1994) In situ reversible promotion of catalyst surfaces via NEMCA: The effect of Na on the CO oxidation on Pt, *J. Catalysis* **146**, 292-305.
26. Harkness, I.R. and Lambert, R.M. (1995) Electrochemical promotion of the NO+ethylene reaction over Pt, *J. Catalysis* 152, 211-214.
27. Alqahtany, H., Chaing, P., Eng, D. and Stoukides, M. (1992) Electrocatalytic decomposition of hydrogen sulfide, *Catalysis Letters* **13**, 289-296.
28. Basini, L., Cavalca, C.A. and Haller, G.L. (1994) Electrochemical promotion of oxygen atom back-spillover from yttria-stabilized ZrO_2 onto a porous Pt electrode, *J. Phys. Chem.* **98**, 10853-10856.

29. Cavalca, C., Larsen, G., Vayenas, C.G. and Haller, G.L. (1993) Electrochemical Modification of CH_3OH oxidation seletivity and activity on a Pt single-pellet catalytic reactor, *J. Phys. Chem.* **97**, 6115-6119
30. Varkaraki, E., Nicole, J., Plattner, Comninellis, C. and Vayenas, C.G. (1995) Electrochemical Promotion of IrO_2 Catalyst for the gas phase combustion of ethylene, *J. Appl. Electrochem.* **25**, 978-981.
31. Pliangos, C., Yentekakis, I.V., Verykios, X.E. and Vayenas, C.G. (1995) Non-Faradaic Electrochemical Modification of Catalytic Activity: 8. Rh-catalyzed C_2H_4 oxidation, *J. Catalysis* **154**, 124-136
32. Ladas, S., Kennou, S., Bebelis, S. and Vayenas, C.G. (1993) XPS investigation of Pt electrodes subject to electrochemical promotion, *J. Physical Chem.* **97**, 8845-8851.
33. Vayenas, C.G., Bebelis, S.I., Yentekakis, I.V. and Neophytides, S. (1996) Chapter 13, *Electrocatalysis and electrochemical reactors* in the CRC Handbook of solid state electrochemistry, P.J. Gellings and H.J.M. Bonwmeester, eds., CRC Press, Boca Raton.
34. Palermo, A., Lambert, R.M., Harkness, I.R., Yentekakis, I.V., Mar'ina, O. and Vayenas, C.G. (1996) Electrochemical Promotion by Na of the Pt-catalyzed Reaction between CO and NO, *J. of Catalysis* **161**, 471-479.
35. Zipprich, W., Wiemhöfer, H.-D., Vohrer, U. and Göpel, W. (1995) In-situ photoelectron spectroscopy of oxygen electrodes on stabilized zirconia, *Ber. Bunseges. Phys. Chem.* **99**, 1406-1413.
36. Palermo, A., Tikhov, M.S., Filkin, N.C., Lambert, R.M., Yentekakis, I.V. and Vayenas, C.G. 91996) Electrochemical promotion of NO reduction by CO and by Propene, *Studies in Surface Science and Catalysis* **101**, 513-522.
37. C.G. Vayenas, *"Electrochemical activation of catalytic reactions"* in *"Elementary reaction steps in Heterogeneous Catalysis"* R.W. Joyner and R.A. van Santen, Eds, NATO ASI Series, Kluwer Academic Publishers, Dordrecht 1993, p. 73.

FABRICATION PROCESSES FOR ELECTROCERAMIC COMPONENTS

J. WILL, R. STADLER, M.K.M. HRUSCHKA, andL.J. GAUCKLER
ETH Zurich
Department of Materials
Nonmetallic Materials
Sonneggstrasse 5
8092 Zurich

This paper is devided into two parts: The first part covers the fundamentals of ceramic processing in general, in the second part some examples of electroceramic components are given. Our overview is limited mostly to components made via the powder route.

1 Ceramic Processing Fundamentals

In the first sections some key aspects of ceramic processing steps are given. After the production of ceramic powders and their characterization are introduced, a section follows on ceramic suspensions and some basic processing technologies for green compacts. The paper closes with remarks on sintering of the ceramic green compact.

1.1 CERAMIC POWDERS

Natural raw materials are nonuniform crude materials from natural deposits to which usually physical separation methods are applied. These require crushing and grinding to obtain a particle size distribution suitable for most ceramic processing techniques. Higher purity starting materials are those produced via the chemical treatment of natural raw materials. They have been beneficiated to remove mineral impurities thereby upgrading their chemical purity and physical properties.

To make the best use of the potential properties of ceramics, new powders with high purity, high homogeneity and fine particle size are necessary. For this reason, new methods have been developed to synthesize ceramic powder materials with special characteristics.

This chapter will introduce new processing techniques for the preparation of ceramic powders. A good overview about these techniques are given by Ring [1] and Ganguli and Chatterjee [2]. Some current processing technologies for special powders, applied in Solid Oxide Fuel Cells (SOFC), are mentioned specifically.

H.L. Tuller et al. (eds.), Oxygen Ion and Mixed Conductors and Their Technological Applications, 165–243.

1.1.1 Solid-Solid Reactions

The most common technique used in the synthesis of new materials are solid-solid reactions between two or more starting materials after conventional mixing and calcination to form a new chemical compound. This type of reaction normally takes place at elevated temperatures and atmospheric pressure. Two major processes are involved in solid-solid reactions:

- Breaking and reconstruction of bonds at the contact region of the starting materials particles, leading to nucleation of the product phase.
- Transport of matter to the contact region by bulk or surface diffusion.

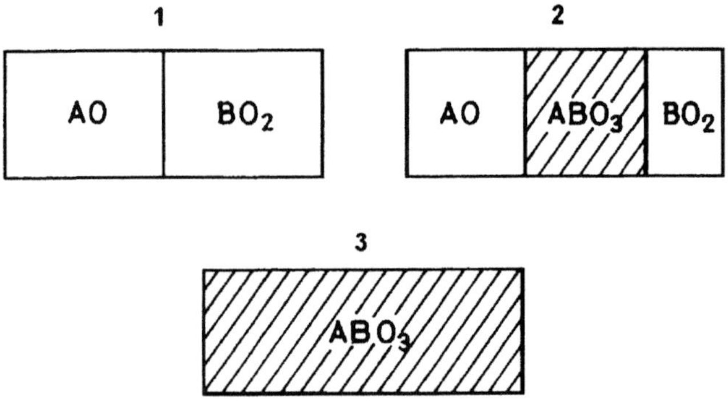

Figure 1: Scheme of solid-solid reactions [2]
(1) reactants (AO and BO_2) in contact at zero time
(2) intermediate stage of reaction
(3) product (ABO_3) after completion of reaction

In practical situations the process of transportation is the slower and rate-controlling step. The transport of a reaction species is described by the rate of diffusion, i.e. the diffusion coefficients for bulk and surface diffusion [3]. In the last years, new techniques for the production of material compounds by solid-solid-reactions have been developed to overcome the slow, rate-controlling diffusion process, which lead to long reaction time and expensive products. One of them is the high energy ball milling process, often called "mechanochemical synthesis". It is carried out by feeding the mixed starting powders into a cylindrical container with the required proportion of the grinding media applying high energy vibration or planetary milling. The powder charge : grinding media weight ration varies between 1 : 10 and 1 : 2. The time of milling is generally 24 to 30 hours. High energy ball milling is an interesting technique for the synthesis of carbides and silicides [4], but it is also used for mechanical alloying and reduction processes (TiO_2 to TiB and ZrO_2 to ZrB_2) [5].

Today, microwaves are being used not only for drying or sintering, but also for synthesis of ceramic materials. This synthesis technique is based on the fact, that some substances can adsorb microwaves and couple with them, so as to obtain the energy required for the solid-solid reaction within the powder mixture [6], [7]. One of the most attractive features is the very short time required to complete the reaction and to obtain a nearly phase-pure compound. Different silicates, oxides ($YBa_2Cu_3O_7$) and non-oxides (β-SiC, AlN) powders of industrial interest have been prepared within some minutes to one hour. The recent advances in microwave synthesis have been reviewed by Rao and Ramesh [8].

SOFC applications:

- Metallic interconnector material ($CrFe5Y_2O_3$), prepared by high energy ball milling (mechanical alloying) [9].
- New anode material (porous Ni-TiO_2 cermet), produced by conventional mixing and calcination [10].

1.1.2 Solution Techniques

All these techniques for ceramic powder production start from solutions of metal salts in water or organic liquids, which are processed, dried and finally calcined. Two of the most important techniques are mentioned here.

Precipitation and Coprecipitation. Precipitation of a solute from a homogeneous solution is a well-known process for obtaining crystalline particles of various compounds used for ceramics. The solubility of a salt in the solvent at a given set of parameters, like pH and temperature, is the maximum limit up to which the salt can remain dissolved in the solvent. The equilibrium saturation concentration c* of a multicomponent solution is given by the empirical relation [2]

$$c^* = s_0 + s_1\,\theta + s_2\,\theta^2 + \ldots \qquad (1)$$

where θ is the solution temperature and s_0, s_1, ... are the coefficients of the components [11]. A supersaturated solution tries to restore the equilibrium condition by precipitating the excess solute as a solid of the same composition. Crystalline salts obtained in this way are thermally decomposed by calcination to convert them to oxides of interest for ceramics.

The precipitants commonly selected are NH_4OH, NaOH, Na_2CO_3 or a suitable mixture of them. The addition of either the precipitating agent to the salt solution or vice versa, i.e. the sequence of addition, is an important point to be considered. In the case of a single cation present in the starting solution, the precipitant is added in excess in order to obtain complete precipitation. This method is called "direct strike". When two or more cations are present in the solution, simultaneous and homogeneous precipitation of the corresponding hydrated species becomes difficult. To avoid local precipitation, leading to inhomogeneity in the final product, the starting mixed solution is added in a stream into an excess of NH_4OH solution. This is called the "reverse strike" method or "co-precipitation".

Agglomeration is a very important problem in the powder precipitation. Some of the procedures, which are recommended to avoid agglomeration, are listed below.

- Vigorous stirring of the solution during and after precipitation.

- Addition of H_2O_2 to the salt solution to form stable peroxo complexes with the metal ions.
- Washing the precipitate with organic solvents of lower surface tension than that of water. Low surface tension of the pore liquid reduces the capillary force in the precipitate, producing soft agglomerates after calcination.
- Drying of the precipitate at high humidity (~ 95%) and temperature (~ 90°C) usually produces agglomerates that are easily crushable.
- Freeze-drying of the precipitate (see below).

SOFC applications:

- Electrolyte material (ZrO_2), zirconia with a large variety of different dopant oxides can be prepared by precipitation [12, 13] and coprecipitation [14].
- Electrolyte material (CeO_2), produced by precipitation [15], [16].
- New electrolyte material ($Ce_{0.67}Y_{0.33}O_{1.83}$), prepared by coprecipitation [17].
- Cathode material ($La_{1-x}Sr_xMnO_3$), prepared by coprecipitation and subsequent calcination [18].
- Anode material (NiO-YSZ composite), also produced by coprecipitation [19], [20].
- Y_2O_3 powder, prepared by precipitation [21].

The Sol-Gel Process. Various processes for synthesis of single or multiple oxide powders start with the preparation of an aqueous or organic sol. The subsequent processing steps vary quite significantly, but the basic chemical principles may remain the same.

As a result, it is difficult to describe a unique powder preparation technique as the "sol-gel process". In this short introduction only sol-gel processes, which use bulk gels as solid precursors for powder preparation are considered.

All Sol-Gel techniques start from a sol, which is defined as a stable dispersion of colloidal particles (< 1μm) in an aqueous, aquo-organic or organic liquid medium. Sols are generally prepared by one of the three main ways:

- dispersing a batch of particles of the desired size range in a liquid medium
- causing nucleation and growth of particles to the desired size range within the solution [22], [23]
- precipitating large particles and agglomerates in the liquid by, e.g., hydrolysis reaction of a premixed precursor solution and peptizing [24] them into much smaller particles of the desired size range

For obtaining a solid matter, the sol must first be destabilized. All methods to achieve this involve a progressive reduction in the distance between the suspended particles. Finally a network structure, a rigid or semi rigid gel, is formed through contact among them. This change from a sol to a gel is called the sol-gel transition.

These bulk gels can now be used for powder preparation by drying and calcining at the required temperature for crystallization. Through milling, finally a powder of desired particle size and composition can be achieved.

The Pechini method [26], which is also known as the citrate gel method, has been used widely for the synthesis of titanate and other powders [27]. A detailed description of the process as applied to the synthesis of $Pb_2MgNb_2O_9$-based dielectric powders is given by Anderson et al [28].

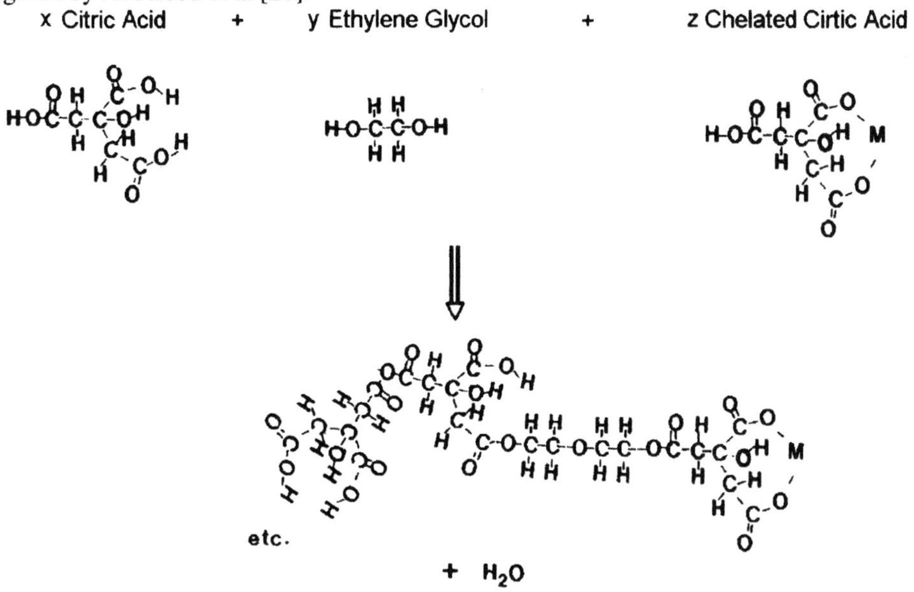

Figure 2: Chemistry of the Pechini process [25]

In case of $SrTiO_3$, the process is described by Budd and Payne [29]. In the first step, a chelate complex had to be formed between mixed cations with a hydroxycarboxylic acid (citric acid is preferred). Various cation salts can be used, such as chlorides, carbonates, hydroxides, isopropoxides, and nitrates (In most cases nitrites are preferred.). The nitrate and citric acid solution is then mixed with a poly(hydroxyl alcohol); (either ethylene or diethylene glycol) and stirred while heating at around 80°C to 110°C until a clear solution is obtained. Subsequent heating to moderate temperatures of 150°C to 250°C causes a condensation reaction with the concurrent formation of a water molecule. A schematic representation of the reactions is shown in Fig. 2.

During heating to moderate temperature, polyesterificaton occurs and most of the excess water is removed, resulting in a solid polymeric "resin". The general idea is to distribute the cations atomistically throughout the polymer structure. Heating of the resin in air or other gases causes a breakdown of the polymer and "charring" at about 400°C. It is assumed that there is little segregation of the various cations that remain trapped in the char. Subsequently, the cations are oxidized to form crystallites of mixed cation oxides at 500°C to 900°C. The process is quite complicated, and it has many changeable experimental variables that affect the final product.

1.1.3 Solvent Vaporization

Simple Evaporation. One of the easiest methods of ceramic powder preparation is the evaporation of a solution to obtain a powdery solid. This method involves the

heating of a soluble salt solution containing one or more components until complete evaporation of the solvent occurs. The dried product is then calcined at a suitable teperature to decompose it into the desired oxide powder (example: PLZT [30]).
The major problem is the lack of control concerning the powder formation, so that the products are generally highly agglomerated particles, which have to be milled and sieved to the desired particle size.

SOFC applications:

- Anode material (NiO-YSZ composite), prepared by single evaporation of a metal salt solution with an addition of glycine and subsequent spontaneous combustion of the dried solid [31].

Spray Drying.

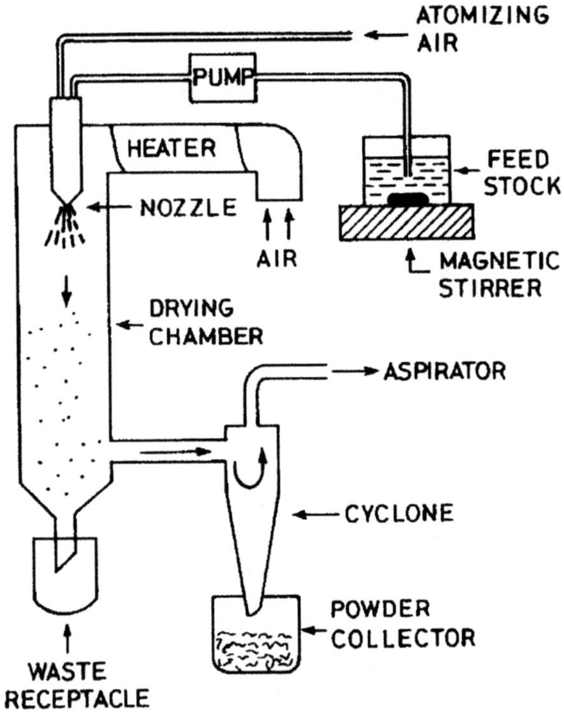

Figure 3: Schematic representation of a laboratory spray dryer [2]

This method is an old and well-known industrial process in order to avoid the problems of the simple evaporation process. A detailed description of the process is given by Masters [32]. The starting suspension of one or more soluble metal salts is sprayed through a special nozzle to atomize it into small droplets. These droplets are instantly dried by heated air between 200°C and 350°C and collected in a cyclone. In a final

calcination step, the precursor powder is decomposed to obtain the desired ceramic powder. Fig. 3 shows a schematic diagram of a laboratory spray dryer system.
There are several different modes of the spray drying process, but the basic steps of this technique are always the same:

- Preparation of the homogeneous solution.
- Spraying (atomizing) of the salt solution.
- Drying of the small droplets to obtain spherical compact particles composed of the solute.
- Calcination and if necessary, sintering of the dried precursor spheres to highly dense (theoretical density, if possible) ceramic particles.

In addition to the preparation of ceramic powders, spray drying can also be used for controlled agglomeration of powders to modify their flowability and packing density.

SOFC applications:

- Electrolyte material (ZrO_2), spherical particles prepared by spray-drying [33].

Spray Pyrolysis.

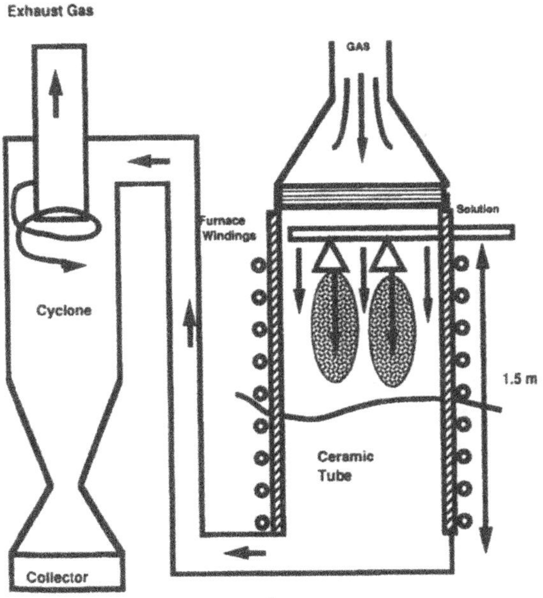

Figure 4: Scheme of a spray pyrolysis apparatus [1]

In the spray pyrolysis process the salt solution is spray dried to compact or hollow spheres as shown in the spray drying technique. In distinction from this technique, the atomized precursor can also undergo a heat treatment by which it decomposes and transforms to the desired ceramic powder directly within the spraying chamber. Fig. 5

shows a schematic diagram of the spray pyrolysis apparatus. The only notable difference to a spray dryer is a much larger drying and reaction chamber to reach the required temperature for the internal decomposition reaction. Spray pyrolysis can be used for the synthesis of both single and multicomponent oxide powders [34]. In principle, the method has also been used for non-oxide powders [35].

SOFC applications:

- Electrolyte material (ZrO_2), prepared by spray-pyrolysis [36].
- Cathode material (La(Sr)MnO_3-YSZ composite), prepared by aerosol flow pyrolysis [37].
- Y_2O_3 powder, prepared by spray-pyrolysis [38].

Freeze Drying.

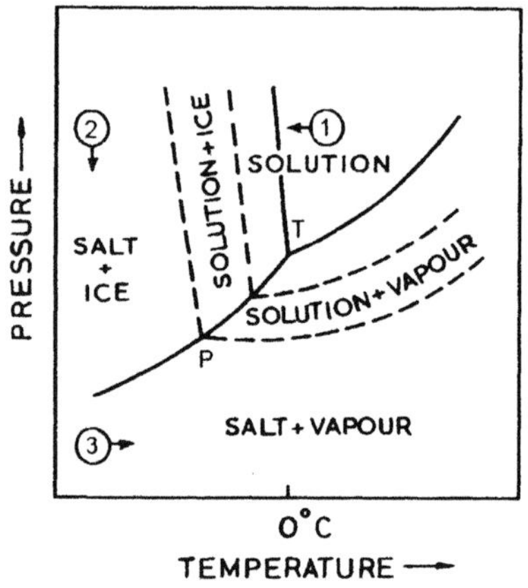

Figure 5: Pressure-temperature relationsships in the freeze drying process [2]

The reaction pathway during freeze drying is illustrated in Fig. 5. The removal of solvents is achieved by sublimation of the solvent [39]. The salt solution **(1)** is atomized and rapidly cooled down below the freezing point of the solve nt **(2)** to convert it to ice and to separate the droplets. By the reduction of pressure **(3)**, below the invariant point P (T indicates the invariant point of the pure solvent) of the salt suspension, the ice sublimates, leaving behind dry powder. Through final calcination the salt powder is decomposed into the desired oxide powder. In the simplest case therefore the freeze drying process consists of the following steps:

- Spraying of the water based salt solution into a liquid of a temperature much below the freezing point of water (e.g. hexane at -40°C, freezing point of hexane: -80°C), freezing the droplets to ice.
- Treatment of the frozen droplets at low temperature under vacuum condition to obtain salt particles as soft agglomerates by sublimation of ice without melting.
- Calcination of the precursor powder to obtain oxide ceramic powders.

1.1.4 Vapour-Phase Techniques

Vaporization-Condensation. Preparation and deposition of fine and ultrafine powders by vaporization and condensation is deduced from the coating techniques for thin films by vacuum deposition [41]. The method can be used to prepare nanosized particles of metals or ceramic materials [42].

Vaporization of the solid material is normally achieved by resistance-heating in evacuated chambers. In most cases, the material melts before vaporization. Ceramic substances are generally not suitable for this process because of their refractoriness (high melting points, high resistivity). One way to avoid this problems with ceramic materials is the preparation of metal particles, which are subsequently oxidized to the desired ceramic powder [43].

Vapour-Solid Reaction. Vapour-solid reactions are well known for the preparation of non-oxide powders. The two major variants for commercial use are:

- Direct synthesis: carbide or nitride formation from metal powders
- Carbothermal reduction of oxide powders for preparation of carbides or nitrides

In the direct synthesis technique, fine (~ 5 to 20 μm) loose packed metal powders react with the reactant gas, normally nitrogen or nitrogen-ammonia gas mixtures, at around 1200 to 1400°C for 10 to 30 hours [44].

To obtain high reaction rates and high quality products several requirements have to be met concerning the quality of the metal powders. The oxide layer, commonly present on metal particle surfaces or their surface nitridation are major obstacles to controlled rapid reaction rates [45]. The direct synthesis is widely used for the production of Si_3N_4 and AlN. The carbothermal reduction of oxides starts from a mixture of the respective oxide powder and a source of carbon which exceeds the stoichiometric amount. The reactions take place in the range of 1100 to 1700°C, depending of the oxide powder, in nitrogen or other inert atmospheres. The carbothermal reduction process is especially important for SiC (Acheson process [46]).

1.1.5 Calcination / Precursor Decomposition

Most of the techniques of powder preparation are based on dissolution and controlled recrystallization of metal salts. To obtain the desired oxide powder these salt particles have to be thermally decomposed by calcination. The calcination step refers to a high temperature treatment of the powder between 800°C to 1600°C to modify the specific characteristics of the powder. The decomposition of hydroxides, carbonates and nitrides into oxide powders is only one modification achieved by this processing step. Also, crystallisation may occur and then coarsening of small particles to large agglomerates by fusing, bonding or crystallite growth.

1.2 CHARACTERIZATION OF CERAMIC POWDERS

Powder characteristics influence the packing uniformity and the development of the microstructures during sintering. For a powder a detailed set of characterization experiments is listed in Table 1. They can be divided into four groups: physical characteristics, chemical composition, phase composition, and surface characteristics.

TABLE 1: Powder characteristics important toceramic processing [44]

Physical characteristics	Chemical composition	Phase composition	Surface characteristics
Particle size and distribution	Bulk composition	Crystallinity	Surface structure
Particle shape	Minor elements	Phases	Surface chemistry
Degree of agglomeration	Trace impurities		
Density and porosity			
Surface area			

In the following sections we introduce some of the characterization methods. The experimental details in performing these methods are not given, they are discussed at length in handbooks (i.e. [44], [47 - 54]). We concentrate on the principles of these methods.

1.2.1 Physical Characterization

In this paper, we adopt the terminology proposed by Onoda and Hench [55]: A *particle* is a discrete, low-porosity unit and can be either a single crystal, a polycrystalline particle, or a glass. If any pores are present, they are isolated from each other. A particle is the smallest unit in the powder with a clearly defined surface. An *agglomerate* is a cluster of particles held together by surface forces, by liquid, or by a solid bridge. Agglomerates are porous. They can be classified into soft agglomerates and hard agglommerates. Soft agglomerates are held together by fairly weak surface forces and can be easily broken down into primary particles. Hard agglomerates consist of particles that are chemically bonded by solid bridges. They can not be broken down easily in particles. Hard agglomerates generally lead to the formation of microstructural defects in the sintered ceramic. Fig. 6 is a schematic diagram of an agglomerate. Normally, *particles* can be viewed as small units that move as separate entities when the powder is dispersed by agitation. They consist of primary particles, agglomerates, or some combination of both. Most particle size analysis techniques refer to such particles. *Granules* are large agglomerates that are deliberately formed by the addition of a granulating agent to the powder, followed by tumbling or spray drying. These large, nearly spherical agglomerates improve the flowability of the powder during filling and compaction operations. *Flocs* are clusters of particles in a liquid. The particles are held together weakly by electrostatistic forces or by organic polymers and can be redispersed by appropriate modification of the interfacial forces through alteration of the solution chemistry. The formation of flocs is undesirable because it decreases the uniformity of the consolidated body. A *colloid* is a system consisting of a finely dispersed phase in a fluid. A colloidal suspension or sol consists of fine particles dispersed in a liquid. The particles, referred to as a *colloidal particles*, undergo Brownian motion. The size of colloidal particles range from about 1nm to 1 μm. An *aggregate* is the coarse constituent in a mixture, it is larger than 1mm.

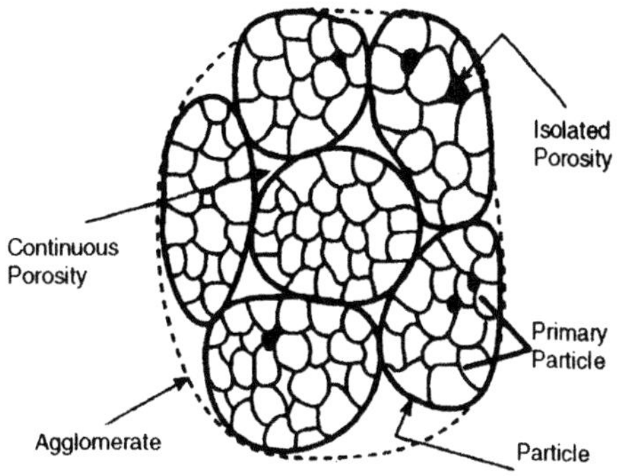

Figure 6: An agglomerate of dense polycrystalline primary particles [44]

Particle Size and Particle Size Distribution The presence of agglomerates, the particle size and its distribution are some of the most important powder characteristics influencing most of the following processing steps. The strength of ceramic green bodies e.g. is inversely proportional to the particle diameter. The sinter activity also depends on the particle size of the starting powder. Typically, the finer the powder and the greater its surface area, the lower is the sintering temperature and time for densification. Ceramic particles normally are far from being round shaped. For particle size measurements, the commonly used size parameter is the equivalent spherical diameter, which is the diameter of a sphere having the same volume or the same surface area as the particle. This is illustrated in Fig. 7. Usually the particle size is defined in a fairly
arbitrary manner in terms of a number generated by one of the measuring techniques. A particle size measured by one technique may therefore be quite different from that measured by another technique [44].

The particle size techniques are summarized in Table 2 with their measuring range.

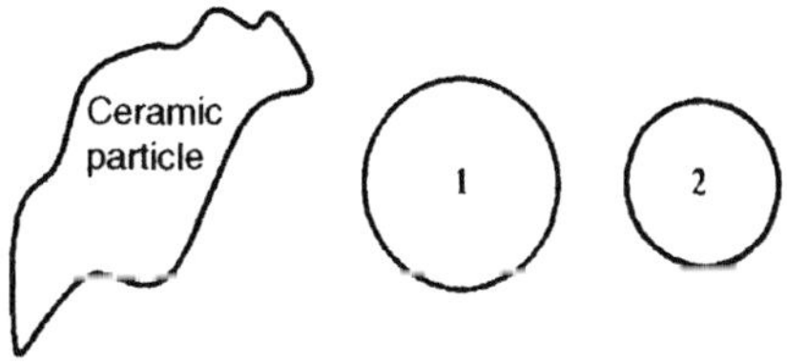

Figure 7: Examples of equivalent sphere concept,
1. Sphere of equivalent surface, 2. Sphere of equivalent volume

TABLE 2: Various classification methods forparticle size measurements(adapted from [69]). The singlemethods are described in [56 - 58].

Classification method	Size Range	Remarks
sieving	> 44 µm	overestimates the fraction of smaller particles
sedimentation	1 - 100 µm	agglomerate
light microscope	1 - 100µm	overestimates the fraction of larger particles
air classification	0.1 - 1000 µm	
coulter counter	0.3 - 100 µm	sphericity assumed
light scattering	0.1 - 10 µm	sphericity assumed
centrifuging	0.05 - 10 µm	
SEM	0.01 - 100 µm	overestimates the fraction of larger particles
TEM	0.001 - 5 µm	
XRD	< 0.05 µm	
Photon correlation spectroscopy	0.005 - 5 µm	

The distribution of the particle sizes is determined either as a distribution by number or by weight. The information may be displayed graphically either as a histogram or as continuos distribution curve as shown in Fig. 8. The median particle size is usually denoted as the d_{50} value. At this point half of the particles are smaller and half are larger.
Any phenomenon which is dependent on the particle size can be used to measure that size. However, different techniques measure different particles sizes and distributions. It is convenient to know the relationship between the measured features and the particle size in the form of an equation, and if that relationship is monotonic. Today a great diversity of commercially equipment is available. They are however based on relatively few physical principles. These techniques are well surveyed and discussed in the book by Allen [56]. By direct particle characterization techniques such as observing a sample of the particles through either a microscope or an electron microscope a wealth of detail about the particle shapes can identified immediately.

Surface Area. The specific surface area is the surface area of the particles per unit mass or volume of material. It is determined by the physical adsorption of a gas. Brunauer, Emmett and Teller used nitrogen to adsorb on the powder surface. For this BET method see [59].
The surface of the powder is degassed at intermediate temperatures (150 - 300°C). At 77 K, the boiling temperature of fluid nitrogen, N_2 gas is adsorbed at the surface and the adsorption isotherm is measured. Each N_2-molecule covers a surface area of about 0.162 nm^2 when it sticks to the surface. Hence, the total surface area can be calculated from the total amount of adsorbed nitrogen according to the BET theory. The measur-

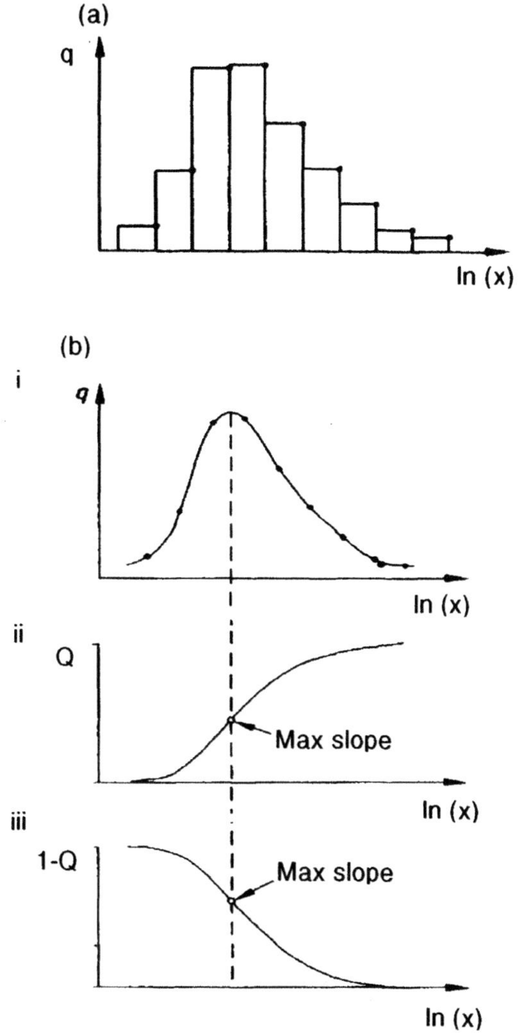

Figure 8: Graphical representations of a particle size distribution as (a) histogram and (b) a continuous distribution (i) or a cumulative distribution (ii) cumulative undersize, (iii) cumulative oversize, x = particle size

able range is 3 - 300 m^2/g [48]. The particle size can be calculated by the assumption of spherical nonagglomerated particles with the formula:

$$d = \frac{6}{\rho S} \tag{2}$$

where d denotes the median primary particle size in μm, ρ the theoretical density in g/cm^3, and S the specific surface area in g/m^3.

As shown in Fig. 9 the specific surface area of a ceramic powder is proportional to the

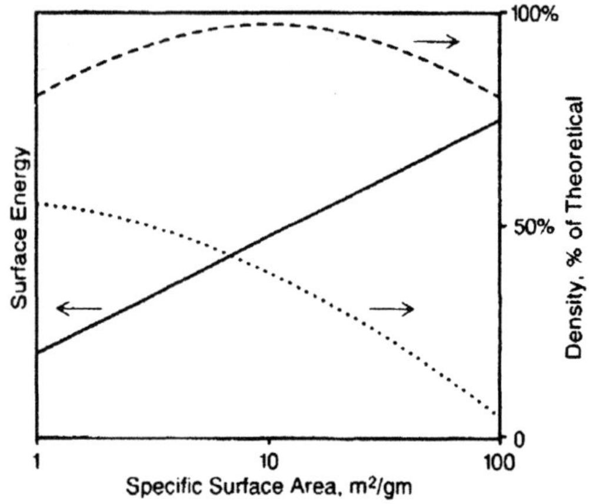

Figure 9: Effects of surface area on green density (dotted curve), interacting with surface energy (solid line), to produce fired density (dashed curve) [60]

surface energy, which is a major driving force for sintering. Therefore, the fired density is expected to depend also on the surface area. However, fine particles having high surface areas tend to pack together poorly because of the formation of agglomerates. The achieved green densities are therefore very low resulting in low sinter densities.

In Fig. 9 the lines of the green density and the surface energy line are added to yield an estimate of sinterability. The optimum surface area can be estimated as 10 square meters per gram [60]. The main advantage of a high surface area is that, when the powder will easily sinter. The main disadvantage of the large specific area is the tendency to agglomerate. High surface areas can raise the viscosity of a suspension requiring more liquid for lowering the viscosity. This decreases solids loading and therefore green density.

1.2.2 Chemical and Phase Composition

TABLE 3: Various chemical characterization methods [48]. The single methods arexplained in [58] and [61].

Method	Principle	Remarks
Wet chemical analysis	dissolve and use various techniques (titration, spectroscopy)	inertness problem, ppm level possible
X-ray fluorescence	record characteristic X-ray lines	nondestructive, calibration necessary
Emmision spectroscopy	vaporize, record optical spectra	ppm level possible
Neutron activation analysis	irradiate, record radioactivity	expensive
Energy dispersive X-ray analysis, EDX	record energy X-ray lines	qualitative, flat surface required, light element problem
X-ray diffraction, XRD	record diffraction angles, compare with JCPDS files	rapid, qualitative, nondestructive, detection limit few percent

1.3 CERAMIC SUSPENSION FORMATION

In ceramic terminology a suspension of solid particles in a liquid is called a slurry. The terms dispersion, suspension and slurry are interchangeable. In [62] the process of forming a suspension is described in detail; it involves the following steps:

1. Wetting the ceramic powder surface by a solvent
2. Breaking up the ceramic agglomerates
3. Colloidal stabilization of the suspension

1.3.1 Wetting and Dispersion

During the wetting process, the liquid spreads onto the surface of the particles and then adhere to it. When a powder is mixed into a liquid, the agglomerates of the powder contain occluded air which must be displaced by the liquid. As the liquid wets the outer surfaces and is drawn into the agglomerate, the air trapped becomes compressed. Wetting, penetration, and pressure buildup continues until the pressure of the occluded air balances the capillary force pulling the liquid into the agglomerate. The powder within the occluded air bubble cannot be wetted unless the agglomerate is broken and the air is released. Agitation is required to draw the powder into the bulk of the liquid, keep it there and break up the agglomerates. Dispersion processes are described in [63 - 65]

1.3.2 Comminution

Comminution by crushing and milling is widely used in ceramic processing to reduce the average particle size of a material, to modify the particle size distribution, to disperse agglomerates, to reduce the maximum particle size, to increase the solids content of colloids, and to modify the particle shape [66]. Communution processes are described by [67] and [68].

Ball milling is extensively used for mixing and milling in the ceramic industry. There is the danger that particles rubbed off the ball or off the brick lining will contaminate the material. Therefore, it is preferable to use ceramic balls and lining material of the of the same kind than the powder. The ball mill functions mainly by compression and shear action. A common laboratory ball mill consists of a cylindrical chamber made of or lined with porcelain, alumina, zirconia or tungsten carbide rotating horizontally on its own axis. Hard ceramic or metal is used as grinding media. Slow horizontal rotation of the chamber causes progressive reduction in size of the material via interaction with the grinding media.

1.3.3 Colloidal Stabilization

Once the primary particles have been wetted and the bonds within the agglomerates have been broken the dispersion has to be stabilized. The balance between the forces of Van der Waals attraction and electrostatic or steric repulsion will determine whether the particles cluster into flocs or whether they keep dispersed. Generally, a stable dispersion is desired. Attractive van der Waals forces exist between particles on short distances If the attractive forces are large enough, the particles will stick together, leading to rapid sedimentation. Repulsive forces are used in order to prevent flocculation. Repulsion may be achieved be electrostatic or steric means. In the textbook of Hunter [70] and other textbooks e.g. [1], [48], [49], [52], [53] the colloidal stabilization of ceramic suspensions is explained in detail.

Electrostatic Stabilization Electrostatic stabilisation occurs when the repulsion of the particles depend on the electrostatic charges on the particles. However, the repulsion is not a simple repulsion between charged particles. The surface charges of the particles are compensated by an equal, but opposite countercharge surrounding them. In this way, an electrical double layer of charge is built up between the particles, and the repulsion occurs as a result of the decrease of entropy due to the overlap of the of these double layers. Two double layers with equal sign of the surface charge repel each other, although as a whole they are electroneutral. The explanation of this phenomenon is one of the merits of the DLVO theory [71].

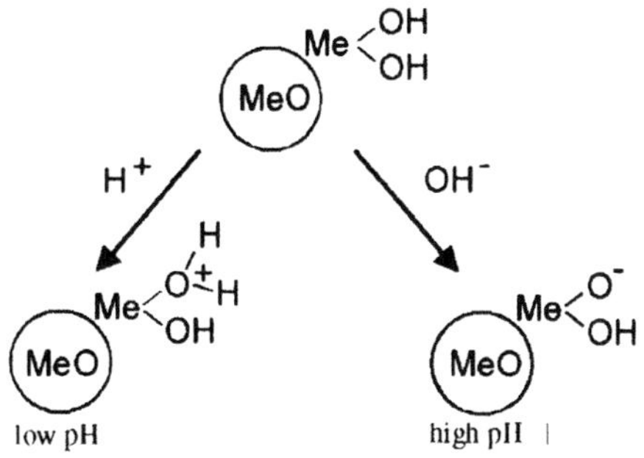

Figure 10: Effect of pH on the hydroxylated oxide particle surfaces, Me corresponds to the metal making up the oxide powder composition

Nearly all ceramic powders dispersed in aqueous solutions bear a charge due to preferential adsorption of certain ionic species, adsorption of charged organics, and/or dissociation of surface groups. These charges can be either positive or negative. Most ceramic oxide powders exhibit hydrated surfaces. In water, they will slightly ionize depending on the pH, thereby leaving the surface either positively or negatively charged (Fig. 10).

Thus, in acid solution, adsorption of H^+ ions will add hydronium ions onto the surface and the charge is positive, whereas in basic solution hydrogen is removed from the surface, forming a negatively charged particle surface. At some intermediate pH, the adsorption of H^+ ions is equal to the adsorption of OH^-. The surface charge will be effectively neutral. This point is referred to as the point of zero charge (PZC) of this powder. Acidic oxides such as SiO_2 have a low PZC and basic oxides such as Al_2O_3 have a high PZC.

Ions and polar molecules in the solution surrounding a particle will respond to the charged surface. Electrostatic forces will repel like-charged ions, but attract unlike charged ions into a region near the surface increasing their concentration relative to that in the bulk. For a positively charged ceramic particle there will be an equal number of opposite counter charges in the solution as shown in Fig. 11.

This layer forms the dense *Stern layer or Helmholtz- layer* around the particle. In total absence of thermal motion, an equal number of counter ions would adsorb on the surface. However, due to thermal motion, the counter ions are spread out in the liquid and form a second layer which is called the *Gouy-Chapman layer.* The Stern layer and the Gouy-Chapman layer together form the *double layer.* There is a rapid change in the concentration of the positive and negative ions in the double layer as we move away from the surface. Therefore, the electric potential also falls off rapidly with distance from the surface. The potential $\Psi(x)$ decays linearly from Ψ_0 just at the particle surface to Ψ_δ inside the Stern layer followed by an exponential decay which can be approximated by the Debye-Hückel equation:

$$\Psi(\xi)=\Psi\delta\,\varepsilon^{(-k(x-d))} \tag{3}$$

with κ = double layer thickness and δ the thickness of the Stern layer.

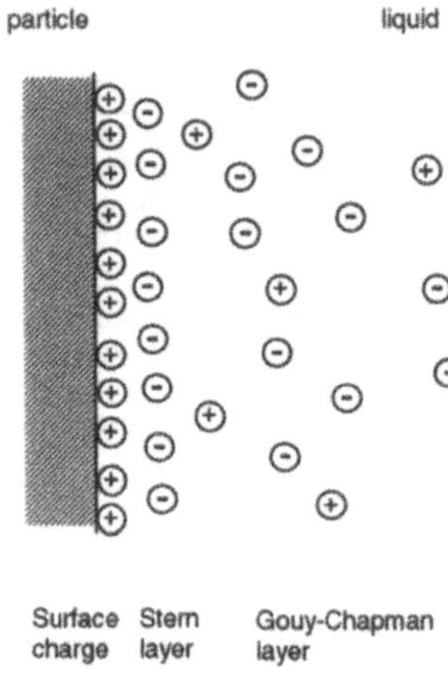

Figure 11: Distribution of the electrical charges

In Fig. 12 the Stern potential is plotted versus the distance from the particle surface. The inverse of the Debye constant, κ is a measure of the thickness of the double layer,

$$\kappa = \frac{8\pi n_0 e^2 z^2}{\varepsilon kT} \tag{4}$$

where n_0 and z are the counter ion concentration and valency, e the electron charge, ε the permittivity of the double layer, k the Boltzmann constant and T the absolute temperature. It follows from this, that the thickness of the double layer decreases with increasing electrolyte concentration and valency, that is, it is controlled entirely by the ionic strength of the solvent. The interactions between double layers are described by the DLVO theory [71].

When a particle moves in the liquid the dense Stern layer will be moving with the particle thus forming an electrokinetic unit. A schematic of this electrokinetic unit is illustrated in Fig. 13. The Zeta-potential represents the potential at this slipping plane. It is a reasonable good approximation for the Stern potential Ψ_δ which can not easily be measured whereas the Zeta-potential can be readily obtained by electrophoretic techniques [70]. In this technique, a known electric potential is applied across a suspension of fine ceramic particles. Then, by using the direction and the speed of movement of the particles, the sign and magnitude of charges can be calculated.

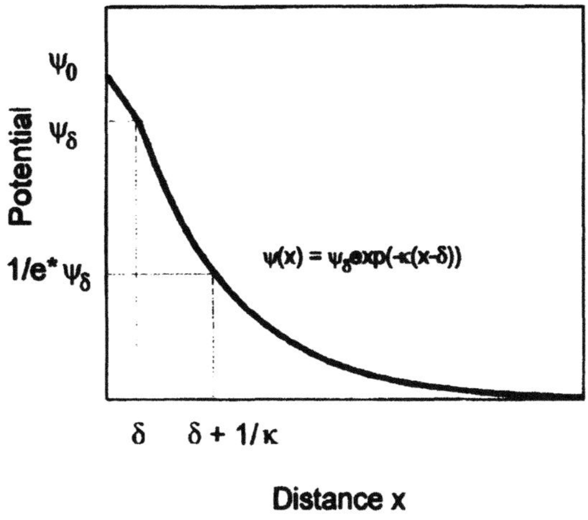

Figure 12: Electrical potential increases with distance of the particle's surface

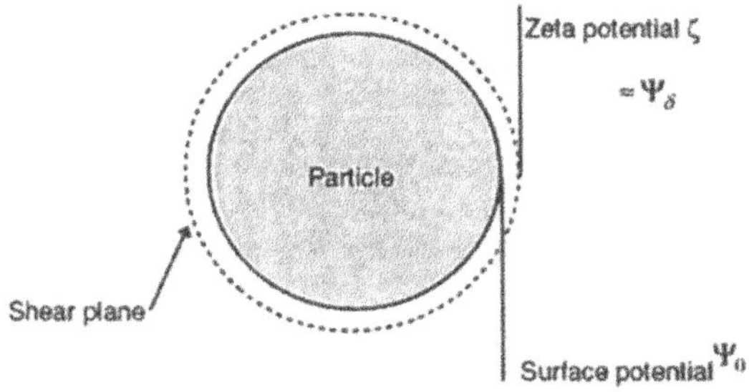

Figure 13: The particle moves with an attached layer of liquid so that a shear plane develops between the moving particle and the liquid. The Zeta-Potential is taken to be the electric potential at the plane of shear between the particle and the electrolyte solution. It is approximately equal to the potential at the surface of the Stern layer [44].

The Schmolouchowki equation relates the electrophoretic mobility to the product of the Zeta-potential and the viscosity of the liquid

$$\mu = \frac{\varepsilon \zeta}{\pi} \qquad (5)$$

μ = electric mobility, ζ = Zeta potential and η= the viscosity of the liquid.

The Zeta potential, ζ, is used to characterize the stability of colloidal suspensions. It controls the repulsive part of the particle interaction. Suspension prepared at pH values close to the isoelectric point (IEP) flocc because the repulsions are not sufficient to overcome the van der Waals attraction. Suspensions prepared with high Zeta Potentials are stable because the repulsive forces are much larger than the attractive van der Waals forces. In practice, for good stability, suspensions are often prepared at pH values comparable to those of the plateau regions of the Zeta potential or electrophoretic mobility curve.

The electric mobility of a $LaCoO_3$-powder is shown in Fig. 14. This powder would be dispersed at pH value of about 6 where the mobility is high in order to obtain a stable suspension. The isoelectric point is at 9.

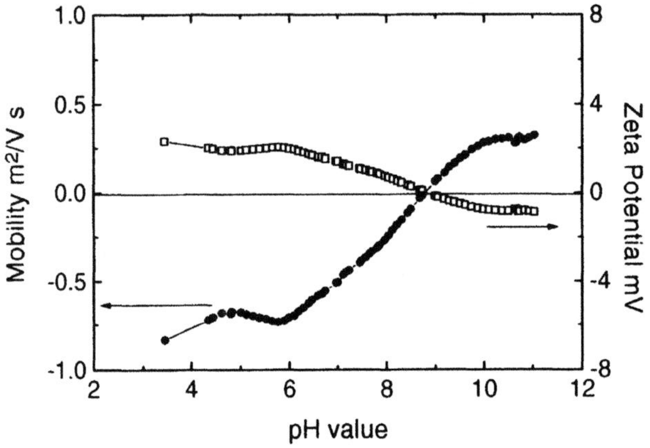

Figure 14: Mobility measurement of a stoichiometric $LaCoO_3$-powder, IEP = 9

The interactions between the particles can be analyzed in terms of a repulsive potential. The ionic concentration at a point in the middle between the two charged surfaces (Point A in Fig. 15) will be higher than that far away from the surfaces (Point B in Fig. 15). This gives rise to an excess osmotic pressure that acts to repulse the two surfaces. In case two particles in the suspension approach each other at close distance, the two double layers are interpenetrating each other. The double layer repulsion opposes the attraction from van der Waals interactions.

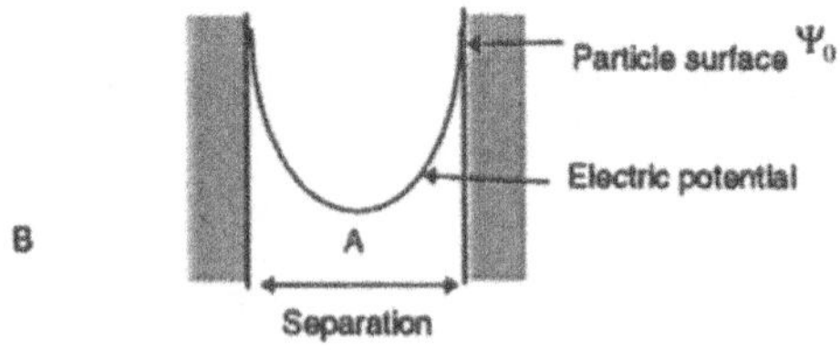

Figure 15: Overlap of the electrical double layer associated with two parallel surfaces leads to an increase in the ionic concentration. The osmotic pressure that results from this concentration increase acts as a repulsive potential.

Using the convention that repulsive potentials are positive and attractive potentials are negative, Fig. 16 shows an example for the van der Waals attraction and the double layer repulsion potentials. The resulting potential shows a deep minimum close to the surface of the particles and a secondary minimum. Two particles approaching each other will not be able to keep separated in case their thermal energy (kT) is comparable to their interaction potential maximum. Flocculation will result, leading to a sediment of loosely packed particles. For a colloid dispersion to be stable for a long time period, the interaction potential should be 3 to 5 kT. Increase of the concentration and the valence of the ions in the solution can compress the double layer and lead to flocculation

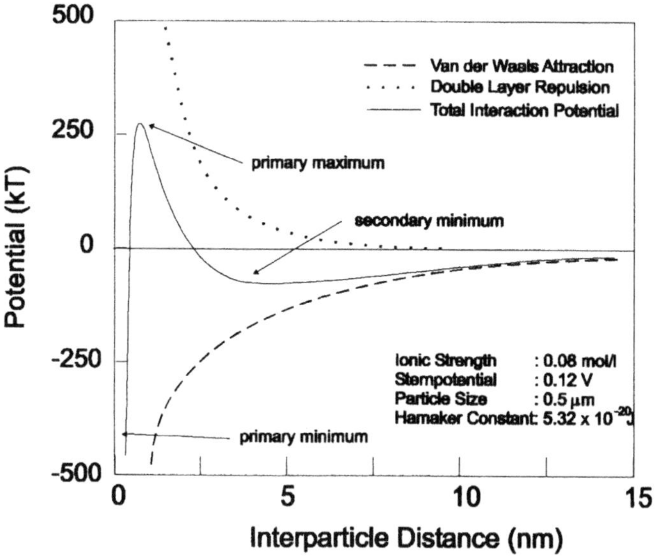

Figure 16: The potential energy between two particles in a liquid resulting from the effects of the van der Waals attraction and the double layer repulsion

Steric Stabilization Steric stabilization refers to the stabilization of particles by organic polymer molecules. The stability of the suspension is achieved by polymer molecules adsorbed or attached to the surfaces of the colloidal particles. The polymer consists of two parts, one part is practically insoluble and anchors on the particle, and the other part is soluble in the liquid. When two particles covered with polymer molecules approach each other (Fig. 17) they start to interact when the distance between the particles is equal to 2L, where L is the size of the polymer molecule. On further approach, the polymer chains may interpenetrate. The concentration of the polymer is increased in the interpenetrating region, and this leads to a repulsion. The repulsion can also be thought of as corresponding to an osmotic pressure. As the interparticle distance decreases below L, the polymer chain on one particle may be compressed by the rigid surface of the other particle. This compression generates repulsion and provides stability.

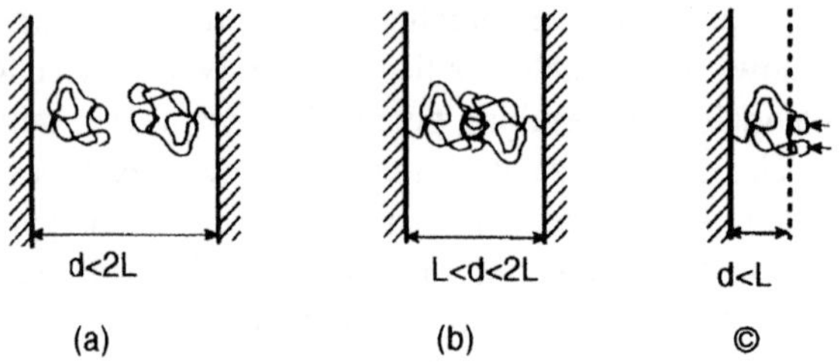

Figure 17: Steric repulsion (a) no interpenetrating for large separation (b) interpenetrating of the coils leads to repulsion, (c) compression of the polymer chain leads to repulsion

1.4 PROCESSING ADDITIVES

A ceramic suspension consists of one or more ceramic powder, one ore more liquid, often a dispersant to stabilize the ceramic powder against agglomeration and often a polymeric binder. The interactions between the liquid, particles, and the additives are wetting, dissolving, and adsorption as shown in Fig. 18.
The choice of all components for this whole system is very complex. The solvent has to be chosen that it wets the powder, the additives should readily dissolve themselves in

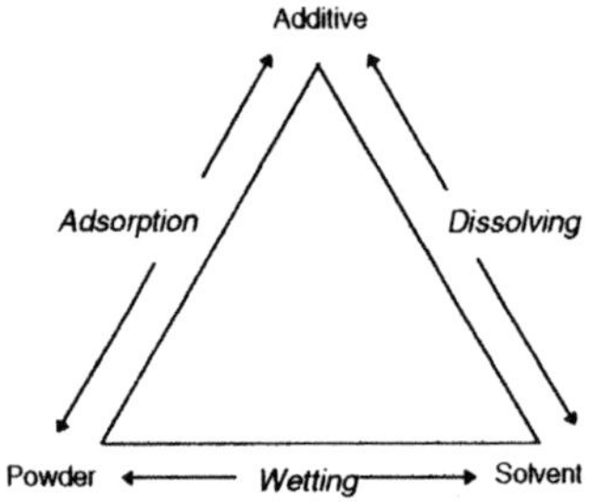

Figure 18: Powder, liquid, and additive components of system and wetting,

the used liquid so that they become homogeneously distributed and adsorb on the surface of the powder. Wetting of the particles by the liquid is essential for particle dispersion. In ceramic processing several different additives are introduced into the batch to meet the requirements of particle dispersion and flow behavior. With the exception of the solvent all these additives are added in small amounts. Most of them are eliminated in later stages of the processing (e.g binder burnout) and do not appear in the final ceramic product. Their ratio has to be carefully balanced as indicated in Fig. 19. Otherwise the green body has a low strength or the suspension will form a gel which can not be further processed. Low solids loading is also a result of poor additive balance which results in low green density, and low sinter density.

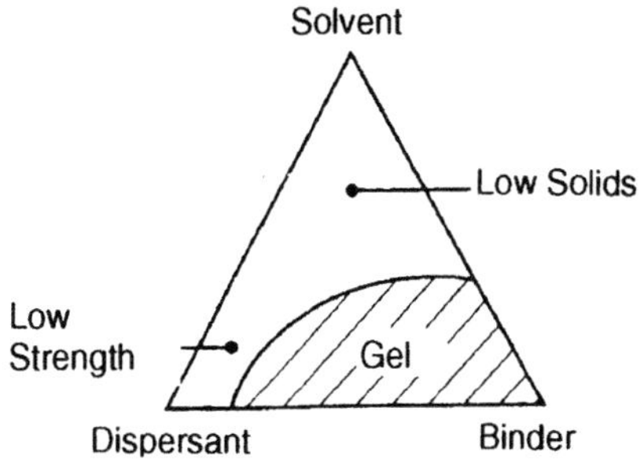

Figure 19: Effects of additive ratio on properties of the suspension [60]

1.4.1 Solvent

TABLE 4: Representative Properties of Liquids [52]

Liquid	Formula	Dielectric Constant	Surface Tension (mN/m)	Viscosity (mPa s)
Water	H_2O	80	73	1.0
Methyl alcohol	CH_3OH	33	23	0.6
Ethyl alcohol	C_2H_5OH	24	23	1.2
n-Propyl alcohol	C_3H_7OH	20	24	2.3
Isopropyl alcohol	C_3H_7OH	18	22	2.4
n-Butyl alcohol	C_4H_9OH	18	25	2.9
n-Octyl alcohol	$C_8H_{17}OH$	10	28	10.6
Ethylene glycol	$C_2H_6O_2$	37	48	20
Glycerol	$C_3H_8O_2$	43	48	20
Tricholorethylene	C_2HCl_3	3		5.5
Methylethyleneketone	C_4H_8O	18	25	0.4

The solvent has to (1) dissolve binders, plasticizers and all the other additives, (2) dispersing the powder particles, and (3) providing suitable viscosity. Water is the main liquid used in ceramic processing due to its low cost, its environmental compatibility and its availability. It can also be treated as a raw material which has to be quality controlled. It is polar, has a high surface tension and is a good solvent of polar and ionic compounds. It can form hydrogen bonds with materials with - OH and - COOH groups on their surfaces. Water also has disadvantages, such as the tendency to react chemically with certain ceramic powders like barium titanate, aluminium nitride, or MgO forming carbonates (due to CO_2 desolved in the water) or hydroxides degrading their properties. Water does not evaporate as quickly as some organic solvents, and therefore, drying is much slower and requires more heating. Nonaqueous liquids such as trichlorethylene, alkohols, ketones are also used as solvents if ceramic powders react with water or when dispersion or drying is a problem. Nonaqueous, organic liquids are less polar and are good solvents for relatively nonpolar substances. Organic liquids have a lower surface tension as water. This helps in the wetting of the solid particles. The initial choice of the solvent determines the choice of chemical family for the other additives. In Table 4 representative properties of common used liquids are given.

1.4.2 Surfactant/dispersant

The surfactant molecular structure consists of one portion that is relatively soluble (lyophilic) in the liquid and a second portion that is relatively insoluble in the liquid (lyophobic). In water, they adsorb at a surface of the ceramic powder with their lyophobic end sticking out in the polar liquid as shown in Fig. 20.

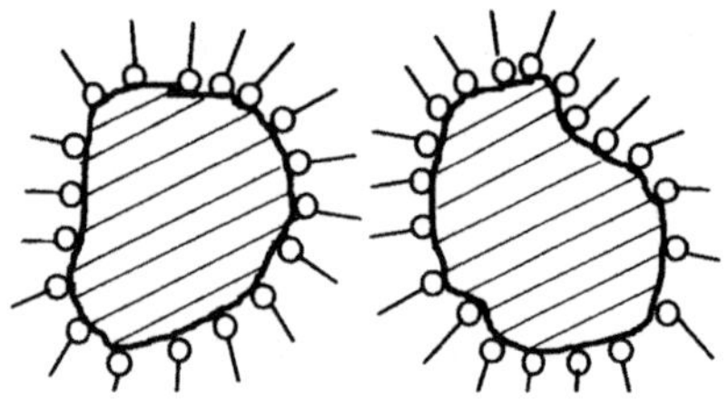

Figure 20: Polar end of surfactant molecule is adsorbed on the surface on an oxide particle [52]

Thereby, the interfacial energy solid/liquid is significantly reduced and the wetting of the suspended ceramic particles is greatly improved. Surfactants may also improve the compatibility of the ceramic particle with the liquid. Some surfactants may also aid in keeping particles from agglomeration and then are called a dispersant.
Surfactants can be classified by the charge they carry on the portion that adsorbs on the particle surface. Anionic surfactants carry a negative charge on their head groups, therefore, they act as dispersants for ceramic particles having positively charged surfaces as indicated in Fig. 21. The ceramic particles will repel each other due to the electrical charges. The suspension is then electrostatically stabilized. Some polymers

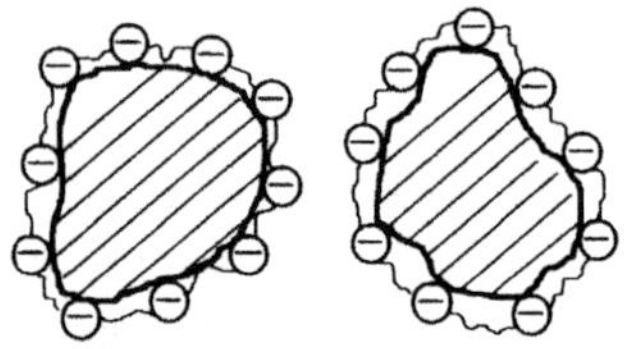

Figure 21: Adsorption of an anionic surfactant [52]

stabilize a suspension by steric hindrance. Neighboring particles are kept at a distance where van der Waals forces are ineffective. The steric layer may consist of loops, tails or trains of polymer segments.
A list of common used surfactants and dispersants used in ceramics can be found e.g.in [72] and [73]. Dispersants are typically added in concentrations of 0.5 to 2 wt.%.

1.4.3 Binders

Binders are polymer molecules or coagulated colloidal particles which may adsorbe on ceramic particles. They may form bridges between these ceramic particles and have therefore a binding action. The most important function of a binder is to improve the green strength of the ceramic. The unfired green body has to be strong enough to be handled for inspection and for loading into the furnace. Binders are available with a wide range of viscosity, which primarily depends on the molecular weight and the strength of molecular bonds of the organic. Binders represent the largest segment of the current market for organic additives in ceramic processing [60]. In Table 5 common binders are given and the applications they are specifically used.

TABLE 5: Common binders used in different applications [60]

Application	Binders	Advantages
Slip Casting	Starch	Inexpensive
	Sodium Lignosulfonate	Inexpensive
	Na Carboxzmethyl Cellulose	Inexpensive
	Sodium Silicate	Strong
	Ammonium Polyarylate	High Solids
Tape Casting	Polyvinyl Butyral (nonaqueous)	Strong
	Methacrylate Sltn. (in MEK)	Easy burnout
	Ammonium Polyacrylate	High Solids
Extrusion	Methyl Cellulose	Heat-Gelation
	Starch	Inexpensive
	Sodium Silicate	Strong
Injection Molding	Wax and Polyethylene (nonaqueous)	Easy burnout
	Epoxy or Phenolic (nonaqueous)	Very strong
Dry Pressing	Polyvinyl Alkohol	Plasticizable
	Methacrylate Emulsion	Easy burnout
Screen Printing	Alginates	Inexpensive
	Gums	Inexpensive
	Ethyl Cellulose (nonaqueous)	Pseudoplastic
	Polyvinyl Butyral	Easy burnout

Most soluble organic binders are long-chain polymer molecules. The backbone of the molecule consists of covalently bonded atoms such as carbon, oxygen, and nitrogen. Attached to the backbone are side groups located at frequent intervals along the length of the molecule. The chemical nature of these side groups determines in part what liquids will dissolve the binder. If the side groups are highly polar, solubility in water is promoted. Binders soluble in polar organic solvent have side groups of intermediate polarity. Solubility in nonpolar liquids is promoted by nonpolar side groups.

While binder is added to provide the necessary green strength, it must also impart the appropriate viscosity to the liquid in the batch material. For slip casting, tape casting, and spray drying, the slip must have a low enough viscosity to carry out the process. In contrast, the liquid present for extrusion processes must have high viscosity [55].

1.4.4 Plasticizers

Plasticizers are small molecules which are distributed among the larger binder molecules thus causing them to pack less densely and reducing the Van der Waals forces holding the binder molecules together. The binders are softened and their flexibility is increased. With the use of a plasticizer the total amount of the binder can be decreased because the binder is effectively stronger. Typical plasticizers are glycerine, ethyleneglycol, and dibutylphthalate for water-based systems and dioctylphthalate, dibutylphthalate, benzylptylphthalate and polyethylene-glycol for nonaqueous solutions [74].

1.4.5 Functional Additives

These additives adsorb on particle surfaces, they can have special functions such as flocculation, homogenization, defoaming, lubrication, wetting, release agents, etc.

1.5 FORMING PROCESSES

Ceramics can be formed by several methods which results in different mechanical and thermal properties of the sintered component. Careful control of the density and microstructure of the green ceramic body is necessary because large defects introduced in forming are usually not eliminated when the product is fired. In the following sections some of the commonly used methods for the fabrication of ceramic green bodies are introduced. This chapter is broken down into different methods of green body manufacturing depending on the initial state of the ceramic material: dry ceramic powders, dilute ceramic suspensions or ceramic pastes. After a detailed description of the various forming methods such as casting, plastic forming, coating and printing, some of the gas assisted processes are described in the last section. Each description of a forming method ends with a summarizing table of the characteristics of the process and the shape of products made via this forming method..

1.5.1 Pressing

Spray Drying This process is described in detail in the textbook of Masters [99] and in the review article of Lukasiewicz [75].
Fine powders usually show a poor flowability. For some forming processes (e. g. pressing) it is necessary to go through a process of granulation before pressing. The granulated powder flows almost like a liquid because the granules are much bigger than the particles and have a round shape. The goal is to make granules that are strong enough to do the filling job but sufficiently weak so that they can still fracture , or at least deform during pressing. The spray drying process used for it was already described in chapter 1.1.3.

Dry Pressing Products produced by pressing include a wide variety of magnetic and dielectric ceramics and spark plugs, engineering ceramics such as cutting tools and refractory sensors, ceramic tile and porcelain products, coarse grained refractories, grinding wheels and structural clay products. Dry pressing is summarized in [100].
Dry pressing can be performed in the axial mode using a rigid die with moving punches or in the isostatic mode employing the pressure via an oil bath to a rubber mold filled with the powder. Usually dry pressing is used for parts thicker than 0.5 mm and parts

with surface relief in the pressing direction. Almost always density variations occur over the compact due to the inhomogeneous filling of the die and the pressing process itself. Feedstocks for dry pressing are spray dried powder granules containing ceramic powder, deflocculant, binder plasticizer and lubricants. Dry pressing is performed by means of punching in hardened metal dies. Powder filling degree, pressure and granule density determine the green density in the pressed part. This might lead to wrappage and possible cracking of the part during sintering.

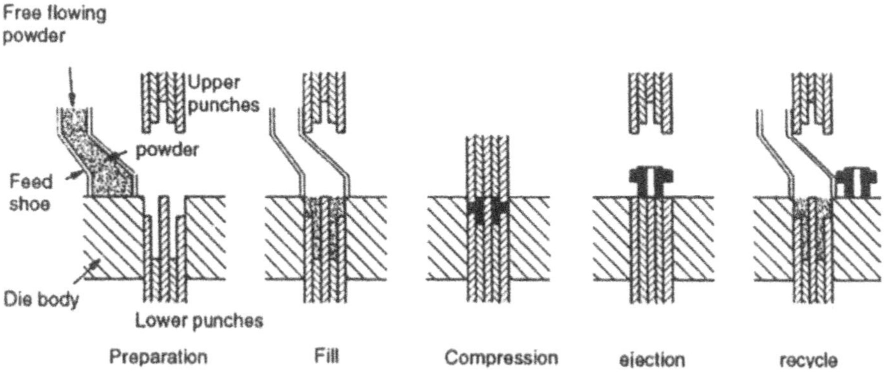

Figure 22: Schematic drawing of a dry press process [66]

Shape of product	Characteristics
small, flat objects, mostly uniform cross sections	cheap / inhomogeneous density, limited in size and shape

Isostatic Pressing Isostatic pressing in flexible rubber molds, commonly called isopressing, is used for producing shapes with relief in two or three dimension, shapes with elongated dimensions such as rods and tubes, and very massive components with a thick cross section.

The process is described e.g. in [52]. In the wet bag process, shown in Fig. 23a, flexible molds are filled and sealed at a separate station. The filled molds are then submerged in a liquid filled pressure chamber and pressed. After decompression, molds are removed, and the part is ejected. Dry bag isopressing (Fig. 23b) resembles dry pressing except that pressure is applied radially by means of the pressurized liquid medium between a flexible mold and a rigid shell. Operating pressures up to 200 MPa are common in isopressing.

Shape of products	Characteristics
rods, tubes, massive components with thick cross section	high uniform density, complex shapes

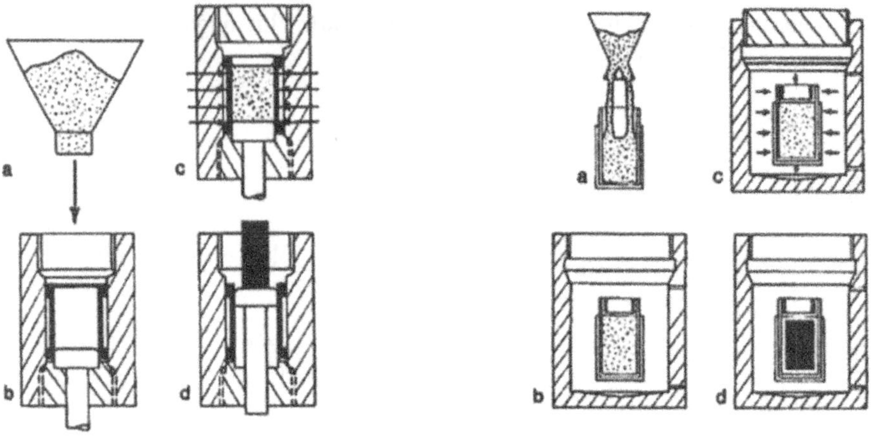

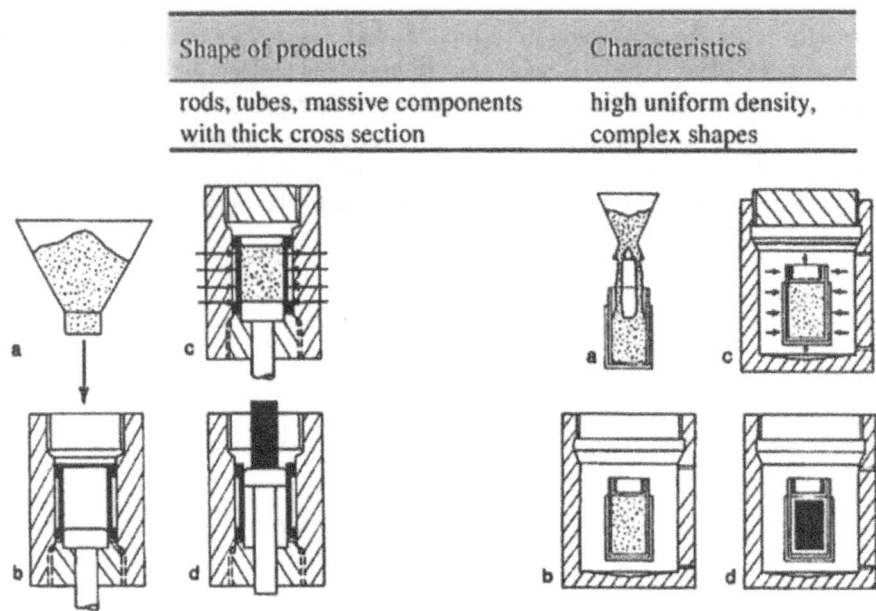

Figure 23: Wet bag (a) and dry bag (b) isostatic pressing [52]

Hot Pressing Hot pressing is described e.g. in [42], [44] and [52]. Dry powder is placed in a mold, made out of carbon, alumina or SiC and high temperature and pressure are simultaneously applied. Thus the product is formed and sintered at the same time. A ceramic may often be produced with a comparable density but a finer grain size

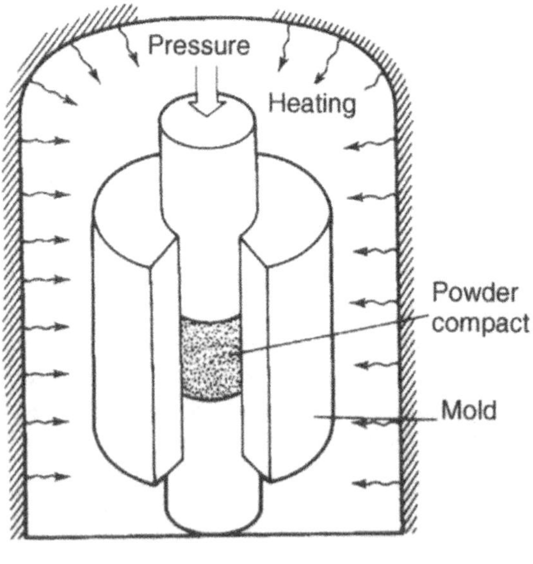

Figure 24: Schematic drawing of hot pressing [42]

or a comparable grain size but a higher density. The industrial use of uniaxial hot

pressing is limited, because its productivity is relatively low and the die maintenance is expensive. It is used to produce some commercial materials when a sintering aid or grain growth inhibitor is unknown, and for producing specimens for research.
Hot isostatic pressing (called HIP, or hipping), in which a product previously sintered to the final stage of sintering to a hot pressurized gas, is used industrially to reduce the size of closed pores in high-performance structural materials. Applied pressures used are in the range of 10 - 200 MPa. Porous materials or products may be hipped if first enclosed in a refractory, compressible enclosure or when the surface is sealed [52].

Shape of product	Characteristics
tubular, columnar, or spherical,	uniform density expensive

1.5.2 Casting

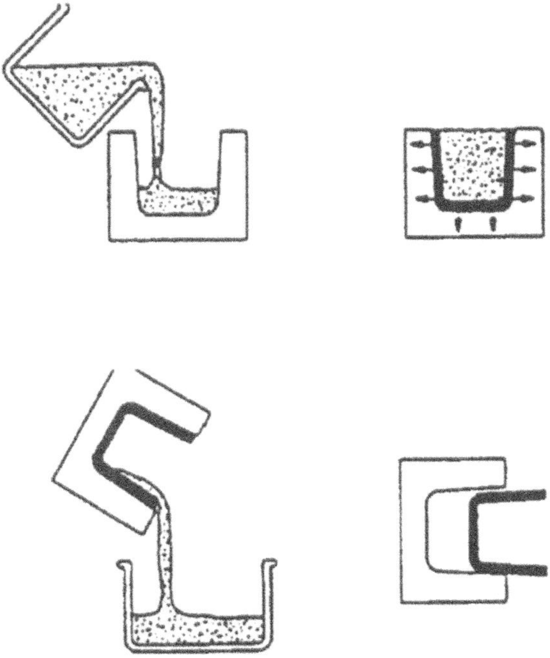

Figure 25 Schematic drawing of slip casting [66]

Slip Casting A detailed description of the process can be found in [76] and all other textbooks mentioned before e.g. [1], [44], [52], and [53]. Slip casting uses a dilute suspension that is poured into a porous mold having the shape of the desired part. The mold is made out of plaster of Paris. Water is soaked up from the contact area into the mold and a hard ceramic green layer is built up. The mold is then removed. The distinctive features of slip casting are the fineness and accuracy of the objects that can be cast in this way, the high rate of contraction, and the considerable time required for drying. Slip casting can be performed with two options, drain casting and solid casting. In drain casting, the mold is filled with a dilute slip typically less than 5% solids, which is dewatered by the porous mold giving a cast layer on the mold wall. The excess suspension is drained from the mold, and the cast layer is allowed to dry and shrink away from the mold. In solid casting, a thick slip is poured into the mold, where it completely dewaters. In both cases, the mold is filled with a ceramic suspensions. A combination of hydrostatic pressure and capillary suction dewaters the ceramic suspension adjacent to the mold as shown in Fig. 25. The thickness of the consolidated layer depends on the time the suspension remains in the mold.

Shape of product	Characteristics
thin walled products with irregular shapes	simple equipment, cheap / segregation in the slip

Tape Casting Reviews of the tape casting process can be found in [76], [78], [79], [80] and [101]. Typical examples for products produced via tape casting are substrates of Al_2O_3 and AlN used for thick and thin film circuiting, $BaTiO_3$ for capacitors, piezoelectric ceramics for actuators or transducers, RBSN and SiC for heat exchangers, mullite substrates for photovoltic solar energy cells and Al_2O_3 and ZrO_2 for the production of composite structures [79].

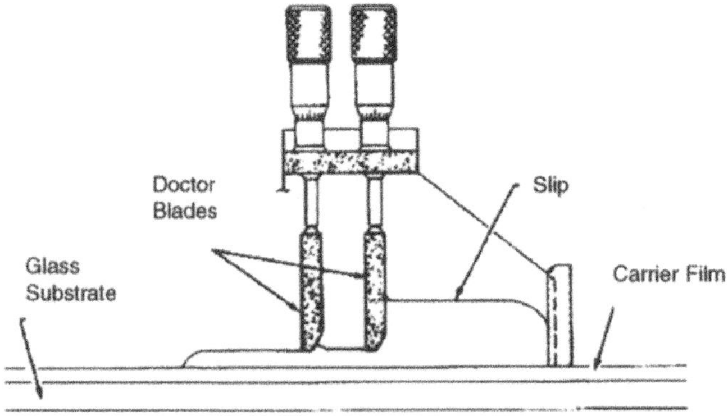

Figure 26: The tape casting process [77] with a flexiblel doctor blade

Tape casting is used generally to produce a green body in form of a thin layer from a ceramic suspension. Tape casting is a suitable processing technique for thin ceramic sheets with smooth surfaces and precise dimensional tolerances. The most important factors determining the properties of the cast product are the properties of the ceramic powder and composition as well as the preparation of the tape casting slurry. This process is widely used for multi-layer ceramic substrates and capacitors for the electronic industry. Casting is performed using a ceramic suspension consisting of powder

dispersed in an organic binder/solvent and a movable doctor blade on a temporary support. After casting, the green ceramic tape is dried and the plastic band can be handled for shaping, such as corrugating, coating or laminating. Multi-layer tapes are fabricated by subsequently casting several layers on top of each other.
After evaporating the solvent, the tapes can be fired. During sintering to full density, shrinkage of about 50 vol.% occurs. During the sintering process the materials' inherent strength is low and the component is crack sensitive. Co-firing of different laminated materials requires perfect matching of shrinkage and shrinkage rate throughout the whole binder burn out and sintering regime. This is the major draw-back of a monolithic stack design. Self supporting tapes commonly are 150 to 200 μm in thickness. However, also non self supporting layers as thin as 2 - 10 μm can be fabricated.

Shape of product	Characteristics
thin sheets	outstanding productivity and good accuracy / expensive equipment, limited to thin sheets, large residual polymer content

Direct Coagulation Casting Detailed description of this new process can be found in [81], [82], and [83]. The method uses an aqueous, electrostatically stabilized ceramic suspension free of agglomerates and of low viscosity. It is cast into a mold and then coagulates forming a stiff, wet green body. Coagulation is performed by changing the pH of the suspension and/or by creating salt directly inside the suspension using a controlled time-delayed reaction. Enzyme catalyzed reactions that are decomposing a substrate, or self-propagating decomposition reactions of a substrate, can be used. After the coagulation reaction in the suspension, the wet green body shows good mechanical properties and can be demolded, dried and sintered.
This new forming process **DCC** (**D**irect **C**oagulation **C**asting) is characterized by near net shape capability of complex shaped parts with high quality and homogeneity in the green as well as in the sintered state. Only small amounts of organic additives which amount to less than 0.5 wt-% based on ceramic powder are needed for the catalytic reactions. Since the forming process takes place without pressure and at ambient temperatures, inexpensive molds and tools can be used. Ceramic bodies with homogeneous, defect poor microstructures can be cast using almost any ceramic powder. Complex shaped components with thin and thick cross sections can be cast as there is no gradient in density and/or temperature during consolidation of the green body.
DCC uses a mechanism of coagulating a suspension by internal change of pH or raising the ionic strength after casting and thus producing rigid green bodies. It has the potential to avoid most of the limitations of conventional shaping techniques and can be applied to a large variety of ceramic powders, sols and polymers or even combinations thereof.

In case we eliminate the repulsive forces, the attractive Van der Waals forces between the particles will cause them to attract each other - the suspension coagulates. This can either be achieved by shifting the pH of a suspension towards the isoelectric point or by increasing the ionic strength in the suspension at a constant pH. The remaining attractive Van der Waals forces between the particles lead to the coagulation of the suspension.

In the following we outline the technical steps necessary to produce ceramic green parts via coagulation of a high solid loading suspension. An agglomerate free aqueous ceramic suspension is made by homogenizing the powder in water. After mixing the low viscous suspension of high solids content with the substrate, the catalytically acting enzyme is added. After additional mixing, the suspension is cast into a mold. The suspension coagulates in 5 to 180 mins leading to a wet, stiff green body. The coagulation reaction kinetics is controlled by temperature and enzyme concentration. The resulting green body is demolded, dried and sintered. The flowchart of the DCC process is shown in Fig. 26.

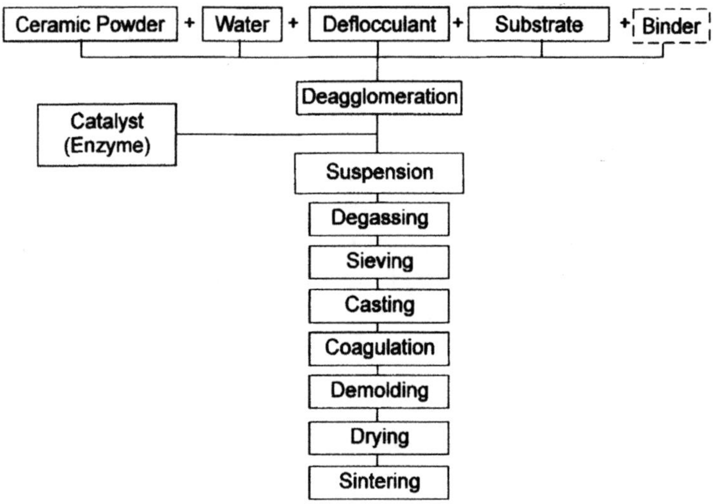

Figure 27: Flowchart of the Direct Coagulation Process

Dip Coating Reviews of the Dip Coating and the Spin Coating process are given by [84], [85], and [86].The first applications of the film and coating produced via dip or spin coating were optical applications. Since then, many uses for caotings have appeared in electronic, protective membrane and sensor applications. The dip coating process can roughly be divided in five stages: immersion, start up, deposition, drainage and evaporation (see Fig. 28). A substrate is dipped into a suspension and pulled out again. It drags an amount of liquid with it which then evaporates and leaves a film of material on the substrate. Dip coating can be done with simple experimental set up, but the film thickness will not be uniform over the whole substrate. It is only possible to coat both sides of a substrate. The thickness of the film is governed by six competing forces:

1. viscous drag on the liquid
2. force of gravity
3. surface tension
4. inertial force on the boundary layer liquid/substrate
5. surface tension gradient
6. disjoining or conjoining pressure (important for film thickness under 1 micron)

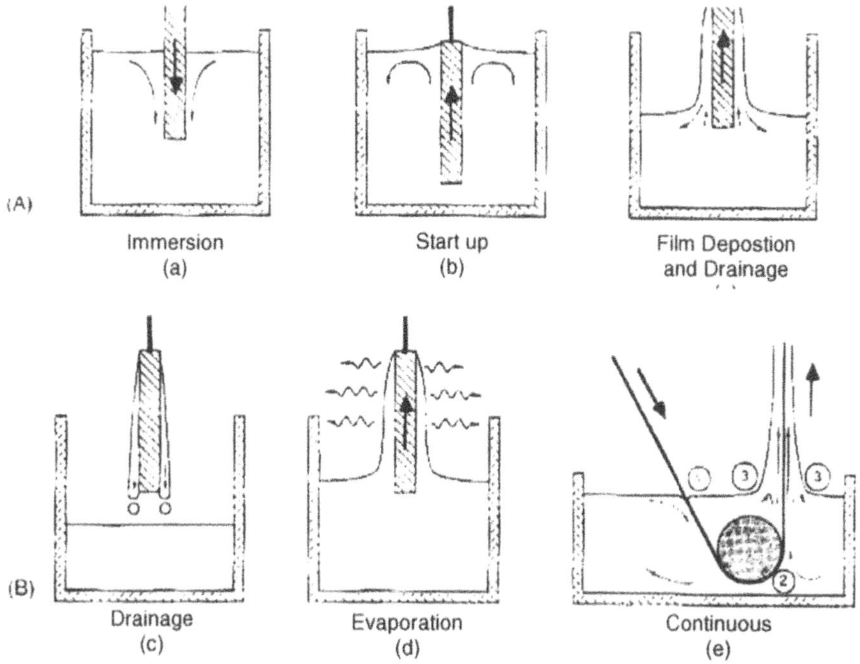

Figure 28: Stages of the dip coating process: (a) Immersion of the substrate into the suspension, (b) start up and withdrawing of the substrate from the suspension © film deposition and beginning of drainage (d) drainage (e) evaporation starting with withdrawal from suspension (f) continuous dip coating process [22]

Shape of product	Characteristics
flat, thin layers, thicknes of one coat 5 - 50 µm	rapid deposition, continuous process possible, variety in geometries, critical drying,

Spin Coating A review of the spin coating process may be found in [102]. The spin coating process may be divided in four stages: deposition, spin up, spin off and evaporation. The evaporation process may take place as soon as the suspension comes in contact with the surrounding atmosphere. The suspension is deposited on the substrate which then is put into rotation. Due to the centrifugal force the liquid will drag to the outside of the substrate, leaving a uniform film. In opposition to dip coating the evaporation of the liquid is more uniform. Two competing forces are responsible for the uniform film formation:

1. centrifugal force to the outside
2. viscous force (friction) inwards

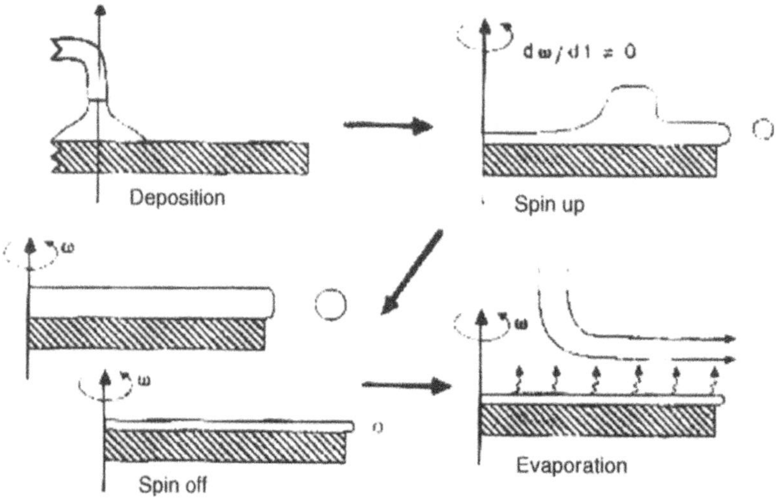

Figure 29: Stages of the spin coating process (a) deposition of suspension in excess on the non rotating substrate (b) spin up and moving of excess suspension to the edge of the substrate © spin off and leaving of excess suspension as droplets from the edge (d) evaporation which may begin already during deposition stage[22].

Shape of product	Characteristics
flat, thin layers, thickness, thickness 0.5 to 50 µm, rotational symmetric	low cost process, easy parameter control, usable for most dispersible suspensions, sensible to viscosity changes, critical drying

Screen Printing Reviews of this techniques may be found in [77] and [88]. The screen printing technique is used to produce porous or dense layers with a thickness from a few microns up to several ten microns. This technique has widely been used to fabricate hybrid circuits. A paste consisting of a mixture of starting powder, an organic binder and plasticizer is screen-printed in a desired pattern onto a substrate. This process is schematically described in Fig. 29. The patterned screen is affixed to the printer, the paste with a high viscosity is placed on top of the screen. The paste is then forced through the openings in the screen by a squeegee which depresses the screen to contact the substrate, as it traverses the pattern. The kind and amount of organic binder as well as the mixing rate of binder and powder should be optimized to obtain continuous and homogeneous microstructure. The thickness of the green film is determined by the thickness of the polymer film. Very often controlled porosity is desirable as well as a good attachment of the layer on the substrate which is achieved by careful control of powder characteristics and sintering conditions.

Shape of product	Characteristics
dense or porous layers layer thickness 3 - 100 μm,	inexpensive, quick , simplicity of fabrication parameters, high throughput, generally discontinuous, limited porosity

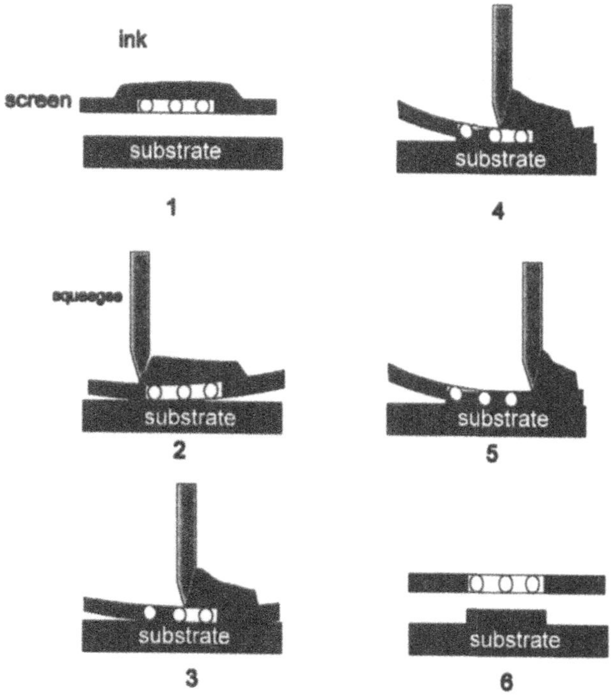

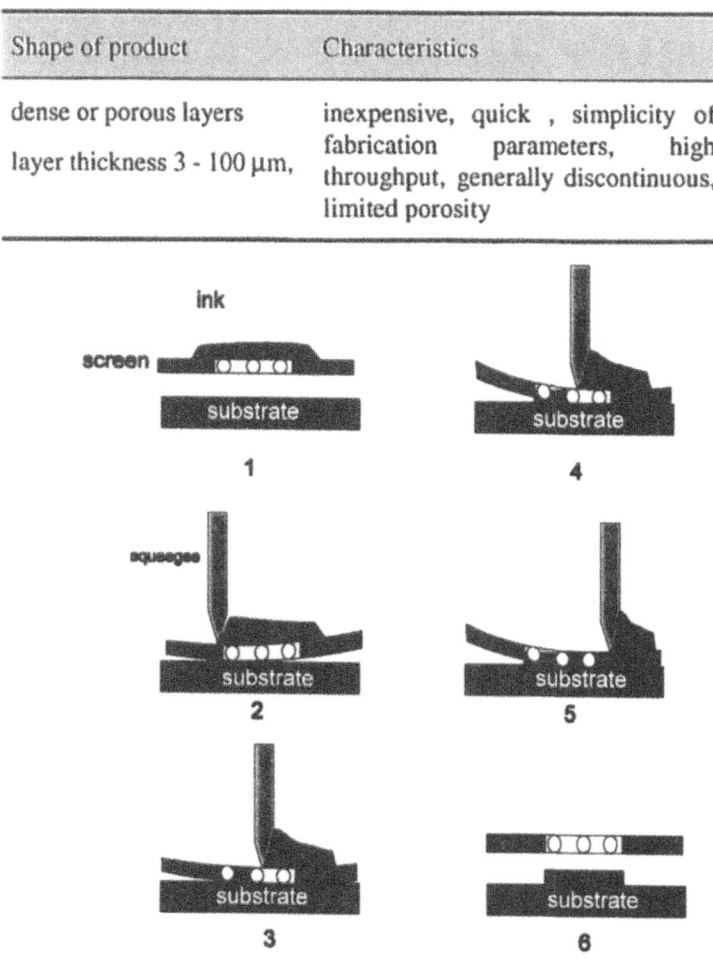

Figure 30: Screen printing [87]

Electrophoretic Deposition Review articles of this process can be found in [92], [93], and [94]. Electrophoretic deposition has been used commercially to manufacture whitewares and structured clay tile. More recently, electrophoretic deposition has been used of components of electrochemical cells, f.e. electrolytes in solid oxide fuel cells.

The formation of a body or coating from a suspension by electrophoretic deposition occurs by the electrophoretic migration of charged particles towards an electrode followed by the compaction of the discharged particles to form a deposit on the electrode EPD process is schematically shown in Fig. 32. Positively charged ceramic particles move toward the negatively charged electrode and form a deposit. EPD has advantages of short formation times, little restriction in the shape of the substrates, simple deposition apparatus and is suitable for mass production.
Electrophoretic disposition (EPD) is a colloidal process wherein ceramic bodies are shaped directly from a stable colloid suspension by a dc electric field. A dc field causes the charged particles to move toward, and deposit on, the oppositely charged electrode. EPD is a combination of two processes: electrophoresis and deposition. Electrophoresis is the motion of charged particle in a suspension under the influence of an electric field. Deposition is the coagulation of particles to a dense mass. In Fig. 31 the EPD process is schematically shown. Positively charged ceramic particles move toward the negatively charged electrode and form a deposit. EPD has advantages of short formation times, little restriction in the shape of the substrates, simple deposition apparatus and is suitable for mass production.

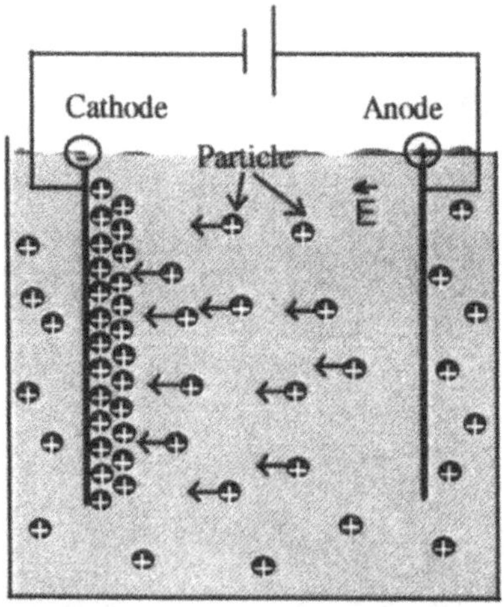

Figure 31: Schematic drawing of an EPD cell showing the process. Positively charged ceramic particles move under the influence of an electric field and deposit on the negatively charged electrode.

Shape of product	Characteristics
dense or porous layers variety of shapes	inexpensive and simple equipment, suitable for mass production, porous or dense substrates easy parameter control, organic (or water based suspensions)

1.5.3 Plastic Forming

Plastic forming involves producing shapes from a mixture of powder and additives that are deformable under pressure. About 25 to 50 vol% organic additive is required to achieve adequate plasticity for forming. A major difficulty in plastic-forming processes is removing the organic material prior to firing. In the following sections, some plastic forming techniques are described.

Extrusion The extrusion process was reviewed by [92] and [93]. Extrusion is used to form long objects with simple cross-sectional forms, such as cylindrical or square rods. In Fig. 32a, some extruded ceramic shapes which can be prepared via extrusion are shown. Extrusion has been used extensively for fabrication of ceramics for furnace tubes, bricks, insulators, pipes, tubular capacitors, catalyst supports, magnets, heat exchanger tubes, and other parts with a constant cross section. Some products are also made by machining the extruded rod shape after it has been dried. In extrusion, a highly viscous, doughlike ceramic paste is continuously pushed through a cylindrical die e.g. to form a drainage tile or a honey comb catalyst support [52]. Stages in extrusion are (1) material feeding, (2) consolidation and flow of the feed material in the barrel, (3) flow through a tapered die or orifice, (4) flow through the finishing tube of constant or nearly constant cross section and (5) ejection.

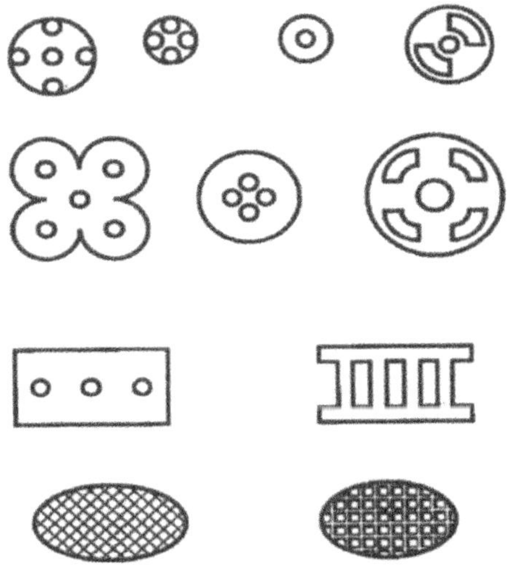

Figure 32a: Extruded ceramic shapes [1]

Because of the high shear rates applied to the viscous mass a slightly increased temperature softens the polymeric binder decreasing the viscosity. Upon leaving the die, the paste is no longer under shear and it solidifies into the desired shape. In some industrial operations where filter pressing is not feasible, plastic material is prepared by mixing feed slurry and spray dried slurry in a high shear mixer. Further processing commonly occurs in a pug mill and extrusion system where paddles or augers mix the feed material and force it through a shredder into a vacuum chamber (Fig. 32b).
Industrial extrusion pressures rang up to about 4 MPa for porcelain bodies and up to about 15 MPa for some organically plasticized materials. Capacities range widely, depending on the product size, but may approach 100 T/h. for large products [52].

Shape of product	Characteristics
long objects in tubular, columnar, or spherical forms with uniform cross sections	low cost method, high productivity, defects such as laminations, longitudinal faults, dark spots, restrictions in terms of materials, great variety of products geometries, need drying step and binder burnout

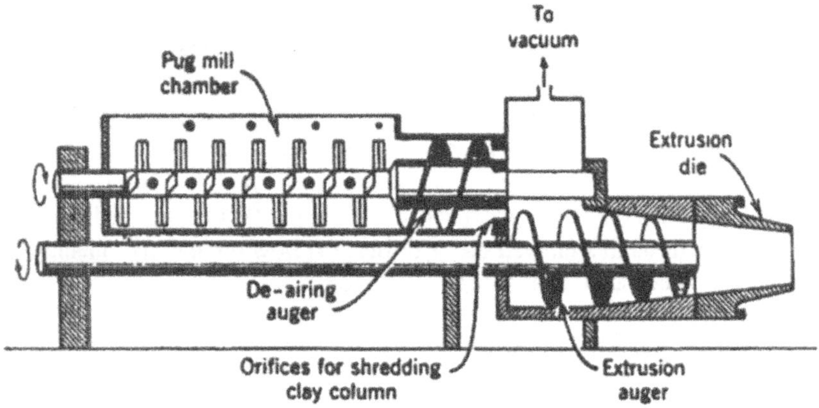

Figure 32b: Industrial pug mill with de-airing chamber and extrusion auger [52]

Injection Molding A good description of this process can be found in [94], [95] and [98]. Injection molding has the potential to produce complex shaped components to near-net shape. Injection molding is a forming operation that injects a heated powder-binder mix into a cooled metal die.

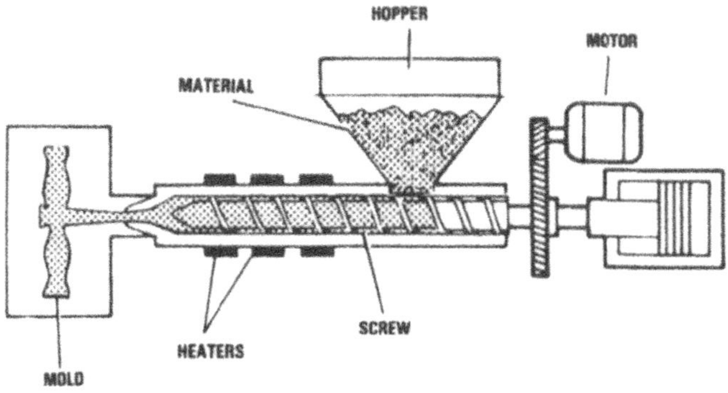

Figure 33: Schematic drawing of the injection molding process

A complete cycle consists of the following operations:
The mold is closed and a locking force is applied. The plunger moves forward, carrying a fresh charge of material into the softening zone of the cylinder and displacing previously softened material through the nozzle, which is in intimate contact with the mold. This softened material flows through the sprue into runners that terminate in gates leading to the mold cavity. As the cavity fills, the material in the cavity begins to cool and contract. During this time the pressure in the mold also increases. Pressure is maintained on the plunger for a period of time during which the material freezes at the gate and no flow occurs through the gate in either direction. The plunger retracts and the mold opens. Some residual pressure remains in the cavity and the part is ejected from the mold. The injection time required to mold a part can vary from several seconds for small components, such as turbine engine blades, to several minutes for turbocharger rotors. After removal from the die, the organic binder between the ceramic grains must be removed by a binder burnout step. This step is accomplished by a thermal cycle that can take as long several weeks to ensure no damage to the part. After this, the part is sintered [53].

Shape of product	Characteristics
complex shapes	mass production possible, high mold cost, difficult burn out process, 40-50 vol% polymer content, limitations in thickness of the products

Calendering The tape calendering is a technique to form a continuous thin sheet or tape of controlled size as shown in Fig 34. Detailed description can be found in [77] and [95]. Ceramic powder, binder and plasticizer are mixed in a high-intensity mixer.

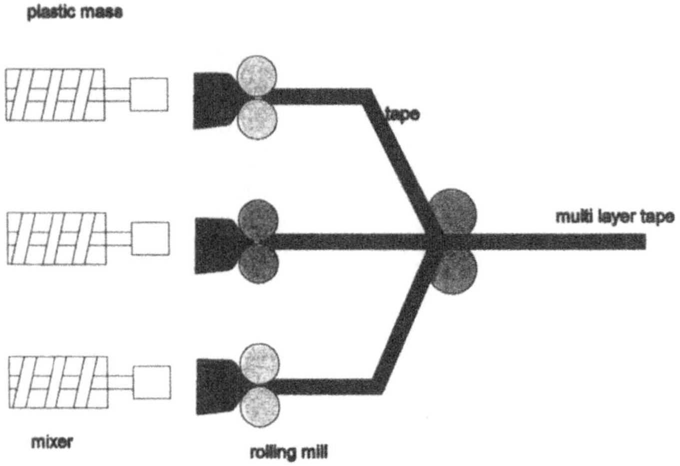

Figure 34: Schematic drawing of the Calendering Process [77]

Heat from the friction during mixing softens the binder to form a flowing plastic mass. The mass is rolled into a thin, flat tape using a two-roll mill, and the tape thickness is controlled by the spacing of the rolls. This technique has been applied to fabricate monolithic fuel cells via a co-sintering process, using multi-layer tapes, where individual tapes are laminated in a second rolling operation. For co-firing processes, the starting powder characteristics especially surface area and particle size should be optimised in order to match firing shrinkage profiles. Otherwise, the individual tapes may be detached due to the shrinkage mismatch.

Shape of product	Characteristics
thin ceramic sheets, multilayer structures, various geometries possible	low cost process, continuous production needs binder burnout, temperature control of role is critical,

1.5.4 Gas Assisted Processes

Chemical vapour deposition (CVD) CVD is a chemical process in which one or more gaseous precursors form a solid material by means of an activation process. The reactant vapours are (1) transported to the surface of a substrate (2) adsorbed on the substrate surface where (3) the chemical reaction leads to a solid product which (4) grows by crystal growth. CVD has been widely used for fabrication microelectronics.

The apparatus used for CVD depends on the type of reaction being used, the reaction temperature, and the configuration of the substrate. In Fig. 35, an example is shown for a CVD reactor. The process variables are flow rates of reactant gases, pressure in the vessel and temperature of the substrate. The pressure in the vessel influences the concentration of the reactant gases, the diffusion toward the substrate and the diffusion

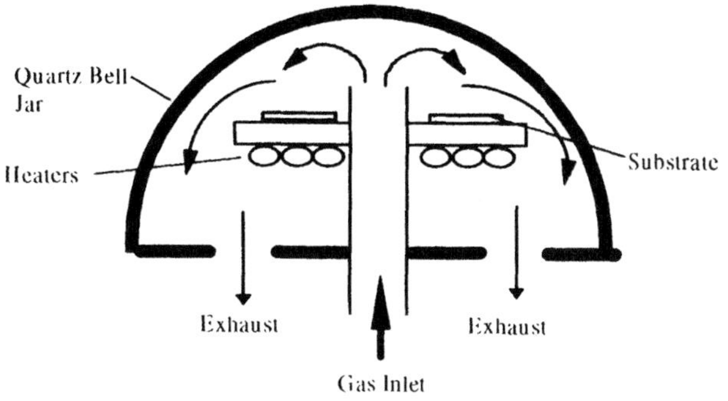

Figure 35: Schematic drawing of an CVD reactor.

of the exhaust products. The temperature of the substrate influences the deposition rate and is the main factor controlling the quality of the deposit. The temperature lays between 800 and 1800°C depending on the reaction being used [44].

Shape of product	Characteristics
films, coatings, monolyts	slow grow rate(5-150 nm/min), easy control of composition and microstructure

Electrochemical vapour deposition (EVD) Electrochemical vapour deposition techniques have been used to fabricate dense thin layers on porous substrates. The process is described in [77] and [95]. In case of the electrolyte for tubular cells, steam and/or oxygen is fed to the interior of the seal-less support tube while metal chloride vapour, hydrogen and argon are fed to the outside. Hydrogen is used to remove the chlorine formed, and argon is used as a flow-regime conditioning gas.
The EVD consists of two steps, schematically described in Fig. 36 and with the following equations:

$$\text{Stage 1: } MeCl_2 + H_2O \longrightarrow MeO + 2\,HCl$$
$$\text{Stage 2: } MeCl_2 + O^{2-} \longrightarrow MeO + Cl_2 + 2e^-$$
$$H_2O + 2e^- \longrightarrow H_2 + O^{2-}$$

The reactant metal chloride and steam (or oxygen) react to form oxides, depositing on the substrates. Once the pores are completely closed, the reactants are no longer in direct contact. Further growth of the film proceeds by oxygen ion diffusion through the film due to the presence of a large oxygen chemical potential gradient across the deposited film. The metal chloride reacts with oxygen ions diffusing through the deposited dense oxides in the presence of electronic conduction. The growth rate is controlled by solid state diffusion in the oxide film and can be described like the Wagner oxidation of metals. The layer thickness is proportional to the square root of the reaction time.

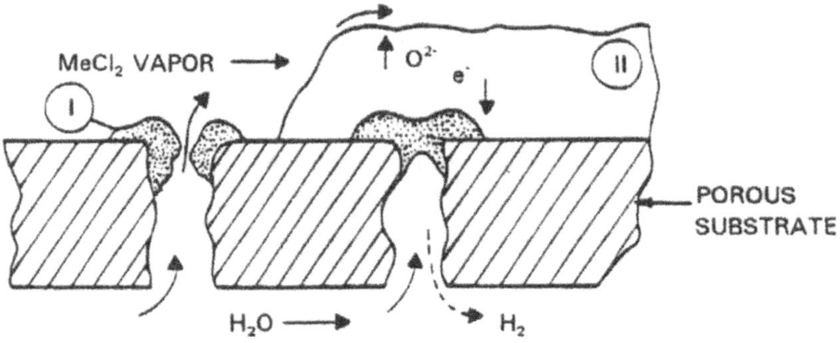

Figure 36: Model of the layer growth in EVD process [77]

Shape of product	Characteristics
very thin layers	fine grained microstructure, virtually zero porosity, very high purity limited to thin walled structures, practical for simple chemistry only

Physical Vapour Deposition (PVD) Description of this process is given in [89 - 91], and [101]. The sputter deposition or physical vapor deposition has become a generic name for a variety of processes. In these processes surface atoms from an electrode surface are knocked out by momentum transfer from bombarding ions to surface atoms. Therefore, sputtering is an etching process. Sputtering produces a vapor of electrode material which is used as a deposition material for a substrate. Modifications of this physical vapor deposition process are diode sputtering, reactive sputtering, bias sputtering, magnetron sputtering, and ion beam sputtering. The microstructure consists of fairly large columnar grains as shown in picture 37a where a ZrO_2 - layer made via the PVD process is shown. Fig. 37b illustrates schematically the PVD process.

Shape of product	Characteristics
very thin layers, dense layer	expensive equipment, complex stoichiometries, Dense or porous substrates, limitations in sample geometry and dimension, complex process control

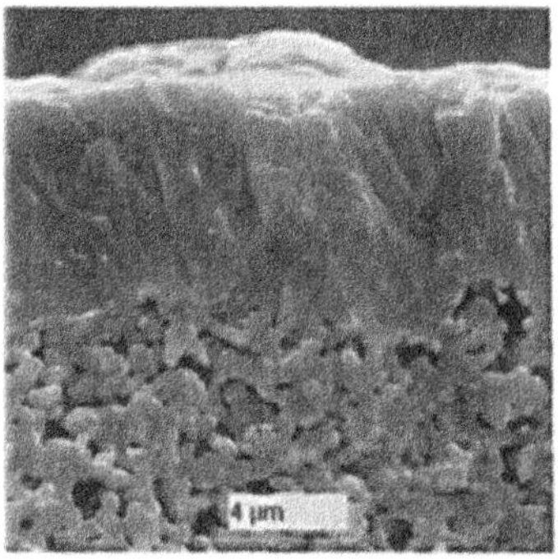

Figure 37a: dense Zirconia layer on porous reactor perovskite substrate made via the PVD process

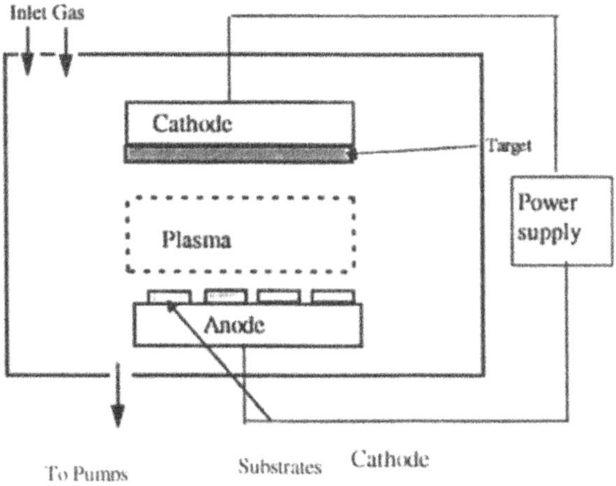

Figure 37b: schematic drawing of the PVD

Thermal Spraying References for this process are given in [95], [97] and [99]. In this deposition process the coating material is fed in the form of a rod or powder into a zone of very high temperature and is projected at high velocity onto a substrate surface. The flame spray technique (FS) uses a combustion flame (H_2-O_2 or C_2H_2-O_2) to melt the spraying material and usually air as a carrier gas. The plasma jet from the combustion gas which passes through a dc arc provides the heat source in plasma arc

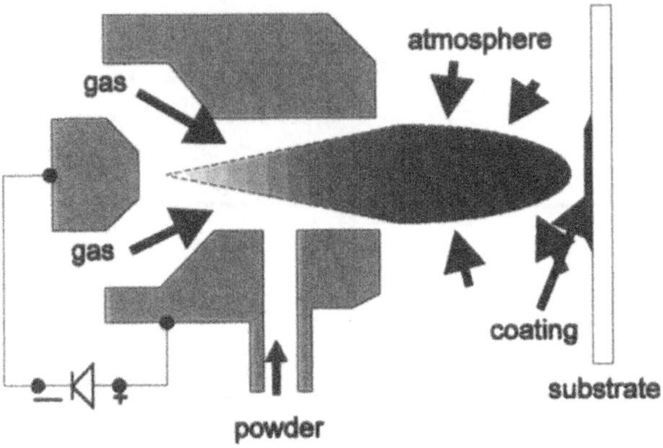

Figure 38: The plasma spray process [97]

spraying. The temperature of the plasma flame can reach up to 16`000°C and is capable of melting all refractory oxides.

The substrate might be heated in order to clean the surface from adsorbed volatile species increasing the adherence of the oxide coating on the substrate.
Common to all thermal spraying processes (Fig. 38) are the high temperatures involved, the rapid quenching of the melted ceramic particles on the substrate during deposition and the thermal treatments required to improve the coating quality. To achieve dense layers, plasma spraying in vacuum is highly promissing.

Shape of product	Characteristics
Coatings e.g. thermal barrier coatings	fast, variety of compositions / expensive, material losses

Chemical Spray deposition Chemical spray deposition of ceramic films (spray pyrolysis, spray hydrolysis, aerosol pyrolysis) involves spraying of a dilute solution of appropriate precursors onto heated substrates. For the **electrostatic spray deposition** methods (corona spray pyrolysis, electrostatic spray pyrolysis (ESP) and aerosol assisted chemical vapor deposition in an electric field (ACVDe)), the aerosol droplets are charged by applying a high voltage and directed onto the substrate using an electrostatic field. The formation of films is due to a chemical reaction (e.g. thermal decomposition, hydrolysis, oxidation, reduction) which occurs on, or in the vicinity of the substrate surface. Wheather or not these processes can be classified as CVD depends on wheather the liquid droplets vaporize before reaching the substrate or react on it after splashing. For too large droplets, the substrate temperature decreases under the film formation temperature due to the solvent vaporization on the surface. On the other hand, too small droplets are already completely vaporized far away from the substrate. This leads to undesirable homogeneous nucleation in the gas phase and results in powder deposition.

The control parameters of the chemical spray deposition are: solution composition, solvent properties (chemical stability, boiling point, specific heat and the heat of vaporization, viscosity, surface tension, wettability), gas and solution flow rates, droplet size distribution, nozzle to substrate distance, electric field, substrate temperature and nature of the substrate.

Films of following compounds have already been prepared by the chemical spray deposition: SnO_2, In_2O_3, ITO, ZnO, CdO, Cd_2SnO_4, $CdSnO_3$, $ZnSnO_3$, TiO_2, SiO_2, Bi_2O_3, Mn_2O_3, Fe_2O_3, RuO_2, V_2O_5, WO_3,. ZrO_2, CSZ, YSZ, CuS, $CuInS_2$, $CuInSe_2$. A typical gas flow of 3-10 l/min and solution flow 5-20 ml/min are commonly used. The molarity of the precursor solution needs to be kept below 0.5 M in order to achieve good results. The growth rate of films varies from about 100 nm/min up to 5µm/min.

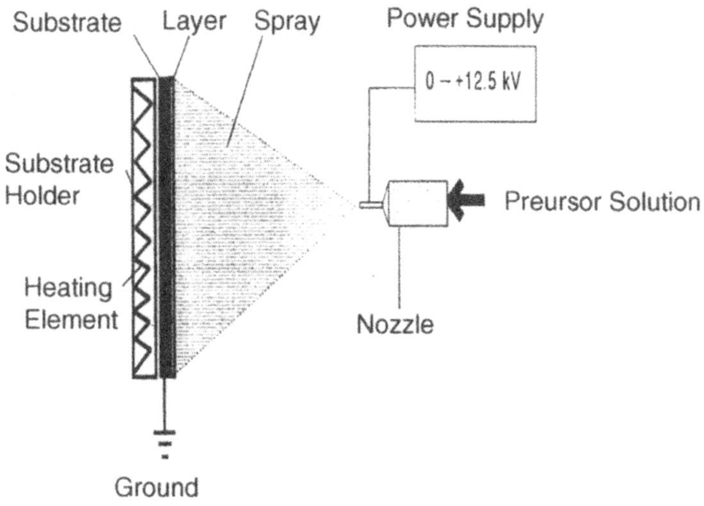

Figure 39: Spray Deposition

Shape of product	Characteristics
Coatings, large area films	low processing temperature, easy control of composition and microstructure, high deposition rate, simple equipment for high speed continuos fabrication, low costs,
	high solution losses due to overspray, aerosol emissions to the environment

1.6 DRYING

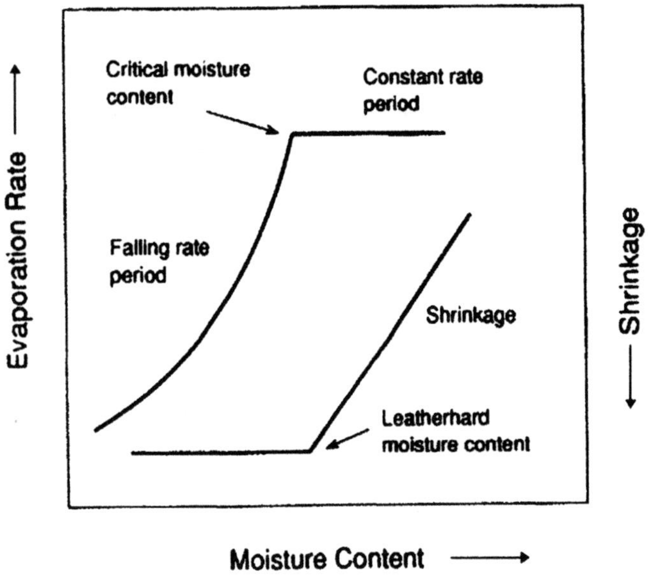

Figure 40: Drying occurs in three stages corresponding to the ranges of liquid content for which the drying rate is increasing, constant and decreasing [52]

Drying is used to remove liquids which were added during mixing or milling or liquids from a green body produced by ceramic suspension methods. The green body is heated or placed in an atmosphere where the solvent evaporates. The weight loss during drying is shown in Fig 40. This figure shows that there exists a constant rate period where the surface of the green body is always wet by the flow of liquid to the surface. At some point, the particle network becomes rigid and no more shrinkage can occur. This rigidity threshold takes place at high volume fraction solids where the particles come into contact. After this critical point, the liquid-vapor interface start to recede into the pores. At this stage, surface tension driven flows in the direction to the free surface attempt to keep a monolayer of solvent on the surface of all the ceramic powder particles. Once

the drying front enters the green body, the drying rate decreases abruptly. This decreasing rate continues until the large pores are essentially free of liquid, except for that which is trapped at the point of contact between the particles as shown in Fig. 41.
During drying, the green body is susceptible to nonuniform stresses that may warp or crack it due to (1) the pressure gradient of the flow of liquid during shrinkage in the constant rate period, (2) the macroscopic pressure gradient of the escaping gasses during the decreasing rate period, or (3) the differential thermal expansion of the ceramic due to temperature gradients in the green body. The rate of green body drying is typically controlled by temperature and humidity in the atmosphere[1], [52].
All ceramic green bodies contain varying amounts of moisture and organic additives - depending on the method of processing and therefore must be dried and subjected to binder burnout before being fired. This stage of "preheating" usually is carried out rather slowly between 100 and 350°C in order to avoid creation of pressures due to the evaporation of moisture and the decomposition of the binders. Sufficient air or oxygen should be supplied in order to achieve complete binder burnout. Special care is required for large, voluminous components and for products with high liquid and/or binder content (e.g. products formed by the injection molding or by tape casting). In most cases, the drying step and/or the binder burnout is carried out in a separate drying chamber or low temperate kiln followed by firing in a high temperature furnace. High rates may lead to temperature gradients in the product and the differences in shrinkage of layers inside the body. These shrinkage separations during drying or binder burnout are one of the main sources of cracks or crack networks in the products.

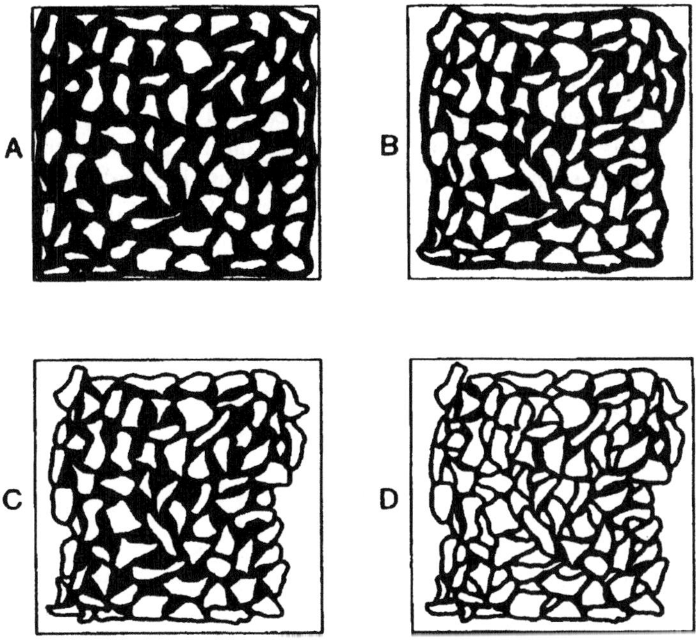

Figure 41: Schematic picture of the drying process, (white: ceramic particles, black: liquid) [52]

The use of microwaves for drying of wet bodies and/or binder burnout is one of the very successful innovations and allows much faster drying and burnout compared with conventional drying.

1.7 SINTERING

Sintering transfers a weakly bonded ceramic green compact into a rigid component with inorganic bonds between the particles. Before sintering all organic binders and residual volatiles have to be removed from the interior of the powder compact by thermal or chemical treatments.

The processes leading to sintering may take place without and with liquid phase enhancing densification. In the following we will concentrate on solid state sintering without the formation of a liquid phase.

During sintering with macroscopic shrinkage the pores are removed from the green body combined with bond formation between particles by redistribution of material. The driving force for sintering is the lowering of the free surface energy of the powder particles as the total interfacial energy of a polycrystalline solid is lower. This process takes place at elevated temperatures. Different transport mechanisms act in series and parallel during heating up and isothermal sintering. Some of them are suited to establish bonds between particles without removing the pores from the body and some of them enable bonding of the particles with removal of pores from the volume. In this latter case densification occurs and macroscopic shrinkage of the compact can be observed.

Macroscopic description:

The dimensional changes of a compact can be measured in a dilatometer at a constant temperature versus time (isothermal sintering) or during a controlled heating ramp versus time (rate controlled sintering). Most of the sintering studies for modeling the sintering process were carried out under isothermal conditions. Fig. 42 shows an example of sintering experiment of an α-Al_2O_3 powder compact. Density increases with time at a constant temperature.

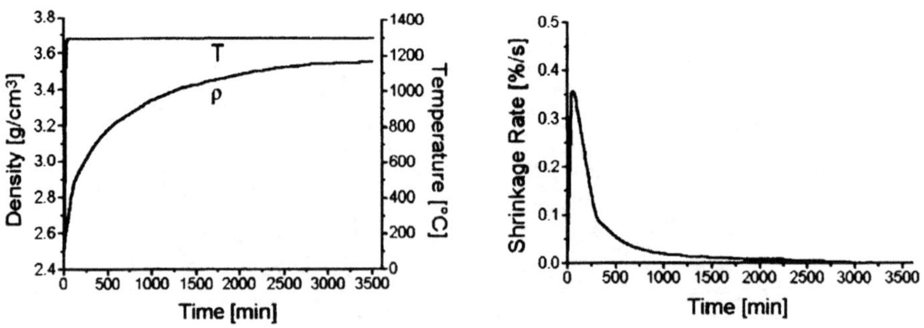

Figure 42: Isothermal sintering of α-Al_2O_3

The shrinkage rate is maximum at the beginning of the process and might lead to detrimental stresses in the compact in case of temperature gradients in the compact during the relative rapid heating period.
As sintering takes place already during the heating up phase experiments with a constant heating rate are performed also.

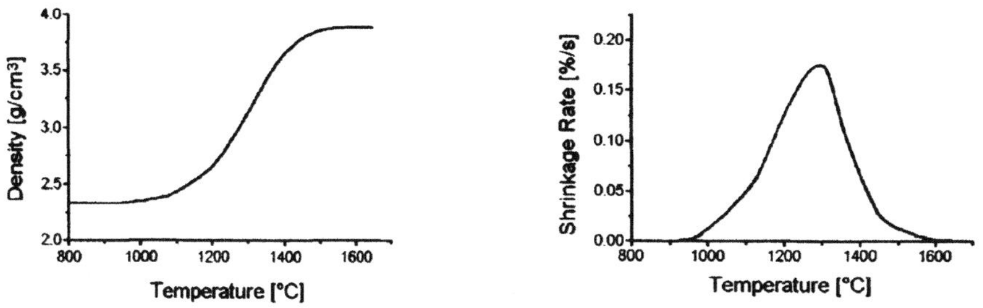

Figure 43: Sintering of α-Al_2O_3 with a constant heating rate

An example with a constant heating rate is shown in Fig.43. Densification occurs during heating up and most of the densification has occurred when reaching the maximum sintering temperature.
The given rate of shrinkage may also be used as master to design special temperature - time schedules in order to minimize stresses and to avoid overheating. Then temperature is controlled as slave and may be a complex function of time (see also rate controlled sintering).

Microscopic description:
The changes of the microstructure during sintering may be grouped in three stages:

- rearrangement of particles in the initial stage and neck formation at particle contacts
- densification during the second stage
- final densification and grain growth in the final stage of sintering

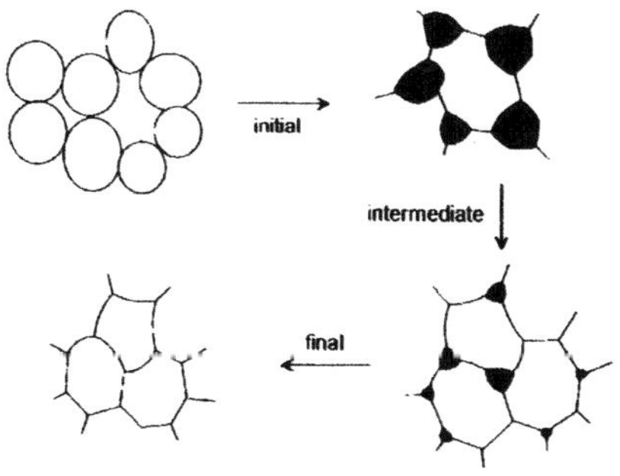

Figure 44: Changes in the microstructure during sintering

Fig.. 44 shows schematically the changes in the microstructure during sintering.

During solid state sintering material is rearranged and transported. Transport of material may occur by evaporation and condensation, viscous flow, surface-, grain boundary and volume diffusion.
Vapor transport changes the pore shape and particle shape but does not result in shrinkage of the compact. The driving force is the difference in vapor pressure of different curved surfaces. Material evaporates from the concave surfaces of the particles and condenses in the convex shaped particle necks. Thus it bonds particles due to neck formation without the particles approaching each other.
Sintering with shrinkage involves grain boundary and volume diffusion. Surface diffusion does not result in shrinkage. The driving force for bulk diffusion is the difference in the chemical potential between the free surface of particles and the interfacial energies of particle/particle interfaces.
Many models have been developed to describe the sintering process. Most of them take in account bulk-, grain boundary and surface diffusion and the chemical potential difference of the atoms at the different locations in the microstructure. Their main differences are in the geometric description of the microstructures during the different sintering stages.

The early stage of sintering:
In a green body with a density of around 50 to 65 % of the theoretical density (TD) each particle is in contact with 4-6 neighbors. During the heating up phase necks will grow at the particle contacts and the particles will rearrange and increase their number of contacts. Local equilibrium is approached and the shape of the particles and pores will change upon further heating. The dihedral angle Ψ between the grain boundary and the pore surfaces will increase according to the ratio of the surface energy γ_{sv} to grain boundary energy γ_{ss}:

$$\cos\frac{\psi}{2} = \frac{\gamma_{ss}}{2\gamma_{sv}}$$

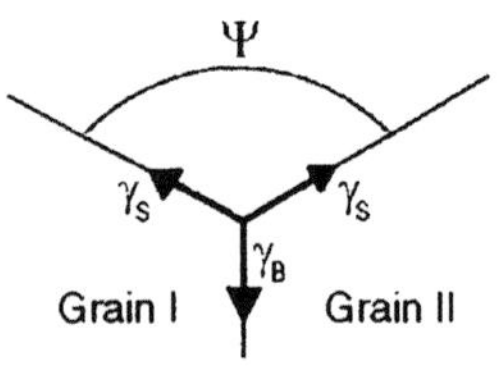

Figure 45: Dihedral angle between grain boundary

In order to approach local equilibrium material is transported to the neck region of the particle / particle contact. The neck will grow and in case the material is transported via bulk and grain boundary diffusion the centers of the particles will approach each other. This results in shrinkage of the compact as shown in Fig. 44.

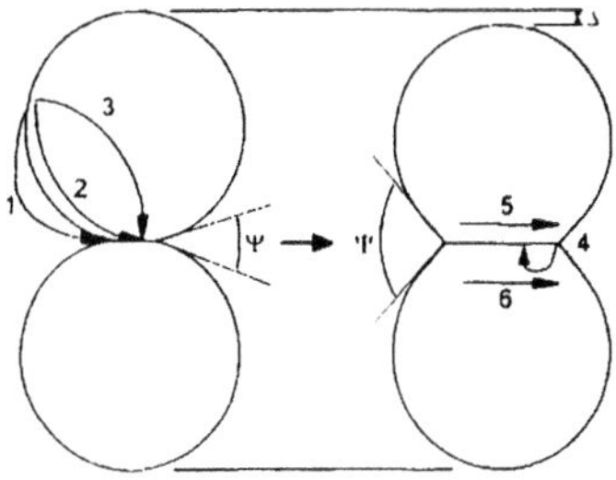

Figure 46: Neck growth and particle approach in the two particle model

Without shrinkage	With shrinkage
1. Vaporisation/condensation	4. Volume diffusion from the grain boundary
2. Surface diffusion	5. Grain boundary diffusion
3. Volume diffusion from the surface	(6. Plastic deformation)

For the macroscopic shrinkage of the compact during the early stage of sintering Coble [105] derived the following expression:

$$\frac{\Delta L}{L_o} = \left[\frac{K D_v \gamma_s s a^3 t}{k_B T d^n} \right]^m$$

from which it is obvious that shrinkage (ΔL/Lo) is increased with smaller particle diameters (d), higher temperature (T) enhanced diffusion ($D \approx e^{-Q/kT}$), high surface energy of the powder and prolonged time. The constant K is depending on the microstructural geometry. The exponent n is in the order of 3, m ranges from 0.3 to 0.5 and is depending on the transport mechanisms involved. This relation is an approximation for the early stage of sintering only, and does not take in account changes in particle positions and geometry.

During this early stage however, particle rearrangement occurs as has been shown by Exner [106] with model experiments using planar copper spheres. He could show the particle rearrangement and the importance of the microstructure and the defects in the green body to have a drastic influence on sintering and the final microstructure of the almost dense component. Fig. 47 shows a series of optical micrographs from the early stage of sintering of copper spheres.

The particles are in contact with more than one neighbor. Therefore more than one neck forms per particle which leads to particle rearrangement.

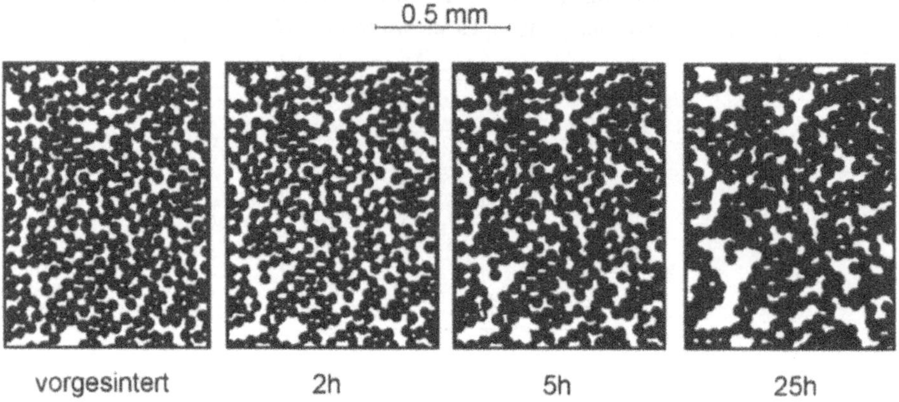

Figure 47: Particle rearrangement and neck formation in the early stage of sintering, copper spheres [106]

The almost equal distribution of copper spheres in the green compact becomes irregular with further neck formation. Densely packed regions densify more and in loosely packed regions open up to large pores with decreased particle contact number. The total shrinkage of the compact is still negligible. The homogeneity of the microstructure decreases and large inhomogeneities are formed. Small pores are eliminated at lower temperatures and earlier in the process, whereas large pores need higher sintering temperatures and longer sintering times. Therefor, large pores will lead to large grain growth in the final stage of sintering and to inhomogenous microstructures. These experiments demonstrate the importance of a uniform packing of particles in the green body in order to achieve locally uniform high shrinkage at low sintering temperatures. Thpresintered ntering finishes at about 75 % theoretical density when the particle coordination number has increased above 6 and no more particle rotations are possible. Now, in the intermediate stage of sintering there is no rotational movement any more of individual particles. Material is transported from the bulk to the neck regions. The pores form a three dimensional network of pipes where three particles meet with their faces ending in cavities where four and more particles meet with their corners.

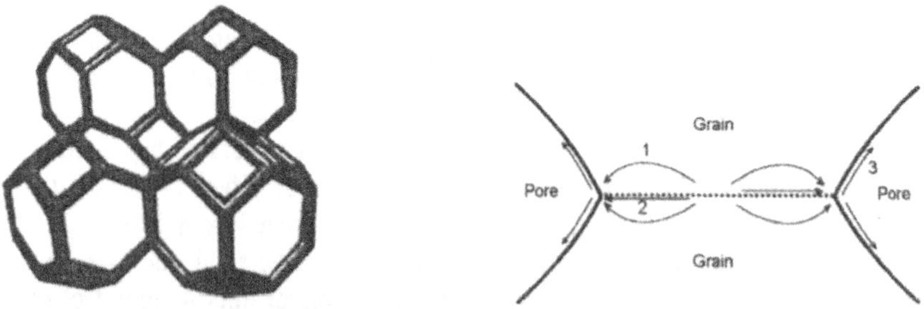

Figure 48: Three dimensional pore network in the intermediate stage of sintering and materials' transport: 1) Volume diffusion 2) Grain boundary diffusion 3) Surface diffusion

The pore volume decreases and grains start to grow. The decrease of the pore volume ΔV_p with time t may be described for the isothermal sintering after Coble by:

$$\Delta V = V_o - V_p = K \ln \frac{t}{t_0}$$

where V_o and V_p are the pore volumes at times t_0 and t respectively. K is a constant and is depending on the material and the geometry of the pore network.
Fig. 47 shows the microstructure of an α-alumina at the end of the intermediate stage of sintering. Some of the pores are already closed and the density is above 95 % TD.

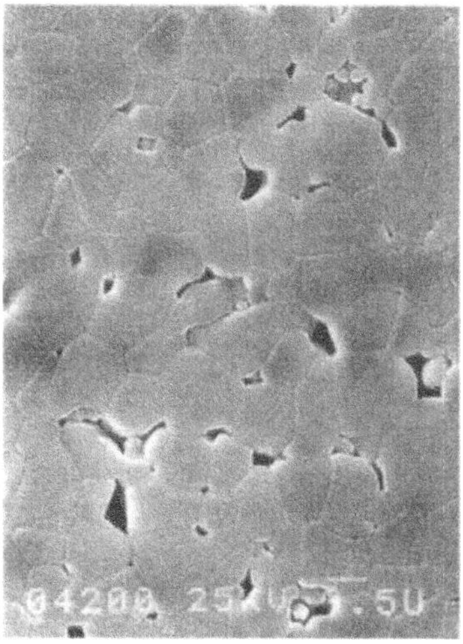

Figure 49: Microstructure of alumina at the end of the intermediate stage of sintering

During the final stage of sintering the large pores formed during green body manufacturing and rearrangement in the early stage of sintering have to be closed. In this stage mainly grain growth occurs due to Ostwald ripening. Insoluble gases entrapped in the closed pores might prevent reaching complete densification. A typical microstructure of alumina in the final stage of sintering is shown in Fig. 50.
Grain growth occurs in all stages of sintering but predominantly in the final stage. Large grains hinder densification and affect adversely the mechanical properties of the component.
Additives such as second phase particles pinning the grain boundaries and slightly soluble elements dissolving preferentially in near grain boundary areas in the grains hinder grain growth.
For a more extended excursion about the fundamentals of sintering the reader is referred to the work of U. Eisele [107].

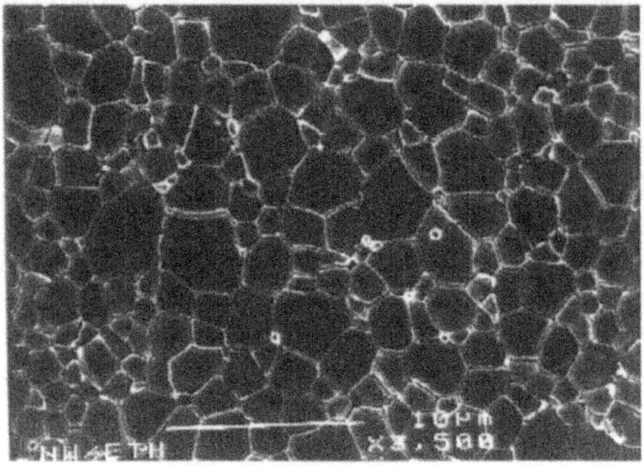

Figure 50: Microstructure of alumina in the final stage of sintering (1500 °C/2h in air).

2. Processing of electroceramic components

In the second part of this articles examples of processing electroceramic are given from the areas of solid oxide fuel cells, inorganic oxygen separating membranes and ceramic oxygen sensors. The interest is focused on ionic and mixed ionic oxygen conductors.

2.1 DESIGN OF A MATERIAL SYSTEM

The design of a materials subsystem is governed by the requirements it should fulfill and the restrictions imposed by the properties of the selected materials and their fabrication possibilities. This concept can be illustrated by manufacturing systems triangles (Fig. 51), which relate product design, process design, and economic analysis. Each discipline of which has its own unique relationship to the other two disciplines [108].

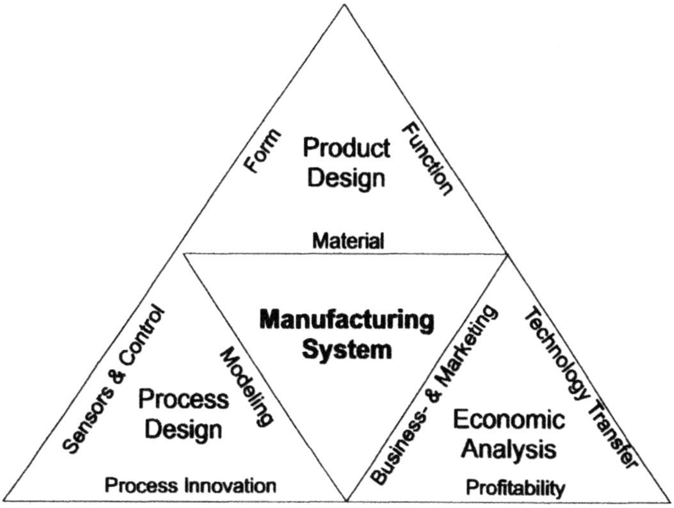

Fig. 51: Materials manufacturing systems triangles [108]

An important aspect for different designs is their feasibility in terms of fabrication as well as their potential of up-scaling from small to large components. The conditions or the environment in any process step should not impose the desired material characteristics of the components. Therefore, each design requires appropriate ceramic fabrication and assembly methods. Fabrication and assembly methodologies must attain the desired structural integrity, shape, electrical properties, and electrochemical performance of every conducting system. The tasks should be obtained at minimum costs and maximum reliability.

2.2 SOLID OXIDE FUEL CELLS

Three basically different designs are used in the development of solid oxide fuel cells:

-Tubular-design (e.g. Westinghouse, USA)
-Hexis-design (e.g. Sulzer-Innotec, Switzerland)
-Planar design (e.g. Siemens, Germany)

Solid oxide fuel cell systems must satisfy electrochemical performance, mechanical integrity and gas manifolding requirements. Electrochemical performance includes high fuel utilization and high power output at low to medium temperatures. Furthermore sufficient mechanical strength during assembling is required. The design has to guarantee good electrical contact between the individual cell components. Leak free gas manifolds and sealings should be guaranteed even at high temperatures. The SOFC-types mentioned above differ fundamentally in gas routing manner and fabrication processes. The shape of the cell-stack must supply the system with air as well as with fuel from the inlet of the stack to each cell and removing unreacted gases and reaction products.
Fabrication processes selected for each design depend on the configuration of the cells within the stack. Characteristics of designs used by industry used today are:

TABLE 6: Comparison of the main characteristics of three different SOFC-designs

	Design		
Feature	Hexis-design	Planar design	Tubular-design
Structural support of the PEN	No	Yes	Yes
Internal electrical resistance in high temperate areas	Low	Low	Medium
Gas sealing	Yes	Yes	No
Power density	High	High	Medium

PEN: Positive air-electrode, Electrolyte, Negative fuel-electrode

Planar cell designs allow a significant reduction in the number of fabrication steps, compared with the much more developed tubular cells. The literature often refers to high fabrication costs as one of the major problems of tubular SOFCs. In addition to the large number of steps required to produce a tubular cell also the processing costs of the components are very high [109]. Planar SOFC design concepts offer an attractive potential for achieving higher power densities compared to the tubular design [110]. The latter, produced by Westinghouse has been operated continuously with minor degradation in performance for more than 30,000 hours while planar stacks have been tested only for thousand hours at much smaller scales.

2.2.1 Seal- less tubular design (e.g. Westinghouse, USA)

The seal-less tubular design, first described in 1980 [111, 112], is the most advanced of the several SOFC concepts proposed. Cells of this design have been operated for thousands of hours. Multikilowatt-size stacks have been constructed and tested with a variety of practical fuels.

Early seal-less tubular SOFC technology used a ZrO_2 support tube of 20 cm long with a 2 mm wall thickness [113]. Nowadays, tubular designs of Westinghouse have a cathode with a thickness which is sufficient to support the cell itself and eliminates the ZrO_2 tube. Due to this special arrangement the possibilities of fabrication methods are limited. In order to improve the power of the fuel cell the length of the tube has been continuously increased currently up to 2 m [77]. Tab. 7 shows the materials and fabrication processes for state-of-the-art Westinghouse tubular fuel cells.

TABLE 7: Materials and fabrication processes for state-of-the-art air electrode supported (AES) cells of Westinghouse [114].

Component	Material	Thickness	Fabrication Process
Air Electrode Tube	Doped $LaMnO_3$	2.2 mm	Extrusion-sintering
Electrolyte	$ZrO_2(Y_2O_3)$	40 µm	Electrochemical vapor deposition
Interconnect	Doped $LaCrO_3$	85 µm	Plasma spraying
Fuel Electrode	Ni-$ZrO_2(Y_2O_3)$	100 µm	Slurry spraying or electrochemical vapor deposition

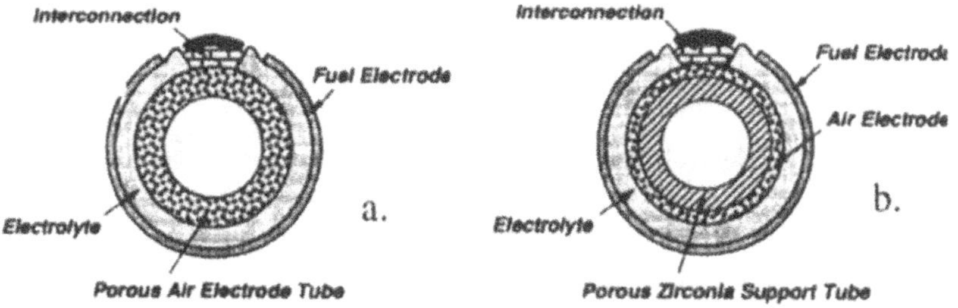

Figure 52: The old (with Al_2O_3 support tube) and new (cathode as substrate) Westinghouse Fuel Cell design

A comparison of the old and the new design of the Westinghouse tubular fuel cell is shown in Fig. 52. The porous supporting ZrO_2-tube (Fig. 52b) has been eliminated and replaced by a thick porous cathode tube (Fig. 52a). Today's research aims at reducing the amount of EVD steps in fabrication due to their high costs. In the near future the fabrication of air electrode supported cells (AES) tubular fuel cell will use the EVD for only the electrolyte [114].

Cathode A doped lanthanum manganite tube (LSM) is used as substrate as well as cathode. The tube is extruded and sintered to about 30 to 35 % porosity and serves as the air electrode to which the other cell components are applied in thin layer form. The increased wall thickness of 2.2 mm accommodates larger pressure drops encountered in longer length of the cell. Presently, the air electrode material is synthesized using high purity component oxides such as La_2O_3 and MnO_2. Over 70% reduction in the cost of the air electrode raw materials are achieved instead of using pure lanthanum compounds to synthesize the cathode. Extrusion is the preferred low-cost method of forming large quantities of tubes. The process is cheap in terms of equipment and maintenance. The most commonly found defects due to the fabrication method are laminations and longitudinal faults. An alternative fabrication method is powder pressing.

Interconnect In the early days the interconnect was made via EVD. Even though EVD provides very high quality thin films [115], it requires capital intensive equipment making the process rather expensive. Today the interconnect is made by plasma spraying [114]. Calcium aluminate containing lanthanum chromite powder is deposited on the porous, doped lanthanum manganite air electrode tube. The calcium aluminate facilitates densification during plasma spraying and subsequent heat treatment [117]. The powder deposition takes place at about 1300°C under vacuum. Powder masking is used to limit the deposition of the interconnect into a strip along the length of the cathode [77].

The plasma spraying has resulted in reduced process cycle time, increased yield, and a major reduction in cell fabrication costs. The sprayed interconnects showed only little performance degradation for periods of time under a variety of operating conditions. Furthermore, the plasma sprayed interconnects are single phase, perovskites which matches much better the thermal expansion of the electrolyte [114].

Electrolyte The electrolyte is applied to the cathode air electrode tube by electrochemical vapor deposition (EVD). 8 mol% Y_2O_3 stabilized ZrO_2 is used as electrolyte material [77]. The electrolyte is deposited over the entire active area of the cell including overlap regions of about 0.5 mm on all sides of the interconnect. The overlap provides gas-tight sealing for the cell. The electrolyte is formed from vapors of YCl_3 and $ZrCl_4$ and an oxygen/steam mixture at about 1200°C under vacuum [77]. The reaction takes place in two stages. The overall reactions for the yttria- and zirconia chloride deposition are explained by equations 2.1 and 2.2.

Stage 1: Pore closure $YCl_3 + 3/2\,H_2O \rightarrow YO_{3/2} + 3HCl$ (2.1)

$$ZrCl_4 + 2H_2O \rightarrow ZrO_2 + 4HCl$$

Stage 2: Scale growth $YCl_3 + 3/2\,O_O^x \rightarrow YO_{3/2} + 3/2\,V_{\ddot{O}} + 3e^- + 3/2\,Cl_2$ (2.2)

$$ZrCl_4 + 2O_O^x \rightarrow ZrO_2 + 2V_{\ddot{O}} + 4e^- + 2Cl_2$$

A scheme of the electrochemical vapor process is explained in detail in 1.1.4.

Important process parameters are the partial pressures of the reactive components, substrate temperature, gas flow rate, and homogeneous distribution of the gas species. Therefore, due to the complexity of the process and the expensive materials, the costs are severe negative factors of EVD. Today's research tries to replace the expensive EVD-process while keeping the characteristics of thin and dense layer fabrication.

Anode The Ni-YSZ fuel electrode deposition by EVD has successfully been replaced by slurry spray deposition. The slurry deposition followed by sintering has yielded electrodes which are equivalent in electrical conductivity to those fabricated by the EVD process. At high fuel utilization the performance of the cell with sintered fuel electrode is comparable to the fuel cell with the EVD fuel electrode [114].

2.2.2 Hexis design (Sulzer Innotec, Switzerland)

The key component of the HEXIS SOFC system is the ceramic/metal hybrid stack with circular planar SOFC elements [118]. Each stack repeat element has a multiple function as electrochemical cell, heat exchanger and afterburner. A detailed scheme of a Hexis SOFC by Sulzer is shown in Fig. 53. Each stack repeat element consists of a PEN-element (PEN: Positive air-electrode, Electrolyte, Negative fuel-electrode) and a metallic interconnect. The interconnect serves as heat exchanger and current collector. The Sulzer system runs thermally self sustaining after electrical heat up. This means that the stack is held at temperature only with the heat from the cells and from the afterburner, which is loaded directly on the outer rim of the stack [118]. The fabrication cost of the cell have to be low in order to compete with traditional energy converting systems. Special attention is given to the following fabrication and cost aspects:

-chemically clean interfaces between functional layers of the cell
-reliable, high yield, automated fabrication process for 200 cm^2 diameter components
-minimum quantities of costly raw materials
-low-cost current collectors

The PEN is manufactured by different companies and delivered to Sulzer. Details of PEN-fabrication are listed in tab. 8 The PEN is stacked on a Al_2O_3 gas manifold. The interconnect is contacted to the cathode with LSC-slurry, while the anode is contacted to the interconnect with a Ni-grid. Both connections are sintered during operation.

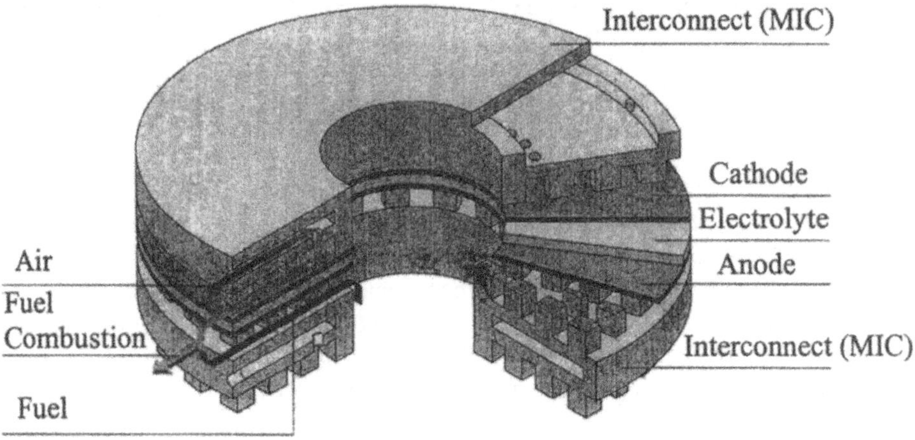

Figure 53: Sulzer Hexis Fuel Cell design [118]

TABLE 8: Material and fabrication processes of the components of the Sulzer Hexis Fuel Cell design.

Component	Material	Thickness	Fabrikation Process
Air Electrode	$LaSrMnO_3$ (LSM)	20-100 μm	Screen printing
Electrolyte	$ZrO_2(Y_2O_3)$ (8YSZ)	120-250 μm	Tape casting
Interconnect	$CrFe5Y_2O_31$		Powder metallurgy
Fuel Electrode	$Ni\text{-}ZrO_2(Y_2O_3)$	20-100 μm	Screen printing

Electrolyte The ZrO_2 (8 mol-% Y_2O_3) electrolyte is manufactured by tape casting. This process is well known and guarantees low costs. Tape casting is used as a fabrication technique for producing large-area, thin and flat ceramic plates. The casting thickness (equal to electrolyte thickness) is controlled by the blade gap. For lower operating temperatures either alternative electrolyte structures or thinner are needed. Future research therefore focuses on thin film electrolytes aiming at thickness of 4-10 μm even for large areas. They show low ohmic losses could be operated at lower temperatures [119].

Alternative fabrication processes are plasma spraying or sol-gel processes. The plasma spraying process is more expensive and the fabrication parameters are complex. With the sol-gel process very thin films can be deposited onto dense or porous substrates.

Anode The NiO/YSZ anode is coated on the electrolyte by screen printing process which provides low costs. Control of the screen mesh and filament diameter, emulsion thickness, and squeegee parameters are required for control of the print thickness. Tape casting is an alternative fabrication method of the anode.

Cathode The cathode material is a perovskite of the lanthanum strontium manganite type (LSM). The cathode is prepared by mixing the powder with solvent and binder into a paste. The cathode is placed onto the electrolyte by screen printing.

Interconnect The interconnect is a metal consisting of 95% Cr, 5% Fe with 3% Y_2O_3 ($CrFe5Y_2O_31$). It is manufactured by powder metallurgy. Further treatments are rolling and electrochemical erosion. This fabrication process can be easily controlled. A key element in the HEXIS design is the metal/ceramic-bonding on the air- and fuel electrode side. The current collector is contacted to the electrodes with LSC-slurry which is applied on the pin of the collector.

2.2.3 Planar design (e.g. Siemens AG, Germany)

The planar design of the Siemens high-temperature fuel cell consists of two metallic end plates and several bipolar plates which direct the process gases to the electrochemically active elements [120]. The Siemens fuel cell combines metallic and ceramic materials with a different design than Sulzer and Westinghouse, and allows high power densities to be achieved. Siemens solved the problem of thermal expansion difference of the interconnect and the electrolyte (8YSZ) by using the newly developed $CrFe5Y_2O_31$-alloy of Plansee (Austria). A characteristic of this design is the multiple cell array concept [120]. It allows the arrangement of several ceramic single cell elements (PEN) parallel in one layer as shown in Fig. 54. With today's fabrication techniques it is possible to fabricate 100 x 100 mm^2 PENs. This leads to a total electrode area of more than 700 cm^2 in one multiple cell array [121]. The sealing of the PENs to the interconnect is realized with deformable glass green tape shaped by a stamping tool which lowers the manufacturing costs.

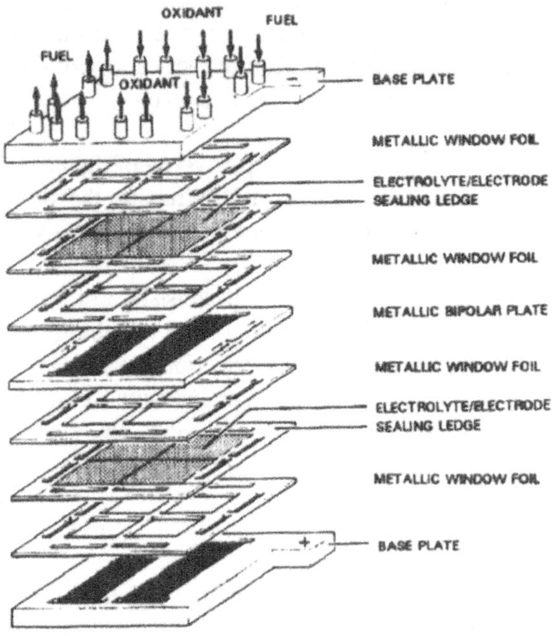

Figure 54: Assembly concept of a stack of Siemens [122]]

The different components and their fabrication methods are listed below in Tab. 9.

TABLE: 9: Materials and fabrication processes of the components of the Siemens Cell design

Component	Material	Thickness	Fabrication Process
Air Electrode	$LaSrMnO_3$ (LSM)	20-100 μm	Screen printing
Electrolyte	$ZrO_2(Y_2O_3)$ (TZP/FSZ)	150-250 μm	Tape casting
Interconnect	$CrFe5Y_2O_31$		Powder metallurgy
Fuel Electrode	$Ni\text{-}ZrO_2(Y_2O_3)$	20-100 μm	Screen printing

Electrolyte The electrolyte is produced by type casting. The material consists 8 mol% yttria stabilized ZrO_2. Siemens uses the same fabrication process for the electrolyte as Sulzer. The alternatives to tape casting are: Screen printing, wet spraying by air brush or spin coating.

Cathode The air electrode is produced by screen printing. The cathode material is $LaSrMnO_3$ (LSM). The advantages and alternatives processes are listed in the description of the Sulzer fuel system.

Anode The anode material is NiO/YSZ. This combination is the state-of-the-art anode material. A screen printed layer of this cermet is used as anode for the SOFC membrane. Alternative fabrication methods of the anode are tape casting or the sol-gel process.

Interconnect The PEN structure of the Siemens is similar to the one of the Sulzer fuel cell design. The two designs differ completely in terms of stack structure and in the design of the collector. In both cases the bipolar plates are key elements for functioning and long term reliability of the solid oxide fuel cell. At Siemens the active cell elements are electrically connected by flat structural $CrFe5Y_2O_31$-alloy plates (interconnects [123]. The requirements of the interconnects are: high temperature mechanical strength, high temperature corrosion resistance, as well as a thermal expansion coefficient adapted to that of the solid electrolyte. Variation of the Fe content adjusts the thermal expansion coefficient to that of the electrolyte/electrode sandwich [110]. The solution hardening of Fe and dispersion strengthening by uniformly distributed Y_2O_3 particles guarantee the mechanical integrity of the interconnect. On the anode side a Ni-grid works as functional layer to improve the electrical contact. On the cathode side $LaCoO_3$ is used for this purpose. The interconnects are fabricated from per-alloyed powders which are compacted to blanks in conventional presses at room temperature. Further densification is achieved by sintering under protective hydrogen atmosphere. Rolling of the blanks provides fully dense interconnect materials. Alternative processes are waterjet cutting or electrochemical machining (ECM). ECM means the controlled partial anodic electrolytical dissolution of material, using a complementary shaped cathode.

2.3 INORGANIC OXYGEN SEPARATING MEMBRANES

There is a large variety of possible structures and materials used as membranes. Depending on the driving force and the physical size of the separated species, membrane processes are classified [125]: microfiltration, ultrafiltration, reverse osmosis, dialysis, electrodialysis, and gas separation. The majority of these membranes have holes discriminating the species to be separated. For oxygen separation however, dense membranes are used where oxygen diffuses through the lattice of the solid dense material. In our further discussion we will give examples of devices and their application.

Membrane processes can be divided into two major modes according to the direction of the feed stream relative to the orientation of the membrane surface: dead-end filtration and crossflow filtration as shown in Fig. 55.

Usually an increased surface to volume ratio is beneficial and therefore flat modules are preferable over tubular systems.

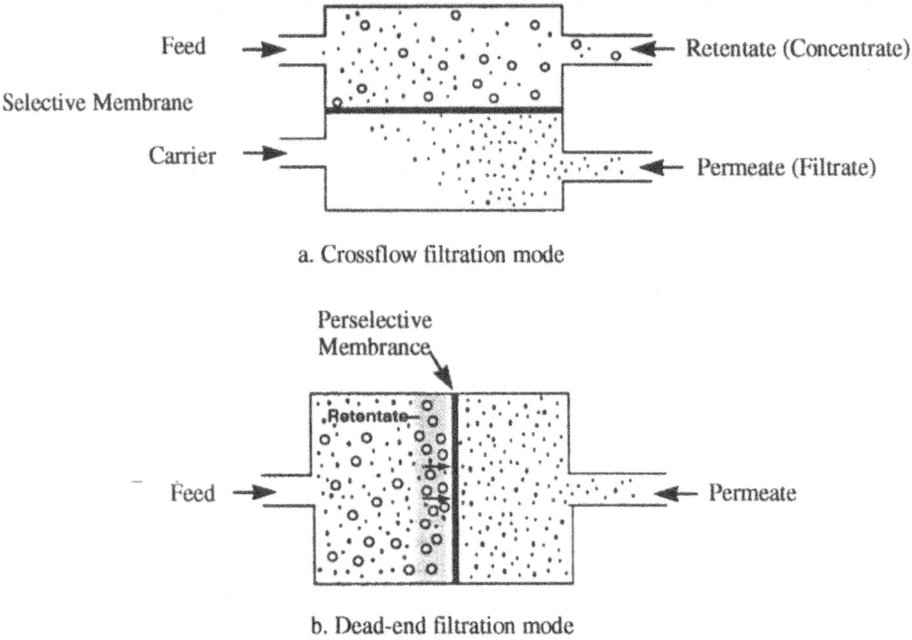

Figure 55: Schematic diagram of a membrane process: acrossflow filtration mode and b. dead-end filtration mode

Dense inorganic membranes with very high selectivity are important for commercial separation of O_2 or H_2 from other gases or for selective feeding into chemical reactors [124]. Two different types of membranes are used: ionic and mixed ionic conductors. The mixed ionic conductors transport oxygen ions but also electrons back from the reactor side to the oxygen/reduction interface. No external electrodes are required and if the driving potential of the transport is sufficient, the partial oxidation reactions should be spontaneous [126]. This reduces the amount of components of the oxygen membrane system, thus the fabrication costs are lowered. On the other hand electrodes are necessary in oxygen separating systems with ionic conductors in order to transport the electrons from the anode to the cathode site. Therefore, and additional fabrication step is needed.

Mixed-conducting oxide membranes have great potential to meet the needs of many segments of the oxygen market. It is further expected that the oxygen flow can be enhanced by thin film deposition on a porous substrate, preferably of the same material to avoid compatibility problems [127]. The applications envisioned range from small-scale oxygen pumps for medical application to large-scale usage in combustion processes, e.g. coal gasification [128]. Another application of mixed-conducting oxide membranes is to be found in the field of chemical processing. Dense inorganic membranes made of metal oxides and metals are used in membrane reactors for decomposition of carbon dioxide to form carbon monoxide and oxygen, oxidation of ammonia to nitrogen and nitrous oxide and oxidation of hydrocarbons such as ethylene, methanol, ethanol, propylene, and butene [125].

Dense ceramic based, oxygen separating membrane processes are divided into two different categories. They may either use a chemical gradient such as different oxygen partial pressures as driving force or an applied electrical potential.

The electrically driven system is inherently more complicated than the pressure gradient driven system because it requires electrodes on each side of the membrane and interconnections. In every membrane system one tries to maximize the membrane area compared to the volume. This can be achieved by sealed-end tubes, extruded honeycombs or planar stack designs. Fig. 56 shows some examples of tubular and monolithic honeycomb shapes of membranes. Fig. 57 shows a multi-tube membrane system. Fabrication methods of inorganic membrane systems and solid oxide fuel cells are very similar.

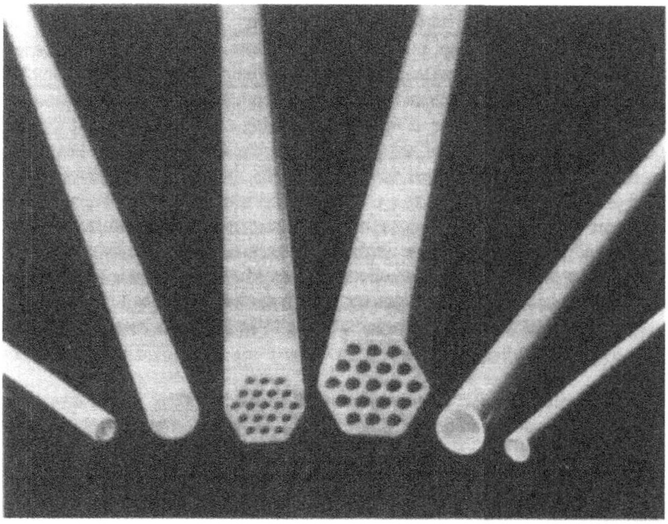

Fig. 56: Photograph of Al_2O_3 membrane elements in tubular and monolithic honeycomb shapes [125]

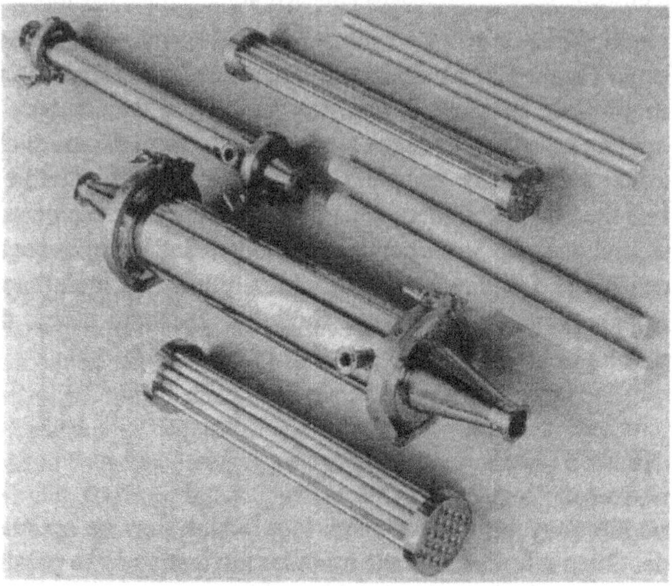

Fig. 57: Photograph displaying how multi-tube modules are assembled [125]

2.3.1 Inorganic dense tubular oxygen membrane

Components for the tubular electrolyte designs are mainly made by extrusion. They have to be fixed by flanges and sealed with glass depending on the reaction temperature and sintering parameters. A very thin dense electrolyte on a ceramic support structure is an alternative membrane design. Tab. 9 lists electrolyte materials used in dense ceramic membranes for tubular design:

TABLE 9: Selected inorganic dense tubular membranes [122]

Manufacturer	Membrane material	Support material
Gaston	ZrO_2	Carbon
NGK	TiO_2	Al_2O_3
TDK	ZrO_2	Al_2O_3
U.S.Filter	ZrO_2	Al_2O_3

Tubular configurations are adapted to tangential filtration. The self supporting structure of the electrolyte demands high mechanical strength and quality. It is possible to fabricate a dense electrolyte tube by powder pressing and sintering.

2.3.2 Dense inorganic stack-design oxygen membrane

Inorganic oxygen separating membranes are used in metallurgy to physically separate molten metals from reducing gases. The ionic current in the membrane is used to control the deoxidation rate. In another application, one side of a ceramic membrane is exposed to molten slags and the other side to a reference gas. Thus,the concentration of oxygen in the slags can be estimated.

Fig. 58 shows a large-scale planar stack design of a ceramic membrane system for metallurgy. Stabilized zirconia plates and zirconia plugs are assembled in such a way that only the alternate openings between the plates are closed by the plugs. A porous Ni-cermet electrode is deposited along the walls of the solid electrolyte plates which are exposed to the reducing gas. Tab. 10 shows two commercial inorganic dense ceramic membrane systems.

TABLE 10: Selected inorganic dense plate design membranes [124]

Manufacturer	Membrane material	Support material
TOTO	ZrO_2	Al_2O_3
Whatman	Al_2O_3	none
Ceramem	ZrO_2	Inocel

Flat plate stacks are mainly fabricated by tape casting. The slurry is spread on a flat support which can be very easily handled because of its plastic characteristics. Thus,complex geometrical shapes can easily manufactured. In industry, tapes can be manufactured up to 1 m of width and 40 m of length. The membrane or electrodes are applied by screen-printing, tape casting or plasma spraying.

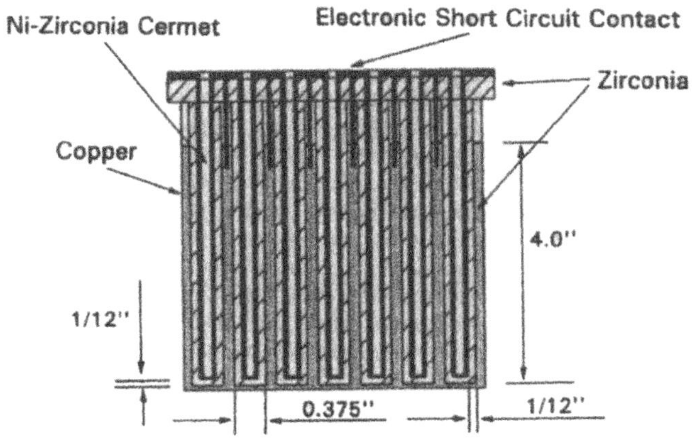

Fig. 58: Schematic of a scale up structure of a oxygen ceramic membrane

2.4 CERAMIC OXYGEN SENSORS

Sensors are devices which convert physical or chemical quantities into electrical signals which are convenient to use. Sensors are getting more and more important in automation and robot technology.
The most important requirements for sensors are[129,131,132]:

-high sensitivity
-reproducibility
-high response rate
- resistance to ambient influence such as heat, vibration, acids, gas, water, etc.

The classification of sensors according to their principles of operation is shown in Figure 59. They can be divided into physical and chemical sensors. According to the materials used, sensors can be divided into the following groups:

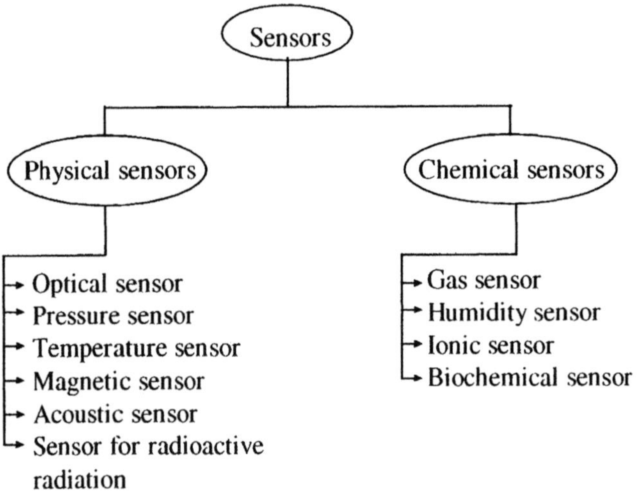

Figure 59: Classification of sensors according to the principle of operation

A large variety of ceramic sensors are known. A small overview is given in Table 11.

This paper focuses mainly on ceramic oxygen sensors. Gas sensors are used in various applications, both in home and industry, e.g. fire detectors, environmental monitoring and control, process control and energy management. Toxic and combustible gas monitoring is important for oil and gas destilling and refinishing, power generation, and in the chemical and petrochemical process industries.

In the following sections some materials and fabrication methods of commercial ceramic-sensors are described. Oxygen sensors are used to detect the presence and to quantify the amount of oxygen in a certain atmosphere. The main application of ionic oxygen sensors is in the automobile exhaust gas analysis (lambda sensor). It is also used in metallurgy, gas production, and food and beverage packaging. Different electrolyte materials used for oxygen ceramic gas sensors are listed in Table 12.

TABLE 11: Ceramic sensors, exploited effects and materials [129]

Type of sensor	Output signal	Effect	Material
Temperature sensor	Change in resistance	Change of the concentration of the temperature carriers	NiO, CoO, FeO, MnO, MnO-NiO-CoO
Temperature sensor	Change in resistance	Thermistor with a positive temperature coefficient (PTC)	$BaTiO_3$
Gas sensor	Change in resistance	Thermal reaction of the catalytic combustion of gases	Pt catalyst/Al_2O_3/Pt conductor
Gas sensor	Electromotive force	Oxgen concentration voltaic cells with a solid electrolyte	ZrO_2-Y_2O_3, CaO, MgO, La_2O_3
Humidity sensor	Change in the capacitance	Change in the dielectric permittivity with humidity adsorption	RuO_2
Optical sensor	Electromotive force	Pyroelectric effect	$SrTiO_3$, $LiNbO_3$, PZT, $PbTiO_3$
Sensors for measuring force, pressure, acceleration	Change in output voltage	Piezoelectric effect	PZT, $PbTiO_3$, $PbZrO_3$

TABLE 12: Different electrolyte materials for oxygen ceramic sensors [129]

Type of oxygen sensor	Electrolyte material
Semiconductor	TiO_2, SnO_2, CeO_2, $BaTiO_3$, $CaTiO_3$, $SrTiO_3$, $Mg_{1-x}Fe_xO$, $Mg_{1-x}Co_xO$
Solid electrolyte	8 mol% stabilized yttria

Semiconductors change their conductivity upon oxygen partial pressure changes in the atmosphere. As this oxidation/reduction reaction is diffusion limited, their response time is larger than sensors based on ionic conductivity solid electrolytes.

Below there are listed three examples of ceramic oxygen sensors produced in industry.

- *Lambda sensor for automobile engines*

The purpose of the lambda sensor is to monitor the exhaust of automobile engines. The sensor uses the oxygen defect chemistry of a dense yttria stabilized zirconia (YSZ) which is used as electrolyte. The Nernst voltage of a stoichiometric exhaust is compared with the potential produced by the effective exhaust. Thus, an online regulation of the fuel injection is possible.
There exist two principle designs of a lambda sensor (1) tubular and (2) flat plate (shown in Fig. 61).

First the ZrO_2 tube of the tubular lambda sensor is fabricated. The powder is pressed and sintered. Afterwards the tube is smoothened. The fabrication processes are mainly different in terms of applying the electrodes. There are two different types:

-Thin films of platinum are deposited (thickness of electrode < 1μm) of platinum by sputtering technique, plasma spraying or chemical vapor deposition.
-A platinum/cermet-mixture is printed on the green sensor, either by tape casting or spraying.

In the platinum/cermet mixture (Pt/YSZ) the triple phase boundary (tpb) of the sensor is increased. Thin film deposition technique on the other hand produces a tpb which is limited to the surface of the electrode. In order to provide long term stability the electrodes have to be protected from the exhaust gas mixture. This can be achieved by plasma spraying or flame spraying of a magnesia-spinel. A different process is co-firing of a protection layer of ZrO_2. The porosity of the protection layer is adjusted by organic additives or Al_2O_3 which is a catalytic sinter additive.

The electrolyte of a flat plate lambda sensor is produced by tape casting. The electrodes, the exhaust manifold or the heating element are applied by screen printing. An example of a planar sensor is shown in Fig. 60.

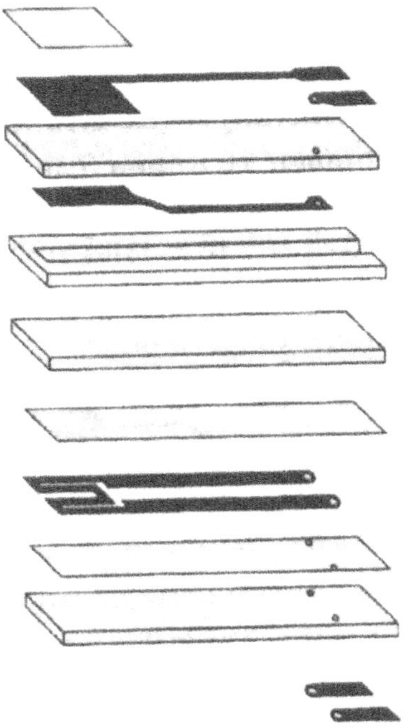

Figure 60: Scheme of packing of a flat plate lambda sensor [130]

2.3.2 - Oxygen sensor on the base of tin dioxide (e.g. by Figaro Engineering Inc.)

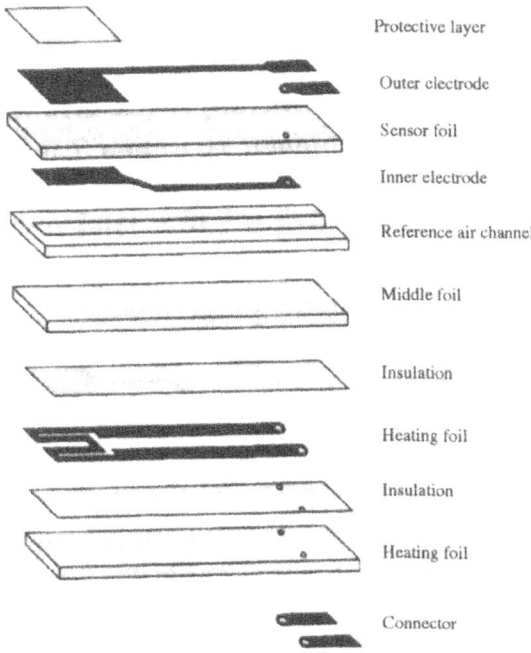

The sensitive element of the gas sensor consists of a tubular ceramic case with electrodes contacted on its outer surface by metallization. The case is fabricated by powder pressing and sintering. The sensitive layer (SnO_2) with catalytic additives is coated over them. It can be applied by electrochemical vapor deposition. This process provides high purity and density of the layer. Disadvantages are the high costs and the complexity of the process parameters. Fig. 61(a) shows a scheme of a sensor construction. A heater filament is enclosed in the ceramic tube. The geometry of the sensing element causes uniformity of the temperature in the sensitive layer. Fig. 61(b) shows the mounting of the sensitive elements. Fig. 61(c) shows a photograph of a ceramic gas sensor. A temperature resistant net made of stainless steel provides an explosion-proof construction [129].

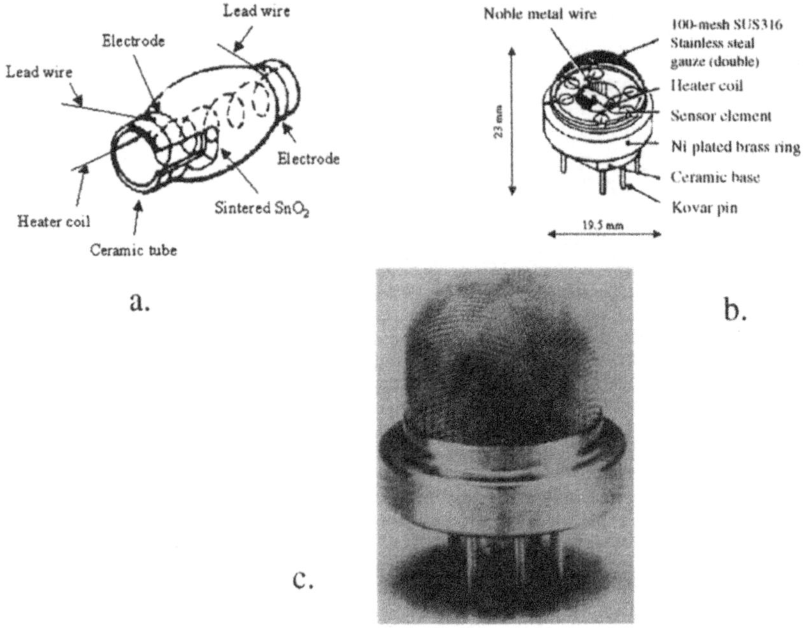

Figure 61: Construction (a), mounting (b) and photograph of a ceramic oxygen sensor (c) made by Figaro Engineering Inc. [127]

Literature

1. Ring, T. A. (1996) Fundamentals of Ceramic Powder Processing and Synthesis, Academic Press.

2. Ganguli, D., Chatterjee, M. (1997) Ceramic Powder Preparation : A handbook,Kluwer Academic Publishers, Boston.

3. Hulbert, S. F. (1996) Models for solid state reactions in powdered compacts : a review, *J. Brit. Ceram. Soc.*, 6, 11-20.

4. Matteazzi, P., Le Cear, g. Bauer-Grosse, E. (1991) Synthesis of advanced ceramics by high energy milling, *Key Eng. Mater.*, **53-55**, 451-456 .

5. Millet, P., Hwang, T., (1996) Preparation of TiB_2 and ZrB_2. Influence of a mechanochemical treatment on the borothemic reduction of titania and zirconia, *J. Mater. Sci.*, **31**,351-355.

6. Sheppard, L. M. (1988) Manufacturing ceramics with microwaves : the potential for economical production, *Am. Ceram. Soc. Bull.*, **67**, 1656-1661.

7. Sutton, W. H. (1989) Microwave processing of ceramic materials *Am. Ceram. Soc. Bull.*, **68**, 376-386

8. Rao, K. J., Ramesh, P. D. (1995) Use of microwaves for the synthesis and processing of materials, *Bull. Mater. Sci.*, **18**, 447-465.

9. Koeck, W., Martinz, H.-P., Greiner, H., Janousek, M., (1995) Development and Processing of Metallic Cr based Materials for SOFC Parts, pp. 841-849 in Solid Oxide Fuel Cells (SOFC-IV),Dokiya, M Yamamoto, O.,Tagawa, H.,. Singhal, S. C (eds) The Electrochemical Society,Pennington.

10. Taimatsu, H.. Kudo, K., Kaneko, H. (1995) Preparation of porous Nickel-Titania Cermets and their Application to Anode Maerials, pp. 706-711 in Solid Oxide Fuel Cells (SOFC-IV),Dokiya, M., Yamamoto, O.,Tagawa, H., Singhal, S. C (eds) The Electrochemical Society,Pennington.

11. Jones, A. G. (1994) Particle formation duringagglomerative precipitation processes, pp.61-94 in Controlled Particle, Droplet and Bubble Formation.,Wedlock, D. J (ed) Butterworth-Heinemann, Oxford, 1994.

12. Itoh, T. (1985) Particle and Crystallite Sizes of ZrO_2 Powder obtained by the Calcination of Hydrous Zirconia, *J. Mater. Sci. Lett.*, **4**, 431-433.

13. Duran, P., Recio, P., Jurado, J. R., Pascual, C., Moure, C. (1989) Preparation, sintering and properties of translucent Er_2O_3-doped tetragonal zirconia, *J. Am. Ceram. Soc.*, **72**, 2088-2093 (1989).

14. van de Graaf, M. A. C. G., Burggraaf, A. J.,(1984) Wet-chemical preparation of zirconia powders : their microstructure and behavior, pp. 744-765 in *Adv. Ceram.*, **12**, Science and technology of Zirconia II.

15. Chen, P.-L.,.Chen, I.-W., (1993) Reactive cerium (IV) oxide powders by the homogeneous precipitation method ,*J. Am. Ceram. Soc.*, **76**, 1577-1583 .

16. Verdon, E., Devalette, M., Demazeau, G. (1995) Solvothermal synthesis of cerium dioxide microcrystallites : effect of the solvent, *J. Eur. Ceram. Soc.*, **15**, 939-950.

17. Van Herle et al. (1995) Preparation of doped Ceria Powder, pp. 1082-1091 in Solid Oxide Fuel Cells (SOFC-IV), Edited by M. Dokiya, O. Yamamoto, H. Tagawa, S. C. Singhal, The Electrochemical Society, Pennington.

18. Zhong-Tai, Z. et al. (1995) Synthesis and Characteristics of Ceramics for Cathode Materials of SOFC, pp. 502-511 in Solid Oxide Fuel Cells (SOFC-IV), Edited by M.Dokiya, O. Yamamoto, H. Tagawa, S. C. Singhal, The Electrochemical Society,Pennington.

19. Macek, J., Marinsek, M. (1996) Coprecipitation Process for Preparation of Nickel Cermet Anodes, pp. 341-350 in Second European Solid Oxide Fuel Cell Forum, Edited by B.Thorstensen, European Solid Oxide Fuel Cell Forum, Oslo.

20. Marinsek, M., Macek, J. (1996) Thermal Processing of Ni-YSZ, pp. 351-360 in Second European Solid Oxide Fuel Cell Forum, Edited by B. Thorstensen, European Solid Oxide Fuel Cell Forum, Oslo.

21. Djuricic, B., Kolar, D., Memic, M. (1992) Synthesis and properties of Y_2O_3 powder obtained by different methods *J. Eur. Ceram. Soc.*, **9**, 75-82 .

22. Brinker C. J.,. Scherer, G. W. (1990) Sol-Gel Science, Academic Press, Boston.

23. Baes, C. F., Mesmer, R. E. (1976) The Hydrolysis of Cations, JohnWiley, New York.

24. Woodhead, J. L. (1984) Sol-Gel processes to ceramic particles using inorganic precursors, J.Mater. Educ., **6**, 888-925.

25. Lessing, P. A. (1989) Mixed-Cation Oxide Powders via Polymeric Precursors *Ceramic Bulletin*, **68**, 5.

26. Pechini, M. (1967) Method of Preparing Lead and Alkaline-Earth Titanates and Niobates and Coating Method Using the Same to Form a Capacitor, U.S. Patent 3 330 697, July 11.

27. Eror, H., Anderson, U. (1986) Polymeric Precursor Synthesis of Ceramic Materials, pp. 571-577 in Better Ceramics Through Chemistry, II, Materials Research Society Symposia Proceedings, Vol. 73. Edited by C. J. Brinker, D. E. Clark, and D. R. Ulrich. Materials Research Society, Pittsburgh, Pa.

28. Anderson, H. U., Pennell, M. J., Guha, J. P. (1987) Polymeric Synthesis of Lead MagnesiumNiobate Powders, pp. 91-98 in Advances in Ceramics, Vol.21, Ceramic Powder Science. Edited by G. L. Messing, K. S. Mazdiyasni, J. W. McCauley, and R. A. Haber. American Ceramic Society,Westerville, OH.

29. Budd, K. D., Payne, D. A. (1984) Preparation of strontiumtitanate ceramics and internal boundary layer capacitors by the Pechini method, pp. 239-44 in Mater. Res. Soc.Symp. Proc., 32, Better Ceramics Through Chemistry. Edited by C. J.Brinker, D. E. Clark , D. R. Ulrich, Elsevier, New York.

30. Thomson, J. Jr., (1974) Chemical preparation of PLZT powders from aqueous nitrate solutions *Am. Ceram. Soc. Bull.*, **53**, 421-24, 433.

31. Kim, S. J., Jung, C. H., Kim, Y. S. (1996) Synthesis of Ultrafine NiO-YSZ Powders by GNP (Glycine Nitrate Process), pp. 321-330 in Second European Solid Oxide Fuel Cell Forum, Edited by B.Thorstensen, European Solid Oxide Fuel Cell Forum, Oslo.

32. Masters, K. (1994) Spray Drying Handbook, J. Wiley & Sons, New York.

33 Chatterjee, M., Ray, J., Ganguli, D., (1992) Spray-drying of hydrated zirconia slurries : a laboratory study, *Brit. Ceram. Trans. J.*, **91**, 159-162 .

34 Messing, G. L., Zhang, S.-C., Jayanthi, G. V. (1993) Ceramic powder synthesis by spray pyrolysis, J. Am. Ceram. Soc., **76**, 2707-2726.

35 Mizutani, N., Liu, T. Q. (1990) Synthesis of spherical Si_3N_4 powders by spray pyrolysis ofpolysilazane, Ceram. Trans., **12**, 59-73, Ceramic Powder Science III, American Ceramic SocietyWesterville OH.

36 Xiaming, D., Quigfeng,. Yuying, L. T(1993) Study of phase formation in spray pyrolysis of ZrO_2 and ZrO_2-Y_2O_3 powders, *J. Am. Ceram. Soc.*, **76**, 760-762 .

37 Oobuchi, T. et al.(1995) Preparation of Electrode Powders by the Aerosol Flow Pyrolysis method and Characteristics of the Electrodes, pp. 759-768 in Solid Oxide Fuel Cells (SOFC-IV), Edited by MDokiya, O. Yamamoto, H. Tagawa, S. C. Singhal, The Electrochemical Society, Pennington.

38 Riluson, A. J., Flagan, R. C. (1994) Synthesis of yttria powder byelectrospray pyrolysis *J. Am. Ceram. Soc.*, **77**, 3244-3250 .

39 Dogan, F., Hausner, H. (1968) The role of Freeze-drying in ceramic powder processing, pp127-34 in Ceram. Trans., **1**, Ceramic Powder Science II, British Ceramic Society, Stoke-on-Trent, U. K. .

40 Rakotoson, L., Paulus, M. (1984) Sintering of a freeze-dried 10mol% Y_2O_3-stabilized zirconia, pp. 727-732 in Adv. Ceram., 12, Science and Technology of Zirconia II. Edited by NClaussen, M. Rühle, A. H. Heuer. American Ceramic. Society, Columbus, OH.

41 Pulker, H. K. (1984) Coatings on Glass, Elsevier, Amsterdam.

42 Ichinose, N., Ozaki, Y., Kashu, S. (1992) Superfine Particle Technology, Springer London.

43 Sigel, R. W. et al. (1988)Synthesis, characterization and properties ofnanophase TiO_2, *J. Mater. Res.*, **3**, 1367-72.

44 M. N. Rahaman, Ceramic Processing and Sintering,Marcel Dekker, New York, 1995.

45 J. M. Haussonne, J. Lostec, J. P. Bertot, L. Lostec, S. Sadou, A new synthesis process forAlN, *Am. Ceram. Soc. Bull.*, **72** (5), 84-90 (1993).

46 Ault, N. N., Crowe, T. J., (1993) Silicon carbide,*Am. Ceram. Soc. Bull.*, **72** (6), 114 .

47. Allen, (1990) Particle Size Measurement, London,Chapman & Hall.

48. Brook, R.J., (1996) Processing of Ceramics, Part I, ed. R.J. Brook in Materials science and Technology Vol. 17, VCH, Weinheim.

49. Ganugli, M. Chatterjee, (1997) Ceramic Powder Preparation: A Handbook,Kluwer, Boston.

50. Technische Keramische Werkstoffe, (1990) J. Kriegesmann (ed) Deutscher Wirtschaftsdienst, Köln.

51. McColm, I.J., Clark, N.J. (1988) Forming, Shaping, and Working of High Performance Ceramics, Blachie, Glasgow .

52. Reed, J. (1996) Introduction to the Principles of Ceramic Processing,. J.Wiley, New York.

53. River, Ch. (1985) Advanced ceramic materials, Noyes Publ. Park Ridge, USA.

54. Stockham, J.D., Fochtman, E.G. (ed), (1978) Particle size analysis, Ann Arbor Science Publishers, Ann Arbor.

55. Otsuka, K. (1993) Multilayer Ceramic Substrate Technology,Elsevier.

56. Allen, T. (1990) Particle Size Measurement, London,Chapman & Hall.

57. Barth, H.G., Sibilia, J.P. (1989) Modern Methods of Particle Size Analysis, J. Wiley, New York,.

58. Sibilia, P. (1988), A Guide to Materials Characterization and Chemical Analysis, VCH, New York.

59. Brunauer, R. P.H. Emmett, E. Teller, (1938), Adsorption of Gases in Multimolecular Layers, *J. Am. Chem. Soc.*, 309ff, **60.**

60. Shanefield, J. (1995), Organic Additives and Ceramic Processing,Kluwer Academic Publishers, Boston.

61. Goldstein, J. (1981), Scanning Electron Microscopy andX-Ray analysis , Plenum Press, New York, 1981.

62. Nelson, R.D., (1988) Dispersing of powders into Liquids,Elsevier, Amsterdam (1988).

63. Valentin, F.H.H. (1967) The Mixing of Powders and Pastes: Some Basic Concepts*Chem.* Eng. 208, 99-104.

64. Parfitt, G.D. (1981) Dispersions of Powders in Liquids,Wiley Interscience, N.Y. .

65. 2. World Congress Particle Technology (1985): part 3 Dispersion and Classification

66. Richerson, W., (1992) Modern CeramicEngineeringDecker, Inc.N.Y..

67. Praher, T. (1987) Crushing and Grinding Process Handbook,Wiley & Sons, New York, 1987.

68. Hogg, R., (1981) Grinding and Mixing of Nonmetallic Powders*Am Ceram. Soc. Bull* ***60.***

69. Adair, J.H (1995) Science, Technology, and Applications of collooidal Suspensions, The Ame. Ceram. Soc.

70. Hunter, R.J.(1981) *Zeta Potential in Colloid Science*, New York, Academic Press, 1981.

71. Derjaguin, B. V. and Landau, J. D. (1941) Acta Physiocochim. URSS, 14, 633.

72. Mistler, R.E., (1991) Tape Casting: The Basic Process for Meeting the Needs of the Electronics Industry, *Ceram. Bull*, **69**, 6.

73. Morse, T., (1979) Handbook of Organic Additives for Use in Ceramic Body Formulation, Montana Energy and MHD Research and Development Institute,Butte, MT

74. Williams, J.C. (1976) Doctor Blade Process in Wang, F.Y (ed) Treatise on Material Science and Technology,**9**, Academic Press, N.Y..

75. Lukasiewicz, S.J. (1989) Spray Drying Ceramic Powders, *J. Am. Ceram. Soc.*, **72** (4) 617-624.

76. Mistler, R.E., Shanefield, D.J., Runk, R.B. (1978) in Ceramic Processing bevor Firing, ed G.Y. Onoda, L.L Hench, J. Wiley Sons. N.Y.

77. Minh, N.Q., Takahashi, T. (1995) *Science and Technology of ceramic fuel cells*, Elsevier, Amsterdam

78. Mistler, R.E. (1991) Tape Casting: The Basic Process for Meeting the Needs of the Electronic Industry, Ceram. Bull. 69.

79. Roosen, A. (1988) Basic Requirements for Tape Casting of Powders in Ceramic Powder Science II, ed. G.L. Messing, E.R. Fuller, H. Hausner, The AmericanCeram. Soc. Ohio.

80. Hyatt, E.P., (1986) Making thin flat ceramic - a review,*Am. Ceram. Soc. Bull.* **65** (4) 637-638.

81. Graule, T.J., Gauckler L.J., and Baader, F.H. (1994) Direct Coagulation Casting - A New Shaping Technique, Part I: Processing Principles, published in Proceedings of the 8th CIMTEC - World Ceramics Congress and Forum on New Materials, Florence, Italy, June 28 - July 4.

82. Gauckler, L.J., and Graule, T.J. (1992) Verfahren zur Herstellung keramischer Grünkörper,Swiss Patent Appl. Nr. 02 377/92-1.

83. Graule, T.J., Gauckler, L.J., and Baader, F.H. (1993) Verfahren zur Herstellung keramischer Grünkörper durch Doppelschichtkompression,Swiss Patent Appl. Nr. 01 096/93-6.

84. Samasundara, P., (1978) Theory of Grinding, in Onoda G.Y., Hench, L.L., (eds) *Ceramic Processing before Firing,* Wiley Interscience N.Y..

85. Ichinose (ed) (1993) Introduction to Fine Ceramics, J.Wiley, N.Y..

86. J. Rao, P. D. Ramesh, Use of microwaves for the synthesis and processing of materials, Bull. MaterSci., **18**, 447-465 (1995).

87. R. Yamaguchi et al.,(1993) *3rd. Intl. Symp. Solid Oxide Fuel Cells* . Ed. by S.C. Singhal and H. Iwahara, The Electrochem. Soc.Proc. Vol 93-4.

88. Taussig, W., (1995) Screen Printing, Clayton Amiline Company, Manchester.

89. Thiele, E.S., Wang, L.S. , Wang, T.O, Barnetz,, S.A.,(1991) Depostion and Properties of Yttria Stabilized Zirconia Thin Films Using Direct Current Magnetron Sputtering *J. Vac. Sci.* Technol. A9.

90. Aita, C.R, Kwok, C.K. (1990) Fundametnal Optical Absorpiton Edge of Sputter Deposited Zirconia and Yttria, J. *Am. Ceram. Soc.*, **73.**

91. Smidt, F.A. (1990), Use of Ion Beam asisted Deposition to Modify the Microstructure and Properties of Thin Films,*Int. Mater. Rev.* **35.**

92. Sarkar P., Nicholson P.S., 1996 Electrophoretic Deposition: Mechanisms, Kinetics, and Applications to Ceramics, *J. Am. Ceram. Soc.*, **79** (8).

93. Ishihara, T. , Sato. K. Takita Y., 1996, Electrophoretic Deposition of y2O3 Stabilized ZrO_2 Electrolyte Film in Solid Oxide FuelCells, J. *Am. Ceram. Soc.* 79 (8).

94. Gani, M.S., (1994), Electrophoretic Deposition: A Review, Ind. Ceram. **14.**

95. Kordesch, K. Simader G. (1996), Fuel Cells and Theri Application, VCH Verlagsgesellschaft, Weinheim.

96. Wright, J.K., Edirisinghe, M.J, Zhang, J.G., Evans, J.R.G., (1990), Particle Packing in Ceramic Injection Molding *J. Am. Ceram. Soc,* **73,** 9.

97. Mwray, P. T. et al., (1987)] *Mat. Lett.*, 5, 250-54.

98. Mutsuddy, B.C., Ford, R.G.,(1995) Ceramic Injection Molding,Chapman & Hall.

99. Singhal S.C. (1997), Recent Progress in Tubular Solid Oxide Fuel CellTechnology, Proc. Of the 5th International Symposium on Solid Oxide FuelCells, Aachen, De. Electrochem Soc..

100. K.Masters, (1994) Spray Drying Handbook, J. Wiley & Sons, New York.,

101. Mehta, K., Hong, S.J., Jue, J.F, Virkar, A.v., (1993) Fabrication and Characterization of YSZ Coated Ceria Electrolyted in Porc. Of the 2nd Int. Symp. On SOFC. S.C. Singhal (ed).

102. Otsuka, K. (1991) Multilayer Ceramic Substrate-Technology for VLSI Package / Multilayer Module, Elsevier.

103. Bornside, D.E., Mcosko, C.W., Scriren, (1987) L.E., *J. Imag. Tech*, 13, 122-129.

104. Somiya(ed) S.(1984) Advanced Technical Ceramics, Academic Press Japan, Inc. Tokyo.

105. Coble, R.L.,(1987) J- Am. Ceram. Soc. 56, 9.

106. Exner, H.E., (1978) Grundlagen von Sintervorgängen.

107. Eisele, U. (1996) Sintering and Hot Pressing, in MaterialsScinece and Technology, 17B,Cahn, R.W., Haasen, R., VCH, Weinheim.

108. Kostorz, G: (ed) (1988) High-tech Ceramics, Academic Press

109. Krist K., Wright J.D., (1993), Fabrication Methods for Reduced Temperature Solid Oxide FuelCells, Proceeding of the 3rd Int. Symposiumon Solid Oxide FuelCells (SOFC), Pennington, USA, p. 782.

110. Quadakkers J., Greiner H., Köck W., (1997), Metals and Alloys for High Temperature SOFC Applications, Proceedings of the 5th Solid Oxide Fuel Cell, Aachen,Ge, Vol. 97-18, p. 525.

111. Isenberg A.O., (1982), National Fuel Cell SeminarAbstracts, Newport Beach, CA, Nov. 14-18, p. 154.

112. Berry D.A., Mayfield M.J., (1988), Fuel cells Technology Status Report, Report DOE/METEC-890266.

113. Westinghouse Electric Corporation, (1984), Solid Oxide Fuel Cell Power Generation SystemThe Status of the Cell Technology-A-Topical Report, DOE/ET/17098-15.

114. Singhal S.C, (1997), Recent Porgress in Tubular Solid Oxide Fuel CellTechnology, Proceedings of the 5th Solid Oxide Fuel Cell, Aachen,Ge, Vol. 97-18, p. 37-45.

115. Yokohama H., Miyohora A., Veyo S.E., (1997), Verification Test of a 25 kW class SOFC Generation System, Proceedings of the 5th Solid Oxide Fuel Cell, Aachen,Ge, Vol. 97-18, p. 94-101.

116. Pal U.B., Singhal S.C., (1990), Electrochemical Vapor Deposition of Yttria-stabilized Zirkonia Films, J. Electrochem. Soc., **137**, 2937.

117. Kao L.H., Vora S.D., Singhal S.C., (1997), Plasma Spraying of LanthanumChromite Filmsfor Solid Oxide Fuel Cell InterconnectionApplication, J. Electrochem. Soc., **80**, in press.

118. Diethelm R., Brun J., (1997), Status of the Sulzer Hexis Solid Oxide Fuel Cell (SOFC) System Development, Proceedings of the V Solid Oxide Fuel Cell, Aachen,Ge, Vol. 97-18, p. 79.

119. Barnett S.A., (1990), Energy, **15**, 1.

120. Blum L., Drenckhahn W., Greiner H., (1995), Mulit-kW-SOFC Developmentat Siemens, Proceedings of the 4th International Symposiumon Solid Oxide FuelCells (SOFC-IV), Pennington, NJ, p. 163.

121. Beie H.J., Blum L., Drenckhahn W., (1997), SOFC Developmentat Siemens, Proceedings of the 5th Solid Oxide Fuel Cell, Aachen,Ge, Vol. 97-18, p. 51.

122. Gellings, P.J., Bouwmeester, H.J.M. (ed) (1977) The CRC Handbook of Solid State Electrochemistry, CRC Press

123. Köck W., Martinz H.P., (1995), Development and Processing of Metallic Cr-based Materials for SOFC Parts, Proceedings of the 4th International Symposium on Solid Oxide FuelCells (SOFC-IV), Pennington, NJ, p. 841.

124. Burggraaf A.J., (1994), Key Points in Understanding and Development of Ceramic membranes, Proceedings of the 3rd International Conference on Inorganic Membranes, Worcester, p. 1.

126. Hsieh H.P.(ed), (1996), Inorganic Membranes for Separation and Reaction, Elsevier.

127. Harold M.P., Lee C., Burggraf A.J, (1994), Catalysis with Inorganic Membranes, MRS Bull., **19**[4], p. 34.

128. Balachandran U., Dusek J.T., (1994), Dense Ceramic Membranes for Partial Oxydation of Methane, Proceedings of the 3rd International Conference on Inorganic Membranes, Worcester, p. 229.

129. Bouwmester H.J.M., Burggraaf A.J.(ed), (1996), Dense Ceramic Membranes for Oxygen Separation, Fundamentals of Inorganic Membrane Science and Technology, Elsevier.

130. Carolan M.F., Dyer P.N., (ed), (1993), Process for Recovering Oxygen from Gaseous Mixtures Containing Water or Carbondioxide which Process Employs Ion Transport Membranes, US Patent 5,261,932.

131. Walker D., Kilner J.A., Steele B.C.H., (1997), Oxygen Separation Using Dense Gadolinia Doped Ceria Membranes, Ceramic Membranes, J. Electrochem. Soc., **24**, p. 48.

132. Nenov T.G., Yordanov S.P., (ed), (1996), Cermic Sensors: Technology and Applications, Technomic.

133. Schaumberg H., (ed), (1995), Sensor Anwendungen,B.G. Teubner Stuttgart.

MATERIALS DESIGN AND OPTIMIZATION

H.L. Tuller
Crystal Physics and Electroceramics Laboratory
Department of Materials Science and Engineering
Massachusetts Institute of Technology
Cambridge, MA 02139, U.S.A.

I. Introduction

The suitability of a particular material depends on the particular application for which it is being considered. In this volume, we deal with applications such as solid oxide fuel cells, gas separation membranes, oxygen pumps and potentiometric and amperometric gas sensors, for which we place a high premium on achieving appropriate levels of ionic and electronic conductivity. At the same time, because of their high temperature use, the components of these devices must simultaneously satisfy a number of demanding criteria including chemical, phase, morphological and dimensional stability in oxidizing and/or reducing environments, while maintaining chemical and thermal-mechanical compatibility with other components. In our earlier chapter [1], we focused on the fundamentals relating to the generation and interaction of ionic and electronic charges in oxides. In this section we begin by examining how structure and chemistry need to be manipulated to achieve the desired transport properties. To assist us in this objective, we define "figures of merit" for various applications. We also examine a number of materials systems which illustrate to what extent the concepts introduced above can be applied towards optimizing such figures of merit. We complete this article by considering a more holistic approach in which we demonstrate means for selecting particularly versatile systems which, upon minor modification, can satisfy the needs for various components of the solid state electrochemical system. Such designs can be expected to substantially minimize degradation modes such as destructive chemical reactions between contacting phases and cracking due to lattice and thermal expansion mismatches.

II. Solid Electrolytes

The solid electrolyte is typically the key component of the solid state electrochemical device. The electrical potential, induced by the imposition of a chemical activity gradient across the electrolyte, is the basis of the potentiometric gas sensor and the solid oxide fuel cell (SOFC). Conversely, the chemical gradient induced by the application of an electrical potential is the basis of the electrochemical pump. The "principal" criteria that the solid electrolyte must satisfy are:

1. Adequate ionic conductivity
2. Negligible electronic conductivity

For potentiometric applications, e.g., oxygen sensor, in which the cell is maintained under open circuit conditions, the magnitude of ionic conductivity need not be high. However, in applications where current is drawn, e.g., during discharge of a battery or fuel cell, the ionic conductivity needs to be high to minimize ohmic losses. It should be noted that in many cells, polarization losses associated with the electrodes dominate. Under these circumstances, higher ionic conductivities of the solid electrolyte do not necessarily improve the performance of the cell.

Electronic conduction degrades solid electrolyte performance in several ways. Because electronic conduction serves as an alternate path for charged species through the electrolyte, it decreases the power that can be dissipated through the load in a battery or fuel cell circuit. Further the short circuiting factor also serves to lower the open circuit potential below the Nernst potential, thereby throwing-off the accuracy of potentiometric sensors and allowing gas permeation even under open circuit conditions. These primary figures of merit are summarized in the context of a

H.L. Tuller et al. (eds.), Oxygen Ion and Mixed Conductors and Their Technological Applications, 245–270.

solid oxide fuel cell (SOFC) in Fig. 1. E represents the potential induced across the cell under open circuit conditions for a given Po_2 gradient, $\hat{t}_i$, is the ionic transference number, E_N, the Nernst potential, R_{INT}, R_C, R_{SE}, and R_A are the internal cell, cathode, solid electrolyte and anode resistance, respectively, Jo_2 is the oxygen permeation flux and L the thickness across which the Po_2 gradient is imposed. All other terms have their normal meanings.

The first challenge is to identify solids which simultaneously possess high oxygen ion and low electronic conductivities. This insures a low R_{SE}, a high value of E and a low Jo_2 through the electrolyte. Typically, this translates to the following desired characteristics:

$$\sigma \geq 10^{-2} S/cm \tag{1}$$

$$t_e = \frac{\sigma_e}{\sigma_i + \sigma_e} \leq 10^{-2} \tag{2}$$

Where σ_i and σ_e are the ionic and electronic conductivities and t_e is the electronic transference number.

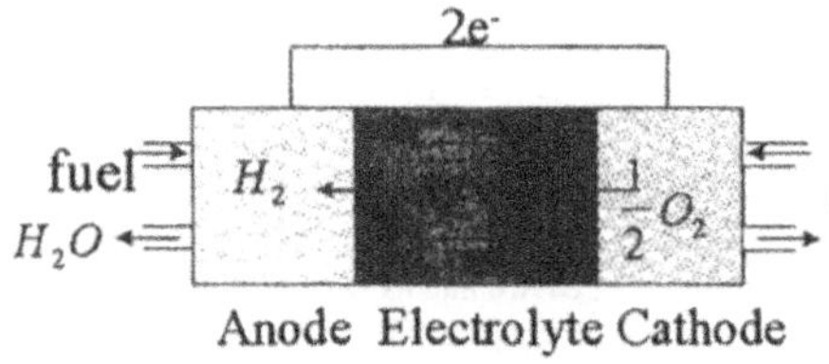

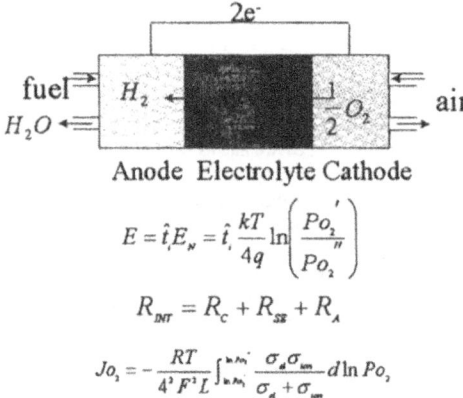

$$E = \hat{t}_i E_N = \hat{t}_i \frac{kT}{4q} \ln\left(\frac{Po_2'}{Po_2''}\right)$$

$$R_{INT} = R_C + R_{SE} + R_A$$

$$Jo_2 = -\frac{RT}{4^2 F^2 L} \int_{\ln Po_2''}^{\ln Po_2'} \frac{\sigma_{el}\sigma_{ion}}{\sigma_{el} + \sigma_{ion}} d\ln Po_2$$

Figure 1. Primary figures of merit for a solid oxide fuel cell. See text for details.

2.1 IONIC CONDUCTIVITY

The oxygen ion conductivity σ_i is given by the sum of the oxygen vacancy and interstitial partial conductivities. In all oxygen ion electrolytes of interest, the interstitial does not appear to make any significant contributions to the ionic conductivity, which can thus be described by

$$\sigma_i = \left[V_o^{\cdot\cdot}\right] 2q\mu_v \tag{3}$$

the product of the oxygen vacancy concentration $\left[V_o^{\cdot\cdot}\right]$ the charge, 2q, and the mobility (μ_v). Optimized levels of σ_i obviously require high charge carrier densities and mobilities.

Classically, high charge carrier densities have been induced in solids by substitution of lower valent cations for the host cations [2]. Implicit in the requirement of high carrier densities are:

1. High solid solubility of the lower valent substituent
2. Low association energies between the oxygen vacancy and dopant

3. No long-range ordering of defects.

As in metals, additives which induce minimal strain tend to exhibit higher levels of solubility. In ionic systems, in addition, the defects formed to maintain charge neutrality must be readily accommodated into the lattice without inducing major crystallographic distortions. The fluorite structure is the most well-known such structure with stabilized zirconia the best-known example. In this case, Y^{3+} substitutes for approximately 10% of Zr in $Zr_{1-x}Y_xO_{2-x/2}$ leading to $\sigma_i \sim 10^{-1}\ S/cm$ at 1000°C and an activation energy of ~1eV. Others include CeO_2 and fluorite related structures such as the pyrochlores $A_2B_2O_7$ [2].

Since the dopant and vacancy are of opposite charge (e.g. Ca_{Gd}' and $V_0^{\cdot\cdot}$) they will tend to associate. Given that cations are much less mobile than oxygen ions in oxide solid electrolytes, this serves to trap the charge carrier. It is of interest to examine how the concentration of "free" mobile carriers depends on the dopant concentration and the association energy. To simplify the analysis, we utilize a modified nomenclature. Consider the neutrality relation representing vacancy compensation of acceptor impurities by

$$N_V = \beta\ N_I \tag{4}$$

where N_V and N_I are the vacancy and impurity densities while β reflects the relative charges of the two species and normally takes on values of 1 (for $A_M^{''}$) and ½ (for $A_M^{'}$). The association reaction is given by

$$(I-V)^{x-y} \Leftrightarrow I^x + V^y \qquad (\beta = x/y) \tag{5}$$

where x and y are the relative charges of the impurity and vacancy respectively. The corresponding mass action relation is then

$$N_I N_V / N_{Dim} = K_A^{\circ} \exp(-\Delta H_A/kT) \tag{6}$$

in which N_{Dim} is the concentration of dimers and N_I and N_V are the corresponding defects remaining outside the complexes. It is straightforward to show that for weak dissociation (low temperatures or high association energies) one obtains the following solutions:

$$\beta = 1: \qquad N_V = (N_I K_A^{\circ})^{1/2} \exp(-\Delta H_A/2kT) \tag{7}$$

$$\beta < 1: \qquad N_V = \left(\frac{1-\beta}{\beta}\right) K_A^{\circ} \exp(-\Delta H_A/kT) \tag{8}$$

The solution for condition β=1 is the more familiar one. As in semiconductor physics [3], the number of free electrons or holes is proportional to the square root of the dopant density at reduced temperature and exhibit an arrhenius dependence with activation energy equal to one half of the association or ionization energy. The solution for condition β<1 is more unusual. Here one predicts that N_V is independent of dopant density! Second the activation energy is predicted to be equal to the association energy.

At sufficiently high temperatures or low association energies, essentially all the dimers are dissociated and

$$N_V = \beta N_I = \beta N_I(total) \tag{9}$$

Nowick and Park [4] have estimated that substantial association should remain in e.g. $Ce_{1-x}Ca_xO_{2-x}$ up to 1000°C for dopant levels as low as ≈2 mol % Ca for association energies ≥ 0.2eV.

$2kT$

Interestingly enough, one rarely experimentally observes the dopant dependencies of σ_i predicted by Eqs. 7 and 8. This is presumably due to the fact that the above equations assume the dilute solution approximation, which is never satisfied in solid electrolytes where high defect densities are the norm. If one examines the dependence of σ_i on dopant concentration in yttria stabilized zirconia, YSZ, one finds a decrease rather than an increase in σ_i for yttrium levels above the minimum necessary to stabilize the cubic phase. This drop in σ_i is attributed to long-range interactions between the defects.

Instead of the zero or linear dependence of σ_i on N_I predicted by Eq. 7 and 8 for $(Gd_{1-x}Ca_x)_2Ti_2O_7$, Fig. 2 shows a nearly exponential dependence on x [5]. This is clearly due to the sharp decrease in activation energy with increasing x as illustrated in Fig. 10 of Ref. 1 which is in disagreement with the dopant independent activation energies assumed in the derivations of Eqs. 7 and 8.

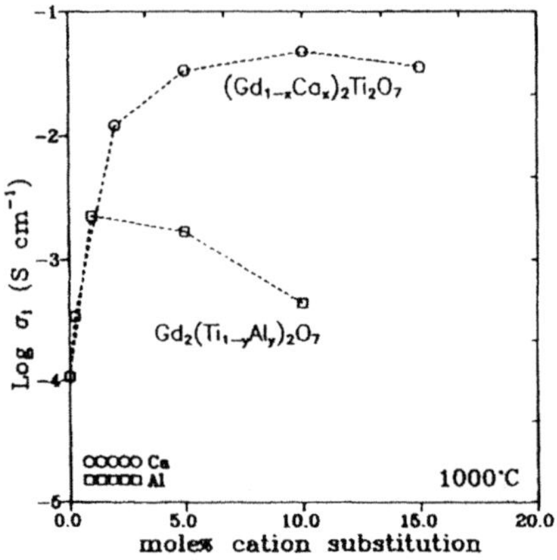

Figure 2. The ionic conductivity of $Gd_2Ti_2O_7$ at 1000°C as a function of acceptor doping on the "A" (Ca'_{Gd}) and "B" (Al'_{Ti}) cation sublattices [5].

An isolated defect complex exhibits a well-defined potential well for the mobile defect. An identical complex, a distance far enough from the first so that there are negligible interactions will be characterized by an identical potential well. However, as N_I and correspondingly N_V increase, the distances between complexes decrease and they begin to perturb each other. As the potential wells begin to overlap, the activation barrier for an ion to move from one site adjacent to a dopant to a near equivalent one can decrease dramatically. Vacancies need no longer fully dissociate from a dopant to move through the lattice but can, in the concentrated regime, percolate between sites adjacent to dopants. From both a practical and a fundamental viewpoint, it then becomes of interest to establish the relative roles of (1) dopant size, (2) dopant charge, (3) dopant concentration and (4) substitutional lattice site in influencing the effective migration characteristics of the ionic charge carriers.

Kilner [6] some years ago reviewed the effect of dopant size in determining the oxygen ion conductivity in fluorite structure oxides. Atomistic defect energy calculations on a number of systems all show a similar trend, i.e., a minimum binding energy for $r_{dopant}/r_{host} \approx 1$ and a steep increase in binding energy as the above ratio decreases below unity. This trend has also been observed experimentally in the CeO_2 system [7] and is reproduced in Fig. 3 together with calculated values [8]. Kilner [6] introduced a lattice parameter map, reproduced in Fig. 4 which compares the lattice parameters of the key fluorite oxides with the pseudo-cubic lattice parameters of the corresponding rare-earth sesquioxides. He suggests that CeO_2-based solid solutions exhibit the highest ionic conductivities given that the easiest matches occur with CeO_2 while the majority of the rare earth radii are either too large for ZrO_2 or too small for ThO_2.

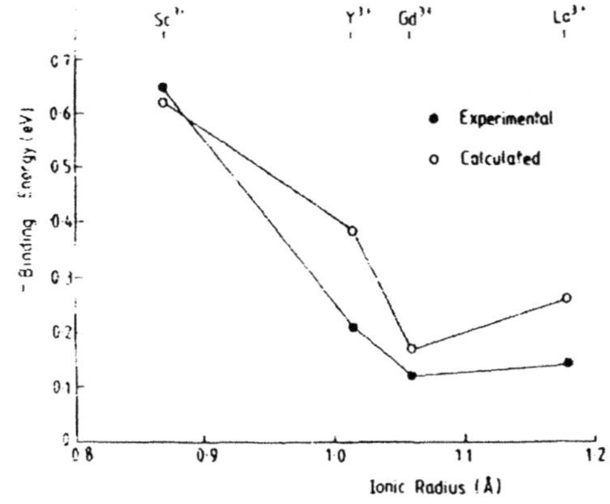

Figure 3. The experimental and calculated binding energy for $Ce(Mf)O_{2-x}$ solid solutions from Refs. 7 and 8 [6].

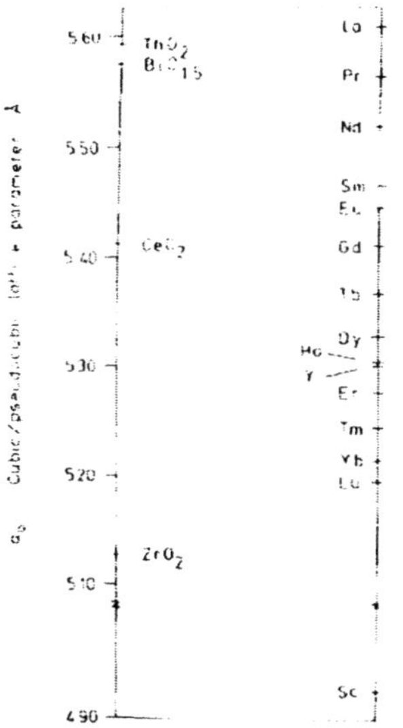

Figure 4. A schematic of the lattice parameter maps for all the fluorite oxides [6].

More recently Chen and co-workers [9] have applied x-ray absorption fine structures (EXAFS) and x-ray near edge structure (XANES) spectroscopy to probe the local structures around Zr and around various additives Fe, Ga, Y, Gd in zirconia solid solutions. They find that the ions Y^{3+} and Gd^{3+} with radii greater than that of Zr, favor 8-fold oxygen coordination leaving the oxygen vacancies preferentially next to the Zr ions. This was previously predicted by computer simulations. On the other hand, undersized ions such as Cr^{3+} and Fe^{3-} favor 6-fold coordination thereby competing with Zr for oxygen vacancies. The latter response, which includes a very large distortion around the undersized dopants is consistent with the much larger association energies generally observed for $r_{dopant}/r_{host} < 1$.

We have recently experimentally reconfirmed this dependence of σ_{ion} on dopant radius by examining the dependence of σ_{ion} of $(Gd_{0.98}A_{0.02})_2Ti_2O_7$ on the r_A/r_{Gd} ratio [10]. The results shown in Fig. 5 clearly show a strong maximum in σ_{ion} for values of $r_A/r_{Gd} \approx 1$.

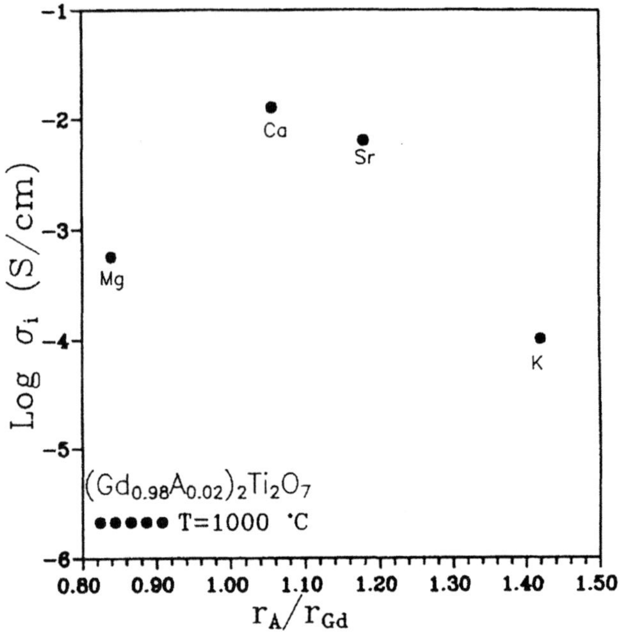

R_A/r_{Gd}

Figure 5. The dependence of the ionic conductivity at 1000°C of acceptor doped $Gd_2Ti_2O_7$ on the acceptor-Gd radius ratio. Note the peak in vicinity of unity radius ratio [10].

Based on simple electrostatic considerations, one would expect association energies to be smallest for dopants with the smallest relative charge. Thus Ca_{Zr}'' or Ca_{Ce}'' should give lower conductivities than Y_{Zr}' or Y_{Ce}'. This is indeed observed. The use of Ca stabilized zirconia is largely based on cost considerations. The results of Fig. 5 are also consistent with this observation, i.e. K_{Gd}'' gives much lower conductivities than Ca_{Gd}' or Sr_{Gd}'. It should be noted, however, that r_K/r_{Gd} is also rather large so it becomes difficult to distinguish effects of size and charge in this case.

In binary compounds such as zirconia or ceria, dopants have only one type of site onto which they can substitute. In ternary or higher order systems such as the pyrochlores, several types of sites may accommodate dopants. For example, in addition to substituting Ca^{2+} on the Gd^{3+} site in $Gd_2Ti_2O_7$ (Ca_{Gd}'), we have also examined the effect on σ_{ion} of the substitution of Al^{3+} on the Ti^{4+} site (Al_{Ti}') [5]. The results are illustrated in Fig. 2. For dopant levels up to 2 fraction %, the results are similar. Above that value, σ_{ion} continues to increase substantially with Ca doping but decreases with Al doping. X-ray analysis shows that the solubility limit is exceeded at much lower levels of Al (~2%) as compared to Ca (~7%). Further studies along these lines are warranted, given the observations of Chen and coworkers (9).

We complete this section by reviewing radius ratio effects in the host lattice. We choose as an example acceptor doped perovskite aluminates with composition $RAl_{0.95}Mg_{0.05}O_3$ with R=La, Sm, Gd, Y [11]. These systems show a pattern of distortion from cubic → rhombohedral → orthorhombic as r_R decreases and/or temperature decreases. Specifically we find the following:

La (rhombohedral → cubic @ 450°C)
Sm (orthorhombic → rhombohedral @ 800°C)
Gd,Y (orthorhombic, all T)

If we compare the conductivities in Fig. 6, we find that the higher the symmetry, the higher the conductivity. This is as expected given that the higher symmetry structures provide more energetically equivalent sites for the mobile ions to move through. Distortions from the cubic tend to trap, e.g. vacancies in deeper potential wells. This leads us next into a discussion of order-disorder phenomena in oxides and their import to ionic conduction.

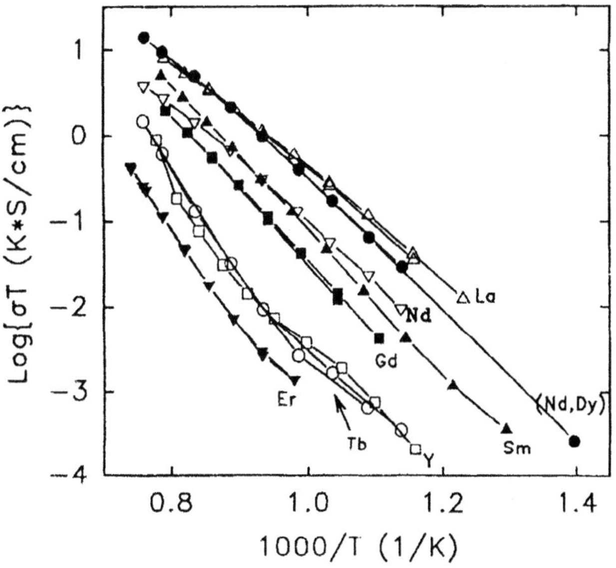

Figure 6. Temperature dependence of the ionic conductivities of $R[Al_{0.95}Mg_{0.05}]O_{3-\delta}$ derived from lattice plus grain boundary resistances. Various A-cations are indicated [11].

2.2 INTRINSIC IONIC CONDUCTORS

The other avenue to high defect densities is via the identification and selection of intrinsically disordered phases. The earliest example of this type of oxygen electrolyte is Bi_2O_3-based, first identified by Takahashi and Iwahara [12] which goes through an order-disorder transformation at 730°C, reaching values of $\square_i$ above 1 S/cm. More recent examples include the highly oxygen deficient perovskite-related phases first investigated by Goodenough and co-workers [13] and the perovskite-Bi_2O_3 intergrowth phases studied by zur Loye and co-workers [14]. Results are summarized in Fig. 7. Since the formation of defects in intrinsically disordered compounds does not require the incorporation of dopants of opposite charge, defect-dopant association, which limits ionic conduction in doped systems at reduced temperatures, is eliminated. Further studies, however, are required to establish the long term stability of the disordered phases and their sensitivity to thermal history.

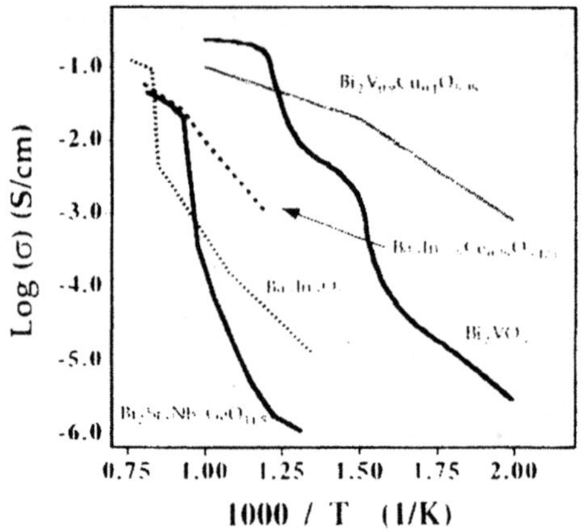

Figure 7. Representative data for a number of oxygen ion conductors with the brownmillerite-perovskite intergrowth structures $(AB'O_{2.5})_x(AB''O_3)_y$ and aurivillius phases $(Bi_2A_{n-1}B_nO_{3n+3}, n=1-5)$ [14].

Our own work on such intrinsically disordered oxides has focused on the pyrochlores particularly since we were able, by going to solid solution systems, to control the degree of disorder nearly continuously from low to high values. By replacing half the Zr^{4+} in ZrO_2 with Gd^{3+}, one out of eight oxygen must be removed to maintain charge neutrality resulting in the chemical formula $Gd_2Zr_2O_7$. At temperatures above ≈ 1550°C, the material is highly defective and remains fluorite. Below this temperature, ordering on both cation and anion lattices occurs leading to the pyrochlore structure.

The pyrochlore structure is a superstructure of the defect fluorite structure with exactly twice the lattice parameter. It has a general molecular formula of $A_2B_2O_6X$. In our compositions, X is also an oxygen ion but with a different site symmetry than the remaining 6 formula oxygens. A diagram of the pyrochlore structure, projected onto the (100) plane, is shown in Fig. 8. The larger A cations occupy the 16c sites and the smaller B cations occupy the 16d sites, forming parallel strings of atoms along the <110> directions. The six oxygens per formula unit occupy the 48f sites while the additional oxygen is found on the 8a site. As is evident from the diagram, the 48f oxygens are displaced from their fluorite positions towards the neighboring empty 8b sites. While oxygen vacancies occur at random throughout the anion sublattice in ideal defect fluorite (e.g. YSZ), they are ordered onto particular sites (8b) in the pyrochlore structure. Thus, one properly views these as empty interstitial oxygen sites rather than oxygen vacancies. As a consequence, ideal pyrochlore oxides are ionic insulators.

Two types of disorder are important in pyrochlores. The first is anti-site disorder on the cation sublattices in which A and B cations switch positions. In the limit of complete randomization of the A and B cations on the 16c and 16d sites, we revert back to the situation in the fluorite case._The second type of disorder, which is relevant to our analysis of ionic transport, concerns a Frenkel-like disorder on the anion sublattice. Here, oxygen ions leave the normally occupied 48f sites and enter the 8b interstitial sites, thereby forming oxygen vacancy-interstitial pairs.

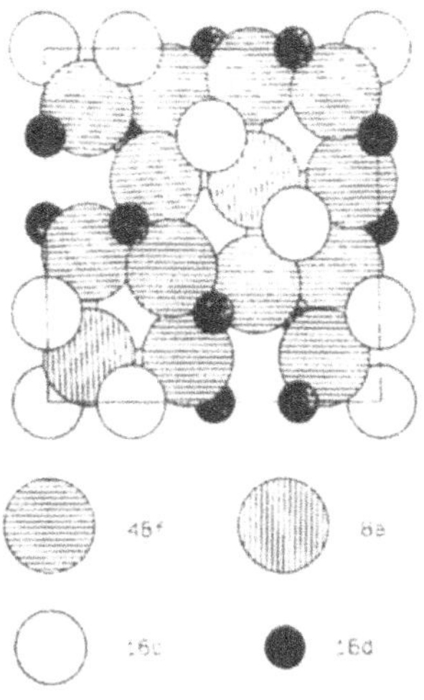

Figure 8. (100) Projection of a portion of one unit cell of the pyrochlore structure. Ionic radii are based on $Gd_2Ti_2O_7$. The empty space in the lattice represents the vacant 8b site [15].

Since neither the 8a nor 8b sites are interconnected, we expect oxygen transport to occur primarily via vacancy motion within the 48f sublattice. This assumption was confirmed by Moon and Tuller [16] in doping experiments performed on $Gd_2(Zr_{0.3}Ti_{0.7})_2O_7$.

A number of earlier studies suggested that there might be a correlation between structural disorder in the pyrochlores and the radius ratio of the A and B cations, r_A/r_B. Intuitively, one might suspect that as the radius ratio approaches unity, anti-site disorder would increase. Further, as the cation environments o the oxygen ions become more homogeneous, 48f-8b exchange would also become more favorable. Moon and Tuller [17] were able to test this hypothesis by studying compositions in the system $Gd_2(Zr_xTi_{1-x})_2O_7$ in which the r_A/r_B ratio was varied from approximately 1.74 to 1.47 as x was increased from 0 to 1. Fig. 9 illustrates the large increase in the ionic conductivity, 4.5 orders of magnitude at 600C as Zr is systematically substituted for Ti. Similar results were obtained in the $Y_2(Zr_xTi_{1-x})_2O_7$ system [17].

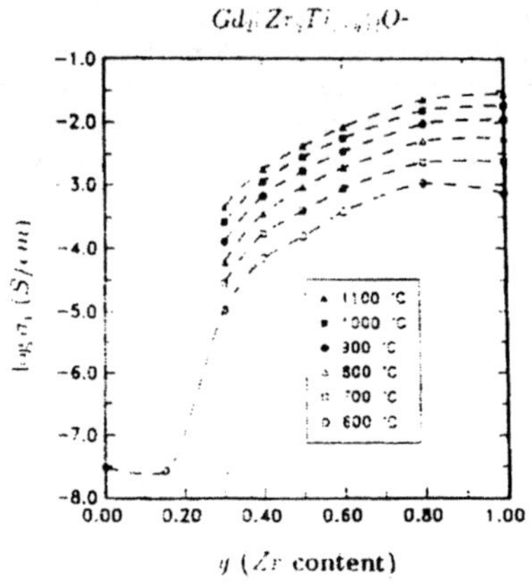

Figure 9. Ionic conductivity $\log(\sigma_i)$ of $Gd_2(Zr_yTi_{1-y})_2O_7$ as a function of y [17].

By examining the dependence of the pre-exponential and exponential terms of $\sigma_i = (\sigma_o/T)\exp(-E_i/kT)$ as a function of x, one finds the following. First, E_i remains relatively insensitive to x. Rather, it is largely increases in σ_0 rather than decreases in E_{ionic} outside of the extrinsic-controlled regime at small x that contributes to the increases in σ_{ionic} for x values above 0.25 as illustrated in Fig. 10.

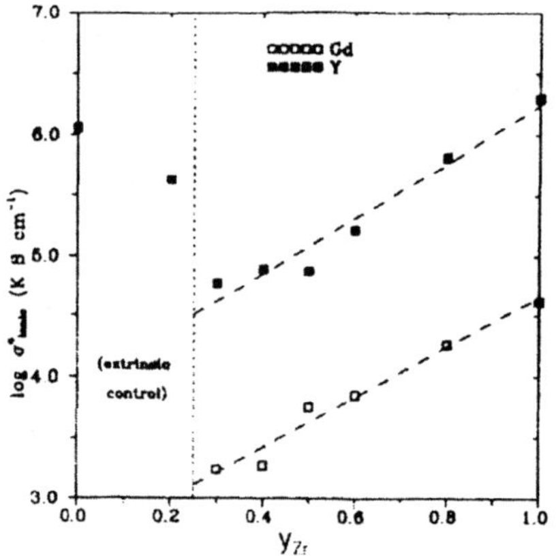

Figure 10. Pre-exponential term for ionic conductivity as a function of y for $Gd_2(Zr_yTi_{1-y})_2O_7$ and $Y_2(Zr_yTi_{1-y})_2O_7$. Dashed lines do not indicate parametric fitting [16, 17].

The related $Y_2(Zr_xTi_{1-x})_2O_7$ system was next studied in order to examine the role of the ratio of the effective ionic radii in the "A" and "B" sites, on structural disorder and transport. As above, the smaller the r_A/r_B ratio, the larger the A-B antisite disorder and correspondingly the larger the oxygen Frenkel-like disorder. Comparisons of the magnitudes of σ_o for YZT and GZT, for the same values of x in Fig. 10, appear to confirm this hypothesis, i.e. ~30

σ_e (GZT). However, because E_{ionic} is 0.3-0.5 eV larger than E_{ionic} (GZT), σ_{ionic} (YZT) remains lower in magnitude below ~1000C. The higher activation energy is presumably de to YZT's smaller lattice parameter.

Neutron diffraction studies on YZT with x=0.3, 0.45, 0.60 and 0.90 by Heremans and Wuensch [18] confirm the systematic disordering of the oxygen sublattices with the oxygen vacancy fraction on the 48f site increasing from 0.006 (x=0.3) to 0.043 (x=0.45) and finally to 0.078 (0.6). At x=0.9 all three oxygen sites (48f, 8a, and 8b) are 1/8 empty, which is consistent with the structure having reverted back to defect fluorite. Surprisingly, the same authors detect significant A-B anti-site disorder only above x=0.45.

While the level of disorder and ultimately the density of ionic charge carriers is critically important in determining the level of ionic conductivity that can be obtained, the mobility of the carriers must also be optimized. We have found that pyrochlores with similar levels of disorder can differ substantially in their ionic conductivities. As an example, we illustrate, in Fig. 11, the effects of substituting Sn^{4+} for either Ti^{4+} or Zr^{4+} [19]. The radius of Sn is nearly that of Zr and thus would be expected to initiate disorder in $Gd_2(Ti_{1-x}Sn_x)_2O_7$ (GTS) with increasing x while it should have little effect on $Gd_2(Zr_{1-x}Sn_x)_2O_7$ (GZS) which should remain largely disordered. Instead we find a sharp drop by ≈ 3 orders of magnitude in GZS as x increases towards unity. This can only be explained on the basis of a drop in mobility. We attribute this to the more highly covalent nature of the Sn-O as compared to the Zr-O and Ti-O bonds. This is consistent with our measured drop in relative dielectric constant in GTS from ≈ 80 at x=0.2 to ≈30 at x=10 [20]. The initial increase in GTS with x is consistent with increasing disorder given that $r_{Sn} > r_{Ti}$ while the subsequent drop for x>0.4 is due, we believe, to the considerably lower ion mobility in the stannate.

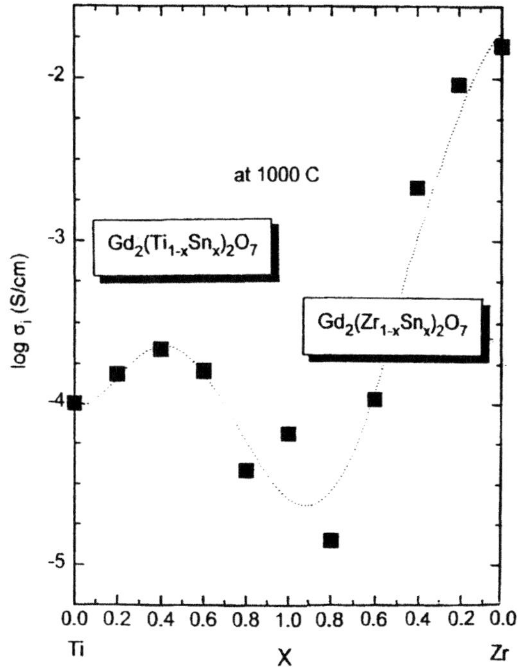

Figure 11. The ionic conductivity of $Gd_2(Ti_{1-x}Sn_x)_2O_7$ and $Gd_2(Zr_{1-x}Sn_x)_2O_7$ systems at 1000°C [19].

2.3 ELECTRONIC TRANSFERENCE NUMBER

The electronic conductivity σ_e which must be minimized, is given by the sum of electron and hole partial conductivities, viz.

$$\sigma_e = ne\mu_e + pe\mu_h \qquad (10)$$

Several key factors control the electron/hole concentrations. The intrinsic levels, which represents the minimum possible carrier densities, are determined largely by the band gap, E_g. For our purposes, the larger E_g, the better. The author has derived expressions for the minimum E_g values necessary to maintain t_e below a certain level, say 10^{-2}, for given levels of ionic conductivity [21]. Fig. 12 shows the results of such an analysis applied to acceptor doped CeO_2. Note that this necessary criteria is easily achieved in this material.

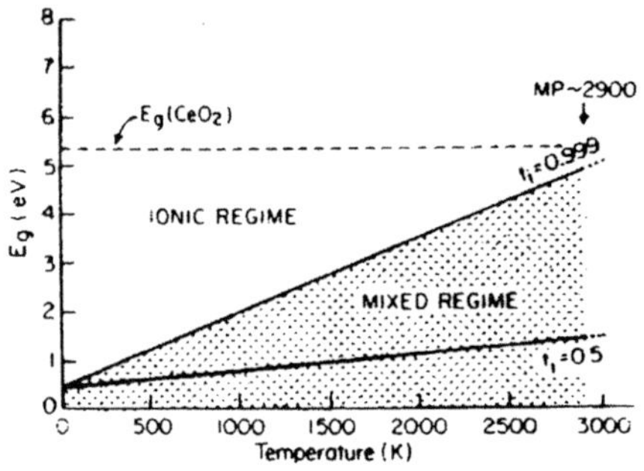

Figure 12 The E_g versus T plot for CeO_2+5mol%Y_2O_3 shows that it remains an electrolyte at all temperatures in the vicinity of the electronic conductivity minimum [21].

Since oxides, like other compounds, deviate from stoichiometry, the electron or hole densities can easily reach levels many orders of magnitude greater than the intrinsic levels given by

$$n_i = p_i = (N_c N_v)^{1/2} \exp(-E_g/2kT) \qquad (11)$$

where N_c and N_v are the conduction and valence band density of states functions. For example the reduction of Ce^{4+} to Ce^{3+} via the reaction

$$2Ce_{Ce}^{x} + O_o \rightarrow V_o^{\bullet\bullet} + 2Ce_{Ce}' + 1/2O_2 \qquad (12)$$

is reflected as a transfer of electrons from the oxygen vacancy donor level to the Ce-4f conduction band. Thus, one may write equivalently

$$n \equiv \left[Ce_{Ce}'\right] \qquad (13)$$

Correspondingly, the Fermi level can be lowered below mid-gap in a given oxide by the formation of acceptor states due to oxidation.

Such oxidation-reduction reactions are minimized in oxides for which the cations remain stable in a given oxidation state e.g., Zr^{4+}. Conversely, transition and some rare earth elements tend to exhibit multiple valence states, which readily lead to nonstoichiometry and the corresponding generation of electronic carriers. We return to such systems in our discussion of MIECs.

The suitability of particular pyrochlore compositions depends, obviously, on the particular applications for which they are being considered. Figures of merit for an electrolyte or an electrode might include the magnitudes of ionic and electronic conductivity, σ_i and σ_e, the ionic transference numbers, t_i and t_e, the leakage or permeation current, j_i, the positions of the electrolytic domain boundaries under oxidizing and reducing conditions, P_p and P_n and their ratio, the domain width, P_p/P_n [22]. As discussed above, for an electrolyte one wishes σ_i and t_i to be as large as possible over a broad electrolytic domain. These factors insure a high EMF, low resistive losses and low permeation currents under open circuit conditions. Ideal electrodes, on the other hand, require high electronic conductivities together with high ionic conductivities to minimize both resistive and polarization losses and to insure low R_A and R_C values. This implies narrow electrolytic domains and high permeation currents.

We have already derived expressions in [1] for $[V_o^{\bullet\bullet}]$ and n under conditions of acceptor doping (Eqs. 43 and 46). Multi0plying each of these carrier densities by charge and mobility gives us the respective conductivities. By setting, $\sigma_i = \sigma_n$, and solving for the Po_2 one obtains an expression for P_n. The lower bound to the electrolytic domain under extrinsic (unassociated) dopant control is thus given by

$$P_n = 4K_R^{2}[A]^{6}(\mu_e/\mu_i)^{4} \tag{14}$$

where μ_i and μ_e are the ion and electron mobilities. The upper bound P_p is found in a similar fashion by setting $\sigma_i = \sigma_p$ and solving for Po_2. Table 1 summarizes the expressions for the figures of merit σ_i, P_n, P_p, and P_p/P_n under the limiting cases of (a) intrinsic Frenkel disorder (b) unassociated acceptor doping and (c) associated acceptor doping [23].

Fig. 13 illustrates the dependence of P_n on temperature for compositions in the YZT system. Typically, the electrolytic region shrinks with increasing temperature since E_n is normally larger than E_i (see Eq. 38 in [1]). One also sees the marked effect of x on extending the electrolytic region, e.g. ~20 orders of magnitude at 800°C as x increases from 0 to 0.6. This decrease is qualitatively consistent with the predictions of Table 1 in that $P_n \propto K_F^{-3}$ which implies that P_n decreases as the intrinsic disorder increases.

Fig. 14 shows the effects of dopant concentration on the domain boundary position as log P_n vs log $[Ca'_{Gd}]$ in $Gd_2Ti_2O_7$ in which Ca varies over nearly two orders of magnitude. The slope of −6.4 is close to the predicted value of −6 (see Table 1).

Examining the factors which optimize the figures of merit, it is instructive to begin with σ_i. Optimum values require high oxygen vacancy densities coupled with high oxygen mobilities (see Table 1). For intrinsic disorder, K_F, should be as large as possible, while for extrinsic disorder the acceptor density $[A']$ needs to be high. Since association lowers the fraction of mobile ions, systems with low association energies should be selected. We find this is usually achieved in the pyrochlores, as in the fluorite oxides, by selecting dopants with low effective charge and good size match with the host ion for which they substitute, as discussed above.

TABLE 1: Expressions for electrolyte figures of merit, σ_i, P_n, P_p, and P_p/P_n, under the special limiting cases of (a) intrinsic Frenkel disorder, (b) unassociated acceptor doping, and (c) associated acceptor doping [23]

	σ_i	P_n	P_p	P_p/P_n
Case #1 Intrinsic	$2e(K_F)^{1/2}\mu_i$	$\frac{K_R^2}{16K_F^3}(\frac{\mu_e}{\mu_i})^4$	$\frac{16K_R^2K_F}{K_i^4}(\frac{\mu_i}{\mu_h})^4$	$\frac{256K_F^4}{K_i^4}(\frac{\mu_i^2}{\mu_e\mu_h})^4$
Case #2 Dopant (unassociated)	$2e(\frac{[A']}{2})\mu_i$	$\frac{4K_R^2}{[A']^6}(\frac{\mu_e}{\mu_i})^4$	$\frac{4K_R^2[A']^2}{K_i^4}(\frac{\mu_i}{\mu_h})^4$	$\frac{[A']^8}{K_i^4}(\frac{\mu_i^2}{\mu_e\mu_h})^4$
Case #3 Associated (concentration independent)	$2e(K_A)\mu_i$	$\frac{K_R^2}{16K_A^6}(\frac{\mu_e}{\mu_i})^4$	$\frac{16K_R^2K_A^2}{K_i^4}(\frac{\mu_i}{\mu_h})^4$	$\frac{256K_A^8}{K_i^4}(\frac{\mu_i^2}{\mu_e\mu_h})^4$

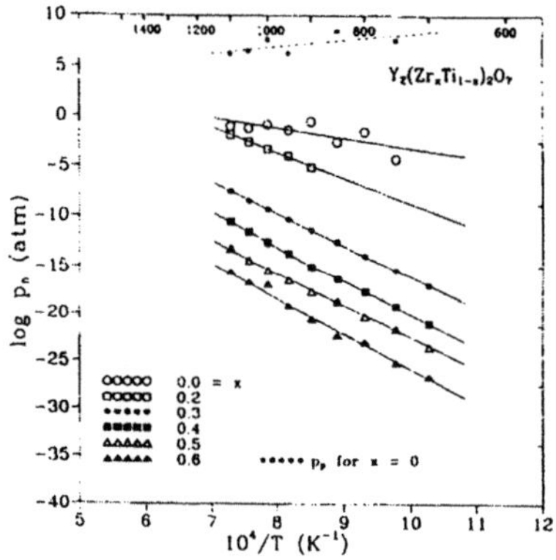

Figure 13 P_n as a function of temperature or $Y_2(Zr_xTi_{1-x})_2O_7$ for a number of values of x. P_p is also included for comparison for x=0 [23].

The ionic mobility of large ions such as oxygen are influenced more by strain than by electrostatic considerations. One tends to select compositions with large lattice parameters, all else being equal. We find GZT compositions offer higher ionic conductivities and lower activation energies than YZT compositions largely because of the higher mobilities. The generality of this observation, however, requires further confirmation.

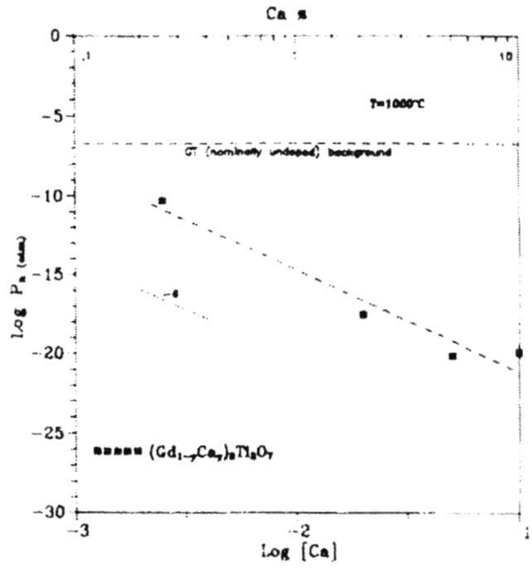

Figure 14 P_n, as a function of $\left[Ca'_{Gd}\right]$ for $\left(Gd_{1-y}Ca_y\right)_2Ti_2O_7$ at 1000C. The horizontal dashed line represents the value measured in nominally undoped GT, due to background impurities. The dashed line with a slope of roughly –6.4 is a linear interpolation of the data while the solid line with a slope of –6 represents te model prediction [23].

The figure of merit, P_n, which we wish to minimize, introduces two additional parameters K_R and μ_e. The same parameters which optimize σ_e are also found in the denominator. We have observed that increasing x in GZT and YZT results in a sharp decreases in σ_e [24]. This results, we believe, from a decrease in both K_R and μ_e induced by the dilution of Ti with Zr which removes the readily reducible species as well as narrowing the Ti derived 3d-like conduction band. Similar arguments could be made for the dilution of Ti by the addition of acceptor dopants such as Al'_{Ti}.

Lastly, we examine the figure of merit P_p/P_n, a measure of the domain width, which is maximized under conditions for which the ion/electron-hole carrier density and mobility ratios are optimized. Notice that, while K_R does not influence the domain width, it does determine its position in Po_2 space. In fuel cell or sensor applications, we wish P_p and P_n to include, within their bounds, the most oxidizing and reducing gases likely to be utilized.

III. Electrodes

The electrodes serve two primary functions in an electrochemical cell such as the SOFC. First, they provide electrons to, or collect electrons from the external circuit (see Fig. 1). Thus they must exhibit a sufficiently high electronic conductivity to minimize ohmic losses in the device. Secondly, they provide reaction sites for the electrochemical reduction of oxygen at the cathode and the corresponding oxidation of the fuel at the anode.

In traditional designs, the active sites at which the reduction or oxidation reactions occur are limited to the "three phase interfaces" at which the electrolyte and electrode both simultaneously intersect the gas phase (see Fig. 15). This follows from the fact that the electrodes solely supply electrons, the electrolyte, ions and the gas phase, gas molecules. This requires that the electrodes be porous and the electrode/electrolyte interface morphology be optimized to provide the maximum possible triple contact line [25].

One obvious approach to broaden the active region is to introduce ionic conduction into the electrode, i.e., rendering it an MIEC solid. In this manner, the reaction sites do not need to be adjacent to the solid electrolyte, since ions can reach any location at the electrode/gas interface by ionic conduction (see Fig. 15) A classic example of how an MIEC can enhance the catalytic activity of the cathode in a zirconia-based SOFC is the performance comparison between $(RE_{1-x}Sr_x)MnO_3$ and $(RE_{1-x}Sr_x)CoO_3$, two closely related perovskite oxides, with RE= rare earth element. Cobaltites are reported to give much lower overpotentials than manganites, e.g., the cathodic overpotential for $100mA/cm^2$ in $Gd_{0.7}Sr_{0.3}CoO_3$ is ~90mV at 800^0C, while it is ~170mV for the corresponding manganate even at 1000°C [26]. Similar observations had earlier been made by Yamamoto et al [27] for the lanthanum systems.

This difference in performance is believed to be largely due to the much larger oxygen ion conductivity in the cobaltite versus the manganite. Steele [28] has pointed out that since Co reduces more readily than Mn, it already exhibits a good deal of oxygen deficiency at Po_2=1atm while the manganate does not (see Fig. 16). Oxygen diffusion measurements confirm this large difference in oxygen ion conductivity [28].

While the cobaltite electrode exhibits considerably better polarization performance, the manganite is nevertheless the material of choice in SOFC devices operating at $900\text{-}1000^0C$. This underlines the importance of "secondary properties". In this case, the difficulty that arises with the cobaltite is its high reactivity with YSZ to form the poorly conducting product phase $La_2Zr_2O_7$. Yamamoto et al [27], for example, found the overpotential at 800^0C increased from 10's of mV initially to hundreds of mV after 5h operation at $100mA/cm^2$. $LaMnO_3$ based electrodes, on the other hand, appear to be relatively stable at these temperatures.

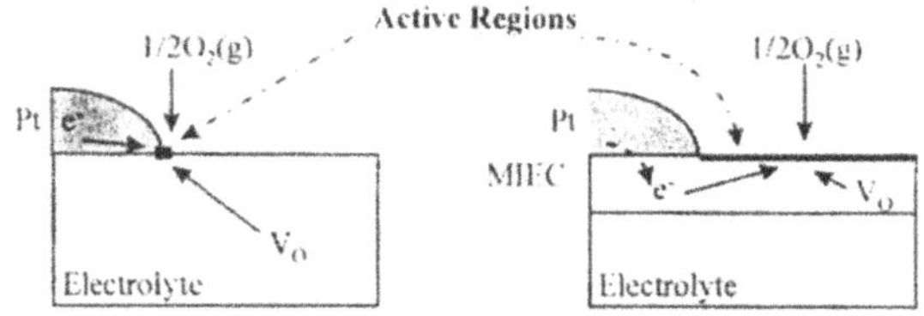

Figure 15. Schematic illustrating the ability of MIEC to extend the active area of the electrode-gas interface [25].

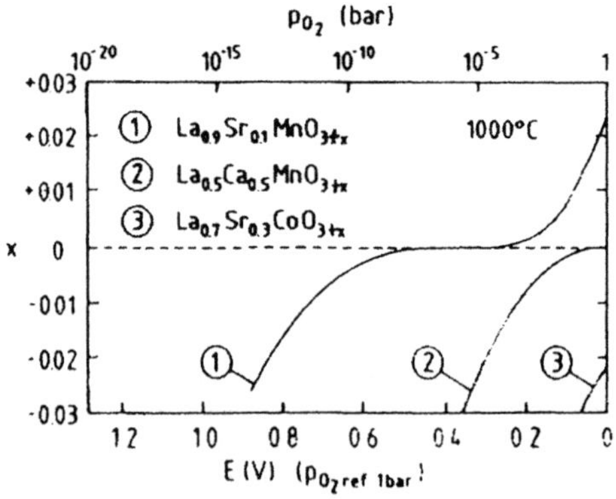

Figure 16 Composition-oxygen partial pressure isotherms for a number of perovskites. Note that the cobaltite is already oxygen deficient at 1atm while the manganates are not [28].

As this last example demonstrates, one cannot consider the properties of the solid electrolyte and the electrode materials in an isolated fashion. Indeed, the solid electrolyte, anode, cathode, and interconnect must remain in intimate contact for extended periods of time (10's of years for SOFC) at elevated temperatures. Thus a more holistic approach to selecting and optimizing materials is desired.

IV. Materials Compatibility

Aside from the "primary" requirements of electrical and catalytic properties, a number of key "secondary" requirements must be satisfied. These include chemical, morphological, and dimensional stability, compatibility with neighboring materials, adequate strength and fabricability at reasonable cost.

4.1 CHEMICAL STABILITY

The electrolyte and the electrodes must remain phase stable over the operating conditions to which they are exposed. Bismuth oxide-based electrolytes, while exhibiting some of the highest reported oxygen ion conductivities, are reported to decompose under the anodic conditions of the fuel cell. The two phase cermet Ni/YSZ used for the anode, which remains stable under reducing conditions would be useless as the cathode since the Ni would oxidize to form insulating NiO.

4.2 MORPHOLOGICAL STABILITY

As discussed above, electrodes which are not mixed conductors, require a certain percentage porosity and a given pore-size distribution for optimum operation. The sintering and densification of such structures with time is thus detrimental to performance.

4.3 DIMENSIONAL STABILITY

Dimensional changes due to temperature or Po_2 excursions can induce stresses of sufficient magnitude to result in fracture of the ceramic component. Mogensen et al [29] has pointed to potential difficulties inherent in the dilation that ceria-based electrolytes are susceptible to due to oxygen deficiency induced under anodic (reducing) conditions. Since the electrolyte also sees high Po_2 at the cathode, this leads to stresses in the bulk of the material.

4.4 COMPATIBILITY

Chemical incompatibilities result in reaction products, as between YSZ and $LaSrCoO_3$. These can lead to current blocking due to formation of insulating intermediates, e.g., $La_2Zr_2O_7$, or fracture or scaling when the products have poor lattice matching to the host materials. Thermal expansion mismatches will also lead to delamination during heating and cooling cycles, even if lattice matching exists at operating temperatures.

V. Monolithic Structures

The author and his co-workers have proposed a monolithic structure for solid state electrochemical cells which resolves many of these compatibility difficulties [30, 31]. The principle is based on the selection of a single crystalline structure/phase which serves as a template for the cell. The composition is then spatially modulated to achieve the desired functionality for that part of the cell. For example, the middle section of the material serves as the solid electrolyte exhibiting high ionic and low electronic conductivities and remains stable at both high and low Po_2. On the other hand, the region at one end, exposed to oxidizing environments, is modified to exhibit both high electronic and ionic conductivities. Likewise, the material at the other end is controlled to exhibit similar properties but under reducing conditions. In this manner a single phase, with only minor spatial variations in lattice parameter and thermal expansion coefficient, can be expected to exhibit long term stability over wide operating conditions, including during heating and cooling cycles. A schematic of such an arrangement is shown in Fig. 17.

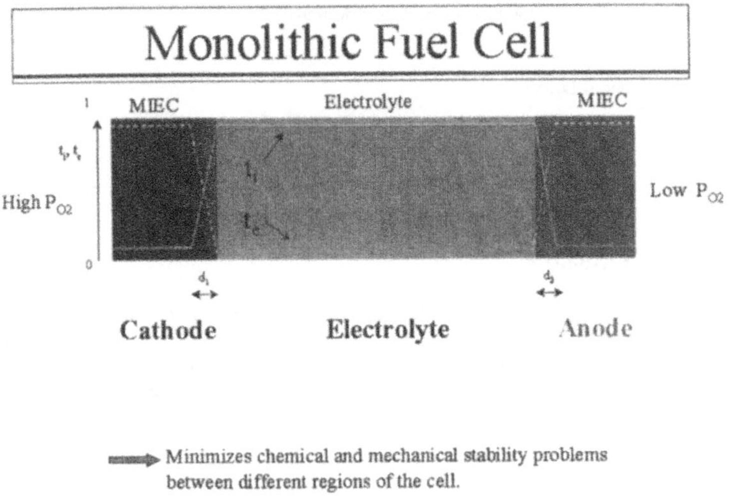

Figure 17. An illustration of a monolithic SOFC structure. The spatial dependence of the ionic and electronic transference numbers depends on the corresponding compositional profiling. The dimensions d_1 and d_2 correspond to possible graded interfaces. Full line:t_i, broken:t_e [25].

5.1 MATERIALS DESIGN

In order to satisfy the above criteria for a monolithic structure, one requires materials systems of the necessary versatility to achieve either high ionic/low electronic conductivity or high ionic/high electronic conductivity without major changes in composition.

Attempts have, over the years, been made to increase the electronic conductivity in fluorite-based solid electrolytes [32]. However, due to a combination of limited solubility of transition metal ions and narrow conduction bands, the levels of electronic conductivity achieved have remained low, typically $\leq 10^{-2}$S/cm in YSZ.

We have focused on the related ternary pyrochlore system, with general formula $A_2B_2O_7$, e.g., $Gd_2Zr_2O_7$. We have demonstrated above that it is possible to achieve high oxygen ion conductivities ($\geq 10^{-2}$S/cm at 1000°C) in two ways. First, by selecting intrinsically disordered pyrochlores, e.g., in the solid solution system $Gd_2(Zr_xTi_{1-x})_2O_7$ with $x > 0.3$. Second, by acceptor doping highly ordered pyrochlores such as the above system in which $x \leq 0.2$.

Since 100% of the fixed valent B-site ion, Zr^{4+}, can be replaced by the variable valent $Ti^{3,4+}$ transition metal ion, reasonably high levels of n-type conductivity can be achieved under reducing conditions [15]. More recently, we have reached metallic levels of electronic conductivity (~100 S/m) in titanate/molybdate pyrochlore solid solutions while apparently retaining high levels of ionic conductivity [33]. This, we believe, is due to the formation of a Mo-conduction band which broadens as the fraction of Mo is increased.

Unfortunately, due to the tendency for Mo to oxidize from the 4+ towards the 6+ oxidation state, the $Gd_2(Ti_{1-x}Mo_x)_2O_7$ pyrochlore decomposes, e.g. above $Po_2=10^{-13}$ atm at 1000°C. In an attempt to stabilize this highly conductive system to higher Po_2, ideally to 1 atm, we investigated the $Gd_2((Mo_{1-y}Mn_y)_xTi_{1-x})_2O_7$ or GMMT system [34]. The approach taken here was to allow the Mo to take the 6+ oxidation state and maintain the pyrochlore oxygen stoichiometry by adding a compensating ion with an oxidation state below 4+. Assuming Mn takes on a 3+ oxidation state then the composition $Gd_2\left(Mo^{\bullet\bullet}_{1/3}Mn'_{2/3}\right)_2O_7$ should be stable and oxygen stoichiometric. This compensation mechanism can also be visualized via the energy band diagram of Fig. 18, in which Mo^{6+} is observed to be compensated either by oxygen interstitials or, when they are present, by lower lying Mn states.

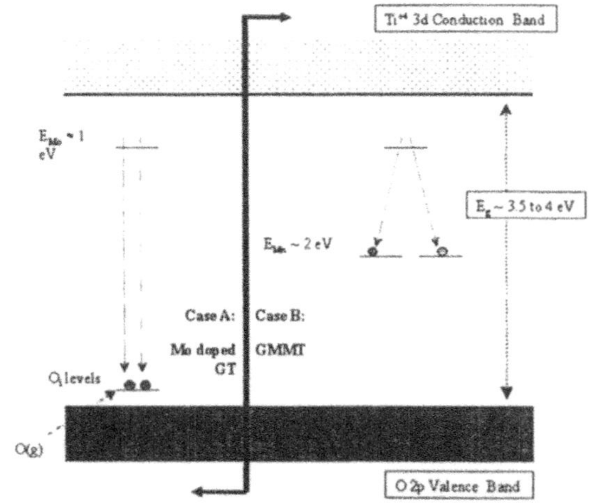

Figure 18. Energy band diagram of $Gd_2(Ti_{1-x}Mo_x)_2O_7$ in which either low lying oxygen interstitial or Mn "acceptor" levels serve to compensate the Mo "donor" states [34].

A series of GMMT specimens with $0 \le x \le 0.3$ were prepared and found to be stable under cathodic conditions. The electrical conductivity of GMMT measured in air at 900°C is plotted as a function of x in Fig. 19. The substitution of $Gd_2\left(Mo_{1/3}Mn_{2/3}\right)_2O_7$ for $Gd_2Ti_2O_7$ is observed to increase the conductivity by over four orders of magnitude.

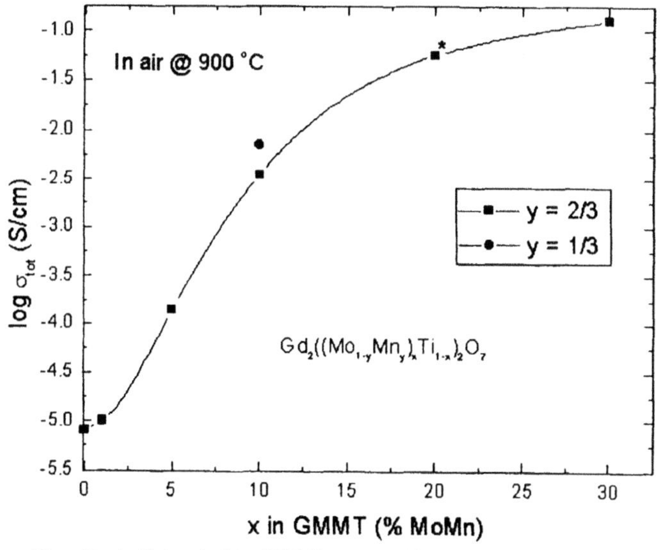

Figure 19. The electrical conductivity of GMMT measured at 900°C in air is plotted as a function of x [34].

Examination of the Po_2-dependence of the conductivity confirms that the electrical conductivity is n-type and that much of the increase with x is due to a decrease in activation energy. This is consistent with Mo and Mn levels forming at low x which broaden into bands with increasing x as illustrated in Fig. 20.

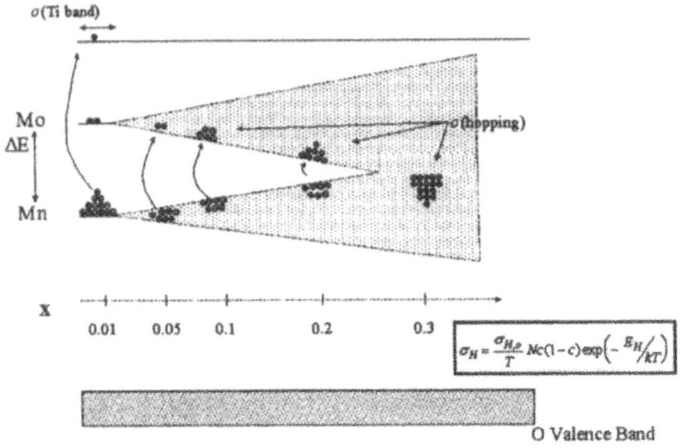

Figure 20. Defect band formation in GMMT [34].

Electron blocking experiments allow one to extract the minority ionic conductivity from a predominantly electronic conductor. Application of this method to GMMT with x=0.1 confirms these materials to support reasonably high levels of ionic conductivity, i.e. $10^{-3} < \sigma_i < 10^{-2}\ S/cm$ at 900°C.

While these results demonstrate the ability to modulate the relative levels of ionic and electronic conductivity in a given host material, the total conductivity, particularly in the GMMT material, is limited to $\approx 10^{-1}$ S/cm. For a cathode, one would desire at least several orders of magnitude higher conductivity.

Another system with promising characteristics is the perovskite structure. We have already mentioned that $(La_{1-x}Sr_x)CoO_3$ exhibits excellent MIEC characteristics but unfortunately reacts with YSZ. However, if a perovskite-based solid electrolyte could be identified, the chemical incompatibility issue could perhaps be resolved.

Interest in oxygen ion conduction in perovskites has been of long standing. Takahashi and Iwahara [35] reported on the study of ionic conduction in a number of perovskite solid solutions, including $La_{1-x}Ca_xAlO_3$ and $CaTi_{1-x}Al_xO_3$. The latter was found to have a substantial oxygen ion conductivity of 3 x 10^{-2} S/cm at 1000^0C. At the same time, the ability of the perovskite structure to accommodate the multivalent Ti ion also contributes to MIEC [35].

In this decade, interest was reignited by the work of Ranløv et al. [36] and Ishihara et al. [37, 38]. In particular, Ishihara et al. [38] demonstrated that substitution of Ga for Al and acceptor doping on both A and B sites (ABO_3) could boost the ionic conductivity to levels above that of zirconia and ceria. e.g., 3 x 10^{-1} S/cm at 850^0C.

In order to take optimum advantage of the 'monolithic" design, we wish to insure optimum compatibility between compositions serving as the solid electrolyte and those serving as the anode and cathode, respectively. Selecting compositions within a given solid solution family enables one, in principle, to retain close thermal-mechanical and chemical compatibility.

Given the exceptionally high oxygen ion conductivites exhibited by acceptor doped $LaGaO_3$, we decided to base our monolithic design concept on this electrolyte. In order to induce high electronic conductivity in this system, the authors formed solid solutions of the type
$La_{0.9}Sr_{0.1}Ga_{1-x}Ni_xO_3$ (LSGN) [39, 40]. The temperature dependence of LSGN measured in air is shown in Fig. 21.

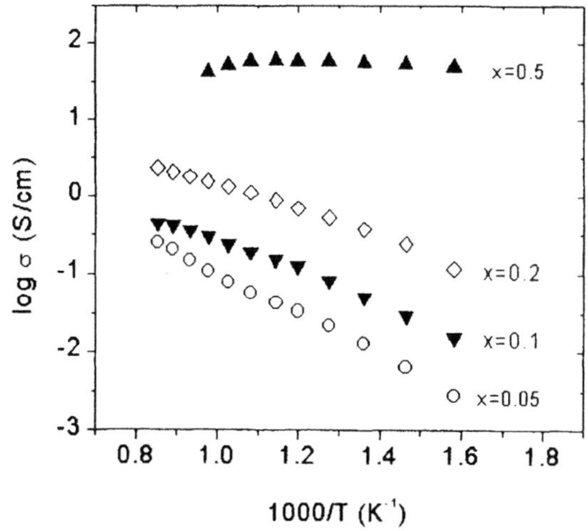

Figure 21. Temperature dependence of conductivity for LSGN measured in Po_2=0.21 atm [40].

One observes a clear shift from semiconducting behavior for $0.05 \le x \le 0.2$ to metallic behavior at x=0.5. Examination of the Po_2-dependence of the conductivity as in Fig. 22 for LSGN with x=0.2 shows a p-type behavior at high Po_2 and a Po_2-independent ionic conductivity-like behavior at low Po_2. Electron blocking experiments performed on cells of the type

$$O_2, M/LSG/LSGN/LSG/M, O_2$$

where LSG is $La_{0.9}Sr_{0.9}GaO_3$ and M are metal electrodes, typically platinum, confirm the ionic conductivity in air at 800°C to be $\sigma_i = 1.2 x 10^{-1} S/cm$, a value very close to that in the low Po_2 regime. We expect the ionic conductivity to be likewise high in LSGN with x=0.5.

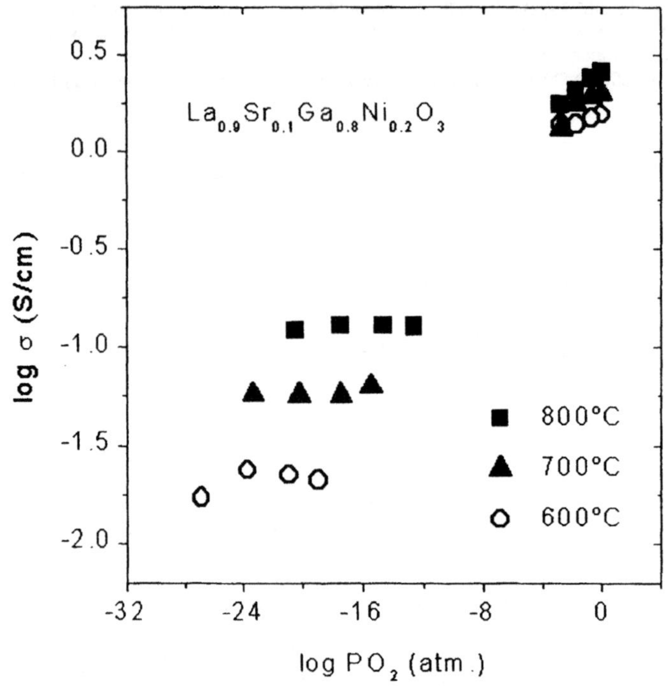

Figure 22. P_{O_2} dependence of electrical conductivity of LSGNCX=0.2 [40].

The above results suggest that LSGN (x=0.5) may be an excellent cathode compatible with LSG. It exhibits an electronic conductivity of ≈ 50 S/cm at 800°C together with an ionic conductivity of ≈ 0.1 S/cm. To confirm this possibility, LSGN (x=0.5) porous electrodes approximately 85 μm thick were applied to $La_{0.9}Sr_{0.1}Ga_{0.8}Mg_{0.2}O_3$ (LSGM) electrolytes and the cathodic overpotentials were studied by the current interruption technique [41]. The results are reported in Fig. 23 and are compared with overpotentials induced under similar conditions for LSM and LSC electrodes.

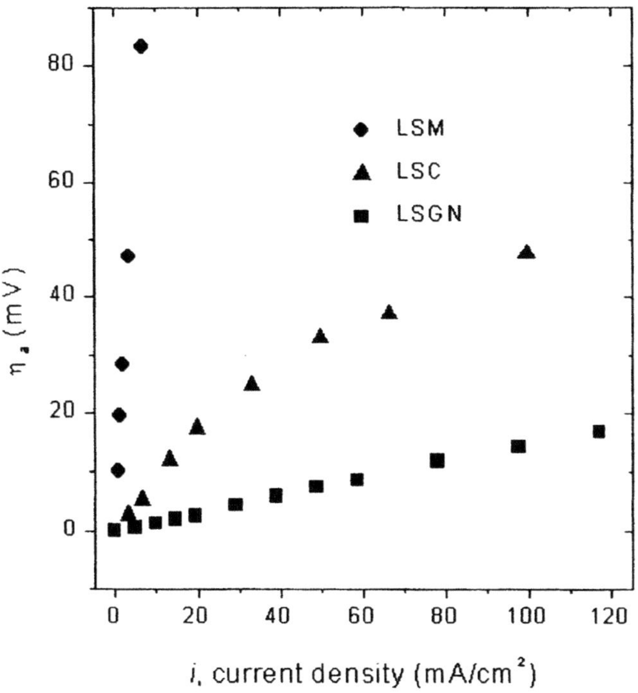

Figure 23. Comparison of cathode overpotentials in air at 800°C for LSM, LSC and LSGN on LSGM electrolyte [41].

These excellent results support the hypothesis that electrodes with high MIEC should exhibit low impedances and thus low overpotentials.

Aside from the primary electrical and electrochemical considerations, we emphasized above the need for satisfying the secondary considerations of thermal-mechanical stability and compatibility in a complex composite structure such as the solid-oxide fuel cell. In Table 2 we compare the thermal coefficients of expansion (TCE) of the LSGN materials for various values of x. In general, TCE increases with increasing x. While the difference in TCE is not large below 700°C, it becomes significant above 800°C which could cause difficulties if LSGN (x=0, 2 or 0.5) is applied to LSG. The large increase in TCE in LSGN for large x may be related to dilation induced by the Ni rich material going-off oxgyen stoichiometry at elevated temperatures. A similar effect was mentioned above for ceria-based electrolytes exposed to reducing conditions [29].

Compound	TEC (300-700°C)	TEC (800-950°C)
$La_{0.9}Sr_{0.1}Ga_{0.95}Ni_{0.05}O_3$	11.0×10^{-6} K^{-1}	15.7×10^{-6} K^{-1}
$La_{0.9}Sr_{0.1}Ga_{0.9}Ni_{0.1}O_3$	12.4×10^{-6} K^{-1}	19.2×10^{-6} K^{-1}
$La_{0.9}Sr_{0.1}Ga_{0.8}Ni_{0.2}O_3$	12.6×10^{-6} K^{-1}	24.2×10^{-6} K^{-1}
$La_{0.9}Sr_{0.1}Ga_{0.5}Ni_{0.5}O_3$	13.2×10^{-6} K^{-1}	25.2×10^{-6} K^{-1}

Table 2. Thermal coefficients of expansion (TCE) of LSGN materials [40].

VI. Summary

The primary requirements for operation of the solid electrolyte and electrodes in solid state electrochemical devices were described and means for achieving them discussed. A novel monoithic structure was proposed which resolves many of the thermal, mechanical and chemical compatibility difficulties typical of present designs. A number of materials systems based on the pyrochlore and perovskite structures were evaluated for their potential in a monolithic design. The materials system $La_{0.9}Sr_{0.1}Ga_{1-x}Ni_xO_3$ was shown to be particularly promising.

Acknowledgements

Support of this work was provided by the Basic Energy Sciences Division of the U.S. Department of Energy under contract #DE FG02 86ER 45261. The work of former students: P. Moon, S. Kramer, M. Spears, T.-H. Yu and J. Sprague and colleagues O. Porat, I. Kosacki, C. Heremans, H. Takamura and B.J. Wuensch are gratefully acknowledged.

References

1. Tuller, H.L. (2000) Defects and transport: implications for solid oxide electrolytes and mixed conductors, Chapter 3, this volume.
2. Tuller, H.L. (1992) Mixed ionic-electronic conduction in a number of fluorite and pyrochlore compounds, *Solid State Ionics* **52**, 135-146.
3. Pierret, R.F. (1996) *Semiconductor Device Fundamentals*, Addison-Wesley Publishers, Reading, MA.
4. Nowick, A.S. and Park, D.S. (1976) Fluorite-type oxygen conductors, in G. Mahon and W. Roth (eds.) *Superionic Conductors*, Plenum Press, New York, pp. 395-412.
5. Kramer, S.A. and Tuller, H.L. (1995) A novel titanate-based oxygen ion conductor: $Gd_2Ti_2O_7$, *Solid State Ionics* **82**, 15-23.
6. Kilner, J.A. (1983) The role of dopant size in determining oxygen ion conductivity in the fluorite structure oxides, in R. Metselaar, H.J.M. Heijligers and J. Schoonman (eds.) *Solid State Chemistry 1982*, Elsevier, Amsterdam, pp. 189-192..
7. Gerhart-Anderson, R. and Nowick, A.S. (1981) Ionic conductivity of CeO_2 with trivalent dopants of different ionic radii, *Solid State Ionics* **5**, 547-550.
8. Butler, V., Catlow, C.R.A., Fender, B.E.F., and Harding, J.H. (1983) Dopant ion radius and ionic-conductivity in cerium dioxide, *Solid State Ionics* **8**, 109-113.
9. Li, P., Chen, I.W., and Penner-Hahn, J.E. (1994) Effect of dopants on zirconia stabilization--an x-ray absorption study: I, trivalent dopants, *J. Am. Ceram. Soc.* **77**, 118-128.
10. Kramer, S.A. (1994) Mixed ionic-electronic conduction in rare earth titanate/zirconate pyrochlore compounds, Ph.D. Thesis, Dept. Materials Sc. & Eng., MIT, Cambridge, MA.
11. J. Ranløv (1995) Perovskite-type metal oxides. Electrical conductivity and structure [Risø-R-796(EN)], Ph.D. Thesis, Risø National Lab, Roskilde, Denmark.
12. Takahashi, T. and Iwahara, H. (1978) Oxide ion conductors based on bismuthsesquioxide, *Mat. Res. Bull.* **13**, 1447-1453.
13. Goodenough, J.B., Ruiz-Diaz, J.E., and Zhen, Y.S. (1990) Oxide ion conduction in $Ba_2In_2O_5$ and $Ba_3In\ 2CeO_8$, $Ba_3In_2HfO_8$, or $Ba_3In_2ZrO_8$, *Solid State Ionics* **44**, 21-31.
14. Kendall, K.R., Navas, C., Thomas, J.K. and zur Loye, H.C. (1995) Synthesis and ionic-conductivity of a new series of aurivillius phases, *Chem. Mat.* **7**, 50-57.
15. Tuller, H.L., Kramer, S., and Spears, M.A. (1993) High temperature electrochemical behaviour of fast ion and mixed conductors, in F.W. Poulsen, J.J. Bentzen, T. Jacobsen, E. Skou and M.J.L. Ostergard (eds.), *Proc. 14th Risø Intl. Symp. on Mats. Sci.*, Riso National Laboratory, Roskilde, Denmark, 151-173.
16. Moon, P.K. and Tuller, H.L. (1988) Ionic conduction in the $Gd_2Ti_2O_7$-$Gd_2Zr_2O_7$ system, *Solid State Ionics*, **28-30**, 470-474.
17. P.K. Moon and H.L. Tuller (1989) Intrinsic fast oxygen ion conductivity in the $Gd_2\ (Zr_xTi_{1-x})_2O_7$ and $Y_2(Zr_xTi_{1-x})_2O_7$ pyrochlore systems, in G. Nazri, R.A. Huggins, and D.F. Shriver (eds.), *Solid St. Ionics, MRS Symp. Proc. Vol. 135*, Pittsburgh, PA, pp. 149-163.
18. Heremans, C. and Wuensch, B.J. (1995) Fast-ion conducting $Y_2(Zr_yTi_{1-y})_2O_7$ pyrochlores: neutron Rietveld analysis of disorder induced by Zr substitution, *J. Sol. St. Chem.* **117**, 108-121.
19. T.-H. Yu and H.L. Tuller, unpublished.
20. Yu, T.-H. (1996) Electrical properties and structural disorder in stannate pyrochlores, Ph.D. Thesis, Dept. Materials Science and Eng., MIT, Cambridge, MA.
21. Tuller, H.L. (1981) Mixed conduction in nonstoichiometric oxides, in O.T. Sorensen (ed.), *Nonstoichiometric Oxides*, Academic Press, New York, pp. 271-335.
22. Tuller, H.L. (1978) Optimized Electrolytic Domain Boundaries in Solid Oxide Electrolytes, in H.S. Isaacs, S. Srinivasan and I.L. Harry (eds.), *Proceedings of the Workshop on High Temperature Solid Oxide Fuel Cells*, Brookhaven National Laboratory, NY, pp. 104-113.
23. Kramer, S., Spears, M., and Tuller, H.L. (1993) Solid electrolyte figures of merit: rare earth titanate pyrochlores, in S.C. Singhal and H. Iwahara (eds.), *Proceedings of the Third Intl. Symp. on Solid Oxide Fuel Cells (Proc. Vol. 93-4)*, The Electrochemical Society, Pennington, NJ, pp. 119-128.
24. Moon, P.K., Spears, M.A., and Tuller, H.L. (1989) The defect chemistry of pyrochlore solid solutions of the type $R_2(Zr_xTi_{1-x})_2O_7$ with R=Gd,Y, in B.C. Larson, M. Ruhle and D.N. Seidman (eds.), *Characterization of the Structure and Chemistry of Defects in Materials (MRS Symp. Proc. Vol. 138)*, Pittsburgh, PA, pp. 157-160.
25. Tuller, H.L. (1996) Materials design and optimization, in F.W. Poulsen, N. Bonanos, S. Linderoth, M. Mogensen, and B. Zachau-Christiansen (eds.), *High Temperature Electrochemistry: Ceramics and Metals*, Risø National Lab, Roskilde, Denmark, pp. 139-153.
26. Yamamoto, O., Watanabe, S., Keno, K., Imanashi, N. Takeda, Y., Sammes, N., and Phillips, M.B. $Gd_{1-x}A_xMnO_3$ and $Gd_{1-x}A_xCoO_3$(A=Ca,Sr) for the electrode of solid oxide fuel cells, in M. Kokiya, O. Yamamoto, M. Tagawa and S.C. Singhal (eds.), *Solid Oxide Fuel Cells IV*, The Electrochem. Soc., Pennington, NJ, pp. 414-423.
27. Yamamoto, O., Takeda, Y., Kanno, R., and Noda, M. (1987) Perovskite-type oxides as oxygen electrodes for high temperature oxide fuel cells, *Solid State Ionics* **22**, 241-246.
28. Steele, B.C.H. (1992) Oxygen ion conductors and their technological applications, in M. Balkanski, T. Takahashi and H.L. Tuller (eds.), *Solid State Ionics*, North Holland Pub., Amsterdam, pp. 17-28.
29. Mogensen, M., Lindegaard, T., Hansen, V.R., and Mogensen, G. (1994) in T.A. Ramanarayanan, W.L. Worrell and H.L. Tuller (eds.), *Ionic and Mixed Conducting Ceramics*, The Electrochemical Soc., Pennington, NJ, pp. 317-324.
30. Tuller, H.L., Kramer, S.A., and Spears, M.A. (1995) Solid electrolyte-electrode system for an electrochemical cell, U.S. Patent No. 5,403,461.
31. Kramer, S.A., Spears, M.A., Pal, U.B., and Tuller, H.L. (1996) Method for making an electrochemical cell, U.S. Patent No. 5,509,189.
32. Han, P. and Worrell, W.L., op. cit. Ref. 29, pp. 317-324..

33. Porat, O. Heremans, C., and Tuller, H.L. (1997) Phase stability and electrical conductivity in $Gd_2Ti_2O_7 - Gd_2Mo_2O_7$ solid solutions, *J. Am. Ceram. Soc.* **80**, 2278-2284.
34. Sprague, J.J. (1999) Mixed conduction and defect chemistry of Mn and Mo substituted gadolinium titanate pyrochlore, Ph.D. Thesis, Dept. Materials Science and Engineering, MIT, Cambridge, MA.
35. Takahashi, T. and Iwahara, H. (1971) Ionic conduction in perovskite-type oxide solid solution and its application to the solid electrolyte fuel cell, *Energy Conv.* **11**, 105-111.
36. Ranløv, J., Poulsen, F.W., and Mogensen, M. (1993) Mixed ionic and electronic conductivity of rare earth aluminates, op cit. Ref. 15, pp. 389-396.
37. Ishihara, T., Matsuda, H., and Takita, Y. (1994) Oxide ion conductivity in doped $NdAlO_3$ perovskite-type oxides, *J. Electrochem. Soc.* **141**, 3444-3449.
38. Ishihara, T., Matsuda, H. and Takita, Y. (1994) Doped LaGaO3 perovskite type oxide as a new oxide ionic conductor, *J. Am. Ceram. Soc.* **116**, 3801-3803.
39. Long, N.J. and Tuller, H.L. (1998) Mixed ionic-electronic conduction in Ni-doped lanthanum gallate perovskites in materials for electrochemical energy storage and conversion II – batteries, capacitors and fuel cells, in D. Ginley, B. Doughty, T. Scrosati, T. Takamura and Z. Zhang (eds.), *MRS Proc. Vol. 456*, Materials Research Society, Warrendale, PA, pp. 129-.
40. Long, N.J., Lecarpentier, F., and Tuller, H.L. (1999) Structure and electrical properties of Ni-substituted lanthanum gallate perovskites, *J. Electroceramics*, in print.
41. Lecarpentier, F., Tuller, H.L. and Long, N. (1999) Performance of $La_{0.9}Sr_{0.1}Ga_{0.5}Ni_{0.5}O_3$ as a cathode for a lanthanum gallate fuel cell, unpublished.

CHEMICAL SPRAY DEPOSITION OF CERAMIC FILMS

P. BOHAC and L. GAUCKLER
Nichtmetallische Werkstoffe, ETH Zürich
CH-8092 Zürich, Switzerland

Abstract

The present overview gives a short comprehensive description of chemical deposition methods and an introduction to novel chemical spray deposition techniques of ceramic films. The mechanism of the chemical spray deposition is critically discussed and some examples of sprayed YSZ films on dense and porous substrates are presented. In contradiction to the common conception of the prevailing CVD mechanism for spray techniques, it is proposed that dense ceramic films are deposited when still liquid droplets spread on the substrate surface. A rapid evaporation of solvents then leads to the formation of an intermediate solid film, followed by a solid state reaction and film densification. The same mechanism rules the sol-gel and MOD wet processing.

1. Introduction

Ceramic films and coatings are used in many areas of technical applications, including electronic and optical devices, fuel cells, cutting tools, protection coatings against corrosion at high temperatures etc. Due to the wide range of ceramic materials, the physical properties of films can be tailored to fulfill the demands for various fields of technology. The principal techniques for the deposition of ceramic coatings can be primary divided in physical and chemical methods. For all chemical film deposition methods in common is the formation of films by a chemical reaction which occurs on, or in the vicinity of the substrate surface, followed by a nucleation on the substrate surface and subsequent film growth. The chemical reaction involved may be a thermal decomposition of starting materials, their hydrolysis, oxidation or reduction as well as any other kind of chemical reaction between the present species. The deposition rate increases in general with temperature due to the higher mobility of atomic particles and high activation energies. However, high temperatures are not a necessary condition for the film deposition and films can be formed even at room temperature.

The chemical spray deposition is a well-known method for coating materials with films of various oxides from precursor solutions. The ability for mass production of large

H.L. Tuller et al. (eds.), Oxygen Ion and Mixed Conductors and Their Technological Applications, 271–294.

area films and multiple coatings, high deposition rates and low processing costs are the characteristic features of this process which offers a great potential for forming ceramic films at atmospheric pressure and at temperatures below 600 °C. Recent investigations of modified techniques of the chemical deposition of ceramic films by spraying of appropriate precursor solutions on heated substrates [10,13,14,27,56,58,61,62,63,71, 72] have indicated new possibilities for the fabrication of films of solid electrolytes and electrode materials for intermediate temperature solid oxide fuel cells. Using the spray film deposition techniques, the thickness of the electrolyte can be reduced and even complete PEN structures for SOFC can be fabricated. However, many questions concerning the mechanism of chemical spray deposition and the most suitable experimental conditions for this technique remain still open.

2. Chemical deposition methods for ceramic films

A number of techniques for chemical deposition of ceramic films has already been developed. Classifications used for chemical deposition processes concern either the type of the chemical reaction (oxidation, reduction, pyrolysis, hydrolysis, thermal decomposition, electrolysis etc.) the relevant process parameters (dip coating, spin coating, brush painting, spraying, LPE, electrodeposition, aerosol, flame, electrostatic field etc.), the environment (vacuum, vapor, plasma, solution, melt) or the structure and properties of the deposits (thin, epitaxial, porous, transparent, conducting) and the nature of depositing species (atoms, molecules, particles, metalorganic). They are not satisfactory since several techniques overlap different categories. Moreover, different authors often use different terms for the very same technique. This is valid especially for chemical spray deposition methods, the mechanism of which is not always well understood.

Ceramic films can be deposited from a homogeneous gas phase or liquid solution, or they can be formed by a solid phase reaction. According to the state of matter of reacting species, chemical deposition methods can be divided into three groups (Table 1).

2.1 DEPOSITION FROM THE GAS PHASE

The chemical vapor deposition (CVD) consists of a chemical reaction at a heated substrate surface involving gaseous species [1,5,9,11,17-19,40-43,50,61,75]. Films can be deposited by pyrolysis, hydrolysis or oxidation of vapors of volatile compounds like metal halides, acetylacetonates, or organometallic compounds e.g. tetramethyltin and diethylzinc. The main control parameters of CVD are: gas flow, gas composition and substrate temperature. A typical deposition rate of CVD is between 30 and 100 nm/min, although rates up to 1 μm/s were also reported.

TABLE 1. Chemical deposition methods

CHEMICAL DEPOSITION METHODS

a) deposition from the gas phase	
CVD (APCVD, LPCVD, MOCVD)	
AACVD	(aerosol)
Cold plasma deposition	(high electric field)
(PECVD, PACVD)	(high electric field)
ACVDe	(aerosol, high electric field)
EVD	(elchem. potential gradient)
FAVD	(flame torch)
Spray deposition	(spraying)

b) deposition from the solution	
Chemical solution growth	
Solution chemistry technique	
Electrodeposition	(electric field)
Electroless deposition	
Thermal decomposition, MOD	
Sol-Gel	(colloidal solution)
Spray deposition	(spraying)

c) deposition by a solid state reaction	
Thermal decomposition, MOD	
Sol-Gel	(colloidal solution)
Spray deposition	(spraying)

A=Atmospheric or Aerosol or Assisted; C=Chemical; D=Deposition; E=Electrochemical or Electrolytic or Electrostatic or Enhanced or Epitaxy; e=electric field; F=Flame; L=Low; M=Metal O=Organic P=Pressure or Plasma or Particle or Precipitation; S=Spray; V=Vapor

Modified forms of CVD are the aerosol assisted CVD (AACVD) [1], chemical vapor infiltration (CVI) [66], particle-precipitation aided CVD (PPCVD) [66], as well as the glow discharge or vapor phase electrolytic deposition (VED) [73], and plasma assisted (PACVD) or enhanced CVD (PECVD) [74]. AACVD enables higher delivery rates of gas-phase precursors by atomization of a precursor solution, followed by precursor evaporation from aerosol particles, which undergo very rapid evaporation compared to bulk material in a sublimation bed. It is a useful technique when precursors with poor volatility or thermal stability are involved. VED is based on electrolytic deposition using a glow discharge plasma as the conductive medium. In PECVD, the plasma reactions in electrical discharges involve excited molecules which are generated through direct collision or by recombination of ions with electrons.

Processing by plasma assisted techniques is being increasingly used in various areas of production and manufacturing. Low pressure plasma, nonequilibrium plasma, cold plasma and glow discharge are some terms used to designate the same process [74,75]. Plasma chemistry takes place under nonequilibrium conditions and the main advantage of plasma assisted processing evolves from considerable lower deposition temperatures than for reactions at thermodynamic equilibrium. Bombardment of the solid surface with energetic plasma particles can also affect the film deposition. It can cause densification of the films, eliminate columnar microstructure and improve adhesion and the quality of films. Ion impact causes enhanced diffusion, collisional mixing and the formation of metastable phases. Diamond like coatings can be obtained only with the assistance of ion bombardment [74]. The substrate temperature required to obtain crystalline coatings can also be lowered. For example, single crystal ZnO films have been obtained at temperatures as low as 200 °C [74].

The electrochemical vapor deposition (EVD) is a modified form of CVD, which utilizes a chemical potential gradient [23-26,28-33,38,46,51-53,65-67]. In the first stage of the process, the substrate pores have to be closed by CVD. The further EVD film growth is only possible for metal oxides with mixed conductivity. The oxygen ions then migrate through the growing film and react at its surface with chloride vapors to oxide film. This method provides very high quality gas-tight electrolyte films on porous substrates, but is rather expensive.

The mechanism of the formation of films by spray methods FAVD (flame assisted VD) and SD (spray deposition) is not known. Films can ensue from the vapor phase as well as from the liquid or the solid phase.

2.2 DEPOSITION FROM THE SOLUTION

For preparation of films of various materials on the substrate surface from homogeneous solutions, the chemical solution growth technique, solution chemistry technique, electro- and electroless deposition are used [11,21,37,49,59,70]. The major parameters which control the deposition process are the composition of the deposition bath, its pH and its temperature. Electroless deposition is an autocatalytic chemical reduction process for the deposition of metal films (Au,Pd,Ag,Cu,Ni,Co) using reducing agents such as hypophosphite, formaldehyde, borohydride, hydrazine etc. Surface activation is a critical step of the film formation. The deposit itself catalyses the reaction, which permits the deposition of relatively thick films

Liquid solutions of film precursors in appropriate solvents are often used for the fabrication of ceramic films by thermal decomposition or chemical reaction. The application of precursor solutions is performed by painting, spraying, dipping or spinning. The liquid film is then dried and decomposed by heat and chemical reaction. Liquid precursor methods in a broad interpretation can thus include sol-gel [79], metal-organic deposition (MOD) [80], aqueous salt deposition and various spray techniques. They are then called wet-processes. However, on thermal treatment the solvent vaporizes and the film formation takes then place either from the solid phase or vapor, depending on process conditions and precursor properties. Various solvents, e.g. water, alcohols (methanol, ethanol, butanol, ethylene glycol), ketones (acetone), butylacetate, acetylacetone, and their mixtures can be used in wet-processes. The concentrations of the precursor vary between 0.01 mol/l and 0.5 mol/l. Multiple component compounds can be obtained from solutions containing mixtures of different precursors.

2.3 SOLID STATE REACTION

Film formation by a solid state reaction includes colloidal deposition by the sol-gel technique [22,35,48,49,69,79], MOD [80], as well as a large part of the aqueous salt deposition processes and spray methods. Precursor chemicals are dissolved in an appropriate solvent, deposited as a homogeneous solid film and then heated to form the desired compound by decomposition or chemical reaction. Mechanical techniques such as brushing, spraying, dipping and spinning are used for the deposition of precursor coatings from liquid media. The dip technique consists of inserting the substrate into a solution and pulling it out at a constant speed. Spin coating uses high speed rotation to distribute the liquid in a thin layer. Spray coating is the most versatile mechanical coating technique, well suitable for high-speed mass production. The deposition process is then followed by drying and subsequent heat treatment. Gelling often occurs in the layer through the formation of a polymeric network. On further heating the coating residue decomposes to produce a low-density amorphous film or network of nanosized particles. At higher temperatures, polycrystalline dense thin films of the desired compound are formed by the densification of ultrafine nanoparticles or by a chemical reaction of the intermediate solid film (pyrolysis, hydrolysis, oxidation, dehydration etc.). The control parameters of the film deposition methods involving a solid state reaction are the nature, concentration, and viscosity of the precursor solution, the pulling or spinning speed, as well as the firing temperature and the heating rate.

In the sol-gel processing of ceramic films [79], also known as wet-chemical processing, colloidal solutions are applied to substrates prior to gelation. The particles in the colloidal sol are then linked to form a gel, which is subsequently dried. Since the ratio of surface area of films to their volume is large, the evaporation takes place very rapidly. Despite high shrinkage on drying cracking is seldom observed for coatings less than about 0.5 μm thick. This is due to the strong adhesion between the film and substrate which prevents relaxation in the film plane. Dry gel films are finely fired to complete the chemical reaction (to decompose intermediate compounds and burn out the residual organic matter or water) and densify. The thickness of dense films amounts usually to only 50-200 nm. Most sol-gel derived films crack upon firing if the film thickness exceeds 0.5 μm. For thicker films, the process of coating and thermal treatment usually must be repeated until the desired thickness of the film is obtained.

The cracking of films occurs due to stresses produced during solvent evaporation, carbon burnout, chemical formation of final product, sintering and crystallization. Stresses due to the high shrinkage during the first stages of processing can be reduced by incorporating organic components into the film structure allowing the gel network to flow and relax. Using viscous precursor solutions in ethylene glycol, crack-free ITO films 1 μm thick have been obtained [37].

The MOD process for deposition of thin films uses metalorganic compounds dissolved in an appropriate solvent [80]. The precursor solution is then deposited on a substrate by any of a variety of possible techniques. The wet film is then dried (without a gel step), heated to remove the residual solvents and finally fired to undergo thermal decomposition to an inorganic film. During the pyrolysis step the film microstructure is developed.

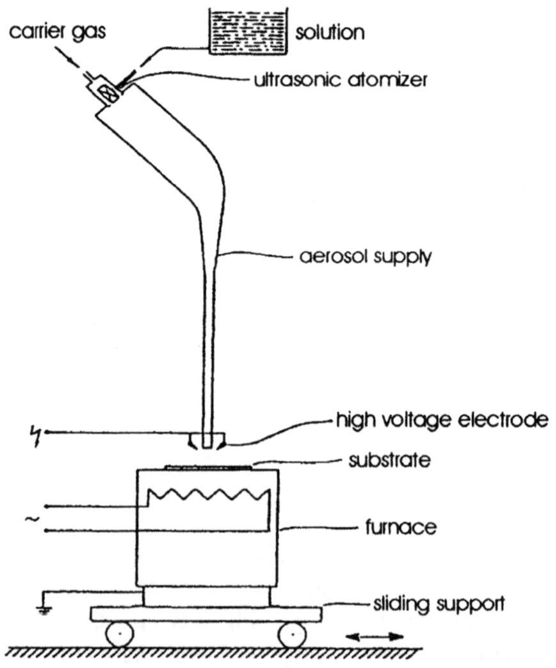

Figure 1. Schematic diagram of equipment for corona spray pyrolysis [10]

The thermal decomposition of the solid precursor occurs without evaporating or melting. The solid state film formation distinguishes the MOD process from the vapor and solution deposition techniques. The volume change ratio of the wet film to the final film is typically between 6 and 30. The thickness of crackfree films is therefore limited to less then 1 μm. To produce thicker films, the deposition and decomposition steps have to be repeated. Compounds suitable to MOD processing are salts of 2-ethylhexanoic acid or neodecanoic acid.

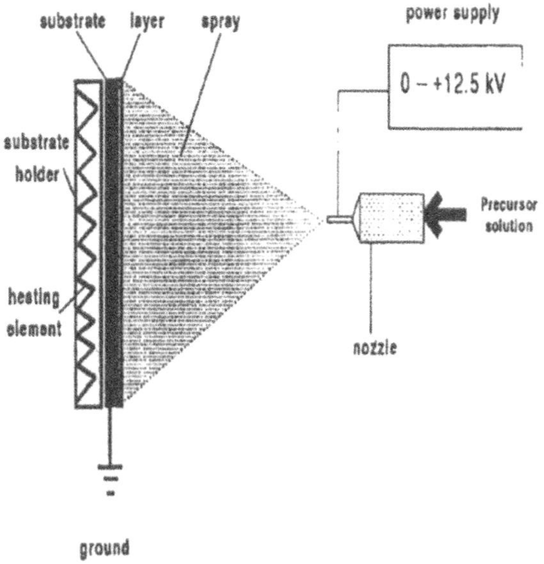

Figure 2. Set-up for electrostatic spray deposition [61]

3 Chemical spray deposition

The various spray methods are called spray pyrolysis [6-12,17,27], spray hydrolysis [9], aerosol pyrolysis, flame assisted vapor deposition (FAVD) [63], corona spray pyrolysis [10,13,14], electrospraying [56], electrostatic spray deposition (ESP) [58,61,62,71], and aerosol assisted chemical vapor deposition in an electric field (ACVDe) [72,77]. They are especially suited to the formation of ceramic films. The spray pyrolysis or hydrolysis involves spraying of a dilute solution of appropriate compounds from an atomizer onto heated substrates, with the aid of a carrier gas. A typical gas flow of 3-10 l/min and solution flow 5-20 ml/min are commonly used. The growth rate varies from about 100 nm/min up to 500 nm/min. The aerosol pyrolysis process uses a mist of very fine droplets of the starting reagents, generated by ultrasonic or electrostatic atomization of liquid solutions. By the corona spray pyrolysis

(Figure 1) and the electrostatic deposition (Figure 2), the liquid is atomized and charged and then directed onto the substrate using an electrostatic field. Spray drops may also be charged during transit through the region of ionization in a corona discharge. Electrode potentials of a few kilovolts and above are used for the last three methods. Electrospraying (ESD) uses the generation of a fine aerosol by applying a high voltage to the surface of a solution [36,58,61,62,71]. The set up for the flame assisted vapor deposition (FAVD) is shown in Figure 3 [63].

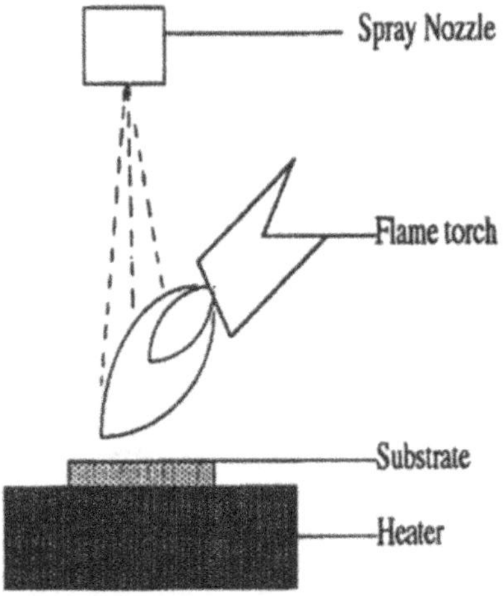

Figure 3: Flame assisted vapor deposition apparatus [63]

The chemical spray deposition methods are summarized in Table 2. The main control parameters of spray deposition methods are the solution composition, the solvent properties (chemical stability, boiling point, viscosity, surface tension and wettability), the temperature, the gas and solution flow rates, the droplet size distribution, the deposition time, the nature of the substrate, the nozzle to substrate distance, and the electric field.

Table 2: Chemical spray deposition methods

CHEMICAL SPRAY DEPOSITION	
Spray pyrolysis	
FACVD	(flame torch)
Corona spray pyrolysis	(high electric field)
ESD	(high electric field)
ACVDe	(high electric field)

Due to the complexity of the system, little is known about the true reaction mechanism of the spray deposition which remains mainly an empirical technique. The deposition may occur both from a gaseous, liquid or solid environment. Some authors [10,13,14,63,72], assume for dense coatings that the film formation and growth takes place from the gas phase. However, whether or not chemical spray deposition processes can be classified as CVD depends on whether the liquid droplets vaporize before reaching the substrate or react on it after splashing. The salient feature of the chemical spray deposition is that the film precursors are transported in solution to the substrate surface. At present, the mechanism of formation of films is not always understood and in most experiments only the starting materials and final products are known. Chemical reactions on the solid surface can be very complicated and only few mechanisms have been studied in detail. Because of the extreme temperature gradient in the vicinity of the substrate (≈50 K/mm [2]), it is even difficult to estimate the exact temperature of the film growth. For liquid solutions, an intermediate gelling can occur through the formation of polymeric network, by hydrolysis of precursors and/or polymerization by a condensation mechanism. In plasma-enhanced processes, positive and negative ions, electrons and radicals can be present in the plasma, in addition to the molecules of starting materials and products.

According to Viguié and Spitz [2], there can be four different deposition processes for spraying volatile precursors and solvents, depending on the size of droplets and substrate temperature (Figure 4). If droplets are too large, the solvent is not fully vaporized during the way to the substrate (process A). The liquid droplets splash onto the surface of the substrate, the temperature of which decreases due to the solvent vaporization under the film formation temperature. This is followed by a condensed-phase reaction. The resulting deposits are often cracked and rough and they exhibit a low adhesion to the substrate.

In the case of process B, the droplets approach the substrate just as the solvent is entirely vaporized. This process leaves dry precursors on the substrate, which form rough deposits by the subsequent chemical reaction. Process C is the true classical CVD process (AACVD). In this case not only the solvent but also the precursor completely vaporizes on reaching the substrate. The chemical reaction takes place in the vapor phase in the vicinity of the surface of the growing film. It is obvious that only those compounds that vaporize without decomposition, e.g. some metal halides and chelates, can undergo this true CVD mechanism [2]. Dense and smooth films can then be obtained. On the other hand, too small droplets in process D are already completely vaporized far away from the substrate. The subsequent chemical reaction leads to nucleation of the solid product in the gas phase and precipitation of powder.

A narrow droplet spectrum is needed to achieve only one of these four possible mechanisms of the deposition. However, it is very difficult to attain a uniform droplet size by atomization. When droplets of different sizes are present in the aerosol, even two of the possible processes can proceed concurrently [2].

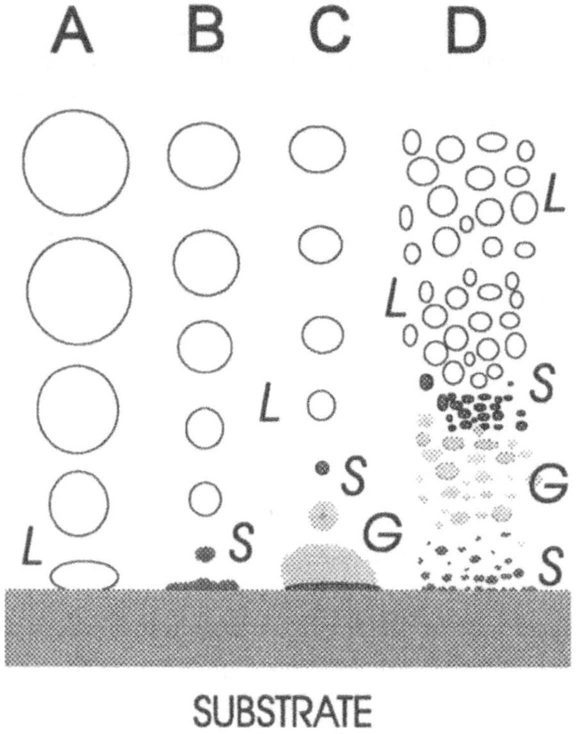

Figure 4. The four possible mechanisms of the spray pyrolysis [2,7,10,14,63]
(*L*=liquid, *S*=solid, *G*=gas)

The model of Viguié and Spitz [2] has been generally accepted in most spray deposition studies and the ideal CVD case C is often assumed as the prevailing mechanism for the film formation [7,10,13,14,63]. However, an empirically determined minimum substrate temperature of 350 °C for the formation of In_2O_3-film [10,14] is in contradiction with the proposed CVD mechanism because of the rather low volatility of $InCl_3$ at this temperature. Moreover, a true CVD process is even suggested for the formation of dense films of zirconia, Fe_2O_3 and $PbTiO_3$ [63] by the Flame Assisted Vapor Deposition (FAVD) technique (Figure 3), despite the striking fact that non-volatile precursors $ZrO(NO_3)_2.xH_2O$, $Y(NO_3)_3.5H_2O$, $Fe(NO_3)_3.9H_2O$ and $Pb(NO_3)_2$ were used.

In Table 3, the deposition temperature, growth rate, and the film thickness of the chemical deposition methods are compared.

TABLE 3. Comparison of chemical deposition methods

Deposition method	T [°C]	Growth rate [nm.min^{-1}]	Film thickness [μm]
CVD	400-1200	5-150	0.1-5
EVD (YSZ)	700-1200	5-500	0.5-50
Spray pyrolysis	80-800	5-150	0.1-1
ESD	250-500	15-100	0.1-5
ACVDe	200-500	≤5000	≤3000
Sol-Gel	600-1400	-	0.1-1

4. Examples of chemical deposition of films from precursor solutions

Transparent conducting thin films of SnO_2, In_2O_3, ZnO, CdO, Cd_2SnO_4, $CdSnO_3$, $ZnSnO_3$, TiO_2, SiO_2, ITO, Bi_2O_3 are produced by hydrolysis or thermal decomposition of metal compounds (chlorides, acetylacetonates, alkoxides, acetates, nitrates) on substrates at 300-800 °C. Films of ZnO deposited from acetate solutions exhibit better optical quality than using $ZnCl_2$ [11]. Two precursors, providing the metal and the non-metal are needed for sulfide and selenide films. $CuInSe_2$ films of thickness ranging from 0.1 to 1 μm were obtained by spray pyrolysis of aqueous solution of corresponding chlorides ($InCl_3$ 0.05 M and CuCl 0.01 M) and selenourea (0.05 M) at 350 °C [12].

A process for spray depositing of coatings of titanium dioxide [6] uses a solution composed of Ti-isopropoxide as Ti-source, n-butyl acetate as diluent solvent, sec-butanol as the leveling agent and 2-ethyl-1-hexanol (33 vol%) as the spraying agent. 2-ethyl-1-hexanol with b.p. of 182 °C promotes wetting and improve the uniformity of the film. The next step is a heat treatment of the liquid film to evaporate solvent at 95 °C and to form TiO_2 by pyrolytic decomposition at 200-450 °C.

Thin films of calcia-stabilized zirconia (CSZ) were fabricated from Zr and Ca acetylacetonato complexes on a (LSM) porous electrode substrate by spray pyrolysis method [27]. The ethanol solution of $Zr(acac)_4$ 0.033 M and $Ca(acac)_2$ 0.0059 M was sprayed by an atomizer on the substrate heated at various temperatures between 80 °C and 200 °C. The spraying step was repeated ten times and then the sample was heated to 600 °C (thermal decomposition step). After every five decomposition steps, the film was submitted to heat treatment at 1000 °C. By repeating the film deposition and heating cycles, a 33 μm thick, dense electrolyte film was obtained. The fuel cell with this electrolyte attained the open circuit voltage of 0.96 V and power density of 500 mW.cm^{-2}. The cracks of CSZ films could be eliminated by coating the film repeatedly, since the voids are filled with the precursor solution by subsequent spraying steps. YSZ coatings were deposited by spin coating from aqueous precursor solution of zirconium acetate and yttrium nitrate and from an isopropanol sol of Zr-propoxide and Y-isopropoxide or Y-nitrate with ethylacetoacetate as a chelating agent, followed by thermal decomposition at 600-1400 °C. The chemical solution growth technique and solution chemistry technique were used for the deposition of various chalcogenide films as well as oxide films (Mn_2O_3, Fe_2O_3, ZnO, SnO_2, RuO_2, V_2O_5, WO_3). Spin coating with precursor solutions of chlorides in ethanol were followed by heat treatment at temperatures between 400 °C and 800 °C [70]. A number of wet-chemical synthesis routes have been reported for making ferroelectric films, e.g. $BaTiO_3$ and $PbTiO_3$, as well as antireflection coatings and planar waveguides for integrated optical circuits [79].

A modified Pechini process has also been successfully applied to deposit ceramic films [62]. When a mixture of metal salt, citric acid, and ethylene glycol is heated, polyesterification occurs and solid resin is formed. Metal ions are distributed in the polymeric network. The use of ethylendiamine as an additional chelating agent improves the film quality [62]. A final thermal treatment converts the resin coatings to dense ceramic film. Dense crack-free thin films of YSZ (0.2-2 μm thick) were deposited on porous and dense substrates using repeated spin coating cycles followed by drying and heat treatment up to 600 °C [57]. Epitaxial zirconia thin films were prepared by spin coating of an aqueous precursor solution of zirconium acetate and yttrium nitrate and heat treatment at 600 °C [49].

A homogeneous ITO film with a thickness of 1 μm without any cracks was obtained by a single spin coating of viscous ethylene glycol solution of indium and tin chlorides, followed by heat-treating at 600 °C for 1 hour [37]. The thickness of the spin coated film depends on the viscosity of the solution. To increase its viscosity, the chloride solution in ethylene glycol was first heated to 200 °C.

The control parameters of the spray deposition are: nature and temperature of the substrate, the solution composition and properties of its components (boiling point, specific heat and the heat of vaporization, surface tension and wettability, viscosity, chemical stability), the gas and solution flow rates, the droplet size distribution, deposition time and the nozzle to substrate distance. The typical droplet size is between 1 μm and a few microns. However, only a limited number of atomizers produce reasonably mono-sized droplets. Most atomizers produce droplets in the range of a few hundred microns down to a few microns.

The experiments showed that the molarity of the precursor solution needs to be kept below 0.5 mol/l in order to achieve good results [70]. The film formation should take place at the substrate surface. The size of droplets and their distribution affect the uniformity of films. The sprayed-on microdroplets must spread locally on the substrate surface and coalescence to form a continuous and homogeneous liquid film. The undesirable homogeneous volume phase reaction in the gas phase affect the deposition rate significantly especially at higher temperatures and may result in a powder deposition. Film grown at low temperatures <300 °C are usually amorphous, higher temperatures yield polycrystalline films. An optimum temperature is 450-500 °C, above this value a large number of cracks and powdered precipitate were formed. The nature of the substrate surface determine many of the factors which control the nucleation and crystal growth and can also have a significant influence on the properties of the films.

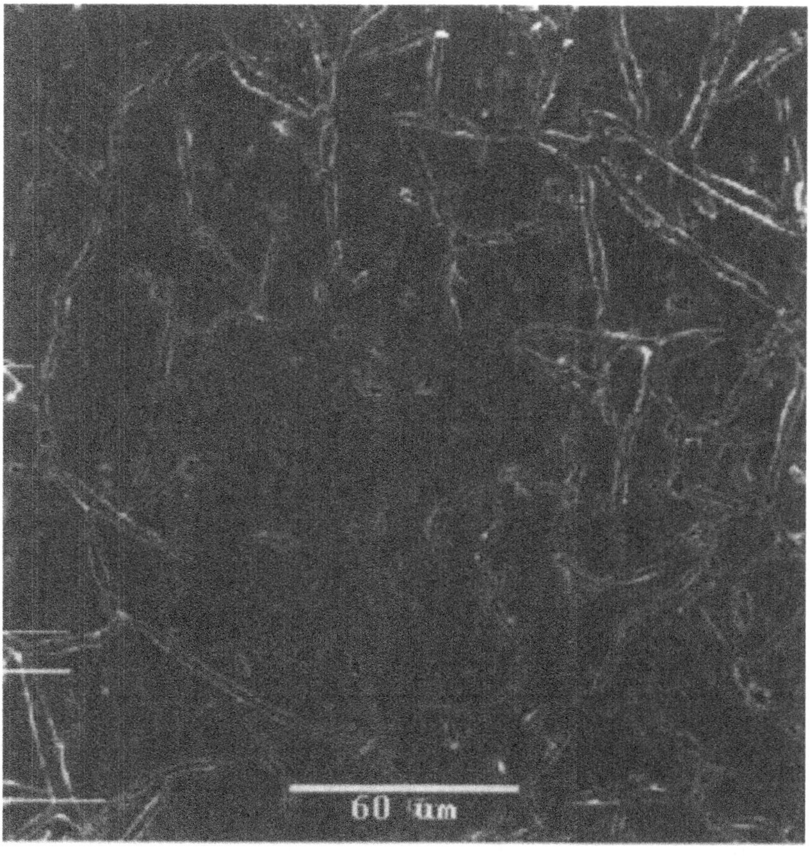

Figure 5. Thin YSZ film on a dense Gd doped ceria substrate

Numerous cracks were found in YSZ films about 3-5 μm thick, fabricated on a dense substrate of Gd doped ceria by 10 times repeated brush coating of the ethylene glycol solution of $ZrCl_4$ and YCl_3 (≈0.1 mol/l YSZ), followed by thermal treatment at 500 °C for 2h (Figure 5). Even more cracks were formed by 20 brush coating/heat treatment cycles of the same precursor solution on a glass substrate (Figure 6). However, some of the initial cracks seal again during the additional steps of the repeated processing. The cross-sectional view of a YSZ film on a porous LSC substrate, fabricated by consecutive (10 times) brush applying of the ethylene glycol solution at 500 °C is shown in Figure 7. It can be seen from this picture that the fine pores of the LSC substrate are covered by a dense YSZ film. However, microcracks cause a partial delamination of the film from the substrate.

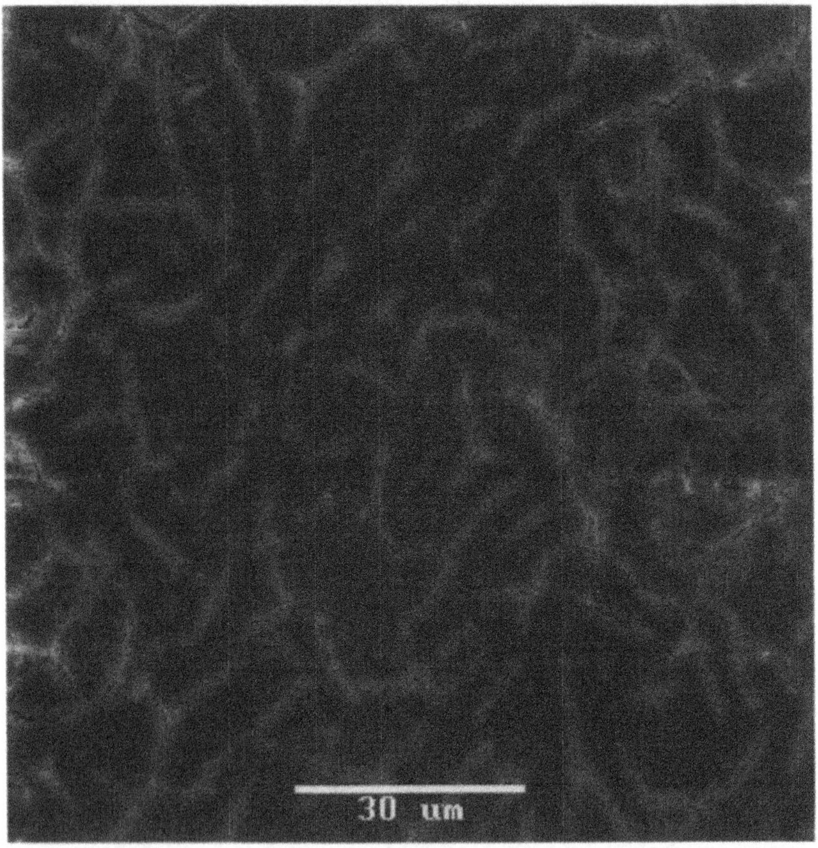

Figure 6. YSZ film on glass

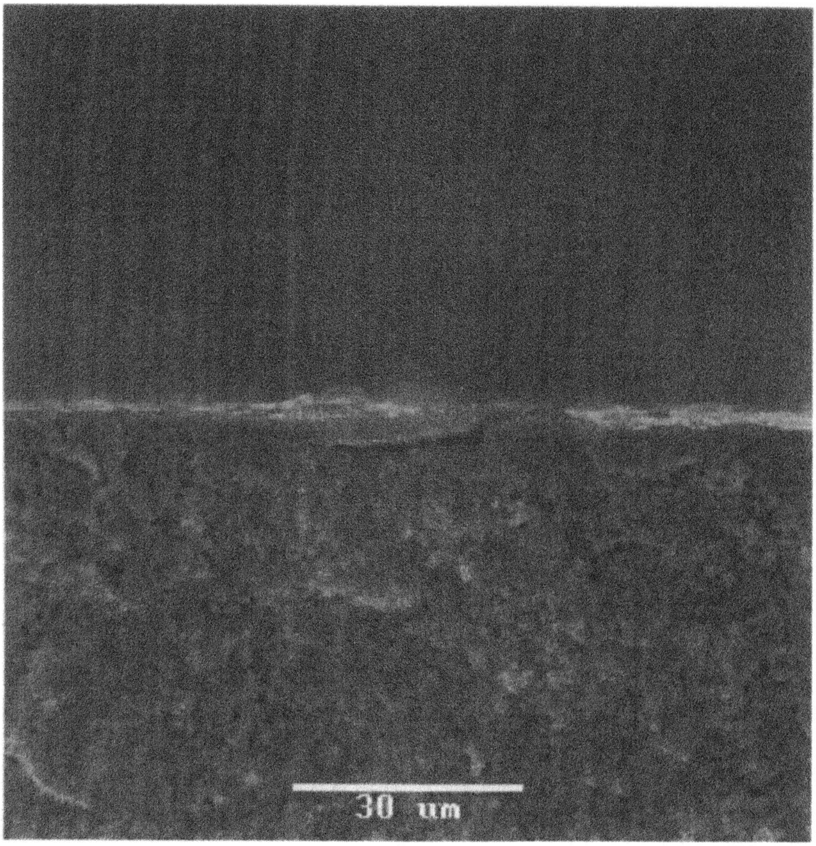

Figure 7. YSZ film on a porous LSC substrate

An YSZ film fabricated at 500 °C on a porous Al_2O_3 substrate (*Inocermic*, open porosity 36.5 %, mean pore size 3.05 μm) by the spray pyrolysis method using an airblast atomizer is shown in Figure 8. Zr-acetylacetonate and YCl_3 precursor solution in methanol/chloroform 1:1 corresponding to 0.1 mol/l YSZ was used for the spray deposition. The majority of pores could be sealed by the spray deposition of YSZ film.

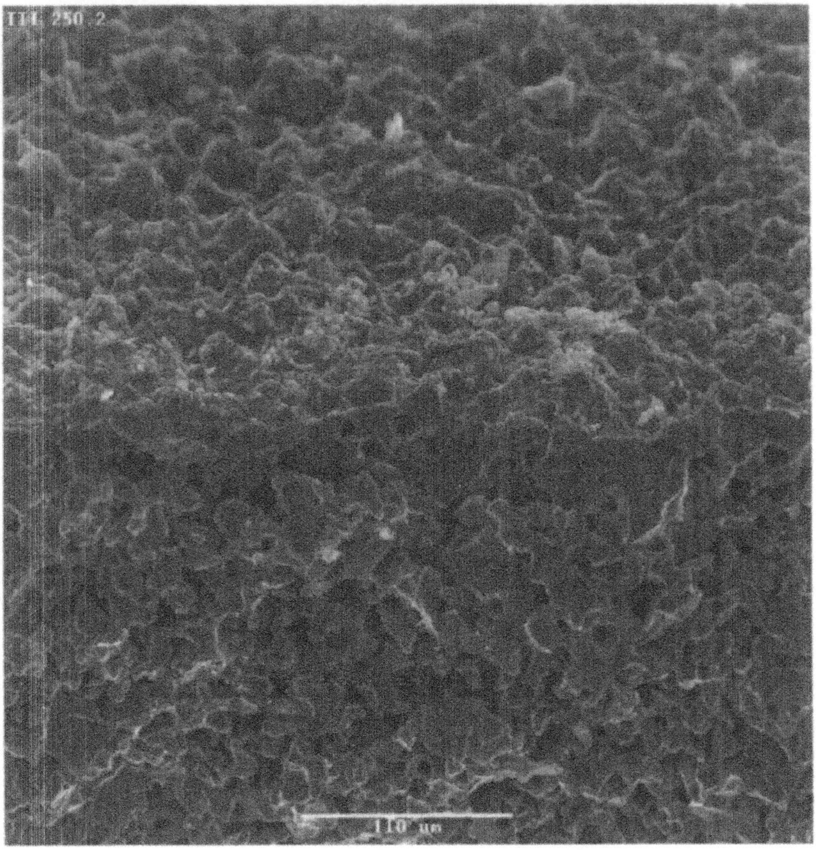

Figure 8. YSZ film on porous Al_2O_3 substrate

5. Advantages of the spray deposition technique

The *chemical spray deposition* offers several advantages for the processing of ceramic films:

a) the simplicity of the method
  -a simple equipment

-low processing temperature
-processing at atmospheric pressure in air

b) the high quality of films
-excellent homogeneity
-good adhesion to substrate
-fine grain size

c) the easy control of composition and microstructure:
-stoichiometry, doping
-porosity, smoothness of the surface, multilayers

d) the technological ability for mass production
-high deposition rates
-large area films
-high-speed continuous fabrication
-multiple coatings (PENs)

e) low costs

The disadvantages of the spray deposition are high solution losses due to overspray and aerosol emissions to the environment. They can be extensively reduced by the electrostatic spraying.

6. Atomization of liquids

Atomizers produce a fine spray of liquid by increasing the surface area of the liquid until it becomes unstable and breaks down. There are a wide variety of experimental arrangements available in the spray atomization [3,4,10,13,14,20,36,54,55,64,76,78]. Spray devices are designated as atomizers or nozzles. In practical spray applications, most atomizers produce a spray having a broad spectrum of droplet sizes with diameters varying between about 1 μm and 500 μm. The size distribution of the spray varies with the distance from the atomizer. Of practical importance is the spray pattern, which gives us an information about the overall spatial distribution of droplets. The technique of atomization determines the droplet size distribution of the spray as well as the droplet velocity and direction (spray angle). The methods are usually grouped according to the source of energy used:

a) **pressure atomizers** use hydraulic forces by forcing the liquid at high pressure through an orifice. When a liquid exceeds a certain velocity, it breaks into small droplets, i.e. it atomizes. The atomized droplets can be sprayed onto a substrate. The airless pressure jet atomizers are available in a very wide range of types and capacities.

b) **blast atomizers** use pneumatic forces by passing a stream of gas (air) at high velocity over the liquid surface of a relatively slow-moving liquid stream. They produce smaller droplets than pressure jet atomizers. Air spray is the principal method used in mass production: a fine stream of liquid leaving the spray gun is atomized by jets of compressed air.

c) **rotary atomizers** by the use of centrifugal forces on a mechanical rotation device. The spinning disc atomizes, i.e., involves the spreading of the liquid sheet on the disc and the formation of droplets at the outer edge of the disc. The droplets are mainly monosized and its size is determined by the liquid flow rate and speed of rotation.

d) **ultrasonic atomizers** by the use of ultrasonic waves. The ultrasonic spray systems produce ultrafine sprays with droplet sizes in the range of 10-100 μm and systems are best suited for clean liquids under a viscosity of 50 centipoise.

Two types of ultrasonic atomizers have been developed. In the first type, the liquid is passed onto a vibrating atomizer surface. An oscillator shears liquid into fine droplets with a relatively narrow size distribution. No high-pressure liquid feed is required. In the second type a low pressure gas passes through a nozzle into a resonator chamber and the resultant high frequency pressure wave into an open cavity in which the liquid is pumped. For its narrow droplet size distribution, the use of ultrasonic spraying for fabrication of coatings of noble metals, metal oxides and sulfides [2,8] was patented under the name "Pyrosol process" (pyrolysis of an aerosol) [8].

e) **electrostatic atomizers** by applying a high potential to a surface of a liquid. The electrostatic atomizers [20,76] produce narrow drop size distributions of almost monodisperse drops of about micron size. In high electric field of about 10^5 V/m the electrostatic spray deposition reduces spray drift (overspray) and enhances the deposition efficiency to more than 80%. A uniform coverage of substrates and minimal solution waste can be achieved.

The Delft electrostatic aerosol generator [33,36] makes use of a strong electric field in which a droplet deforms into a conical shape known as Taylor cone. From the tip of the cone highly charged droplets are generated at a frequency of 10^8-10^{10} droplets per second. The droplet size distribution is narrow and the medium size of droplets depends on the properties of the liquid and is of the order of one micron. However, submicron droplets as small as 0.01 μm can be achieved by the dilution with volatile solvents and evaporation. The electrohydrodynamic spraying (EHD) in cone-jet mode uses a metallic capillary supplied with fluid under low pressure and produces fine droplets of approximately uniform size (monodispersed).

f) **hybrid atomizers** use a combination of two or more techniques to improve the efficiency of spraying. The electrostatic spray guns utilize drop charging by induction and conduction. The spray nozzle is at a high potential with respect to an earthen target. With the aid of an electric field superimposed on the aerosol stream, the droplets are charged and accelerated due to electric force to the substrate. When the aerosol droplets are loaded and transported in an electric field towards the substrate, the coulombic forces influence drop trajectories and velocities. The spray deposition can be easily controlled and the enhancement of the deposition efficiency can be achieved. There are many different types of electrostatic hybrid systems, the most widely used are air-assisted rotary/electrostatic guns using rotating bells or discs.

7. Conclusions

A brief outlook has been given over the chemical deposition of ceramic films, and the novel chemical spray deposition techniques using high electrostatic fields to reduce overspray and to achieve fine atomization with a narrow droplet size distribution have been introduced. Moreover, the present literature survey allows conclusions about the mechanism of the formation of inorganic films by the chemical spray deposition.

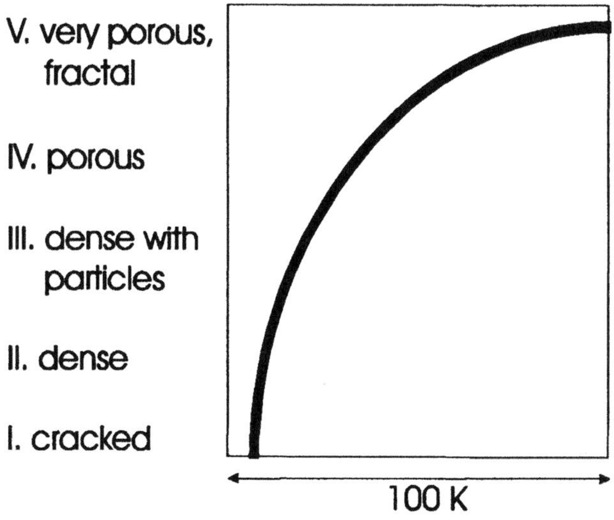

Figure 9. Dependence of the morphology of ceramic films prepared by ESD on the substrate temperature [71]

The general dependence of the morphology of ESD fabricated ceramic films on the deposition temperature was investigated by Stelzer et al. [71] and is shown in Figure 9.

As can be seen, dense ceramic films can be formed only in a narrow temperature window (II and III) the temperature of which depends on the properties of both solvent and precursor, as well as on the droplet size and the deposition rate. At lower temperatures (I) the films are cracked and at higher temperatures (IV and V) porous or powdered precipitates are formed. Moreover, denser layers are obtained if a solvent with higher boiling point and higher applied voltage are used. This is due to the increase of still wet incoming droplets at the substrate surface [71].

From this experimental dependence, the following mechanism for the spray deposition of dense films can be derived:

i) Spreading of a thin liquid film of precursor solution over the hot substrate surface
ii) Rapid formation of an intermediate solid film by solvent evaporation and/or thermal decomposition of precursor solution
iii) Solid state chemical reaction and, if necessary, burnout of organic residue accompanied with nucleation and growth of the ceramic film

Only the proposed mechanism of the solid state film formation can explain the fact that non-volatile precursors (nitrates, acetates) are frequently used for the spray pyrolysis and even the volatility of most metal halides remains too low for true vapor deposition at relevant temperatures. According to this conception, fine droplets of spraying solution are already completely converted into ceramic powder before arriving at the substrate when the substrate temperature is too high (region V in Figure 9). The solvent vaporization followed by the formation of solid precipitates prevails also in the case of porous deposits (region IV), but is accompanied with film deposition from liquid droplets. A clear transition between the formation of solid particles and dense films is then observed at still lower temperatures (region III). Dense films are formed from liquid droplets only in a narrow temperature region II. In this region, the droplets evaporate rapidly on the substrate surface forming an intermediate solid film which then undergoes a solid state reaction. At too low substrate temperature (region I), the liquid droplets on the substrate surface evaporate too slowly. The thickness of the liquid film consequently exceeds the value for the formation of crackless deposits.

Naturally, when compounds of sufficient volatility, i.e. some metal halides, acetylacetonates, and alkoxides, or rather high substrate temperatures are used for spray deposition of ceramic films, a continuous transition from the solid state reaction to CVD is also possible.

Cracks in ceramic films can be eliminated by controlling the substrate temperature and by a continuous spray coating which fills the voids repeatedly with precursor solution. Organic resins in which metal ions are homogeneously distributed in the high viscosity matrix can increase the flexibility of the intermediate solid film during processing and reduce stresses. The novel chemical spray deposition techniques offer new possibilities for the fabrication of ceramic films, especially PEN elements for solid oxide fuel cells and rechargeable batteries.

Acknowledgment: This work was financially supported by the Swiss Aluminium Foundation. The authors would like to express sincere thanks to Dr. Leo Dubal from Swiss Federal Office of Energy for valuable discussions and to M.Kristiansen for REM investigations.

8. References

1. Xia, C., Ward, T.L. and Schwartz, R.W. (1966) Aerosol-assisted chemical vapor deposition of CeO_2-doped Y_2O_3-stabilized ZrO_2 films on porous ceramic support for membrane applications, *Chem. Vap. Deposition* **2**, 48-51
2. Viguié, J.C. and Spitz, J. (1975) Chemical vapor deposition at low temperatures, *J. Electrochem. Soc.* **122**, 585-588
3. Landers, E.U. (1975) *Raumladungsfelder bei hoher Gleichspannung*, Dissertation TU München
4. Williams, A. (1976) *Combustion of sprays of liquid fuels*, Elek. Science, London
5. Vossen, J.L. (1977) Tranparent conducting films, in G.Hass, M.H.Francombe and R.W.Hoffman (eds.), *Physics of Thin Films* Vol.9, Academic Press, N.Y., pp. 1-71
6. Kern, W. and Tracy, E. (1980) Titanium dioxide antireflection coating for silicon solar cells by spray deposition, *RCA Review* **41**, 133-297
7. Blocher, J.M. (1981) Coating of glass by chemical vapor deposition, *Thin Solid Films* **77**, 51-63
8. Blandenet, G., Court, M. and Lagarde, Y. (1981) Thin layers deposited by the Pyrosol process, *Thin Solid Films* **77**, 81-90
9. Jarzebski, Z.M. (1982) Preparation and physical properties of transparent conducting oxide films, *phys. stat.sol. (a)* **71**, 13-41
10. Siefert, W. (1982) *Entwicklung eines Spray Pyrolyse Beschichtungsverfahrens mit erheblich gesteigertem Abscheidegrad*, Dissertation TH Karlsruhe
11. Chopra, K.L., Major, S. and Pandya, D.K. (1983) Transparent conductors - a status review, *Thin Solid Films* **102**, 1-46
12. Agnihotri, O.P., Raja Ram, P., Thangaraj, R., Sharma, A.K. and Raturi, A (1983) Structural and optical properties of sprayed $CuInSe_2$ films, *Thin Solid Films* **102**, 291-297
13. Siefert, W. (1984) Corona spray pyrolysis: a new coating technique with an extremely enhanced deposition efficiency, *Thin Solid Films* **120**, 267-274
14. Siefert, W. (1984) Properties of thin In_2O_3 and SnO_2 films prepared by corona spray pyrolysis and a discussion of the spray pyrolysis process, *Thin Solid Films* **121**, 275-282
15. Dietrich, G. and Schäfer, W. (1984) Advances in the development of thin-film cells for high temperature electrolysis, *Int. J. Hydrogen Energy* **9**, 747-752
16. Yamane, H. and Hirai, T. (1987) Preparation of ZrO_2-film by oxidation of $ZrCl_4$, *J. Mat. Sci. Letters* **6**, 1229-1230
17. Haefer, R.A. (1987) *Oberflächen- und Dünnschichttechnologie. Teil I: Beschichtung von Oberflächen*, Springer-Verlag, Berlin, New York, pp. 213-214
18. Carolan, M.F. and Michaels, J.N. (1987) Chemical vapor deposition of yttria stabilized zirconia on porous supports, *Solid State Ionics* **25**, 207-216
19. Yamane, H. and Hirai, T. (1989) Yttria stabilized zirconia transparent films prepared by chemical vapor deposition, *J. Cryst. Growth* **94**, 880-884
20. Bailey, A.G. (1988) *Electrostatic Spraying of Liquids*, John Wiley & Sons Inc., New York
21. Miller, K.T. and Lange, F.F. (1989) Single crystal zirconia thin films from liquid precursors, *Mat.Res.Soc.Symp.Proc.*, Vol. **155**, 191-197
22. Papet, P. Le Bars, N., Baumard, J.F., Lecomte, A. and Dauger, A. (1989) Transparent monolithic zirconia gels: effects of acetylacetone on gelation, *J.Mat.Sci.* **24**, 3850-3854
23. Lin, Y.S., de Haart, L.G.J., de Vries, K.J. and Burggraaf, A.J. (1989) Thin electrolyte layers for SOFC via modification of ceramic membranes by CVD and EVD", in S.C.Singhal (ed.), *Proc. 1st Int. Symp. on SOFC*, High Temp. Mat. Div. Proc. Vol. **89-11**, The Electrochem. Soc. Inc., Pennington, NJ, pp. 67-70

24. Dekker, J.P., Kiwiet, N.J. and Schoonman, J. (1989) Electrochemical vapor deposition of SOFC components, S.C.Singhal (ed.), *Proc. 1st Int. Symp. on SOFC*, High Temp. Mat. Div. Proc. Vol. **89-11**, The Electrochem. Soc. Inc., Pennington, NJ, pp. 57-66
25. Kiwiet, N.J. and Schoonman, J. (1990) Electrochemical vapor deposition: theory and experiment, in P.A.Nelson, W.W.Schertz and R.H.Till (eds.), *Proc. IECEC-90*, Vol.**3**, Am. Inst. Chem. Eng., New York, pp. 240-245
26. Lin, S., de Haart, L.G.J., de Vries, K.J. and Burggraaf, A.J. (1990) A kinetic study of the electrochemical vapor deposition of solid oxide electrolyte films on porous substrates, *J. Electrochem. Soc.* **137**, 3960-3966
27. Setoguchi, T., Sawano, M., Eguchi, K. and Arai, H. (1990) Application of the stabilized zirconia thin film prepared by spray pyrolysis method to SOFC, *Solid State Ionics* **40/41**, 502-505
28. Carolan, M.F. and Michaels, J.N. (1990) Growth rates and mechanism of electrochemical vapor deposited yttria-stabilized zirconia films, *Solid State Ionics* **37**, 189-195
29. de Haart, L.G.J., Lin, Y.S., de Vries, K.J. and Burggraaf, A.J. (1991) On the kinetics study of electrochemical vapour deposition, *Solid State Ionics* **47**, 331-336
30. de Haart, L.G.J., Lin, Y.S., de Vries, K.J. and Burggraaf, A.J. (1991) Modified CVD of nanoscale structures in and EVD of thin layers on porous ceramic membranes, *J. Europ. Ceram. Soc.* **8**, 59-70
31. Lin, Y.S. and Burggraaf, A.J. (1991) Preparation and characterization of high-temperature thermally stable alumina composite membrane, *J. Am. Ceram. Soc.* **74**, 219-224
32. Schoonman, J., Dekker, J.P., Broers, J.W. and Kiwiet, N.J. (1991) Electrochemical vapor deposition of stabilized zirconia and interconnection material for solid oxide fuel cells, *Solid State Ionics* **46**, 299-308
33. Dekker, J.P., van Dieten, V.E.J. and Schoonman, J. (1992) The growth of electrochemical vapor deposited films, *Solid State Ionics* **51**, 143-145
34. Peshev, P. and Slavkova, V. (1992) Preparation of yttria-stabilized zirconia thin films by a sol-gel procedure using alkoxide precursors, *Mat. Res. Bull.* **27**, 1269-1275
35. Kueper, T.W., Visco, S.J. and De Jonghe, L.C. (1992) Thin-film ceramic electrolytes deposited on porous and non-porous substraters by sol-gel technigues, *Solid State Ionics* **52**, 251-259
36. Meesters, G. (1992) *Mechanisms of droplet formation*, Thesis TU Delft, Delft University Press
37. Yamamoto, O., Sasamoto, T. and Inagaki, M. (1992) Indium tin oxide thin films prepared by thermal decomposition of ethylene glycol solution, *J.Mater.Res.* **7**, 2488-2491
38. Lin, Y.S., de Vries, K.J., Brinkman, H.W. and Burggraaf, A.J. (1992) Oxygen semipermeable solid oxide membrane composites prepared by electrochemical vapor deposition, *J. Membrane Sci.* **66**, 211-226
39. Meesters, G.M.H., Vercoulen, P.H.W., Marijnissen, J.C.M. and Scarlett, B. (1992) Generation of micron-sized droplets from the Taylor cone, *J.Aerosol Sci.* **23**, 37-49
40. Lin, Y.S. and Burggraaf, A.J. (1992) CVD of solid oxides in porous substrates for ceramic membrane applications, *AIChE Journal* **38**, 445-454
41. Brinkman, H.W., Cao, G.Z., Meijerink, J., de Vries, K.J. and Burggraaf, A.J. (1993) Modelling and analysis of CVD processes for ceramic membrane preparation, *Solid State Ionics* **63-65**, 37-44
42. Cao, G.Z., Brinkman, H.W., Meijerink, J., de Vries, K.J. and Burggraaf, A.J. (1993) Pore narrowing and formation of ultrathin yttria-stabilized zirconia layers in ceramic membranes by chemical vapor deposition/electrochemical vapor deposition, *J. Am. Ceram. Soc.* **76**, 2201-2208
43. Cao, G.Z., Brinkman, H.W., Meijerink, J., de Vries, K.J. and Burggraaf, A.J. (1993) Deposition of YSZ on porous ceramics by modified CVD, in S.P.S.Badwal, M.J.Bannister and R.H.J.Hannink (eds), *Science and Technology of Zirconia V*, Technomic Publishing Company, Inc.,Lancaster, PA, pp. 803-810
44. Brinkman, H.W., Cao, G.Z., Meijerink, J., de Vries, K.J. and Burggraaf, A.J. (1993) Morphology of very thin layers made by electrochemical vapour deposition at low temperatures, in S.P.S.Badwal, M.J.Bannister and R.H.J.Hannink (eds.), *Science and Technology of Zirconia V*, Technomic Publishing Company, Inc.,Lancaster, PA, pp. 811-818
45. Cao, G.Z., Meijerink, J., Brinkman, H.W. and Burggraaf, A.J. (1993) Permporometry study on the size distribution of active pores in porous ceramic membranes, *J. Membrane Sci.* **83**, 221-235
46. Brinkman, H.W., Cao, G.Z., Meijerink, J., de Vries, K.J. and Burggraaf, A.J. (1993) Kinetics of the EVD process for growing thin zirconia/yttria films on porous alumina substrates, *J. De Physique IV, Colloque C3*, Vol. **3**, 59-66
47. Cao, G.Z., Brinkman, H.W., Meijerink, J., de Vries, K.J. and Burggraaf, A.J. (1993) On the kinetics of modified CVD in porous ceramics, *J. De Physique IV, Colloque C3*, Vol. **3**, 67-74
48. Sakurai, C., Fukui, T. and Okuyama, M. (1993) Preparation of zirconia coatings by hydrolysis of zirconium alkoxide with hydrogen poroxide, *J.Am.Ceram.Soc.* **76**, 1061-1064

49. Miller, K.T., Chan, C.J., Cain, M.G. and Lange, F.F. (1993) Epitaxial zirconia thin films from aqueous precursors, *J.Mater.Res.* **8**, 169-177
50. Van Dieten, V.E.J., Dekker, J.P., Zomeren, A.A. and Schoonman, J. (1993) Chemical vapor deposition techniques for thin films of solid electrolytes and electrodes,in B.Scrosati, A.Magistris, C.M.Mari and G.Mariotto (eds.), *Fast Ion Transport in Solids*, Kluwer Academic Publishers, Dordrecht, pp. 231-257
51. Dekker, J.P., van Dieten, V.E.J. and Schoonman, J. (1993) The growth of electrochemical vapor deposited YSZ films, in S.P.S.Badwal, M.J.Bannister and R.H.J.Hannink (eds.), *Science and Technology of Zirconia* **V**, Technomic Publishing Company, Inc.,Lancaster, PA, pp. 786-802
52. Sasaki, H., Yakawa, C., Otoshi, S., Suzuki, M. and Ippommatsu, M. (1993) Reaction mechanism of electrochemical-vapor deposition of yttria-stabilized zirconia film, *J. Appl. Phys* **74**, 4608-4613
53. Han, J. and Lin, Y.S. (1994) An improved analysis on kinetics of electrochemical vapor deposition, *Solid State Ionics* **73**, 255-263
54. Cloupeau, M. (1994) Recipies for use of EHD spraying in cone-jet mode and notes on corona discharge effects, *J.Aerosol Sci.* **25**, 1143-1157
55. Kelly, A.J. (1994) On the statistical, quantum and practical mechanics of electrostatic atomization, *J. Aerosol Sci.* **25**, 1159-1177
56. van Zomeren, A.A., Kelder, E.M., Marijnissen, J.C.M. and Schoonman, J. (1994) The production of thin films of $LiMn_2O_4$ by electrospraying, *J.Aerosol.Sci.* **25**, 1229-1235
57. Chen, C.C., Nasrallah, M.M. and Anderson, H.U. (1994) Synthesis and characterization of YSZ thin film electrolytes, *Solid State Ionics* **70/71**, 101-108
58. Kelder, E.M., Nijs, O.C.J. and Schoonman, J. (1994) Low-temperature synthesis of thin films of YSZ and $BaCeO_3$ using electrostatic spray pyrolysis (ESP), *Solid State Ionics* **68**, 5-7
59. Chung, B.W., Brosha, E.L., Brown, D.R. and Garzon, F.H. (1995) Vapor deposition of thin film Y-doped ZrO_2 for electrochemical device applications, in G.A.Nazri, J.M.Tarascon and M.Schreiber (eds.), *Solid State Ionics IV*, Mat.Res.Soc.Proc. Vol. **369**, Mat.Res.Soc., Pittsburg, pp. 623-628
60. Visco, S.J., Wang, L.S., Souza, S. and De Jonghe, L.C. (1995) Solid oxide fuel cells, in G.A.Nazri, J.M.Tarascon and M.Schreiber (eds.), *Solid State Ionics IV*, Mat.Res.Soc.Proc. Vol. **369**, Mat.Res.Soc., Pittsburg, pp. 683-695
61. Chen, C.H., Buysman, A.A.J., Kelder, E.M. and Schoonman, J. (1995) Fabrication of $LiCoO_2$ thin film cathodes for rechargeable lithium battery by electrostatic spray pyrolysis, *Solid State Ionics* **80**, 1-4
62. Liu, M. and Wang, D. (1995) Preparation of $La_{1-z}Sr_zCo_{1-y}Fe_yO_{3-x}$ thin films, membranes and coatings on dense and porous substrates, *J.Mater.Res.* **10**, 3210-3221
63. Choy, K.L. (1995) Fabrication of ceramic coatings using flame assisted vapour deposition, in W.E.Lee (ed.), British Ceramic Proceedings No. **54**, *Ceramic Films and Coatings*, The Institute of Materials, London, pp. 65-74
64. Raetzo, T. (1995) *Charakterisierung von Düsen zum Zerstäuben von Flüssigkeiten*, Diss. ETH Nr. 11 223, Zürich
65. Brinkman, H.W. and Burggraaf, A.J. (1995) Ceramic membranes by electrochemical vapor deposition of zirconia-yttria-terbia layers on porous substrates, *J. Electrochem. Soc.* **142**, 3851-3858
66. Van Dieten, V.E.J., Dekker, J.P. and Schoonman, J. (1995) Thin film fuel cells, in G.A.Nazri, J.M.Tarascon and M.Schreiber (eds.), *Solid State Ionics IV*, Mat.Res.Soc.Proc. Vol. **369**, Mat.Res.Soc., Pittsburg, pp. 669-681
67. Brinkman, H.W., Meijerink, J., De Vries, K.J. and Burggraaf, A.J. (1996) Kinetics and morphology of electrochemical vapour deposited thin zirconia/yttria layers on porous substrates, *J. Europ. Ceram. Soc.* **16**, 587-600
68. Agashe, C. and Major, S.S. (1996) Effect of heavy doping in SnO_2:F films, *J. Mat. Sci.* **31**, 2965-2969
69. Xi, X.M. and Yang, X.F. (1996) Sintering behaviour of Y_2O_3-ZrO_2 gels, *J.Mat.Sci.* **31**, 2697-2703
70. Tressler, J.F., Watanabe, K. and Tanaka, M. (1996) Synthesis of ruthenium dioxide thin films by a solution chemistry technique, *J.Am.Ceram.Soc.* **79**, 525-529
71. Stelzer, N.H.J., Chen, C.H., van Rij, L.N. and Schoonman, J. (1997) Electrostatic spray deposition of doped YSZ electrode materials for a monolitic solid oxide fuell cell design, in A.J. McEvoy and K. Nisancioglu (eds.), *Materials and Processes,* 10th SOFC Workshop, Les Diablerets Jan. 97, Vol. II, Int. Energy Agency, pp. 236-247
72. Choy, K.L. and Bai, W. (1997) Novel fabrication of $La(Sr)MnO_3$/YSZ/NiO-YSZ PEN cells, in A.J. McEvoy and K. Nisancioglu (eds.), *Materials and Processes,* 10th SOFC Workshop, Les Diablerets Jan. 97, Vol. II, Int. Energy Agency, pp. 252-254
73. Ogumi, Z., Uchimoto, Y., Tsuji, Y. and Takeharai, Z. (1992) Properties of thin yttria-stabilized zirconia layers by vapor phase electrolytic deposition, *Solid State Ionics* **58**, 345-50

74. Grill, A. (1994) *Cold Plasma in Materials Fabrication*, IEEE Press, New York
75. Smith, D.L. (1995) *Thin-Film Deposition*, McGraw-Hill, Inc, New York
76. Michelson, D. (1990) *Electrostatic Atomization*, Adam Hilger, Bristol and New York
77. Choy, K.L., Bai, W. and Steele, B.C.H. (1997) Fabrication and properties of new materials for low temperature operation, in A.J. McEvoy and K. Nisancioglu (eds.), *Materials and Processes*, 10th SOFC Workshop, Les Diablerets Jan. 97, Vol. II, Int. Energy Agency, pp. 233-235
78. Lefebvre, A.H. (1989) *Atomization and Sprays*, Series Combustion (ed. N.Chigier), Hemisphere Pub. Corp., New York
79. Fabes, B.D., Zelinski, B.J.J. and Uhlmann, D.R. (1993) Sol-gel derived ceramic coatings, in J.B.Wachtman and R.A.Haber (eds.), *Ceramic films and coatings*, Noyes Pub., Park Ridge N.J., pp. 224-283
80. Vest, R.W. (1993) Electronic films from metallo-organic precursors, in J.B.Wachtman and R.A.Haber (eds.), *Ceramic films and coatings*, Noyes Pub., Park Ridge N.J., pp. 303-347

ELECTROSTATIC SPRAY DEPOSITION (ESD)

-- *A novel technique for ceramic thin film fabrication*

CHUNHUA CHEN[(1)] AND JOOP SCHOONMAN[(1,2)]

(1) *Laboratory for Applied Inorganic Chemistry, Delft University of Technology, Julianalaan 136, 2628 BL Delft, The Netherlands*
(2) *Department of Materials Science and Engineering, Massachusetts Institute of Technology, Cambridge, MA 02139, USA*

1. Introduction

Electrostatic atomization of liquids has been investigated for many years. It has been applied in crop spraying [1] and for painting. However, recently it has been used for the preparation of particles of metal oxides by pyrolysing solution spray droplets [2,3]. Compared with other spray techniques, e.g. ultrasonic or mechanical atomization, it has the advantage that monosized droplets may be produced under proper conditions. More recently, it was developed for preparing a number of ceramic thin films, such as $LiMn_2O_4$, Y_2O_3 stabilized ZrO_2, $BaCeO_3$, $LiCoO_2$, $LiNiO_2$, Li_3PO_4, $Li_{0.1}BPO_4$, CoO, MnO_2, TiO_2, $LaCoO_3$, SnO_2•MnO_2 [4-13], and a CdS-polymer composite film [14]. This technique is referred to as Electrostatic Spray Deposition (ESD), and in aerosol science called Electrostatic Spray Pyrolysis (ESP) in our previous reports. In addition to the very simple set-up, which is a common advantage of spray film fabrication techniques over techniques using vacuum systems, e.g. physical vapour deposition, the high deposition efficiency attained using this technique appears to be another attractive feature. This is mainly due to a well-defined trajectory of spray droplets directed towards the substrate forced by the electric field. In this respect, it is similar to the so-called corona spray technique [15,16]. However, in the ESD process, the charged aerosol is generated more directly and usually consists of monodispersed primary particles, while the corona spray technique produces its spray by other means, e.g. ultrasonically combined with electrical discharge.

H.L. Tuller et al. (eds.), Oxygen Ion and Mixed Conductors and Their Technological Applications, 295–321.

Like other spray deposition techniques, the ESD technique usually atomizes a precursor solution into an aerosol, which is then directed to a heated substrate to form a thin layer. The ESD process to deposit thin films has been developed only recently, hence only a few reports exist concerning the deposition mechanism. Here we present as an example the results of $LiCoO_2$ thin layers prepared by ESD of ethanol solutions containing lithium and cobalt precursors. The focus will be on the control of the morphology of the layers by controlling the deposition parameters. Furthermore, a number of different morphologies produced by this technique will help us to find a unified deposition mechanism model.

2. Processes involved in ESD

There are several physical and chemical processes involved in the ESD of layers, occurring either sequentially or simultaneously. Possible sequential steps are (viz. Fig.1):

(1) spray formation;
(2) droplet transport, evaporation, disruption;
(3) preferential landing of droplets;
(4) discharge, droplet spreading, penetration of droplet solution, drying;
(5) surface diffusion, reaction.

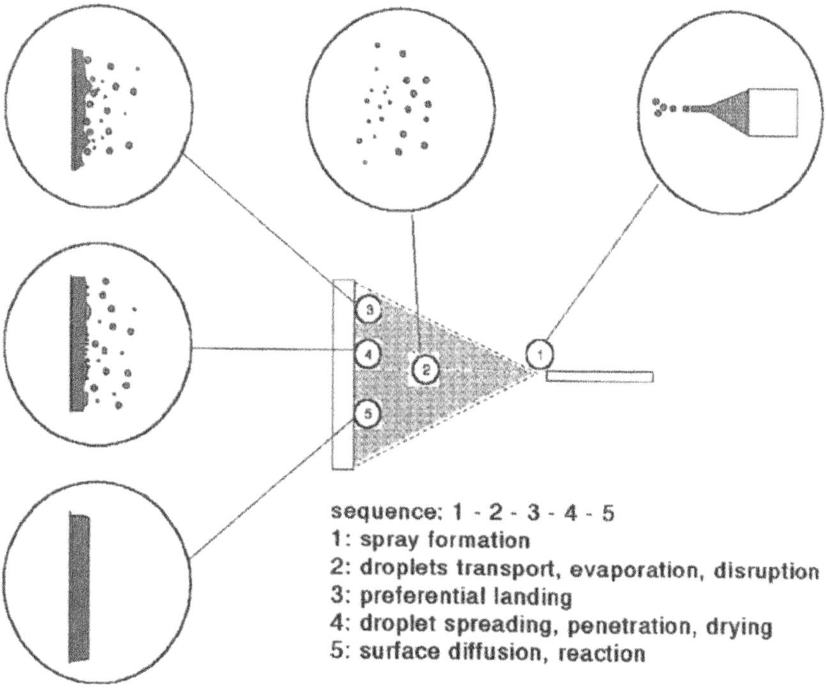

Figure 1. Processes involved in ESD

All of these processes can influence the morphology of the deposited layer. They are described below separately.

2.1 SPRAY PRODUCTION

In the ESD technique a capillary - plate configuration is usually adopted [17]. The precursor solution is placed in a container, which is connected to a metal capillary tube. When a voltage is applied to the capillary, an electrostatic field is immediately set up across the capillary and the grounded plate. This field also penetrates the liquid surface and acts on ions in the solution. In the case of a positive potential at the capillary, positive ions move to the surface of the solution at a rate which depends upon the electrical relaxation time constant t which in turn depends upon the electrical conductivity k and the absolute permittivity of the solution e (= e_re_0 , where e_r is the relative permittivity of the solution and e_0 is the absolute permittivity of free space), as shown in eq.(1):

$$\tau = \frac{\varepsilon}{\kappa} \tag{1}$$

For ethanol solutions, the permittivity is approximately 2 x 10^{-10} F/m [18]. The resistivity of an ethanol solution containing a salt has typical values between 10^{-2} and 1 S/m, therefore, its electrical relaxation time is 2 x 10^{-8} to 2 x 10^{-10} seconds, which means that the surface charge can develop fully in n 1 ms. This surface charge completely shields the bulk of the solution so that there is no free charge inside the solution. The surface charge density s is given by $s = e_0E$ where E is the electric field strength. The surface charge causes an outward electrostatic pressure on the solution, which is opposite to the inward directed pressure from the surface tension. This leads to surface instabilities, which are normally called Rayleigh-Taylor instabilities [19]. Taylor has shown that when the electric field is strong enough, the electrostatically stressed liquid surface can be distorted into a stable conical shape (Taylor cone) [20]. The cone surface is equipotential but the net electric field E_{surf} that exists at the surface of a Taylor cone is given by eq.(2):

$$E_{surf} = 300\left(\frac{8\pi\gamma \cot\alpha}{r}\right)^{1/2} \tag{2}$$

where g is the surface tension of the liquid with respect to the surrounding gas, a the semi-vertex angle of the cone, and r the radius.

An ideal Taylor cone has a semi-vertex angle of 49.3°. At the very apex of a Taylor cone the liquid surface is unstable since according to eq.(2) the electric field would tend to be infinite with r close to zero. In order to compensate for this "infinite" electric field, the Taylor cone will emit charged droplets immediately after the application of an electric potential. In the so-called cone-jet mode [21] (as shown in Fig.1) these primary charged particles are usually monodispersed. Having considered capillary equilibrium, liquid continuity, and moment and charge continuity at the jet, Ganan-Calvo deduced the following relation for polar liquids (eq.(3)) [22]:

$$d \propto \varepsilon_r^{1/6} \left(\frac{Q}{\kappa} \right)^{1/3} \tag{3}$$

where d and Q are the diameter of the droplets emitted at the jet (the primary droplet size) and the feed rate or flow rate of the solution, respectively. As a result, the primary droplet size depends on the flow rate, the conductivity, and the permittivity of the solution.

2.2 AEROSOL TRANSPORT

An electrostatically produced charged droplet of mass m will be attracted towards a grounded substrate by a Coulombic force qE_{sp}, where q and E_{sp} are the droplet charge, and the electric field strength in the travelling space, respectively. Simultaneously, a gravitational force, mg (where m is the mass of the droplet and g is the gravitational acceleration constant), and a viscous drag force 3pdhvC (where h is the dynamic viscosity of air, v is the drop velocity, and C is a correction coefficient) also act on the droplet. The trajectory as well as the flight time taken from the nozzle to the substrate for this droplet will be determined mainly by these forces. The gravitational force may be neglected in the case of electrostatic spraying because droplets produced in this way are very small.

Assuming a homogeneous electric field in the travelling space (i.e. a constant E_{sp}) and a short nozzle-to-substrate distance L, the drag force and the solvent evaporation can be neglected, and the flight time t can be calculated according to eq.(4):

$$t \approx \left(\frac{2\,L\,m}{q\,E_{sp}} \right)^{1/2} \tag{4}$$

On the other hand, if L is long enough, the equilibrium between the Coulombic force and the drag force determines the terminal velocity of the droplet, i.e.

$$v = \frac{q\,E_{sp}}{3\pi\,d\,\eta\,C} \tag{5}$$

and

$$t \approx \frac{3\pi d \eta C L}{q E_{sp}} \tag{6}$$

In addition, there are space-charge forces arising from the repulsive interaction between charged droplets. Moreover, the real situation is further complicated by: (i) the non-uniform temperature profile and the resulting thermophoresis force and (ii) the evaporation of the solvent and the resulting possible droplet disruption (see below). These factors also change the flight speed and time.

2.3 SOLVENT EVAPORATION AND DROPLET DISRUPTION

Alcohol solutions have been used frequently in the ESD process. Solvent evaporation during the flight of a solution droplet is inevitable, especially under heating conditions. The evaporation rate for small volatile drops can be calculated by using eq.(7) [23]:

$$\frac{dd}{dt} = \frac{4D_v M}{R\rho d}\left[\frac{P_\infty}{T} - \frac{P_d}{T_d}\right]\left[\frac{2\lambda + d}{d + 5.33(\lambda^2/d) + 3.42\lambda}\right] \tag{7}$$

where d is the droplet diameter, D_v the diffusion coefficient of the vapour of the solvent in air, M the molecular mass of the solvent, R the gas constant, r the density of the solution, P_4 the partial vapour pressure of the solvent away from the droplet, P_d the partial vapour pressure at the droplet surface, T the ambient temperature, T_d the droplet temperature which is normally < T due to cooling by evaporation, and l the mean free path of air. For a charged droplet, it is not clear whether the evaporation time is influenced by the charge. Even if it is negligible the calculation of the evaporation time is still difficult because the system is not isothermal, and thus T and T_d change from place to place.

Evaporation of the solvent results in shrinkage of the droplet, keeping the total charge the same [17]. A charged droplet may be disrupted into a few smaller droplets, after reaching a maximum attainable charge density, q_R, for a liquid droplet with a radius a. This is the so-called Rayleigh limit [25], which can be expressed as shown in eq.(8):

$$\frac{q_R}{m} = \frac{q_R}{4\pi a^3 \rho / 3} = \frac{6}{\rho}\left[\frac{\gamma \varepsilon_0}{a^3}\right]^{1/2} \tag{8}$$

The disruption of a droplet ("mother droplet") usually occurs with the ejection of a few highly charged, very tiny drops ("daughter droplets").

Therefore, for solutions with rather volatile alcohols as solvents and/or a long nozzle-to-substrate distance and/or a high deposition temperature the effect of droplet disruption should be taken into account. In that case, there is no longer a monodispersed particle size distribution.

In contrast, for solutions with a relatively non-volatile solvent and/or a short nozzle-to-substrate distance and/or a low deposition temperature, the monosized distribution may remain during droplet flight.

2.4 PREFERENTIAL LANDING OF DROPLETS ON THE SUBSTRATE

In the strong electrostatic field, induced charges exist on the surface of the grounded substrate, with a sign opposite to that of the droplets or the nozzle. The charge distribution generally is not uniform, but depends on the position relative to the nozzle and, in particular, on the local curvature of the surface. The charges concentrate more at the places where the curvature is greater. Therefore, the electric field is the strongest there. When a charged droplet approaches the surface, it will be attracted more towards these more curved areas. This is referred to as "preferential landing". This action will cause agglomeration of the particles, especially when the incoming droplets are small (see below). Also, this means that the roughness of the substrate surface may influence the morphology. An increase in the surface roughness will lead to more particle agglomeration.

2.5 DISCHARGE, SPREADING AND PENETRATION OF SOLUTION DROPLETS ON THE SURFACE

As soon as a charged droplet gets in contact with the surface of the substrate or the formed layer, it is discharged by transferring its charge to the grounded substrate either immediately or through the layer to the substrate. This process is very fast according to eq.(1) if the electronic conductivity of the substrate (usually a metal in ESD) and the deposited layer is relatively high. In this case, the discharge process is not expected to determine the morphology of the layer. However, in the case of using insulating substrates or depositing insulating layers, the discharge may proceed slowly and, hence, will influence the morphology. Nevertheless, when wet droplets reach the substrate surface (see below), the discharge process could also be completed through electrical conduction in the concentrated solution on the surface.

If the evaporation of all of the solvent has not been completed when a droplet reaches the surface of the heated substrate, the solution wets the surface of the substrate or the earlier deposited layer. This is usually the case when a high-boiling point solvent is used or deposition occurs at low temperatures. The type and dynamics of spreading depends strongly upon the so-called spreading coefficients (eq.(9)) [26,27]:

$$S = \gamma_{sv}\ \gamma_{sl}\ \gamma_{lv} \tag{9}$$

where g_{sv}, g_{sl} and g_{lv} denote the interfacial tensions between the substrate and ambient gas, between the substrate and the droplet liquid, and between the droplet liquid and ambient gas, respectively. If $S < 0$ only partial wetting occurs with equilibrium reached at a finite contact area. If $S < 0$ the droplet spreads until it completely covers the surface. The value of S is intimately related to the spreading rate. For $S = 0$, r^9 % t, where r is the radius of the contact circle and t is the time, so the rate of evolution of the drop slows rapidly. If $S > 0$, r^4 % t. Therefore, the choice of the substrate will initially affect the spreading rate of the liquid droplet, and may finally affect the morphology of the layer. In addition, the spreading rate is also influenced by the viscosity of the liquid. Qualitatively, the spreading rate decreases with increasing viscosity. However, even if $S > 0$ in an ESD process, the spreading may not be complete if the simultaneous drying process proceeds rapidly. This is why many lamellar particles are usually formed in an ESD layer.

If cracks or pin-holes are formed in the earlier deposited layer, the subsequently arriving solution droplets may penetrate into these defects by capillary action. Therefore, the earlier formed defects are "repaired" in this way, and a crack-free layer is easily obtained. This appears to be another advantage of the ESD technique.

2.6 DECOMPOSITION, REACTION AND SURFACE DIFFUSION OF THE SOLUTE(S)

The decomposition and reaction (either partial or complete) of the solute(s) may have occurred before the droplets reach the substrate, which is expected if the surrounding temperature is high enough and dried droplets have been formed. Rearrangement of these dry particles on the substrate surface by surface diffusion is not expected at moderate deposition temperatures lower than 500°C used in the ESD experiment. In this case, a grain-like structure is expected to be formed instead of a very dense morphology.

On the other hand, at a relatively low temperature the spreading of solution droplets on the surface and the subsequent process, which is actually a wet chemical process of an alcohol solution of metal salt precursors, determines the layer morphology. For the spreading process, the viscosity change of the solution droplets is important. For the wet chemical process, there are many factors which influence the morphology. Specifically, among them are the solution chemistry including the solvation state, for instance, whether there is complexation of the metal ions by the alcohol, evaporation and/or reaction with ambient gas of the solvent on the heated substrate, nucleation and precipitation of the solutes, and dissociation and chemical reaction of the solutes. This may be the only way to form a relatively dense morphology except at an extremely high surrounding temperature at which the whole droplets will be vaporized before reaching the substrate surface. Other morphologies, like a unique porous structure found in our studies, can also be formed by using different solutes and/or different solvents. Furthermore, unlike a normal wet chemical process such as a sol-gel process, this method proceeds via the continuous repetition of many small steps. Therefore, a crack-free layer is more easily formed due to the aforementioned defect-repairing mechanism.

In general, the final morphology of the layer depends upon the relative rate of spreading, precipitation, decomposition, and reaction. For example, if the spreading is slow but the precipitation and the decomposition are fast, the morphology will be granular. The ideal conditions for the formation of a dense layer include at least that: (i) the particles arriving at the substrate are still wet (solutes) (ii) the solubilities of the solutes in alcohol are sufficiently large and (iii) the spreading of the solution droplets is rapid.

3. Experimental considerations

3.1 SET-UP

An ESD set-up consists mainly of the following parts: (1) a precursor solution container; (2) a nozzle, (3) a device for transferring the solution from the container to the nozzle, (4) a high DC-voltage power supply, and (5) a substrate holder provided with heating element able to control the temperature. The schematic illustrations of two ESD set-ups used in our studies are shown in Fig.2. They are horizontal (Fig.2a) and vertical (Fig.2b) configurations, respectively. The nozzle is a hollow metal (e.g. stainless steel) needle with a diameter usually smaller than 1.0 mm. The solution transport may be accurately controlled with a peristaltic pump (Fig.2b) or a syringe pump. Transport can also be achieved by using compressed air or a hydraulic pressure difference (Fig.2a), but in these cases the flow rates cannot be controlled accurately. The DC power supply should be capable of providing voltages up to 15 - 30 kV. The substrate holder is usually made of stainless steel in close contact to a heating element. Although we have found that the polarity of the electric field has no effect on the structure of deposited thin films, the substrate holder is usually grounded for safety reasons.

3.2 PRECURSORS

One of advantages of the ESD technique is the extremely large choice of precursors. In our studies, a variety of chemicals has been used as precursors, such as hydrated metal nitrates (e.g. $Co(NO_3)_2{\bullet}6H_2O$, $Mn(NO_3)_2A4H_2O$), hydrated acetates (e.g. $Li(CH_3COO){\bullet}2H_2O$, $Co(CH_3COO)_2{\bullet}4H_2O$), chlorides (e.g. $SnCl_4$, $TiCl_4$), metallorganic compounds (e.g. $Ti[OCH(CH_3)_2]_4$, $Zr(O_2C_5H_7)_4$), and even non-metal compounds (e.g. H_3BO_3, P_2O_5). The basic considerations here are that: (1) a precursor must be relatively soluble in an alcoholic solvent, (2) it must be decomposed and converted into a desired product at the deposition temperature, and (3) no contaminant or impurity will be introduced to the deposited thin films.

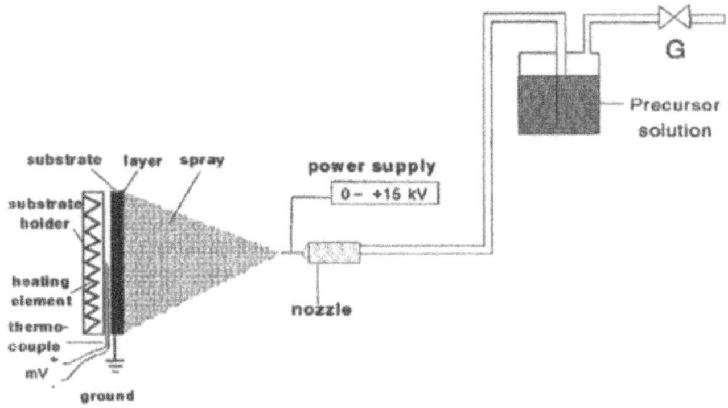

(a)

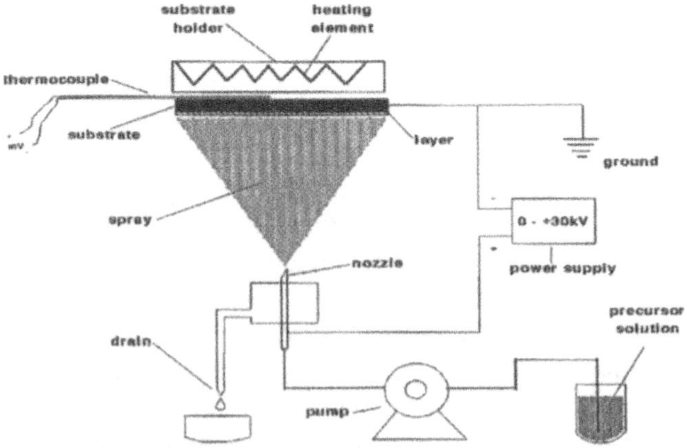

(b)

Figure 2. ESD set-ups with horizontal (a) and vertical (b) configurations

3.3 SOLVENTS

According to eqs. (1)-(3) and (8), a solvent with proper physico-chemical parameters is very important in the atomization of the precursor solution as well as in controlling the droplet size and also the structure of the deposited films because it determines to a great extent the corresponding parameters of the precursor solution. In order for the ESD to work in the cone-jet mode, the conductivity and the surface tension of the solution are the most important parameters. The lower limit of the conductivity varies between 10^{-8} and 10^{-11} S/m, whereas the upper limit can be up to 10^{-1} S/m [28]. The surface tension cannot be too high, otherwise a triggering of corona discharges prevents electrostatic atomization [29]. This is why an aqueous solution is usually not used as the precursor solution. Many organic solvents used in electrospraying and their physical parameters have been summarized by Grace and Marijnissen [17].

Our studies have shown that, in order to control the morphology of the deposited thin films, the boiling point is the most important parameter. The boiling point of the most commonly used solvent, ethanol, is 78°C. To increase the boiling point, another high-boiling-point alcohol may be mixed with ethanol. For instance, butyl carbitol of boiling point 231°C may be used for this purpose. The parameters of other carbitol-related alcohols are listed in Table 1.

TABLE 1. Physical and chemical data of carbitol-related compounds[a]

	$C_3H_8O_3$	$C_4H_{10}O_3$	$C_5H_{12}O_3$	$C_6H_{14}O_3$	$C_8H_{18}O_3$	$C_{10}H_{22}O_3$	C_2H_6O	H_2O
M.W.	92.09	106.12	120.15	134.18	162.23	190.29	46.07	18.015
m.p.(°C)	18.2	-10.4	-70	-55 H	-68	-40	-114.1	0
b.p. (°C)	290	245.8	193	196(202)	231	260	78.2	100
d (g/cm³) [a]	1.2613^{20}	1.1197^{15}	1.035^{13}	0.9885^{20}	0.9553^{20}	0.935	0.789^{20}	1.000^{4}
Solubility [a]	H_2O3, EtOH3, eth2;bz1	H_2O3, eth3, EtOH3, chl3	H_2O5, EtOH4, eth4, ace5	H_2O5, EtOH5, eth4, ace5	H_2O5, EtOH5, eth5, ctc2		H_2O5, EtOH5 eth5, ace5	
n_D	1.4157^{20}	1.4438^{19}	1.4385^{20}	1.4300^{20}	1.4198^{20}	1.4381^{20}	1.361^{20}	1.333
Viscosity[a] (mPa$s)	934^{25}, 152^{50}, 39.8^{75}, 14.8^{100}	30.2^{25}, 11.11^{50}, 4.917^{75}, 2.505^{100}	3.48^{25}, 1.61^{60} H	3.71^{25} H	4.76^{25} H		1.786^{0}, 1.074^{25} 0.694^{50} 0.476^{75}	1.787^{0}, 0.89^{25}, 0.55^{50}, 0.38^{75}
Surface tension [a] (mN/m)	63.14^{17}, 62.5^{25} H	44.77^{25}, 42.57^{50}, 40.37^{75}, 38.17^{100}	34.8^{25}, 29.9^{75} H	31.8^{25}, 27.2^{75} H	30.0^{25} H		23.22^{10}, 21.97^{25}, 19.89^{50}	74.2^{10}, 72.0^{25}, 67.9^{50}, 58.9^{100}
Permitivity [a] (e)	46.53^{20}	31.82^{20}					25.3^{25}	78.54^{25}
Dipole moment D [b]	2.68 (D, 25)	2.69 (D, 25)					1.71 (B, 25)	1.71 (B, 25)
Price (hfl)[c]	98.50 /2.5L	42.20 /2.5L	41.60 /2.5L	33.10 /2.5L	58.70 /3kg	210.70 /3L	154.50 /2L	

a. Unless indicated, all data in the table are from: *D.R. Lide, Handbook of Chemistry and Physics (75th, edn.) (CRC Press, London, 1994)*. The numbers at the superscripts are the values of the temperatures (in °C). For the solubility column, EtOH-ethanol, eth-ethyl ether, bz-benzene, ace-acetone, ctc-carbon tetrachloride, chl-chloroform; 5-miscible, 4-very soluble, 3-soluble, 2-slightly soluble, 1-insoluble.

b. Dipole moment data are from: *A.L. McClellan, Tables of Experimental Dipole Moments (W.H. Freeman and Company, San Francisco, 1963)*. The measurement conditions are shown in the brackets, D-dioxane, B-benzene, 25-25°C.

c. The prices of these compounds are from: *Aldrich Catalogue Handbook of Fine Chemicals (1994-1995) (Aldrich Chemie, 1994)*. H. Data from: *J.A. Dean, Lange's Handbook of Chemistry (14th edn.) (McGraw-Hill Inc., New York, 1992)*.

3.4 SUBSTRATE TEMPERATURE

The current ESD technique is in principle a low-temperature deposition technique, although it may be modified to become a high-temperature technique by, for instance, putting a high temperature furnace between the nozzle and the substrate holder. Therefore, in our studies the substrate temperature is the deposition temperature, since all of the pyrolysis processes and solid state reactions take place at or near the substrate surface. Generally speaking, this temperature should be at least higher than the decomposition temperatures of the precursors.

If a high-temperature furnace is used, the ESD process can be transformed into a Chemical Vapour Deposition (CVD) process. The process sequence: evaporation of solvent, reaction of precursors to solid particles, melting and evaporation of the solid particles schematically describes the change from ESD to CVD.

4. Morphological control of esd derived thin films [7]

In this section, the ESD results of thin films of a model material, $LiCoO_2$, will be demonstrated to show effects of many deposition parameters on the surface morphology of thin films.

Ethanol (C_2H_5OH)(100%) solutions of $Li(CH_3COO)\bullet 2H_2O$ and $Co(NO_3)_2\bullet 6H_2O$ were prepared separately. The solutions were mixed to a molar ratio Li : Co = 1 : 1 and this solution was used as the precursor solution. The Li (or Co) concentrations range from 0.003M to 0.05M. To investigate the effect of solvent, mixtures of alcohols, i.e. ethanol (C_2H_5OH) and butyl carbitol ($C_4H_9OC_2H_4OC_2H_4OH$), were used. A horizontal ESD set-up with capillary C plate configuration was used in most cases(Fig.2a). Stainless steel, platinum or aluminium disks (1.4cm in diameter) were used as substrate. A positive high voltage up to +15kV was applied to the nozzle through which the precursor solution was forced to flow by the pressure difference between the top level of the precursor solution and the solution at the needle orifice. A positively charged spray was generated. The flow rate was controlled using valve G. The nozzle-to-substrate distance was 6 cm. Another set-up with a vertical configuration (Fig.2b) was also used to obtain special morphologies.

To investigate the effect of the substrate, an unpolished alumina square plate partly covered with aluminium foil was used as the substrate, in order to produce the same conditions, necessary for a good comparison. Besides, a thin (0.1mm thick) smooth yttria-stabilized zirconia (YSZ or cubic $(YO_{1.5})_{0.16}(ZrO_2)_{0.84}$) was also used as a substrate under similar deposition conditions.

4.1 GENERAL

It was previously reported that an almost proportional relationship between layer mass and deposition time exists for ESD derived layers [6]. Furthermore, the growth rate in terms of the layer mass per unit time is found to be independent of the deposition temperature. Therefore, over-spray is apparently not a problem in ESD. The overall composition of the layer should be the same as that of the precursor solutions. Elemental analysis by atomic absorption spectroscopy (AAS) for an $LiCoO_2$ layer deposited at 340°C confirmed this conclusion. In fact, this is one of the advantages of the ESD technique in comparison with other spray and non-spray deposition techniques.

In addition to a unique porous microstructure, four types of layer morphologies were observed and they are schematically shown in Fig.3. Type I is a relatively dense layer, type II a relatively dense layer with some particles incorporated, type III consists of a relatively dense bottom layer containing some lamellar particles and agglomerates of lamellar particles on top of this dense layer, forming a porous and sometimes fractal-like structure, while type IV is a very porous structure made of fractal agglomerates of tiny particles. These four types of morphologies are formed under specific deposition parameters, as will be described below.

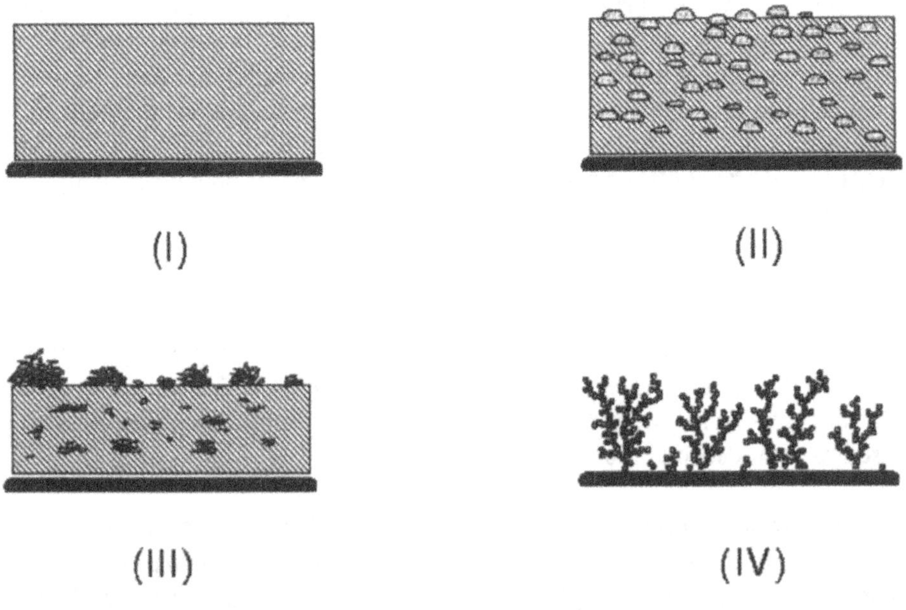

Figure 3. Four types morpholog obtained by ESD. I, dense layer; II, dense layer with inorporated particles; III, porous top layer with dense bottom layer; IV, fractal-like porous layer

4.2 EFFECT OF DEPOSITION TIME

The effect of deposition time on layer morphology deposited at 340°C is shown in Fig.4. It can be seen that the layer deposited within 1 h (ca. 1.5mm thick) is relatively dense, representing type II (Fig.4a). With increasing deposition time and thus increasing layer thickness the morphologies of the layers are shifted to type III (Fig.4b). With a deposition time of 6 h the top section of the layer is very porous (Fig.4c).

This morphology development can be explained by considering the competitive effect between the rates of evaporation, decomposition, and spreading. Probably most of the spray droplets arriving at the substrate are still wet. Initially these drops spread on the metal substrate surface at high speed, because the surface tension of a metal (g_{sv} in eq.(9)) is usually much greater than that of a metal oxide. Therefore, the solution droplets can spread rapidly. Also, the solubilities of $Li(CH_3COO)\cdot 2H_2O$ and $Co(NO_3)_2\cdot 6H_2O$ in ethanol (100 g at 12.5°C and 21.5 g at 25°C, respectively) are sufficiently high. Combining these two factors, a continuous layer is formed. In the mean time, evaporation of ethanol and decomposition of the acetates $Li(CH_3COO)$ and $Co(NO_3)_2$ takes place, producing a relatively dense morphology. It has been established by X-ray diffraction that $LiCoO_2$ is formed at this deposition temperature, i.e. 340EC [6]. Therefore, the reaction between the lithium intermediates such as Li_2O and the cobalt intermediates such as CoO may also contribute to the formation of this dense morphology. The submicron-sized particles incorporated into the layer could be from the disruption of larger "mother droplets" owing to the evaporation of ethanol during the droplet flight and subsequently surpassing the Rayleigh limit. When these disrupted particles arrive at the surface they are likely to be dry. They are, therefore, incorporated into a continuous layer.

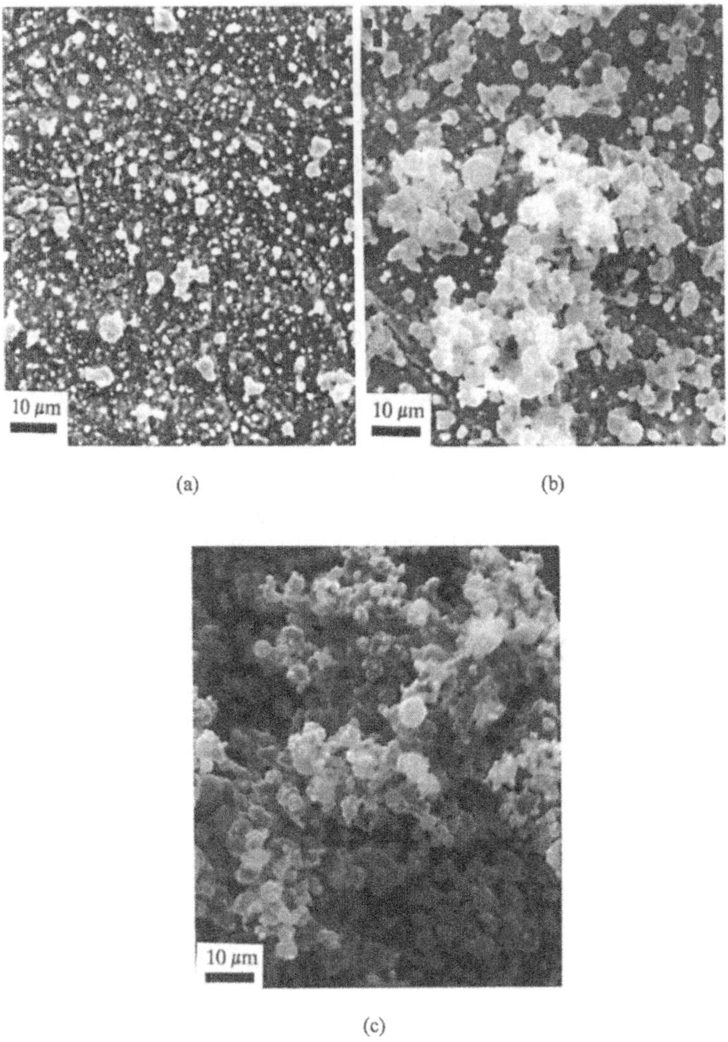

Figure 4. Surface morphologies of layers deposited at 340°C for different deposition times: (a) 1 hour; (b) 3 hours; (c) 6 hours. Precursor solution, 0,04M $Li(CH_3COO){\cdot}2H_2O + Co(NO_3)_2{\cdot}6H_2O$ ethanol solution; substrate, stainless steel; applied voltage, 11kV

With the increase of deposition time and layer thickness, spreading of the solution droplets will occur on the surface of the $LiCoO_2$ layer, which usually has a smaller surface tension than a metal (stainless steel here). In other words, the wetability of the ethanol solution on the $LiCoO_2$ surface is less than that on a metal surface. Therefore, discrete particles may be formed on the surface owing to the slow spreading. Some extent of spreading of the solution leads to the lamellar particles. These discrete particles also increase the surface roughness, which enhances the possibility of preferential landing and agglomeration.

In addition, no crust or hollow particles have been observed when ethanol is used as the solvent. This suggests that the evaporation of ethanol and precipitation of the solutes proceed homogeneously.

4.3 EFFECT OF DEPOSITION TEMPERATURE

Fig.5 shows the differences in morphology of layers deposited at different substrate temperatures. As already shown in Fig.4(c) the layer deposited at 340°C for the same deposition time is quite porous and consists of discrete agglomerates of lamellar particles, belonging to the type III morphology. At lower deposition temperatures (Figs.5a and b) the morphology of the layers is less porous and belongs to type II, but consists of many large particles (4 - 20mm), most of which are also lamellar and "buried" in an amorphous matrix. The continuous "semi-transparent" matrix appears to be formed from the spreading of large droplets. This provides further evidence that the particles landing on the substrate surface are still wet drops. At these relatively low temperatures the incoming droplets are larger and heavier than those at a higher deposition temperature owing to less solvent evaporation. Therefore, their movement direction cannot be changed considerably by the attraction of induced charges at the substrate surface to form agglomerates. There are hardly agglomerates in the layer at the deposition temperature of 230°C (Fig.5a), while minor agglomeration occurs at the deposition temperature of 280°C (Fig.5b). In addition, slower precipitation at low temperatures also favour the formation of the relatively dense morphology. With increasing deposition temperature, the agglomeration extent increases by increasing the effect of preferential landing. The morphology of the layer deposited at 400°C belongs to type III. The agglomeration is substantial but the agglomerates still constitute small lamellar particles (Fig.5c), implying that even at this temperature the incoming droplets are not completely dry. It seems that the incoming droplets are completely dried at 500°C, because lamellar particles are no longer observed and drying traces are absent (Fig.5d). Actually, the morphology of the layer is fractal-like, belonging to type IV. Therefore, with increasing deposition temperature from low to high values the morphology of the deposited layer changes from type II to type IV, i.e. from relatively dense to highly porous.

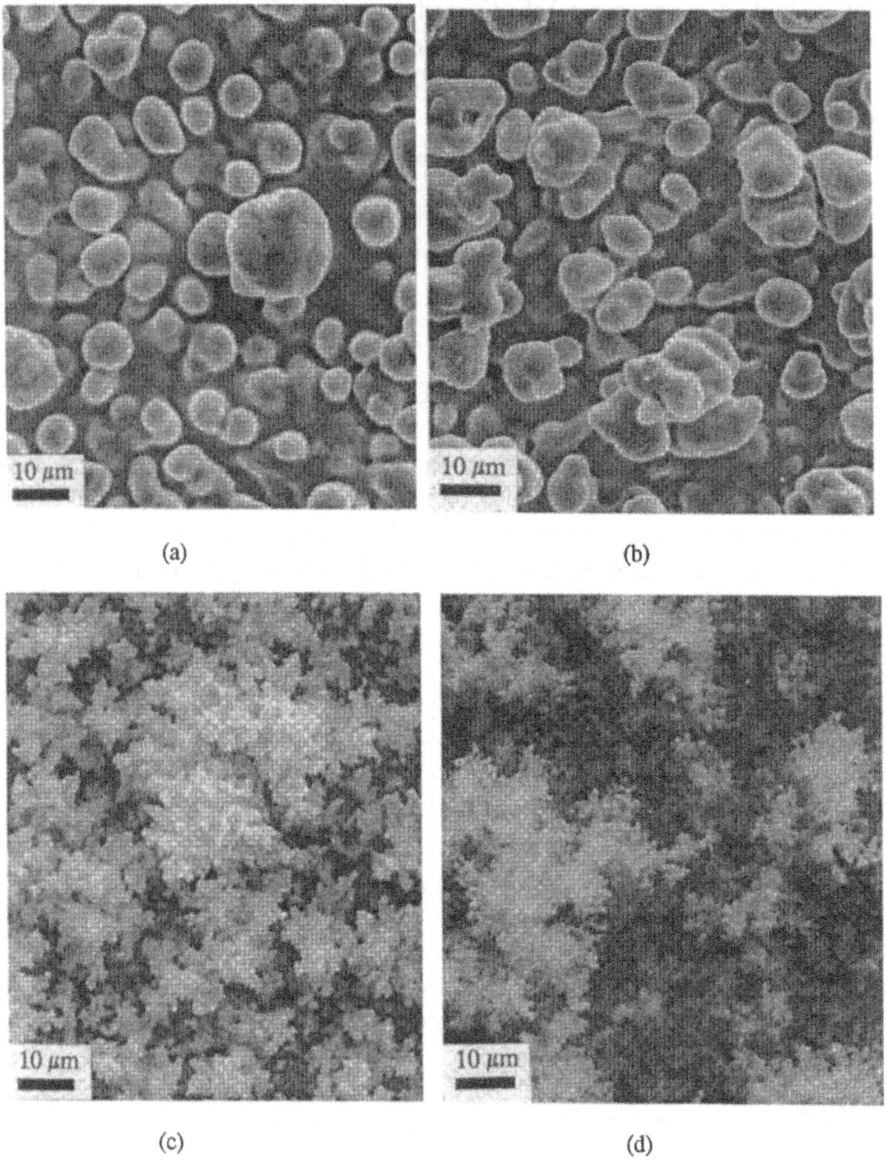

Fig.5 Surface morphologies of layers deposited at different temperatures for 6 hours: (a) 230°C; (b) 280°C; (c) 400°C; (d) 500°C. Precursor solution, 0.04M $Li(CH_3COO)\cdot 2H_2O + Co(NO_3)_2\cdot 6H_2O$ ethanol solution; substrate, stainless steel in (a), (b) and (c), Pt in (d); applied voltage, 11kV.

4.4 EFFECT OF CONCENTRATION OF PRECURSOR SOLUTION

Fig.6 shows two layers, both with type III morphology, prepared with two concentrations of precursor solution. The deposition times are different but the layer thicknesses are similar, i.e. about 0.8mm, which equals the thickness of a layer deposited in 1 h from a 0.04M precursor solution (Fig.4a). Therefore, the influence of the concentration on the morphology is not remarkable as long as the layer thicknesses are similar. However, more scattered agglomerates and lamellar particles are present in the layer obtained with a 0.0038M solution (Fig.6a) compared to that with a 0.01M solution (Fig.6b) and a 0.04M solution (Fig.4a). This is ascribed to the fact that a longer deposition time will result in an increased roughness, as shown in Fig.4. Interestingly, there is no large variation in the particle sizes when precursor solutions with different concentrations are used. According to eq.(3) a lower concentration, and hence a smaller conductivity, will result in a larger primary particle size. However, the solid particle size after drying should also increase with the concentration of the precursor solution. The combination of these two opposing factors may lead to comparable final particle sizes for different concentrations.

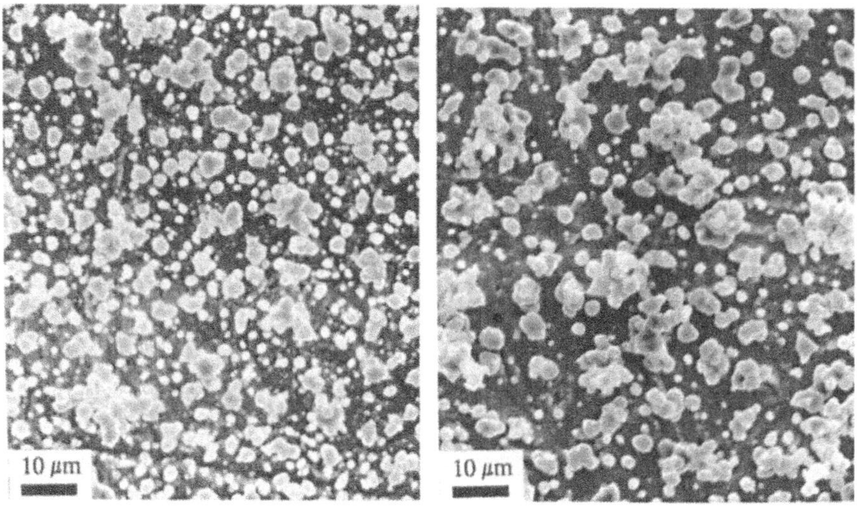

(a) (b)

Figure 6. Surface morphologies of layers deposited at 350°C for 2 h with solutions of different concentrations: (a) 0.0038M; (b) 0.010M. Precursor solutions, $Li(CII_3COO)2H_2O + Co(NO_3)_2{\bullet}6H_2O$ ethanol solution; substrate, stainless steel; applied voltage, 11kV

4.5 EFFECT OF ELECTRIC FIELD STRENGTH

Fig.7 shows two layers deposited by applying 8kV and 15kV, respectively, to the nozzle. Their morphologies both belong to Type III. However, it can be seen that the extent of particle agglomeration increases with decreasing electric field. Therefore, the layer from a stronger electric field (Fig.7b) looks denser than that from a weaker electric field (Fig.7a). Also, the particle size using a weak electric field is smaller. This can be attributed to a shorter flight time of droplets under the stronger field according to either eq.(4) or eq.(6), and, hence, this results in less solvent evaporation and larger incoming droplets at the substrate surface. Another reason might be a stronger preferential landing effect existing in a stronger electric field.

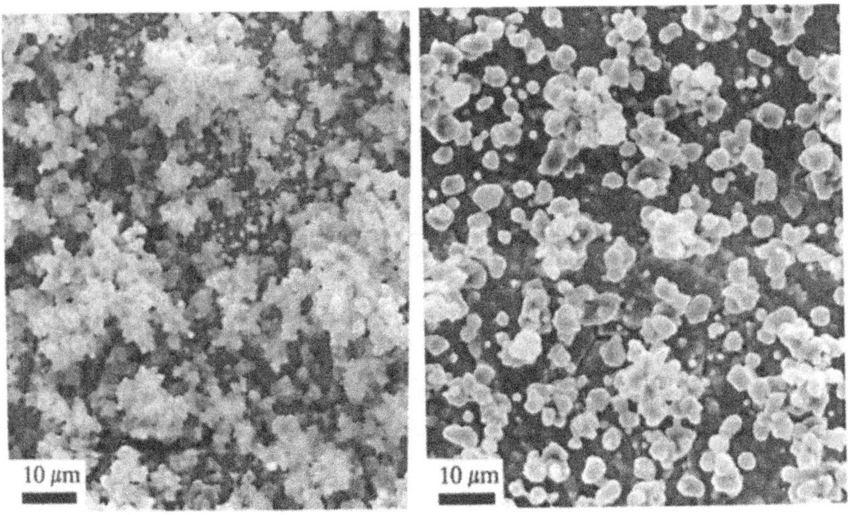

(a) (b)

Figure 7. Surface morphologies of layers deposited at 350°C for 4 h with different applied high voltages: (a) 8 kV; (b) 15kV. Precursor solution, 0.04M $Li(CH_3COO){\cdot}2H_2O + Co(NO_3)_2{\cdot}6H_2O$ ethanol solution; substrate, stainless steel

4.6 EFFECT OF SUBSTRATE

Three layers deposited on different types of substrate are shown in Fig.8. For the two layers simultaneously deposited on aluminium and alumina, the morphologies are quite different. The layer on aluminium (Fig.8a) is rather dense whereas that on alumina (Fig.8b) is not. The latter (Fig.8b) consists of more agglomerates than that on the aluminium substrate (Fig.8a). Note that originally some cracks or cavities are present in the alumina substrate, which enhance the preferential landing effect and accordingly lead to the formation of more agglomerates. In addition, a large difference in the dielectric property between aluminium and alumina may also cause the electric field to be stronger near aluminium than near alumina. This might also contribute to the different morphologies obtained on these substrates. However, under similar deposition conditions, a rather dense layer can also be formed on a thin (0.1 mm thick) and smooth YSZ substrate (Fig.8c). This suggests that the presence of cracks and cavities in the alumina substrate used in Fig.8b is the main reason for the formation of agglomerates. The effect of the thickness of the ceramic substrate on the layer morphology is unclear and requires further study.

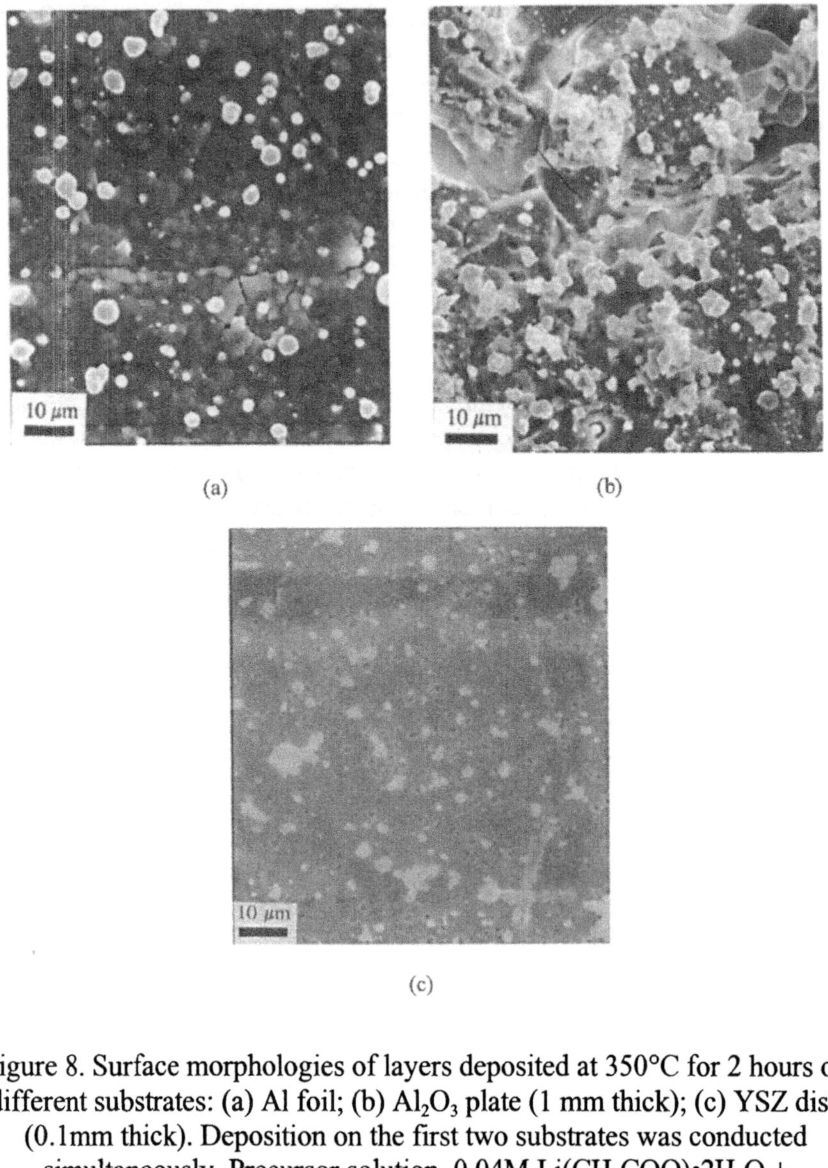

(a) (b) (c)

Figure 8. Surface morphologies of layers deposited at 350°C for 2 hours on different substrates: (a) Al foil; (b) Al_2O_3 plate (1 mm thick); (c) YSZ disk (0.1mm thick). Deposition on the first two substrates was conducted simultaneously. Precursor solution, 0.04M $Li(CH_3COO){\cdot}2H_2O$ + $Co(NO_3)_2{\cdot}6H_2O$ ethanol solution; applied voltage, 11kV

4.7 EFFECT OF SOLVENT

As discussed above the morphology of a deposited layer is largely determined by the droplet size and physical properties, such as boiling point and spreading behaviour on the substrate, of incoming droplets, and in particular the solubilities of the precursor solutes. By changing the solvent composition, the layer morphology may also be modified. Fig.9(a) and (b) show that the morphology changes from type IV to type III, when a mixture of 67 vol% ethanol + 33 vol% butyl carbitol is used as the solvent instead of 100% ethanol. The boiling point of butyl carbitol is ca. 230°C, while that of ethanol is only 78°C. Therefore, at 450°C the droplets arriving at the substrate are probably dried particles when pure ethanol is used as solvent, but they are still wet in the instance of mixtures with a higher boiling point. When using 50vol% ethanol + 50vol% butyl carbitol as the solvent at 250°C the layer (Fig.9c) is relatively dense, and hardly any particle can be discerned. It belongs to the type I morphology. Compared with the layers formed using pure ethanol solution (Fig.5a or b) the morphology is denser. This is due to slower evaporation of solvent, both during transport to and spreading on the substrate surface, and accordingly, a slower precipitation. Therefore, by using high boiling point solvents the morphology of a layer becomes denser.

Fig.9(d) shows the unique morphology of a layer deposited at 230°C using a vertical set-up with a solution containing 15vol% ethanol + 85vol% butyl carbitol as solvent. Note that cobalt acetate instead of nitrate was used in this case. The layer is a highly porous and three-dimensional interconnected structure with a narrow pore size distribution. The pore size is ca. 8mm. It appears to be a stable structure as the network remains unchanged after annealing at 450°C. The formation mechanism for this unique reticulate structure is not yet clear. However, the precipitation step must play a crucial role, because it is found that such a structure cannot be obtained by using cobalt nitrate. According to our experience, the solubility of cobalt nitrate in the solvent is much larger than that of cobalt acetate. Therefore, this morphology is probably formed during the spreading of wet droplets. In addition, there could be chelation between cobalt acetate and butyl carbitol, as the colour of the precursor solution is dark green, rather than pink which is the colour of cobalt acetate and the precursor solution using cobalt nitrate. The possible chelating effect might cause an increase in the viscosity of the solution during the spreading step. This may also contribute to the formation of the structure. Owing to the unique structure and its potential application in many electrochemical devices, it has been studied in more detail [8].

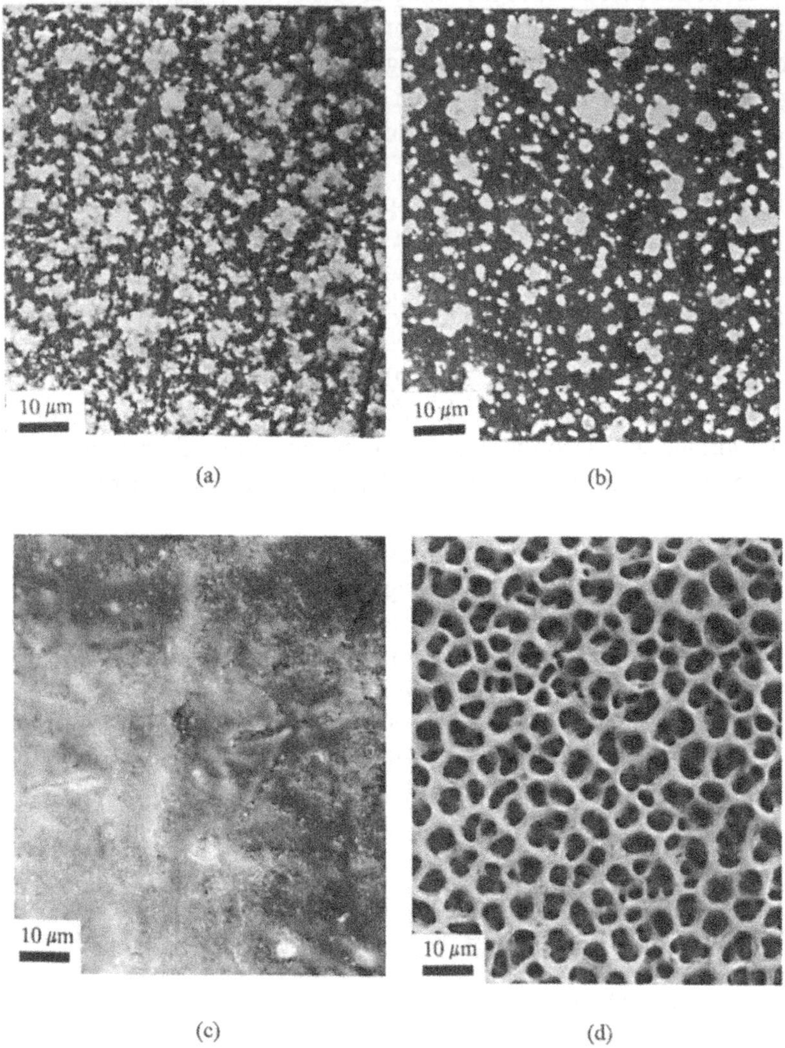

Figure 9. Surface morphologies of layers deposited using precursor solutions with different solvent compositions: (a) 100 vol% ethanol solution (0.04M), at 450°C for 2 h; (b) 67 vol% ethanol + 33 vol% butyl carbitol solution (0.04M), at 450°C for 2 h; (c) 50% ethanol + 50 vol% butyl carbitol solution (0.02M), at 250°C for 4 h; (d) 15 vol% ethanol + 85 vol% butyl carbitol solution (0.005M), at 230°C for 2 h. (a)-(c): $Li(CH_3COO){\bullet}2H_2O + Co(NO_3)_2{\bullet}6H_2O$ precursor, using the horizontal set-up; (d) $Li(CH_3COO){\bullet}2H_2O + Co(CH_3COO)_2{\bullet}4H_2O$ precursor, using the vertical set-up

5. Conclusions

The electrostatic spray deposition (ESD) technique opens up the opportunity to control the morphology of a layer. The morphology is influenced by the spray droplet size (especially the size of the incoming droplets), the deposition temperature, the spreading rate of solution droplets on the substrate, and the solution chemistry including the precipitation process at low deposition temperatures or utilizing a high-boiling point solvent, and the pyrolysis or reaction of the solutes.

However, the main factor that determines the layer morphology is the substrate temperature. The higher the substrate temperature, the more porous the layer. The layer deposited at elevated temperatures with ethanol as solvent is very porous and has a fractal morphology. At moderate temperatures and in an early stage the deposited layer on a metal substrate is relatively dense, but becomes porous with increasing thickness. The concentration of the precursor solution has a minor effect on the layer morphology. Basically, it is easier to obtain a denser layer using a higher concentration compared to using a lower concentration. The electric field strength can influence the flight time of the charged particles. Above the onset voltage the higher the applied voltage, the denser the layer. For a ceramic substrate, its surface roughness (or the smoothness) can affect the morphology of a deposited layer. Cracks or cavities present in the substrate will lead to the formation of more agglomerates. At the same deposition temperature, a layer formed using a precursor solution with a high boiling point solvent is generally denser than one obtained using a low boiling point solvent. A unique porous structure, however, can also be obtained with a high boiling point solvent.

Acknowledgements

The Foundation for Chemical Research in the Netherlands (SON) under the Netherlands Organization for Scientific Research (NWO) is acknowledged for financial support. One of the authors (JS) is grateful to Prof. Dr. H. Tuller of MIT for kind hospitality during a sabbatical leave. Dr. Agnes van Zomeren is acknowledged for critically reading the manuscript.

References:

1. Coffee, R.A. (1981) Outlook Agr., **10**, 350.
2. Slamovich, E.B. and Lange, F.F. (1988) Mater. Res. Soc. Symp. Proc., **121**, 257.
3. Vercoulen, P.H.W., Camelot, D.M.A., Marijnissen, J.C.M., Pratsinis, S. and Scarlett, B., in *Synthesis and measurement of ultrafine particles*, eds. J.C.M. Marijnissen and S. Pratsinis, Delft University Press, Delft, 1993, p. 71.
4. Zomeren, A.A. van, Kelder, E.M., Marijnissen, J.C.M. and Schoonman, J., J. Aerosol Sci., 1994, **25**, 1229.
5. Kelder, E.M., Nijs, O.C.J. and Schoonman, J., Solid State Ionics, 1994, **68**,
6. Chen, C.H., Buysman, A.A.J., Kelder, E.M. and Schoonman, J., Solid State Ionics, 1995, **80**, 1.
7. Chen, C.H., Kelder, E.M., van der Put, P.J.J.M., and Schoonman, J., J. Mater. Chem. 1996, **6**,765.
8. Chen, C.H., Kelder, E.M., and Schoonman, J., J. Mater. Sci. 1996, **31,** 5437.
9. Chen, C.H., Kelder, E.M., Jak, M.J.G., and Schoonman, J., Solid State Ionics 1996, **86-88**, 1301.
10. Chen, C.H., Nord-Varhaug, K., and Schoonman, J., J. Mater. Synthesis and Processing 1996, **4**, 189.
11. Chen, C.H., Kelder, E.M., Put, P.J.J.M. van der, and Schoonman, J., in Proc. Symp. on *Exploratory research and development of batteries for electric and hybrid vehicles*, eds. W.A. Adams, B. Scrosati, and A.R. Landgrebe, PV96-14, pp.43.
12. Chen, C.H., Kelder, E.M., and Schoonman, J., J. Power Sources, in press.
13. Chen, C.H., Kelder, E.M., and Schoonman, J., J. Electrochem. Soc., in press.
14. Salata, O.V., Dobson, P.J., Hull, P.J. and Hutchison, J.L., Thin Solid Films, 1994, **251**, 1.
15. Siefert, W., Thin Solid Films, 1984, **120**, 267.
16. Siefert, W., Thin Solid Films, 1984, **120**, 275.
17. Grace, J.M. and Marijnissen, J.C.M., J. Aerosol Sci., 1994, **25**, 1005.
18. Weast, R.C., *Handbook of Chemistry and Physics*, 56th edn., CRC Press, 1975, E-56.
19. Bailey, A.G., *Electrostatic spraying of liquids*, John Wiley & Sons, New York, 1988.
20. Taylor, G.I., Proc. Roy. Soc. London, A, 1964, **280**, 383.
21. Cloupeau, M., and Prunet-Foch, B., J. Electrostatics, 1990, **25**, 165.
22. Ganan-Calvo, A.M., J. Aerosol Sci., 1994, **25**, Suppl.1, S309.
23. Hinds, W.C., *Aerosol technology*, John Wiley and Sons, New York, 1982.
24. Abbas, M.A. and Latham, ‚J., Fluid Mech., 1967, **30**, part 4, 663.
25. Rayleigh, L., Phil. Mag. Series, 1882, **5**, 184.
26. Davies, J.T. and Rideal, E.K., *Interfacial Phenomena*, 1961, Academic, New York.
27. Beaglehole, D., in *Fluid Interfacial Phenomena*, ed. C.A. Croxton, John Wiley

& Sons, New York, 1986, p.523.
28. Cloupeau, M. and Prunet-Foch, B., J. Electrostatics, 1989, **22**, 135.
29. Burayev, T.K. and Vereshchagin, I.P., Fluid Mech., Sov. Res. 1972, **1**, 56.

DENSE CERAMIC ION CONDUCTING MEMBRANES

B.C.H.STEELE
Imperial College,
London, SW7 2BP, UK

1. Introduction

Inorganic membranes can be divided into porous and dense membranes. Useful surveys of porous inorganic membranes and their applications are available in a variety of publications [1]. Dense (nonporous) membranes can be subdivided into ceramic membranes, metal membranes, and liquid-immobilised membranes. The third category consists of a porous support in which a semipermeable liquid is immobilised and completely fills the pores (cf. molten carbonate fuel cell). Other examples include molten salts that are semi-permeable to oxygen and ammonia [1].

The present survey is concerned with dense ceramic ion conducting membranes (CICM), and benefits from an excellent review recently published [2] on this topic. Most examples of CICM materials are mixed conducting oxides exhibiting high electronic and oxygen ion conductivities.These requirements can be satisfied by non-stoichiometric perovskite oxides, and most investigations so far have involved these systems following the pioneering work of Teraoka et al. [3].

Many market opportunities exist for CICM materials assuming they can be manufactured economically, and exhibit sufficient reliability and durability. Initially CICM technology is expected to be exploited for the small-scale generation of oxygen for medical applications and aerospace life-support systems, and for the production of useful chemicals[4] including the production of syngas from methane using ceramic partial oxidation reactors.

This survey commences with an examination of the parameters influencing oxygen transport through CICM materials, and then proceeds to examine briefly some of the technological applications of this exciting class of ceramics.

H.L. Tuller et al. (eds.), Oxygen Ion and Mixed Conductors and Their Technological Applications, 323–345.

2. Transport Properties

2.1 IONIC TRANSPORT

The phenomenological expression for ionic conductivity (σ_i) is given by a relationship of the type:

$$\sigma_i\,(T) = A\exp(-\Delta E/RT), \qquad (1)$$

where A represents the pre-exponential factor, and ΔE is an activation energy for ionic transport.

A compilation of ionic conductivity values for selected oxide electrolytes and ceramic membranes is provided in Fig.{1} Attention is drawn to the large partial ionic conductivities that can be exhibited by mixed conducting oxide perovskites in the La-Sr-Co-Fe-O system. Furthermore the area specific resistivity (ASR) equals L/σ_i, and so for a given target value (eg. $0.15\Omega cm^2$), the corresponding thickness (L) of the electolyte or membrane can be evaluated.

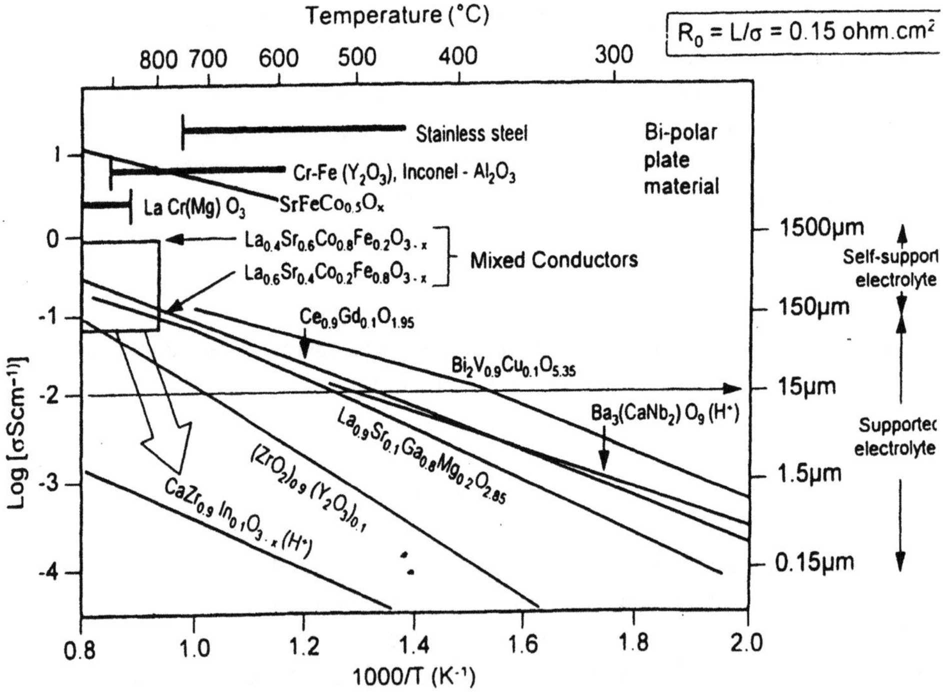

Fig. 1 Specific ionic conductivity values for selected ceramic oxide ion conducting membranes as a function of reciprocal temperature

To examine the parameters controlling the magnitude of oxygen ion conductivity in mixed conductors it is useful to start with the Nernst-Einstein equation:

$$\sigma_i = \frac{(zF)^2 D^* C_O}{RT} \tag{2}$$

where D* is the self-diffusion coefficient of oxygen ions, and C_O, the molar concentration of oxide ions. Making use of the relationship:

$$D^* C_O = D_v C_v \tag{3}$$

the Nernst-Einstein equation can be re-written as:

$$\sigma_i = \frac{(zF)^2 C_v D_v}{RT}, \tag{4}$$

Inspection of equation (4) indicates that for large oxygen fluxes the product $C_v.D_v$ should be as large as possible. However for perovskite oxides (ABO_3), complilations of D_v data [eg.,5,6] indicate that the magnitude of this parameter for a given crystallographic structure is comparable for different compositions, and so it is essential that the value of C_v is optimised. This is usually accomplished by using high concentrations of the aliovalent dopant ion (eg., Sr^{2+}) and a B cation (eg., Co^{3+}) whose valency can easily accommodate to changes in oxygen partial pressure.

The classical Wagner expression [7] for mass transport through a mixed conducting oxide exposed to a chemical potential gradient is:

$$j = \frac{RT}{(zF)^2 L} \int_{\ln p'_{o2}}^{\ln p''_{o2}} \frac{\sigma_e \sigma_i}{\sigma_e + \sigma_i} d\ln p_{o2} \tag{5}$$

For steady state conditions, and assuming $\sigma_e >> \sigma_i$, with σ_i constant over the relevant oxygen partial pressure range, it follows:

$$j = \frac{RT}{(zF)^2} \frac{\sigma_i}{L} \int_{\ln p'_{o2}}^{\ln p''_{o2}} d\ln p_{o2} \tag{6}$$

This expression can be further transformed by substitution of equation (4) for σ_i, viz:

$$j_{o2} = \frac{D_v}{4L} \int_{\ln P'_{o2}}^{\ln P''_{o2}} C_v d\ln P_{o2} \tag{7}$$

For small changes in oxygen partial pressure the variation of C_v with P_{O2} can often be expressed by a simple relationship, $C_v \; \alpha \; P_{O2}^n$, as a single predominant defect equilibria exists and so the modified Wagner expression (7) can be integrated assuming a constant value for D_v. This has been carried out, for example, on $La_{0.9}Sr_{0.1}FeO_{3-x}$ [8], and reasonable agreement was obtained on thick samples as shown in Fig.{2}. The situation is more complicated when large $\Delta\mu_{O2}$ gradients are imposed across a dense oxide membrane, as different defect equilibria will exist over the relevant oxygen partial pressure range making integration more difficult [9,10].

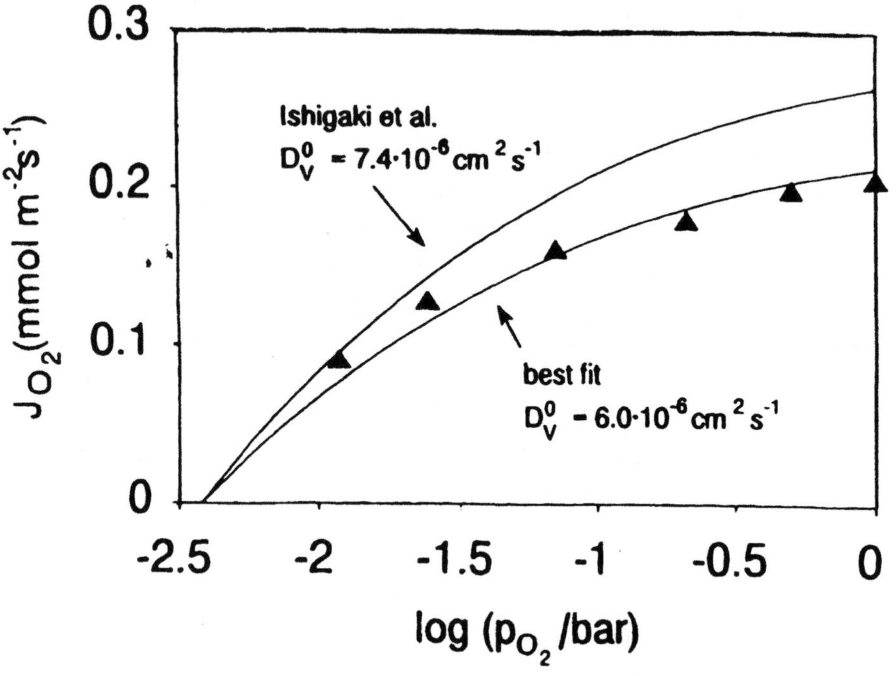

Fig.2 Theoretical fit of feed side P_{O2} dependence of oxygen permeation through $La_{0.9}Sr_{0.1}CoO_{3-\delta}$ at 1000C. The best fit is obtained when D_v equals $6x10^{-6}$ cm^2s^{-1}, which deviates slightly from the corresponding value obtained from isotopic exchange.(Reprinted from J.E.Ten Elsof, H.J.M.Bouwmeester,and H.Verweij, *Solid State Ionics*, 1995, **81**, 97-109

The need for large concentrations of anion vacancies introduces other problems as defect interactions are likely, leading to ordered structures and microdomain textures which effectively trap a significant fraction of the oxygen vacancies [11] resulting in lower fluxes. Moreover when additional vacancies are generated as the oxygen partial pressure is lowered the unit cell volume expands. This process can produce sufficient local strains in an oxide subjected to an oxygen partial pressure gradient to cause structural failure [12,13].

Reliable data for D_v are rather sparse but it does appear at present that values of this parameter are not a strong function of composition or degree of non stoichiometry. However measurements can sometimes be difficult to interpret owing to the possibility of trapping of the mobile species [14], and the role of grain boundaries which can depend upon the purity and processing route used to prepare the samples under examination. The grain boundaries can act as barriers or routes for fast ion transport as reported [15] for the composition $La_{0.6}Sr_{0.4}Co_{0.2}Fe_{0.8}O_{3-x}$ (LSCF) at intermediate temperatures. Fortunately the distribution of ^{18}O can now be revealed by modern dynamic SIMS instruments with enhanced spatial resolution.

2.2 SURFACE OXYGEN EXCHANGE

For most technological applications it is desirable that that oxygen fluxes through CICM materials attain values corresponding to at least 1 Acm^{-2}. This value translates into 2.6μmoles $O_2.cm^{-2}.s^{-1}$, or1.56 x 10^{18} molecules $O_2.cm^{-2}.s^{-1}$, or 3.5 $cm^3O_2.cm^{-2}.min^{-1}$. This flux requirement can be placed into perspective by considering how many O_2 molecules (n) per unit area collide with an oxide surface, assuming temperature and pressure values of 1000K and 1 bar, respectively.

$$n = p/(2\pi mkT)^{1/2} = 2.2 \times 10^{24} cm^{-2} s^{-1} \qquad (8)$$

This implies that approximately 1 in 10^6 of the molecules striking the surface have to be adsorbed which requires high values for the accomodation coefficient which is not normally a feature of high temperature gas/solid reactions. Moreover if we assume 10^{14}-10^{15} oxygen sites cm^{-2}, then the site turnover rate is about $10^3 s^{-1}$. Again this value represents a very high rate for an heterogeneous reaction. It is not surprising therefore that the surface exchange reaction can often limit the oxygen flux through CICM materials.

Although the oxygen surface reaction involves a series of individual reaction steps including; adsorption, dissociation, charge transfer, surface diffusion of intermediate surface species (eg. O_{ad}, $O_2^-{}_{ad}$, $O^-{}_{ad}$, $O^{2-}{}_{ad}$,etc), and finally incorporation into a vacancy in the surface layer, the overall reaction can be represented by the equation:

$$\tfrac{1}{2}O_2 + V_o^{\cdot\cdot} + 2e- = O_o^x, \qquad (9)$$

In subsequent discussions it is important to realise that the above equation is not a charge-transfer reaction as it involves the neutral combination of charged species.

Near equilibrium the oxygen flux (j_O) across a gas/solid interface will be governed by a relationship [16] of the type:

$$j = j_{eq}^{0}\left[\exp\left(\frac{\alpha_f}{RT}\Delta\mu\right) - \exp\left(\frac{\alpha_b}{RT}\Delta\mu\right)\right] \tag{10}$$

where j_{eq}^{0} (mols. of oxygen atoms cm^{-2} s^{-1}) is a 'neutral exchange flux density', $\Delta\mu$ is the chemical potential driving force across the gas/solid interface, with α_f and α_b representing constants that depend upon the specific forward and backward reaction mechanisms.

Isotopic $^{18}O/^{16}O$ exchange experiments [17,18], which yield a value for the oxygen surface exchange coefficient (k $cm.s^{-1}$), provide a means of measuring the magnitude of j_{eq}^{0} as it has been shown [19] that:

$$j_{eq}^{0} = kC_O = k / V_m, \tag{11}$$

where C_O is the molar concentration of oxide ions in the ceramic, and V_m represents the associated molar volume.

This expression, involving a measurable parameter, the oxygen surface exchange coefficient (k), enables the relative importance of the oxygen surface reaction and bulk ionic conductivity to be analysed in a simple manner.

2.3 OXYGEN FLUXES THROUGH **CICM** COMPONENTS

A useful semi-quantitative approach incorporating values for the oxygen surface exchange coefficient has been used by Steele [20,21,22] for the interpretation of the kinetics of porous oxide cathodes and for the selection of dense ceramic membranes, and is summarised below.

The classical Butler-Volmer relationship for electrode kinetics is given by:

$$i = i^{o}\left[\exp\left(\frac{\alpha_{ef} zF}{RT}\eta\right) - \left(\frac{\alpha_{eb} zF}{RT}\eta\right)\right] \tag{12}$$

By analogy between equations (10) and (12) it is possible to suggest the relationship:

$$i^{0} = j_{eq}^{0} zF = kC_0 zF \tag{13}$$

For small overpotentials (<50mV) the Butler-Volmer equation may be simplified to:

$$i = \frac{i^{0} zF\eta}{RT} \tag{14}$$

which allows an electrode resistance, R_e, to be defined for the linear portion of an I-V plot,

$$\text{ie,}\quad R_e = \frac{\eta}{i} = \frac{RT}{zF}\frac{1}{i^0} = \frac{RT}{zF}\frac{1}{kC_0 zF} = \frac{RT}{(zF)^2}\frac{1}{kC_0} = \frac{D^*}{\sigma_i k} \tag{15}$$

using the Nernst-Einstein relationship for ionic conductivity,

$$\sigma_i = \frac{D^*(zF)^2 C_o}{RT}$$

The analogy between equations (10) and (12) is discussed by Adler et al.[16] and care should be taken in stretching the analogy too far. A charge transfer reaction may proceed with a purely electrical driving force in which the composition and thus i^0 retain their equilibrium values. A chemical reaction, however, proceeds only when the composition is displaced from equilibrium. Thus in general, j^0_{eq} , will be a complex function of both the concentrations and driving forces.

Using the above approach it is now possible to construct the following equivalent circuit {Fig.3}to represent the oxygen flux (current) through a dense ceramic membrane. For this initial analysis we shall assume that the two interfacial resistances, R_e, are equal, and steady state conditions apply.

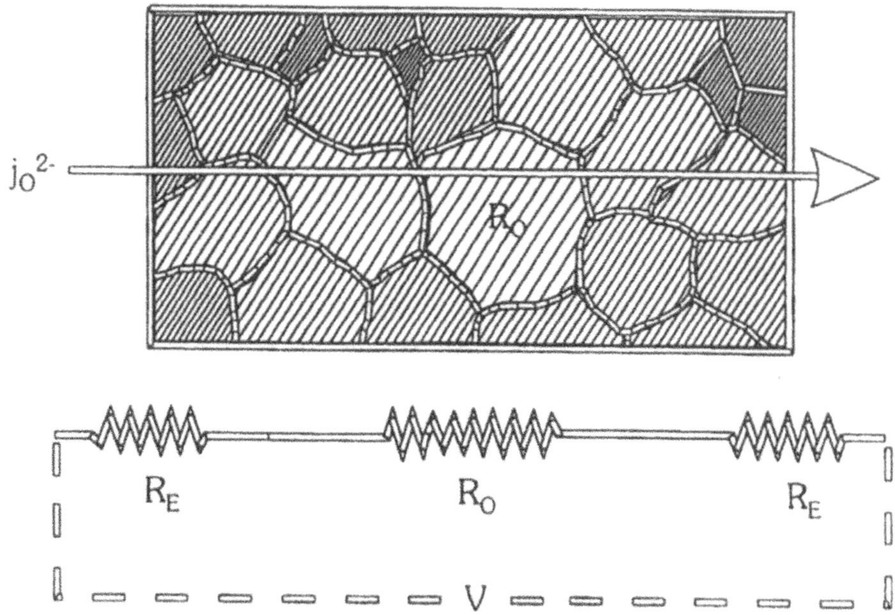

Fig.3 Schematic equivalent circuit for oxygen flux through dense ceramic oxide membrane

From Ohm's law,

$$i = \frac{\eta}{R_t} = \eta \Big/ (R_o + 2R_e), \tag{16}$$

$$\eta = \frac{RT}{zF} \ln p''_{o2} / p'_{o2}$$, and making substitutions for R_o and R_e,

$$j = i / zF = \frac{RT}{(zF)^2} \ln p''_{o2} / p'_{o2} / \Big/ \left(\frac{L}{\sigma_i} + \frac{2D^*}{\sigma_i k} \right) \tag{17}$$

Clearly the changeover from a situation in which the oxygen flux is controlled by the surface reaction to one controlled by bulk diffusion is given by $R_o = R_e$, for a single interface:

ie, $L/\sigma_i = D^*/\sigma_i k$, and a critical thickness, L_c, can be defined which equals D^*/k. The parameter, L_c, is simply the reciprocal of the quantity, h, (k/D^*) which is involved in the interpretation of $^{18}O/^{16}O$ isotopic exchange/SIMS measurements [23].

Using this parameter it follows that:

$$j = \frac{RT}{(zF)^2} \frac{\sigma_i}{L} \ln p''_{o2} / p'_{o2} \cdot \frac{1}{(1 + 2L_c / L)} \tag{18}$$

Inspection of equations (6) and (18) indicates that the diffusional flux through a mixed conductor is reduced by a factor $(1+2L_c/L)^{-1}$ when interfacial reactions are taken into account.

Experimental data for oxygen fluxes through dense ceramic membranes can be interpreted in terms of the above simple equations which may also be used for preliminary evaluations regarding the feasibility of technological applications. However comparison between calculated and measured fluxes often indicates that more complicated models are required to reproduce the experimental data. It is appropriate, therefore, to examine the various relevant parameters in more detail.

2.4 CHARACTERISTIC MEMBRANE THICKNESS (L_c)

The most direct way of measuring L_c is by the isotopic exchange-diffusion profile (IEDP)technique [17,] pioneered at Imperial College. A recent survey [18] of the method outlines the principles and discusses the experimental criteria that can influence the accuracy and range of D* and k values that can be obtained with confidence. Whilst D* can be measured with great precision and reproducibility, there are more problems in obtaining good data for k, particularly as the surface properties can be influenced by segregation, space charges, processing route, and various adsorbates. Analysing published data Kilner [24] has noted correlations between D* and k, and has emphasised that large concentrations of both oxygen vacancies and electronic species should be present to ensure high values for k. Solid electrolytes, for example, generally have low k values [25] with high activation energies due to the lack of electrons even though the concentration of anion vacancies is high. A selection of D* and k data are provided in Table 1 together with the corresponding values for L_c, and R_e. The relative influence of D* and k on the oxygen flux through a membrane interface as a function of thickness for a variety of mixed conductors is clearly evident in Fig {4}.

TABLE 1. Tracer Diffusion Coefficient, D*; Surface Exchange Coefficient, k; Characteristic Thickness, L_c or δ^a; Interfacial resistance, R_e or $R_{chem}{}^a$, for selected perovskite oxides.

Material	T	D*	k	L_c	δ	R_e	R_{chem}
	K	cm^2s^{-1}	$cm\ s^{-1}$	μm	μm	Ωcm^2	Ωcm^2
$La_{0.5}Sr_{0.5}MnO_{3-x}$	973	$2x10^{-15}$	$1x10^{-8}$	0.002	0.02	241	9000
	1173	$3x10^{-12}$	$9x10^{-8}$	0.3	0.2	32	90
$La_{0.8}Sr_{0.2}CoO_{3-x}$	973	$1x10^{-8}$	$3x10^{-6}$	30	2	0.8	0.2
	1173	$4x10^{-8}$	$2x10^{-5}$	20	3	0.1	0.05
$La_{0.6}Ca_{0.4}Co_{0.8}Fe_{0.2}O_{3-x}$	973	$2x10^{-8}$	$4x10^{-6}$	50	3	0.6	0.1
	1173	$3x10^{-7}$	$4x10^{-5}$	70	4	0.07	0.01
$La_{0.6}Sr_{0.4}Co_{0.8}Ni_{0.2}O_{3-x}$	973	$3x10^{-8}$	$2x10^{-6}$	200	4	1.2	0.2
	1173	$4x10^{-7}$	$2x10^{-6}$	2000	18	1.4	0.05
$La_{0.6}Sr_{0.4}Co_{0.6}Ni_{0.4}O_{3-x}$	973	$2x10^{-9}$	$7x10^{-7}$	30	2	3.4	1.0
	1173	$3x10^{-7}$	$3x10^{-6}$	1000	13	0.9	0.05
$La_{0.6}Sr_{0.4}Co_{0.4}Ni_{0.6}O_{3-x}$	973	$1x10^{-8}$	$3x10^{-7}$	300	7	8.0	0.7
	1173	$6x10^{-7}$	$2x10^{-6}$	3000	22	1.4	0.04

a) calculated according to model proposed in ref. [16]

For thick film membranes Fig {4} indicates that the surface exchange reaction often controls the magnitude of the oxygen flux, and so it is sometimes feasible to increase the value of k by dispersing catalysts (eg., Pt) on the surface [26,27]. Alternatively the effective interfacial surface area can be increased by coating the membrane with a graded porous layer of the same material. This latter approach has been shown to be effective by Thorogood et al. [28], and this situation has been modelled by Deng et al.[29,30] who derived an expression for the enhancement factor, ξ, :

$$\xi = \sqrt{L_c S(1-\theta)/\tau_s} + \theta \qquad (19)$$

where S is the pore wall surface area per unit volume, θ the porosity, and τ_s the tortuosity of the solid phase in the porous structure.

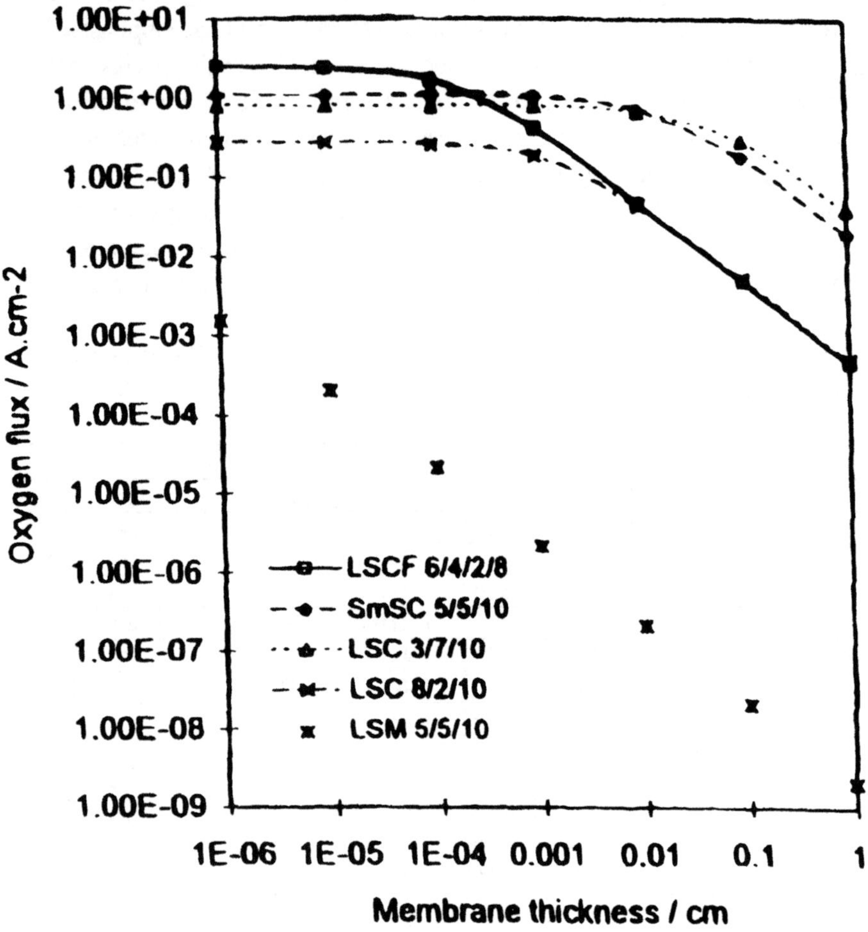

Fig.4 Calculated oxygen fluxes using equation (18) through selected gas/solid interfaces to illustrate influence of critical thickness,Lc

Finally it should be noted that information can also be obtained about k from simple isotopic exchange measurements [31], and from relaxation type measurements [32] which involve observing the change in electrical conductivity or weight as the sample re-equilibrates after a rapid change in the surrounding oxygen activity. However as the actual oxygen diffusion profile is not measured, derivation of the relevant k value is more difficult, and generally the values obtained are less accurate.

An assumption in the derivation of equation (18) is that the value of the oxygen surface exchange coefficient (k) derived using the IEDP technique under equilibrium conditions is appropriate for the actual conditions when a net flux is passing through the interface. Providing the relevant interfacial 'resistance' (equation 15) is not too high,(eg., $< 0.1\Omega cm^2$), which must be the situation for fluxes of technological interest, then the use of the equilibrium k value is probably justified. It must be remembered however that the k value relates to the specific surface under examination and different processing routes, annealing procedures, etc., can create different surface textures and compositions. It is also important to examine the influence of other relevant gases ,eg,H_2O,CO_2,etc, upon the value of k, and there is no reason in principle why this cannot be investigated using the IEDP technique.

2.5 THERMOCHEMICAL AND THERMOMECHANICAL STABILITY OF PEROVSKITES

It is obviously necessary to ensure that the mixed conducting phase is thermodynamically stable over the oxygen partial pressure gradient to be imposed across the mixed conducting membrane. Whilst limited data do exist for some systems, better models are required to predict the likely stability of transition metal cations in specific structures as a function of oxygen partial pressure. For example,estimates for the thermodynamic stability of $Sr_2Co_2O_5$ and $Sr_2Fe_2O_5$ are provided in the Ellingham diagram shown in Fig.{5}, which also contains information about the range of oxygen chemical potentials likely to be encountered in the partial and complete oxidation of methane. This diagram suggests that CICM compositions based on doped $Sr(Fe,Co)_{1.5}O_x$ (10) are unlikely to exhibit long term stability in reactors designed to produce syngas from natural gas feedstocks. However it has to be recognised that even if a multicomponent nonstoichiometric phase is judged to be thermodynamically stable it can decompose in an oxygen chemical potential gradient due to kinetic demixing, a term introduced by Schmalzreid [33,34]. If the mobilities of the slower moving cations are different then concentration gradients are developed in the oxide such that the membrane interface exposed to the higher oxygen partial pressure becomes enriched with the faster moving cation species.The degradation of $SrCo_{0.8}Fe_{0.2}O_{3-\delta}$ membranes, for example, has been attributed [35] to this phenomenon.

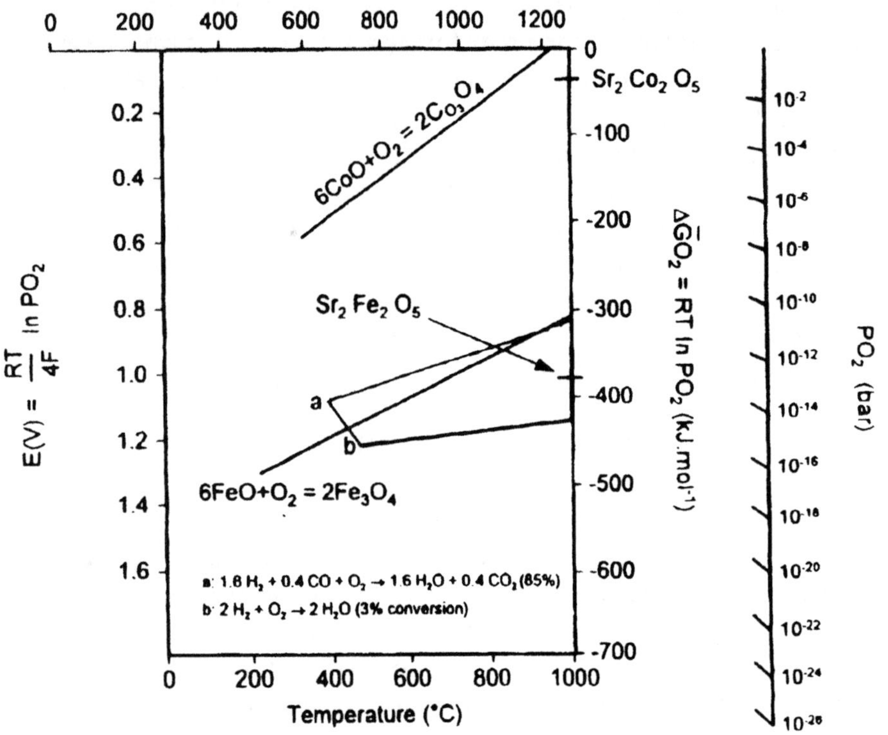

Fig.5 Modified Ellingham diagram depicting relative stabilities of Co and Fe based brown-millerite structures

Performance degradation can also occur due to the formation of oxyhydroxides, oxycarbonates, etc. Some of the constituent oxides (eg, SrO, La_2O_3) used in the perovskite memranes are very basic, and can form these compounds even if the concentration of H_2O or CO_2 is relatively low in the feed gases if their thermodynamic activity is high in the membrane . This can occur due to segregation processes at interfaces or may be an intrinsic property of the doped oxide. Hilpert [36], for example, has shown that the activity of SrO in the bi-polar plate material $La(Sr)CrO_3$ is relatively high, and that $SrCO_3$ can be expected to form under the operating conditions likely to be encountered in solid oxide fuel cells using fossil fuels. Even though significant quantities of $SrCO_3$ and other degradation products may not be detected by X-ray diffraction, incipient formation of these products could produce stress corrosion at crack tips eventually leading to structural failure.

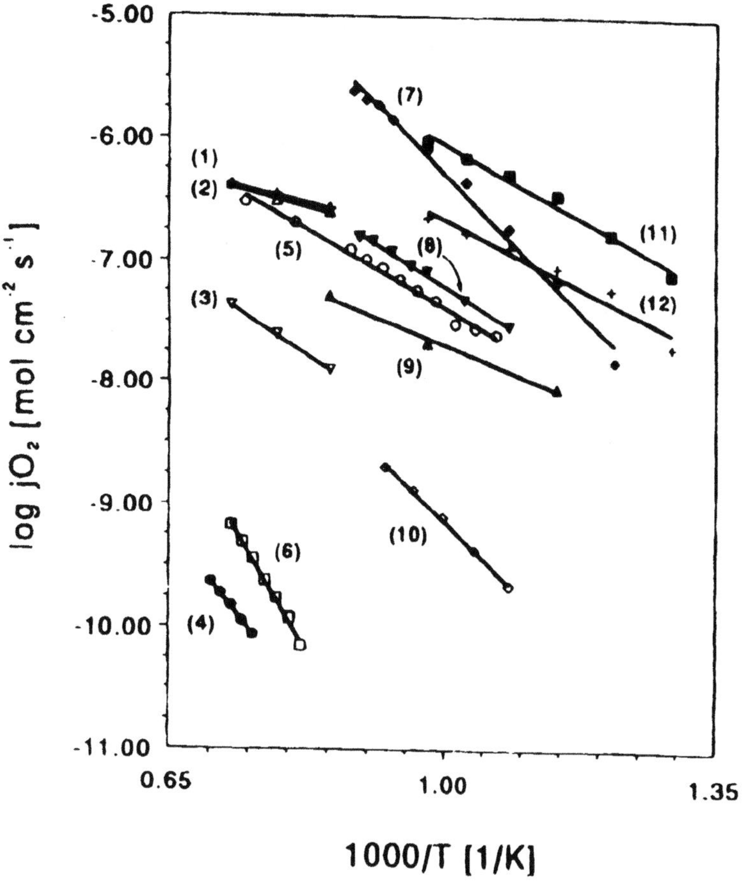

Fig. 6 Arrhenius plots of oxygen permeation for: (1) $La_{0.3}Sr_{0.7}CoO_{3-\delta}$, (2) $Ba0._9Y_{0.1}CoO_3$-δ, (3) YSZ-Pd (40vol%)-continuous Pd phase, (4) YSZ-Pd (30vol%)-discontinuous Pd phase, (5) $La_{0.6}Sr_{0.}5CoO_{3-\delta}$, (6) $(ZrO_2)_{0.7}$-$(Tb_2O_{3.5})_{0.228}$- $(Y_2O_3)_{0.072}$,(7) $SrCo_{0.8}Fe_{0\cdot 2}O_{2.5+\delta}$, (8) BE25-Ag(40vol%), (9) $Ba_{0.66}Y_{0.33}CoO_{3-\delta}$, (10), $(Bi_2O_3)_{0.75}$-$(Er_2O_3)_{0.25}$-[BE25], (11) BY25-Ag(35vol%)-[thickness 90μm], (12) BY25-Ag35(vol%)-[thickness 1.5μm] . For original data see ref. [2].

2.6 DUAL PHASE COMPOSITE MEMBRANES

The comments in the preceeding section indicate that it is difficult to select a single phase oxide material that can satisfy all the design criteria, including long term structural reliability, when the oxygen chemical potential gradient across the membrane is large.It is appropriate, therefore, to consider separating the ionic and electronic transport functions and to fabricate a dual -phase composite membrane with two interpenetrating percolation pathways for the oxygen ions and electrons.

Most investigators in this area have used a combination of a ceramic electrolyte with a noble metal (Ag,Pt,etc), although preliminary measurements have been conducted on dual oxide systems such as: YSZ-$LaMnO_3$ [37], YSZ ITO [38], and another obvious candidate is CGO-LSCF. Encouraging results have been reported [39] for the dual phase system $(Bi_2O_3)_{0.75}(Y_2O_3)_{0.25}$-Ag (35 vol %) as shown in Fig {6},which also contains data for other membrane systems for comparison. It is apparent that it should be possible to obtain larger oxygen fluxes when the microstructure is optimised, and the surface activated to promote the oxygen surface exchange reaction. At present dual phase membranes have been fabricated by conventional ceramic processing routes without taking advantage of alternative routes that could optimise the distribution of the two phases.

3. Applications

3.1 OXYGEN GENERATION

Two approaches are being investigated for oxygen generation. One concept uses a voltage driven system incorporating a ceramic electrolyte which functions as an electrochemical pump transferring oxygen from air to a pure oxygen anode compartment as indicated schematically below:

$$O_2,N_2\ (P_{O2}'),\ C(-)\left|\ \xrightarrow{\quad O^{2-}\quad}\ \right|A(+),\ O_2,\ (P_{O2}'')$$

The applied voltage(E) equals:

$$E = E_0 + IR + \eta_1 + \eta_2\ ,$$

where E_0 is the Nernst voltage, $E_0 = RT/4F\ln(P_{O2}''/P_{O2}')$, IR represents ohmic losses, η_1 is the activation polarisation losses at both electrodes,C,A; and similarly η_2 represents concentration polarisation losses at both electrodes. One advantage of the electrochemical production of oxygen is that the pure oxygen can be generated at higher pressures than the air supplied to the cathodes.Clark et al. [40] have described the performance of a multistack YSZ -based generator, but CeO_2 or Bi_2O_3 electrolyte based units operating at lower temperatures [41]are expected to produce oxygen more cheaply. Applied voltages of 0.5V can generate currents around 0.5Acm^{-2}. This corresponds to 1.5kW/kg O_2 which should be suitable for the small-scale generation of O_2 for medical and aerospace applications. Tubular, or planar, stack configurations have been considered, and Lawless [42] has suggested the use of ceramic honeycombs, which is similar to the concept proposed by Steele[43] for solid oxide fuel cells.

3.2 CERAMIC PARTIAL OXIDATION REACTORS

The use of ceramic oxide electrolytes to study heterogeneous reactions was first diccussed in 1970 [44,45] and this topic is the subject of another contribution in the present book (see relevant chap. by Prof.Vayenas), and other relevant reviews [46,47] are available. It was recognised that useful chemicals could also be produced using oxygen-ion conducting membranes/electrolytes, and an early example is provided by the oxidation of NH_3 to NO [48]. However by the mid 1980's attention had focussed on other partial oxidation reactions [4,49,] used by the chemical industry, and more recent surveys are available [50,51]. The relatively low oxygen fluxes through dense ceramic membranes at intermediate temperatures (~ 500C) has meant that most investigations have been restricted to the following processes which involve products that are stable at high temperatures (700-1000C):

$$2CH_4 + 1/2O_2 \Rightarrow C_2H_4 + H_2O$$

$$CH_4 + 1/2O_2 \Rightarrow CO + 2H_2 \text{ (syngas)}$$

Interest in the production of ethylene (C_2H_4) is declining as the reaction mechanisms proposed for the formation of this compound involve the radical $CH_3^{\bullet}$ in homogeneous gas phase reactions. The use of CICM materials therefore appears to have little influence on the yield and selectivity of C_2H_4. In contrast very encouraging results have been reported by Balachandran et al. [52] for the production of syngas from CH_4 using oxygen fluxes through ceramic tubes fabricated from mixed conducting oxides. Excellent methane conversion levels and CO and H_2 selectivity have been reported {Fig.7}. World -wide consortia lead by Air Products and Praxair, repectively, have been formed to exploit this technology. At present the production of syngas is mostly accomplished by steam reforming which is a very energy and capital intensive process as the reaction is highly endothermic. Dense oxygen ion conducting membranes offer potential solutions to several problems in methane conversion. The associated reduction in plant size is also attractive to the oil industry as it should be possible to convert surplus methane on oil rigs to syngas and subsequently to liquid fuels such as methanol, or hydrocarbons (via Fischer-Tropsch reactions).

For syngas production the ceramic membrane has to operate under severe chemical potential gradients with air on one side and natural gas on the other. The initial work of Balachandran using mixed conducting perovskite oxides in the system La-Sr-Co-Fe-O was unsuccessful due to decomposition of this material when exposed to the reducing environment containing methane. An alternative material based on Co doped $Sr_4Fe_6O_{13}$ has been developed [10] which exhibits a high partial oxygen ion conductivity and kinetic stability, at least, in methane. However doubts have already been expressed in section 2.5 about the long term stability of this material, and it may be that dual-phase composite structures will eventually offer the best solution, Further information about the structure of $Sr_4(Fe,Co)_6O_{13}$ has been provided by Guggilla and Manthiram [53] and it appears that this orthorhombic intergrowth structure incorporates perovskite layers alternating with $Fe_2O_{2.5}$ blocks.

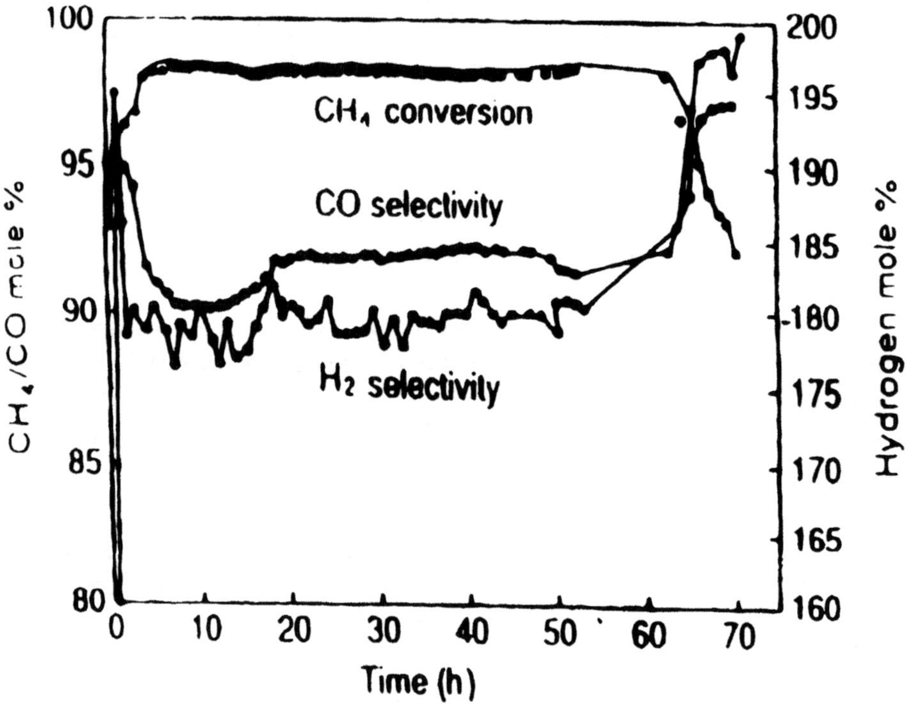

Fig.7 Methane conversion with CO and H2 selectivity in $Sr(Fe,Co)_{1.5}O_x$ membrane reactor with reforming catalyst.Conditions: feed (80% methane,20% argon: flow 2.5cm^3/min.; minimum temperature 850C;atmospheric pressure;membrane surface area 10cm^2.reproduced from ref. [52]

Finally it should be noted that many industrial hydrogenation and de-hydrogenation processes can in principle be carried out using protonic conducting dense membranes [54] However, for protonic conductors, it has been difficult so far to combine good thermochemical and thermomechanical stability with technologically useful levels of conductivity [55].

3.3 CERAMIC MEMBRANES IN SOLID OXIDE FUEL CELLS

The synergy [56,57] between ceramic partial oxidation reactors (CPOR) and solid oxide fuel cells (SOFC) could allow these two devices to be combined in a single system shown schematically in Fig. {8}.

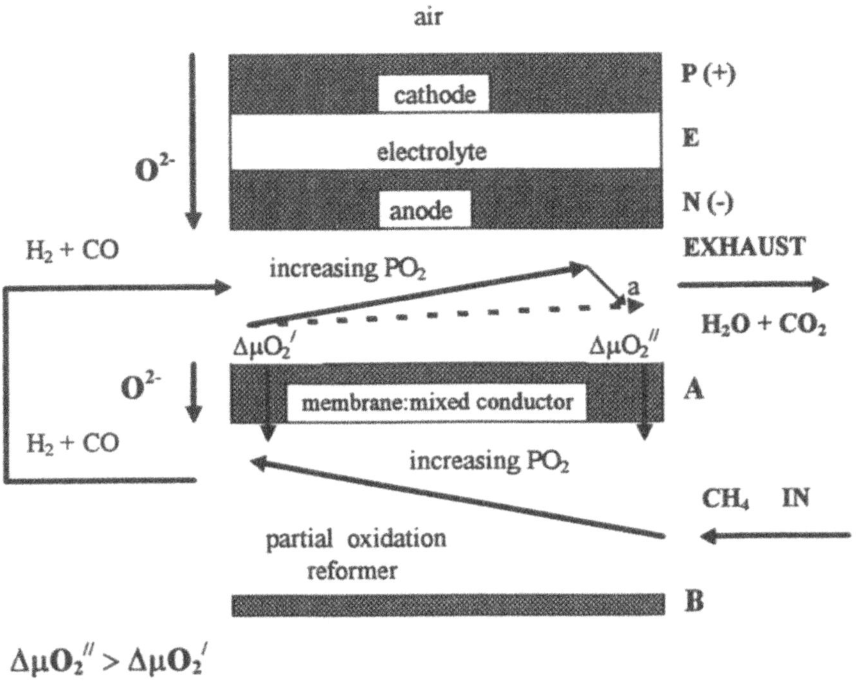

Fig.8 Schematic diagram showing PEN cell and ceramic membrane partial oxidation reformer integrated within a SOFC structure

In any fuel cell the OCV decreases across the cell as the oxygen partial pressure of the anode gas rises from the inlet side to the outlet region due to oxidation of the fuel. For example the change in value of the OCV at 700C in going from gas compositions of H_2/H_2O(90/10) to H_2/H_2O (10/90) is almost 200mV ,which is also indicated in Fig. {5}. Accordingly the fuel cell exhibits a mixed potential, the value of which is dependant upon the rate of the anode reaction as a function of gas composition and the in-plane electronic conductivity of the porous anode. It is very difficult to calculate the value of the mixed potential [58], and most operators use the empirical value obtained experimentally for a given set of operating conditions.

In principle the introduction of a ceramic membrane {A in Fig.8} could reduce the ΔPO_2 gradient across the anode compartment as oxygen from the H_2O and CO_2 generated by the anode reaction could pass through the mixed conducting membrane wall and contribute to the partial oxidation of the incoming CH_4 fuel. This is indicated in Fig. 8 by the arrow 'a'. The oxygen flux through the membrane will vary along the length of the membrane due to the lateral variation of ΔPO_2 but the overall result should be an increase in the mixed potential OCV exhibited by the fuel cell, and thus a greater overall system efficiency. The lower overall oxygen chemical potential in the anode compartment should also reduce the tendency of Ni -based composite anodes to form NiO at lower temperatures and and lower operating stack voltages [22]. It may also be easier to design alternative oxide anode systems based on Ti^{3+}/Ti^{4+}, Mn^{2+}/Mn^{3+}, Mo^{4+}/Mo^{3+} redox couples, as the target range of thermodynamic stability will be reduced [22]. The ceramic membrane has to function at relatively low oxygen partial pressures and so compositions in the Ce-Y-Nb-O system [59] ,for example, could be examined to ascertain whether the magnitude of the oxygen fluxes are sufficient. In this connection it should be noted that the chemical potential driving force across the membrane is relatively small, and so to accomplish effective reforming additional oxygen will probably have to be added to the fuel in the form of H_2O, CO_2 (including recycled anode exhaust gas), air, or possibly O_2 introduced by another ceramic membrane {B in Fig.8}.Finally it should also be noted that if the oxygen fluxes through the SOFC PEN structure and ceramic membrane A {Fig.8} are comparable then the overall process is equivalent to the direct anodic oxidation of methane which was investigated by Steele et al. in 1988 [60] .

4. Acknowledgements

Discussions with members of the Ceramic Ion Conducting Membrane (CICM) Group at Imperial College are gratefully acknowledged , together with financial suport from the Leverhulme Trust for a Senior Research Fellowship.

5. References

1. Bhave, R.R., Ed.,(1991),*Inorganic Membranes, Synthesis,Characterisation and Applications* , Van Nostrand Reinhold, New York
2. Bouwmeester, H.J.M., and Burggraaf, A.J.,(1997), Dense ceramic membranes for oxygen separation, in P.J.Gellings and

H.J.M.Bouwmeester (eds.) *Solid State Electrochemistry*, CRC Press, New York, pp. 481-553.

3. Teraoka, Y.,Zhang, H.M., Furukawa,S., and Yamazoe,N., (1985). Oxygen permeation through perovskite-type oxides,*Chem.Lett.*1743-46
4. Steele , B.C.H.,(1986), Applications of ceramic electrolytes in electrochemical engineering systems, *Chem. and Ind.,* **10**, 651-656
5. Mizusaki, J. (1992), Nonstoichiometry,diffusion, and electrical properties of perovskite-type oxide electrode materials, *Solid State Ionics*, **52**, 79-91
6. Doshi,R., Routbort,J.L., and Alcock,C.B. (1995), Diffusion in mixed conducting oxides-a review, *Defects and Diffusion Forum,*
7. Wagner,C., (1933), *Z.Phys.Chem.*, **B21**, 25
8. Ten Elshof, J.E., Bouwmeester, H.J.M., and Verweij, H., (1995), *Solid State Ionics*, **81**, 97-109.
9. Van Hassel, B.A. Kawada, T., Sakai, N., Yokokawa, H., and Dokiya, M., (1993), Oxygen permeation modelling of perovskites, *Solid State Ionics*, **66,** 295-305
10. Ma, B., Balachandran,U., and Park, J-H. (1996),Electrical transport properties and defect structure of Sr Fe $Co_{0.5}O_x$, *J.Electrochem.Soc.*, **143**,1736-44.
11. Van Doorn, R.H.E., Bouwmeester H.J.M., and Burggraaf, A.J.(1997), Phase related oxygen permeation through Sr-doped $LaCoO_{3-\delta}$,in *Proc. 1st Int. Conf. on Ceramic Membranes*, eds. H.U.Anderson, A.C.Khandkar, and M.Liu, ECS Proc Vol. 95-24, pp.138-45
12. Armstrong, T.R., Stevenson, J.W., and Pederson L.R. (1996), Dimensional instability of cathode and interconnect materials in reducing environments, in *Proc.2nd European SOFC Forum*,ed B.Thorstensen, U.Bossel (ISBN 3-922 148-19-0), Zurich, Switzerland, pp.521-530
13. Atkinson,A., (1997), Chemically-induced stresses in gadolinium-doped ceria solid oxide fuel cell electrolytes, *Solid State Ionics*, **95**, 249-258
14. Maier, J.,(1993) Mass transport in the presence of internal defect reactions-concept of conservative ensembles: I, Chemical diffusion in pure compounds. *J.Am.Ceram.* Soc. **76**,1212-17, II, Evaluation of electrochemical transport measurements,*ibid.*, **76**,1218-22, III, Trapping effect of dopants on chemical diffusion, *ibid.*,**76**,1223-27, IV, Tracer diffusion and intercorrelation with chemical diffusion andion conductivity, *ibid*,**76**,1228-32.
15. Benson,S.J.,Chater,R.J., and Kilner J.A.,(1997), Oxygen diffusion and surface exchange in the mixed conducting perovskite La0.6Sr0.4Co0.2Fe0.8O3-δ, in Proc.3rd Intl. symp. ionic and mixed

conducting oxide ceramics, eds. T.A. Ramanarayanan,W.L.Worrell, and H.L.Tuller,*ECS Proc.Vol. 97-24*, New Jersey,

16. Adler S.B., Lane, J.A. and Steele,B.C.H.,(1996) Electrode kinetics of porous mixed conducting oxygen electrodes, *J.Electrochem.Soc.* **143**, 3554-64
17. Kilner,J.A., Steele,B.C.H., and Ilkov, L., (1984), Oxygen self-diffusion studies using negative-ion secondary ion mass spectrometry (SIMS), *Solid State Ionics*, **12**, 89-97
18. Kilner,J., and De Souza R.A., (1996), Measurement of oxygen transport in ceramics by SIMS in *17th Riso Intl. Symp. on Mat.Sci., High Temperature Electrochemistry: Ceramics and Metals,* eds. F.W.Poulsen et al. Riso National Lab., (ISBN 87-550-2199-9), Denmark., pp. 41-54
19. B.C.H.Steele, Kilner,J.A., Dennis,P.F., McHale, A.E., Van Hemert,M., Burggraaf,A.J. (1986), Oxygen surface exchange and diffusion in fast ionic conductors, *Solid State Ionics*, **18/19**, 1038-44.
20. Steele, B.C.H.. (1992), Oxygen ion conductors and their technological applications, *Mater. Sci.Eng.*, **B13**, 79-87.
21. Steele, B.C.H., (1995), Interfacial reactions associated with ceramic ion transport membranes, *Solid State Ionics*, **75**, 157-165.
22. Steele,B.C.H., (1996), Survey of materials selection for ceramic fuel cells, II: Cathodes and Anodes, *Solid State Ionics*, **86/88**, 1223-1234
23. Chater, R.J., Carter,S, Kiner,J.A., Steele,B.C.H., (1992), Development of a novel SIMS technique for oxygen self-diffusion and surface exchange coefficient measurements in oxides of high diffusivity, *Solid State Ionics*, **53/56**, 859-867
24. Kilner, J. (1994), Isotopic exchange in mixed and ionically conducting oxides, in *Proc. 2nd Int.Symp. Ionic and Mixed Conucting Oxide Ceramics,* eds. T.A. Ramanarayanan, W.L.Worrell, H.L.Tuller, ECS Proc.Vol. 94-12,New Jersey, pp. 174-190
25. Kilner,J.A., Sirman,J.D., Manning,P.S., Oxygen exchange and diffusion in materials with the fluorite structure, in in Proc.3rd Intl. symp. ionic and mixed conducting oxide ceramics, eds. T.A. Ramanarayanan, W.L.Worrell, and H.L.Tuller, *ECS Proc.Vol. 97-24*, New Jersey,
26. Sirman,J., and Kilner J.A.,(1996), Surface exchange properties of $Ce_{0.9}Gd_{0.1}O_{1.95}$ coated with $La_{1-x}Sr_xCo_{1-y}Fe_yO_{3-\delta}$, *J.Electrochem.Soc*, **143,** L229-31.,T.,
27. Simam,J.D.,Lane,J, Kilner J.A. (1997), Comparison of dense and porous SOFC perovskite cathodes, in in Proc.3rd Intl. symp. ionic and mixed conducting oxide ceramics, eds. T.A. Ramanarayanan,

W.L.Worrell, and H.L.Tuller, *ECS Proc.Vol. 97-24*, New Jersey,

28. Thorogood, R.M. Srinivasan, R.,Yee, T.F. and Drake,M.P.,(1993),Composite mixed conductor membranes for producing oxygen, U.S.Patent 5,240,480
29. Deng,H., Zhou,M., Abeles, B.,(1994), Diffusion-reaction in porous mixed ionic-electronic solid oxide membranes, *Solid State Ionics*, **74**, 75-84.
30. Deng,H., Zhou,M. and Abeles,B., (1995),Transport in solid oxide porous electrodes: effect of gas diffusion, *Solid State Ionics*, **80**, 213-22.
31. Boukamp,B.A.,Vinke,I.C.,de Vries,K.J. and Burggraaf, A.J.(1989), Surface oxygen exchange properties of bismuth oxide-based solid electrolytes and electrode materials, *Solid State Ionics*, **32/33**, 918-23.
32. Novotny,J.and Wagner, J.B.Jr.,(1981) Influence of the surface on the equilibration kinetics of nonstoichiometric oxides,*Oxid.Met*.,**15**,169-90.
33. Schmalzreid,H., Laqua,W., and Lin,P.L.(1979), Crystalline oxide solid solutions in oxygen patential gradients, Z.Naturforsch.,**34A,**192-99
34. Schmalzreid,H. and Laqua,W., (1981), Multicomponent oxides in oxygen potential gradients, *Oxid.Met*.,**15**, 339-.
35 Qui,L., Lee,T.H,. Lie,L-M, Yang,Y.L.,and Jacobson,A.J.,(1995), Oxygen permeation studies of $SrCo_{0.8}Fe_{0.2}O_{3-\delta}$, *Solid State Ionics*, **76**, 321-29.
36. Peck,D.H., Miller,M., Nickel H., Das,D., Hilpert,K., (1995), The SrO-Cr2O3-La2O3 phase diagram and volatility of $La_{1-x}Sr_xCrO_3$ (x = 0-0.3) in comparison to Cr base interconnect alloys, in Solid Oxide Fuel Cells IV, eds. M.Dokiya et al., *ECS Proc.Vol. 95-1*, New Jersey, pp.858-68
37. Costa,A.D.S., Labrincha, J.A.,and Marques F.M.B.,(1994), Composite (YSZ + LSM) ceramic filters with oxygen electrochemical permeability, in *Electroceramics IV,vol.2,*eds., R.Waser, S.Hoffman, D.Bonnenberg, Ch.Hoffman, Augustinus Buchhandlung,Aachen, 781-784
38. MazenecT.J., Cable, T.L. Frye, J.G.,(1992),Electrocatalytic cells for chemical reactions, *Solid State Ionics*, **53/56**, 111-118.
39. Shen,Y.S., Liu,M., Taylor,D., Bolagopal,S., Joshi,A., Krist,K.,(1994), Mixed ionic-electronic conductors based on Bi-Y-O-Ag metal ceramic system, in Proc.2nd Int.Symp.on ionic and mixed conducting ceramics,eds, T.A.Ramanarayanan, W.L.Worrell, and H.L.Tuller, *ECS Proc.Vol., 94-12*, pp.574-95.
40. Clark D.J., Losey R.W.and Suitor,J.W.,(1992), Separation of oxygen by using zirconia solid electrolyte membranes, *Gas.Sep.Purif.*,**6**, 201-5.
41. Waller,D., Kilner,J.A., and Steele,B.C.H., (1995), Oxygen separation using dense gadolinia doped ceria membranes, in Proc.1st Int.symp. on

ceramic membranes,eds. H.U.Anderson, A.C.Khandkar, and M.Liu, *ECS Proc.Vol. 95-24*, pp. 48-64

42. Lawless,W.N.,(1992), Solid state generation of oxygen using ceramic honeycombs, Solid State Ionics, **52**, 219-224.
43. B.C.H.Steele, (1984), *Science of Ceramics*, **12**,697
44. Wagner, C, (1970), *Adv.Catal.*, **21**, 323 .
45. Steele, B.C.H., (1970), Electrochemical studies of heterogeneous kinetics at elevated temperatures, in Heterogeneous Kinetics at Elevated Temperatures, eds G.R.Belton and W.L.Worrell, Plenum Press,New York, pp. 135-163
46. Vayenas, C.G., Jaksic, M.M., Bebelis,S.I., Neophytides,S.G. (1996), The electrochemical activation of catalytic reactions, Modern Aspects of Electrochemistry, No.29, eds. J.O'M.Bockris et al., pp57-202
47. Metcalfe,I. A., (1994), Stabilised zirconia solid electrolyte membranes in catalysis, *Catal.Today*, **20**, 283-294.
48. Farr,R.D., and Vayenas ,C.G., (1980),Ammonia high temperature solid electrolyte fuel cell, *J.Electrochem. Soc.*, **127**, 1478-83
49. Sokolovskii V.D., (1990), Principles of oxidative catalysis on solid oxides, *Catal.Rev.-Sci.Eng.*, **32**, 1-49
50. Mazenec,T.J., (1994), Prospects for ceramic electrochemical reactors in industry, *Solid State Ionics*, **70/71**,11-19.
51. Eng,D. and Stoukides,M.,(1991), Catalytic and electrocatalytic methane oxidation with solid oxide membranes, *Catal.Rev.-Sci. Eng.*, **33**, 375-412
52. Balachandran,U.,et al, (1995), Methane to syngas via ceramic membranes, *Am.Ceram.Bull.* **74**, 71-75.
53. Guggilla,S., and Manithram, A., (1997), Crystal chemical characterisation of the mixed conductor Sr(Fe,Co)1.5Oy exhibiting unusually high oxygen permeability, *J.Electrochem.Soc.*, **144**, L120-122
54. Iwahara H., (1995), Technological challenges in the application of proton conducting ceramics, *Solid State Ionics*, **77**, 289-298
55. Norby,T.,and Larring,Y., (1997), Concentration and transport of protons in oxides, *Current Opinion in Solid State & Materials Science*, **2**, 593-99.
56. Steele, B.C.H. (1996), Ceramic ion conducting membranes, *Current Opinion in Solid State & Materials Science*, **1**, 684-691.
57. Steele,B.C.H., (1996),Synergistic applications of SOFC technology, in *Proc.2nd European SOFC Forum*,ed B.Thorstensen, U.Bossel (ISBN 3-922 148-19-0), Zurich, Switzerland, pp.795-810.
58. Tanaka,T., (1989), research with on-site internal reforming molten

carbonate fuel cells, in Proc. intl. gas research conf., pp.257

59. Liddicott K.M., (1994),High temperature materials chemistry of doped cerium oxide ceramics, Ph.D.Thesis, University of London.

60. Steele,B.C.H., Kelly,I.E., Middleton,P.H. and Rudkin,R.A., (1988), Oxidation of methane in solid state electrochemical reactors, *Solid State Ionics*, **28/30**, 1486-89

THERMAL AND ISOTHERMAL EXPANSION

A. BIEBERLE, L.J. GAUCKLER
ETH Zürich
Department of Materials
Nonmetallic Materials
Sonneggstrasse 5
8092 Zürich, Switzerland

1 Introduction

SOFCs are designed in a layered structure with four elements that are in intimate contact with each other (i.e. the electrolyte, the cathode, the anode, and the interconnector). The components are all made of different materials (ceramics, metals, cermets). During fabrication and operation, a SOFC stack is subjected to non-uniform temperature distributions and thermal cycles. Therefore, thermal expansion mismatch of different elements is a critical issue for the selection of suitable materials for a SOFC stack [1,2,3,4].

In addition to expansion due to temperature changes, one has to consider isothermal expansion [5,6]. Isothermal expansion means that a material changes its dimensions, while the temperature is kept constant. This type of expansion is due to changes of the stoichiometry and oxidation state in the materials, e.g. changes in oxygen partial pressure.

2 Thermal Expansion of SOFC Elements

Thermal expansion coefficients of materials typically used in SOFCs are listed in TABLE 1 (the appendix contains a table going into more detail). The thermal expansion values are mean values that are given for a certain temperature range in which the thermal expansion changes almost linearly with temperature. The exact value for the thermal expansion is strongly depending on the stoichiometry and the fabrication method of the composition as well as on the set-up of the thermal expansion measurement. Comparing thermal expansion data, these points have to be considered.

TABLE 1 illustrates already that the state-of-the-art anode material is the most critical SOFC material concerning thermal expansion mismatch with other components.

H.L. Tuller et al. (eds.), Oxygen Ion and Mixed Conductors and Their Technological Applications, 347–358.

TABLE 1: Thermal expansion coefficients of typical SOFC materials (temperature 20-1000°C).

Component	Material	Thermal Expansion Coefficient 10^{-6}/K	Reference
Electrolyte	ZrO_2 + 8wt.% Y_2O_3	10.8	[2,8]
Cathode	$LaMnO_3$	11.2	[4]
Anode	Ni/YSZ Cermet (30 vol.% Ni, 30% porosity)	12.5	[12]
Interconnector	$CrFe5Y_2O_31$	11.3	[15]
Glass Seal	$SrO\text{-}La_2O_3\text{-}Al_2O_3\text{-}B_2O_3\text{-}SiO_2$	8-13	[16]

2.1 ELECTROLYTE

The most commonly used electrolyte material is based on ZrO_2 having a thermal expansion coefficient between $10\text{-}11*10^{-6}$/K [7] in the polycrystalline structure. The thermal expansion coefficient of commercially available polycrystalline ZrO_2 depends in some way on the supplier (TABLE 2).

TABLE 2: Thermal expansion of commercial polycrystalline ZrO_2 materials (temperature 20-1500°C) [7].

Company	Thermal Expansion Coefficient (10^{-6}/K)
PSZ, Nilsen Sintered Products, Northcote, Australia	10.36
TTZ, Feldmühle Plochingen, FRG	11.03
ZDY, Coors Porcelain Co.	10.45
Coors Porcelain Co.	10.59

A considerable amount of dopant, such as Y_2O_3, is normally added to ZrO_2 in order to stabilize the crystal structure and to introduce oxygen ionic conductivity. This dopant increases the thermal expansion coefficient of ZrO_2, however, increasing amounts of Y_2O_3 do not affect the thermal expansion on a large scale [4]. Components made of the most commonly used Tosoh 8YSZ powder (8mol% Y_2O_3 in ZrO_2) have a thermal expansion coefficient of $10.8*10^{-6}$/K [2,8].

In the last years, alternative electrolyte materials, such as doped CeO_2, were studied [6]. The thermal expansion coefficients of doped CeO_2 electrolytes are higher compared to those of ZrO_2 electrolytes (APPENDIX).

In case of an electrolyte supported cell design, the electrolyte is typically selected as the basic material concerning thermal expansion in the stack. The thermal expansion coefficients of other cell materials are then modified in order to match the thermal expansion coefficient of the electrolyte material. The reason is that the electrolyte is sandwiched between two different electrode materials, i.e. the cathode and the anode.

In addition, the electrolyte is in intimate contact with the interconnector and the glass seal.

2.2 CATHODE

Most cathodes are based on the perovskite $LaMnO_3$ with a thermal expansion coefficient of $11.2*10^{-6}$/K [4]. Other perovskites, e.g. $LaCoO_3$, often have higher thermal expansion coefficients ($15\text{-}20*10^{-6}$/K). These materials are therefore not suitable, even though they have good electrical conductivity.
Lanthanum deficiency and oxygen nonstoichiometry lower the thermal expansion of $LaMnO_3$ [3,9]. The thermal expansion coefficient of $LaMnO_3$ fits quite well to that of 8YSZ (α_{YSZ} = $10.8*10^{-6}$/K) [2,8]. The thermal expansion coefficient difference is smaller than $0.5*10^{-6}$/K.

For better electrical conductivity of the cathode, La is often substituted by Sr. Sr increases the thermal expansion coefficient of the perovskite (APPENDIX) because of the lower bonding strength of Sr-O compared to La-O which in turn is due to the lower coulombic attraction from the doubly charged Sr^{2+} compared to La^{3+}. The more Sr is added, the more the thermal expansion coefficient increases (APPENDIX) [4,10]. Cations with smaller ionic radius, such as Ca or Y, lower the thermal expansion coefficient approaching the value of the solid electrolyte [2,4,11].
If Mn is substituted by Co, the thermal expansion coefficient is substantially increased (APPENDIX). Figure 1 illustrates this effect for two different La/Sr ratios. The thermal expansion coefficients range for YSZ and typical alloys for interconnector materials are sketched in order to show the limits for SOFC application. In the case that Co is partially substituted by Fe, the thermal expansion coefficient decreases to lower values (APPENDIX) [10].

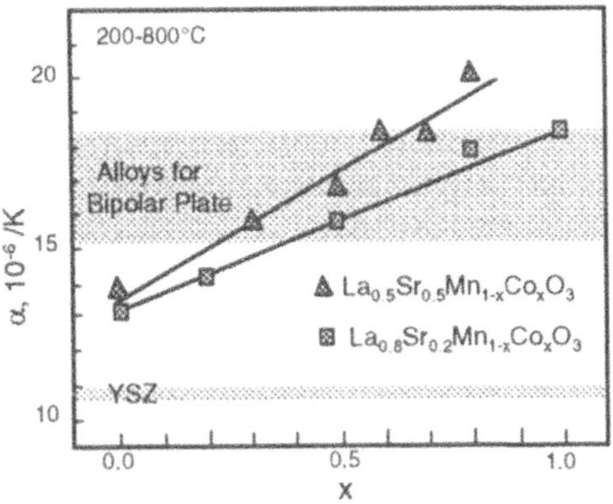

Figure 1: Mean thermal expansion coefficients of ceramic cathode materials between 200-800°C (α(8YSZ) = $10.8 \cdot 10^{-6}$/K, α(alloys for bipolar plate) = $15\text{-}18.5 10^{-6}$/K)[2].

2.3 ANODE

Ni-YSZ cermet anodes generally have a higher thermal expansion coefficient than the YSZ electrolyte and the cathode materials (>$11*10^{-6}$/K [2,12]) (APPENDIX). Cermet anodes contain metallic nickel in the structure which has a higher thermal expansion coefficient than ceramics. Therefore, with increasing Ni content the thermal expansion coefficient of the Ni/YSZ cermet increases (APPENDIX) (Figure 2). The increase will not only depend on the composition (Ni and YSZ), but also on the distribution and the penetration of the two phases into each other.
In order to obtain sufficient conductivity, the anode should contain more than 30 vol.% Ni. Anodes with this composition usually show a thermal expansion coefficient which is almost $2*10^{-6}$/K higher than that of the electrolyte 8YSZ. In order to minimize the thermal expansion mismatch between the Ni-YSZ cermet anode and the electrolyte, the thickness and thickness ratio of the cell components are varied [4].

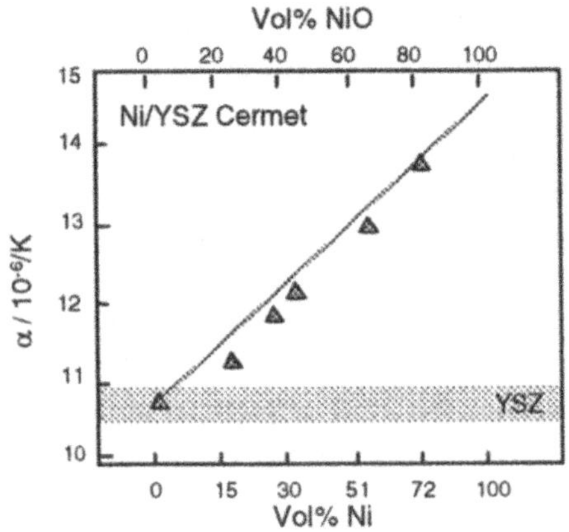

Figure 2: Thermal expansion coefficient of cermet anode as a function of Ni content [4].

2.4 INTERCONNECTOR

Interconnectors are made either of ceramic (doped $LaCrO_3$) or metal (chromia forming alloy $CrFe5Y_2O_31$). In both cases, the thermal expansion coefficient can be varied by changing the composition of the material.
The ceramic material $LaCrO_3$ exhibits different phases depending on temperature. From room temperature to about 240°C, $LaCrO_3$ has a orthorhombic structure ($\alpha = 6.7*10^{-6}$/K [1]). Between 240-290°C, the phase transforms into a rhombohedral structure ($\alpha = 9.5*10^{-6}$/K [1]). At about 1000°C, the structure changes to a hexagonal one. The first phase transformation can be clearly seen in Figure 3 where the thermal expansion coefficient of $LaCrO_3$ is plotted vs. temperature. The thermal expansion changes abruptly at the phase transformation temperature.

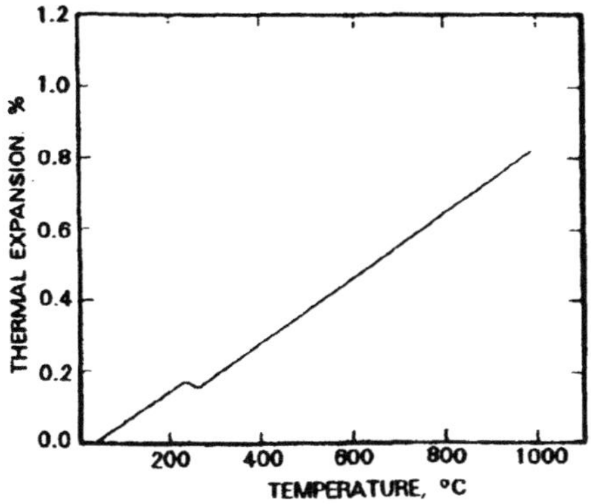

Figure 3: Thermal expansion of $La_{1-x}CrO_3$ (-0.1 < x < 0.1) [4].

The thermal expansion of $LaCrO_3$ can be tailored by the substitution of cations on both the A and the B sites. Dopants, such as Sr, Ca, Co, Ni, and Mg, increase the thermal expansion coefficient of $LaCrO_3$ by up to $4*10^{-6}$/K (APPENDIX) [1]. Exactly the same thermal expansion coefficient as that of 8YSZ was obtained with $La_{0.7}Ca_{0.3}Cr_{0.85}Co_{0.005}Fe_{0.005}Ni_{0.005}O_3$ [14].

The thermal expansion coefficients of the ceramic interconnector $LaCrO_3$ and of the electrolyte YSZ are quite similar; the thermal expansion mismatch is less than $0.2*10^{-6}$/K (Figure 4). From this point of view, $LaCrO_3$ would be an excellent interconnector material. However, compared to metallic interconnectors, ceramic interconnectors have less mechanical stability and the processing of pieces of large dimensions is much more expensive. Therefore, metallic alloys are a good alternative material.

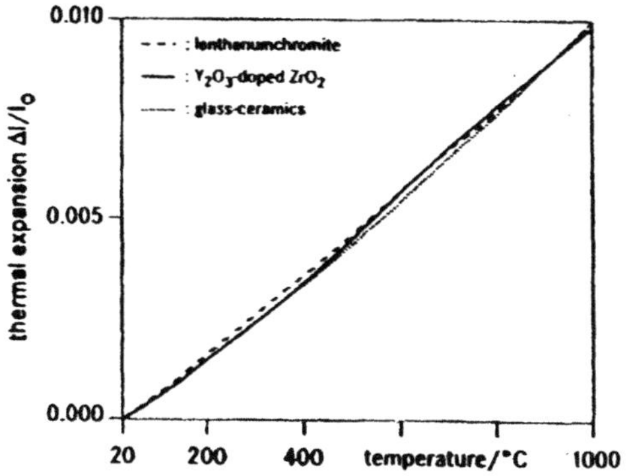

Figure 4: Thermal expansion of YSZ, $LaCrO_3$, and glass ceramic as a function of temperature [15].

Many metals have larger thermal expansion coefficients than ceramics because of the different bonding in both materials. For example, the metallic alloys FeNiCr, NiCrAl, NiCrW, FeCrAl (APPENDIX) cannot be used as interconnector material in SOFCs, as their thermal expansion coefficients are too high in the temperature range of SOFC application (700-1000°C): $\alpha = 15\text{-}18*10^{-6}$/K [2] (Figure 5).

Therefore, much effort was undertaken in the last years to find a suitable metallic interconnector material that does not only have an appropriate thermal expansion coefficient, but can also be easily produced in large dimensions. The best choice for high temperature application seems to be the so-called ODS (= Oxide Dispersion Strengthening) materials, such as $CrFe5Y_2O_31$. Figure 5 demonstrates the similar thermal expansion behavior of $CrFe5Y_2O_31$ and ZrO_2.

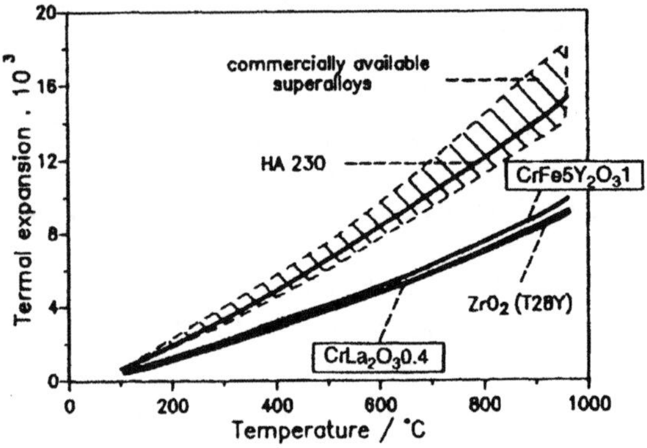

Figure 5: Thermal expansion of metallic components for SOFC application [16].

2.5 GLASS SEAL

In the planar SOFC design, the oxidizing and reducing gas chambers are separated gas-tight by a glass seal. The glass seal is in direct contact with the electrolyte on one side and with the interconnector on the other side. In this way, the glass seal has to compensate the thermal expansion mismatch between the electrolyte and the interconnector material. It should have a thermal expansion coefficient in the range of $10\text{-}11*10^{-6}$/K [8].

Glass seals investigated for SOFC application are listed in the appendix. Borate based glasses, such as Pyrex, are not suitable for large sealing areas because of their very low thermal expansion coefficients ($\alpha_{Pyrex} = 3.2*10^{-6}$/K) [17]. Soda lime glasses show appropriate matching thermal expansion coefficients ($9.5*10^{-6}$/K [17]), but cannot be used because of reactions with the cathode material $LaMnO_3$ and because of their low viscosity (< 10^4 dPas at 1000°C [17]). Mica glass-ceramics (mica = three layer mineral, e.g. $K_2O \cdot 3Al_2O_3 \cdot 6SiO_2 \cdot 2H_2O$), e.g. Macor or Photoveel [18], seem to be practicable concerning their thermal expansion behavior (APPENDIX).

Glasses show a characteristic thermal expansion behavior due to their amorphous structure. Up to a certain temperature, a glass expands more or less linearly with temperature. At temperatures of around 400-600°C (transformation temperature T_g), the thermal expansion coefficient increases suddenly due to the transition from the glassy state to a viscous fluid [17].

If the glass sealant is a viscous fluid at the cell operating temperature, thermal expansion mismatch of the sealant and the other cell components can be tolerated. As the structure cools down to room temperature, significant stress develops below the

glass transformation temperature T_g. The total stress due to thermal expansion mismatch is considerably less than if the stress starts to develop at operating temperature. In order to minimize stresses, T_g should be as low as possible. As T_g roughly corresponds to a viscosity of 10^{13}dPas, the viscosity of the sealant should be higher than 10^{13}d Pas at operating temperature [19].
Since glass is metastable, it crystallizes with time at high temperature. Increasing crystallinity changes the thermal expansion coefficient in long term applications [17].

3 Isothermal Expansion

Dimensional changes of components may not only arise due to changes of temperature. Nonstoichiometric oxides exhibit an expansion behavior depending on oxygen stoichiometry due to reduction or oxidation upon changes in oxygen partial pressures [5,6,20,21]. Examples are doped ceria electrolytes [5,6] and $La_{1-x}Sr_xCrO_{3-\delta}$ interconnectors [20,21].
SOFC electrolytes and interconnectors are exposed to different oxygen partial pressures at the anode and cathode side, respectively. An expansion behavior depending on oxygen nonstoichiometry can therefore lead to different expansions at both sides of the electrolyte and the interconnector. Bending and mechanical failure may result [20,21]. In case of ceria based electrolytes, an additional effect can be observed: Ceria based electrolytes become more or less reduced depending on the operating condition of the fuel cell.

The isothermal expansion behavior of pure nonstoichiometric CeO_{2-x} with a lattice constant increasing with rising oxygen nonstoichiometry x [5,20] can be explained using XRD and dilatometry data. It was found that the predominant defect in reduced ceria are oxygen vacancies due to the reduction of Ce^{4+} to Ce^{3+}

$$O_O^x + 2Ce_{Ce}^x \leftrightarrow V_{\ddot{O}} + 2Ce'_{Ce} + \frac{1}{2}O_2(g) \qquad (1)$$

The ionic radius of Ce^{3+} (1.143Å [22]) is larger compared to Ce^{4+} (0.97Å [22]). Therefore the lattice constant increases with increasing Ce^{3+} concentration.

Isothermal expansion characteristics of three doped ceria oxides, i.e. $Ce_{0.8}Sm_{0.2}O_{1.9}$ (CSO), $Ce_{0.8}Gd_{0.2}O_{1.9}$ (CGO), and $Ce_{0.9}Ca_{0.1}O_{1.9}$ (CCO), were compared [6]. The expansion of CCO as a function of the partial pressure of oxygen is plotted for different temperatures in Figure 6. The expansion is independent of $p(O_2)$ up to a temperature of about 700°C. At temperatures higher than 700°C, the differential lengths of the specimens increases significantly at oxygen partial pressures below 10^{-5}atm.
The increase in differential length change is more pronounced in calcia doped ceria than in samaria and gadolinia doped ceria [5,6].

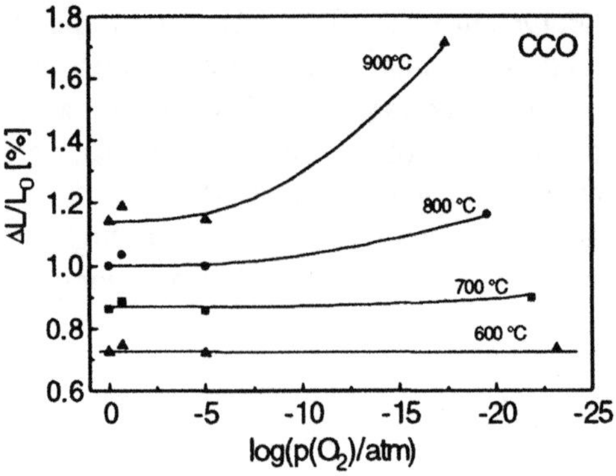

Figure 6: Differential length change of $Ce_{0.9}Ca_{0.1}O_{1.9-x}$ as a function of temperature and $p(O_2)$. L_0 is the length of the specimen at 50°C in air [6].

The oxygen content of the fuel gas and the oxidizing gas are not the only contributions which determine the oxygen partial pressure. The $p(O_2)$ at the electrode/electrolyte interface depends also on the electrode overpotential. By changing the load current of the fuel cell, the $p(O_2)$ at the electrode/electrolyte interface can change quite fast. Changes of the expansion due to different $p(O_2)$ at these interfaces can lead to cracks in ceria membranes and to possible delamination of the electrodes.
Therefore, reduction-oxidation cycles were studied in [6] (Figure 7). It can be noted that the reduction reaction of the sample is much slower than the oxidation reaction.

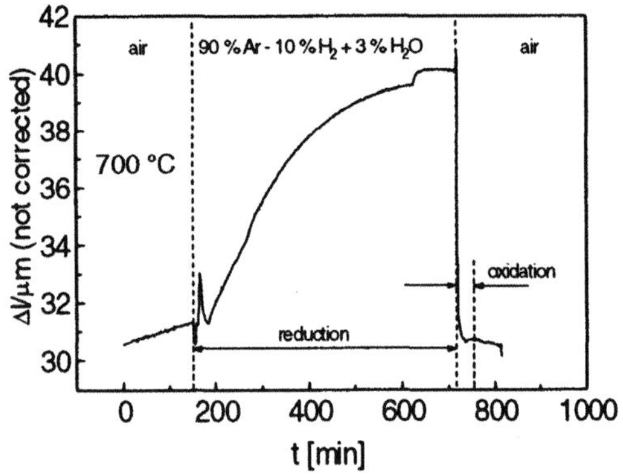

Figure 7: Typical oxidation/reduction cycle of a CSO sample at 700°C (the length change ΔL was not corrected for thermal expansion of reference, sample holder and pushrods) [6].

4 Summary

Small differences in the thermal expansion coefficients of cell components in SOFCs can produce large stresses during fabrication and operation and cause cracking and delamination of bonded elements. Therefore, the thermal expansion coefficient of the cell components must be tailored and modified in order to reduce thermal expansion mismatch. For state-of-the-art SOFCs, the thermal expansion mismatch is most critical concerning the anode material Ni-YSZ cermet. Satisfying materials for the electrolyte, cathode, interconnector, and glass seal exist from a thermal expansion point of view. Isothermal expansion can be observed in oxygen-deficient nonstoichiometric oxides, such as doped ceria electrolytes or $La_{1-x}Sr_xCr_{3-\delta}$ interconnectors as a function of oxygen partial pressure. It is either due to different oxygen content in the applied gases or due to different overpotentials at the electrodes.

5 References

1. Srilomask, S., Schilling, D.P., Anderson, H.U. (1989) Thermal Expansion Studies on Cathode and Interconnect Oxides, in S.C. Singhal (ed.), *Proc. 1st Int. Symp. SOFC*, The Electrochem. Soc., Pennington NJ, 129-40.

2. Ivers-Tiffée, E., Wersing, W., Schießl, M., Greiner, H. (1990) Ceramic and Metallic Components for a Planar SOFC, *Ber. Bunsg. Phys. Chem.* **94**, 978-81.

3. Yamada, H. Nagamoto, H. (1993) Thermal Expansion Coefficient and Electrical Conductivity of Mn-Based Perovskite-Type Oxides, in S.C. Singhal, H. Ihawara (eds.), *Proc. 3rd Int. Symp. SOFC*, The Electrochem. Soc., Pennington NJ, 213-19.

4. Minh, N.Q., Takahashi, T. (1995) *Science and Technology of Ceramic Fuel Cells*, Elsevier, Amsterdam, Netherland.

5. Chiang, H.-W., Blumenthal, R.N., Fournelle, R.A. (1993) A High Temperature Lattice Parameter and Dilatometer Study of the Defect Structure of Nonstoichiometric Cerium Dioxide, *Solid State Ionics* **66**, 85-95.

6. Gödickemeier, M. (1996) Mixed Ionic Electronic Conductors for Solid Oxide Fuel Cells, PhD Thesis No. 11348, Swiss Federal Institut of Technology, ETH Zürich.

7. Adams, J.W., Nakamura, H.H. (1985) Thermal Expansion Behavior of Single-Crystal Zirconia, *J. Am. Ceram. Soc.* **68**, C-228-231.

8. Larsen, P.H., Bagger, C., Mogensen, M., Larsen, J.G. (1995) Stacking of Planar SOFCs, in M. Dokiya, O. Yamamoto, H. Tagawa, S.C. Singhal (eds.), *Proc. 4th Int. Symp. SOFC*, The Electrochem. Soc., Pennington NJ, 69-75.

9. Takeda, Y., Nakai, S., Kojima, T., Kanno, R., Imanishi, N., Shen, G.Q., Yamamoto, O., Mori, M., Asakawa, C., Abe, T. (1991) Phase Relation in the System $(La_{1-x}A_x)_{1-y}MnO_{3+z}$ (A = Sr and Ca)", *Mater. Res. Bull.* **26**, 153-62.

10. Sasaki, K. (1993) Phase Equilibria, Electrical Conductivity, and Electrochemical Properties of ZrO_2-In_2O_3, PhD Thesis No. 10331, Swiss Federal Institut of Technology, ETH Zürich.

11. Fandel, M., Höschele, J., Schäfer, W., Schmidberger, R. (1989) Properties of Ca-Doped $LaMnO_3$, in G. de With, R.A. Terpstra, R. Metselaar (eds.), *Euroceramics Vol.2, Properties of Ceramics*, Elsevier Applied Science, London, UK, 2.236-40.

12. Majumdar, S., Claar, T., Fandermeyer, B. (1986) Stresss and Fracture Behvior of Monolithic Fuel Cell Tapes, *J. Am. Cer. Soc.* **69**, 628-33.

13. Mori, M., Yamamoto, T., Itho, H., Inaba, H., Tagawa, H., (1997) Considerations on Thermal Expansion of Nickel-Zirconia Anode in SOFC During Fabrication and Operation, in: U. Stimming, S.C. Singhal, H. Tagawa, W. Lehnert (eds.), *Proc. 5th Int. Symp. SOFC*, The Electrochem. Soc., Pennington NJ,869-78.

14. Gordes, P., Christiansen, N. (1993) Synthesis and Characterization of a New Perovskite Material for SOFC Applications, in S.C. Singhal, H. Ihawara (eds.), *Proc. 3rd Int. Symp. SOFC*, The Electrochem. Soc., Pennington NJ,414-20.

15. Stolten, D., Monreal, E., Schäfer, W. (1994) Soft Glass Ceramic Sealing for Gastight SOFC Stacks, in U. Bossel (ed.), *Proc. 1st Europ. SOFC Forum*, Vol.2, Lucerne, Switzerland, 517-24.

16. Greiner, H., Grögler, T., Köck, W., Singer, R.F. (1995) Chromium Based Alloys for High Temperature SOFC Applications, in M. Dokiya, O. Yamamoto, H. Tagawa, S.C. Singhal (eds.), *Proc. 4th Int. Symp. SOFC*, The Electrochem. Soc., Pennington NJ,879-88.

17. Ley, K.L., Krumpelt, M., Kumar, R., Meiser, J.H., Bloom, I. (1996) Glass-Ceramic Sealants for Solid Oxide Fuel Cells: Part I. Physical Properties, *J. Mater. Res.* **11**, 1489-93.

18. Yamamoto, T., Itho, H., Mori, M., Mori, N., Watanabe, T. (1996) Compatibility of Mica Glass-Ceramics as Glass-Sealing Materials for SOFC, *Denki Kagaku* **64**, 575-81.

19. Tomsia, A.P., Pask, J.A., Loehman, R.E. (1991) Glass/Metal and Glass-Ceramic/Metal Seals, in S.J. Schneider, J.R. Davis, G.M. Davidson, S.R. Lampman, M.S. Woods, T.B. Zorc (eds.), *Ceramics and Glasses Vol.4*, ASM International, 493-501.

20. Armstrong, T.R., Stevenson, J.W., Pederson, L.R., Raney, P.E. (1995) Instabilities in Doped Lanthanum Chromite in Reducing Environments, in M. Dokiya, O. Yamamoto, H. Tagawa, S.C. Singhal (eds.), *Proc. 4th Int. Symp. SOFC*, The Electrochem. Soc., Pennington NJ,944-51.

21. Hendriksen, P.V., Carter, J.D., Mogensen, M. (1995) Dimensional Instability of Doped Lanthanum Chromites in an Oxygen Pressure Gradient, in M. Dokiya, O. Yamamoto, H. Tagawa, S.C. Singhal (eds.), *Proc. 4th Int. Symp. SOFC*, The Electrochem. Soc., Pennington NJ,934-43.

22. Shannon, R.D. (1976) *Acta Cryst. A* **32**, 751.

23. Nasrallah, M.M., Carter, J.D., Anderson, H.U., Koc, R. (1991) Low Temperature Air-Sinterable $LaCrO_3$ and $YCrO_3$, in F. Grosz, P. Zegers, S.C. Singhal, O. Yamamoto (eds.), *Proc. 2nd Int. Symp. SOFC*, Commission Europ. Communities, Luxembourg, 637-44.

24. Höfer, H.E., Kock, W.F. (1993) Crystal Chemistry and Thermal Behavior in the $La(Cr, Ni)O_3$ Perovskite System, *J. Electrochem. Soc.* **140**, 2889-94.

25. Tai, L.W., Nasrallah, M.M., Anderson, H.U. (1993) $La_{1-x}Sr_xCo_{1-y}Fe_yO_3$, A Potential Cathode for Intermediate Temperature SOFC Applications, in S.C. Singhal, H. Ihawara (eds.), *Proc. 3rd Int. Symp. SOFC*, The Electrochem. Soc., Pennington NJ,241-51.

26. Deutschschweizerische Mathematik- und Physikkommission (1984) *Formeln und Tafeln*, Orell Füssli Verlag Zürich, Schweiz.

6 Appendix

Thermal expansion coefficients of SOFC materials.

Component	Material	Temperature Range / °C	Therm. Expansion Coefficient 10^{-6} K^{-1}	Reference
Electrolyte	ZrO_2 (single crystal)	20-1180	8.12	[7]
	ZrO_2 (polycrstal)		10-11	TABLE 2
	ZrO_2 + 8 wt.% Y_2O_3	100-1000	10.8	[2, 8]
	$Ce_{0.8}Gd_{0.2}O_{1.9}$	50-900	12.9	[6]
	$Ce_{0.8}Sm_{0.2}O_{1.9}$	50-900	13.3	[6]
	$Ce_{0.9}Ca_{0.1}O_{1.9}$	50-900	14.0	[6]
Cathode	$LaMnO_3$	25-1000	11.2	[4]
	$La_{0.9}Sr_{0.1}MnO_3$	25-1000	12.0	[4]
	$La_{0.5}Sr_{0.5}MnO_3$	25-900	12.6	[10]
	$La_{0.9}Ca_{0.1}MnO_3$		10.1	[11]
	$La_{0.5}Ca_{0.5}MnO_3$		11.4	[11]
	$La_{0.4}Y_{0.1}Sr_{0.5}MnO_3$		10.5	[4]
	$La_{0.5}Sr_{0.5}Mn_{0.5}Co_{0.5}O_3$	200-800	16.8	[2]
	$La_{0.5}Sr_{0.5}CoO_3$	25-900	22	[10]
	$La_{0.8}Sr_{0.2}CoO_3$	100-1000	19.7	[25]
	$La_{0.8}Sr_{0.2}FeO_3$	100-1000	12.9	[25]
	$La_{0.8}Sr_{0.2}Co_{0.2}Fe_{0.8}O_3$	100-1000	15.1	[25]
Anode	Ni/YSZ (30vol.% Ni, 30% porosity)		12.5	[12]
	8YSZ	200-1000	10.8	[13]
	Ni	0-100	12.8	[26]
	NiO	200-1000	14.2	[13]
	10mol% NiO	200-800	11.2	[2]
	50mol% NiO	200-800	13.3	[2]
Interconnector				
	$LaCrO_3$ (orthorhombic)	RT-240	6.7	[1]
	$LaCrO_3$ (rhombohedral)	240-1000	9.5	[1]
	$La_{0.9}Sr_{0.1}CrO_3$	240-1000	10.7	[1]
	$La_{0.8}Ca_{0.2}CrO_3$	240-1000	10.0	[4]
	$LaCr_{0.9}Mg_{0.1}O_3$	240-1000	9.5	[1]
	$LaCr_{0.9}Co_{0.1}O_3$	240-1000	13.1	[23]
	$LaCr_{0.9}Ni_{0.1}O_3$	240-1000	10.1	[24]
	$La_{0.7}Ca_{0.3}Cr_{0.85}Co_{0.05}Fe_{0.05}Ni_{0.05}O_3$	240-1000	10.8	[14]
	FeNiCr	20-950	18	[2]
	NiCrAl	20-950	18	[2]
	NiCrW	20-950	16	[2]
	FeCrAl	20-950	15	[2]
	$CrFe5Y_2O_31$	20-1000	11.3	[16]

Glass Seal			
Pyrex		3.2	[17]
Soda Lime Glass		9.5	[17]
SrO-La_2O_3-Al_2O_3- B_2O_3- SiO_2 glass and glass ceramics	50-600	8-13	[17]
Macor glass ceramic	400	9.5	[18]
	800	12.3	[18]
Photoveel glass ceramic	400	8.5	[18]
	800	10.5	[18]

TEMPERATURE LIMITATIONS IN THE PROCESSING SEQUENCE OF SOLID OXIDE FUEL CELLS

C. Kleinlogel and L.J. Gauckler
ETH Zurich
Department of Materials
Nonmetallic Materials
Sonneggstrasse 5
8092 Zurich, Switzerland

1 Introduction

In the fabrication of a SOFC, different consecutive processing steps must be carried out according to the design, the functional requirements and materials of the different cell components. The processing steps must conform to the following boundary conditions:

- *mechanical integrity:* for thin-film deposition methods, a substrate is required with a mechanical strength and a microstructure (e.g. maximum pore size, surface roughness) suitable for thin-film deposition processes. Also screen-printing, as thick-film deposition method, can only be applied on a suitable substrate. Mechanical and thermal stresses between cell components must be kept to a minimum to prevent cracking, delamination, or detachment of the components during fabrication.
- *chemical integrity:* co-sintering of different cell components can lead to interdiffusion or chemical reactions (e.g. zirconate formation [Mitterdorfer96], Mn diffusion in YSZ [Kawada92]). Thin-film deposition by CVD or ECVD is problematic on perovskite substrates due to acid formation during the deposition process.
- *thermal integrity:* As the single cell components are subsequently built and sintered together to maintain good electrical contact, in general the processing and sintering temperature of a single cell component is therefore limited to a maximum temperature. Exeeding this temperature may lead to several undesired effects such as: coarsening of the fine microstructure (e.g. increase of pore size and grain growth in the bulk material), delamination of single components due to a mismatch in the thermal expansion coefficients, interdiffusion and interfacial reactions.

One important aspect in SOFC cell design is its fabricability. Any design requires appropriate ceramic fabrication and assembly methods to incorporate the materials into the cell configuration. The fabrication and assembly processes must ensure that no condition

H.L. Tuller et al. (eds.), Oxygen Ion and Mixed Conductors and Their Technological Applications, 359–374.

or environment in any process step destroys desired material characteristics of individual components.

In the following outline, special focus is given on the maximum processing temperature which cannot be exceeded due to preceding fabrication steps. This is the main limitation in selecting an appropriate fabrication process for a specific cell design. Typical processing steps for cell components with corresponding temperature limitations will be described; the concequences in fabrication will be shown for the well known electrolyte supported cell, the tubular design and for two different planar electrode supported cell designs which have received much attention recently.

2 Temperature limitation in processing

In Table 1, the most important fabrication methods of SOFC cell components and typical temperature ranges for deposition (T_{dep}), sintering (T_{sinter}) and operation (T_{oper}) are listed. It can be seen from this table, that ceramic methods such as screen-printing, tape-casting or sol-gel route need the highest sintering temperatures depending on the material (composition, mean particle size, doping) and on the desired microstructure. Table 2 gives an overview of typical materials and corresponding sintering temperatures used for the electrodes, electrolyte and interconnector of a SOFC.

If deposition methods such as PVD, CVD or VPS are applied for electrolyte formation, dense layers are obtained already during deposition at very low temperatures (~300°C). These layers remain stable up to a maximum temperature (highest possible operation temperature). Further increase of the temperature will lead to intensified sintering, grain coarsening and to pin-holes in the originally dense layer (Figure 1).

Usually, the cell components are formed sequentially as thin or thick layers on a dense or porous supporting substrate; thus, the fabrication conditions cannot be selected independently for each component. For each successive layer it must be selected such that the thermal or chemical environment of this step does not change the properties of the previously formed layers.

Table 1: Typical fabrication methods for SOFC components. The sintering temperature (T_{sinter}) is limiting for subsequent processing steps. T_{max} means the temperature which cannot be exceeded in subsequent processing steps

Shaping process	T_{dep} [°C]	T_{sinter} [°C]	T_{max}, T_{oper}
ceramic methods (screen-printing, tape calendering, slip-casting, sol-gel, electrophoresis etc.)	room temperature	1000 -1600[a]	< T_{sinter}
electrostatic spray deposition (ESD)	200-400	1000 -1100[a]	<T_{sinter}
vacuum plasma spraying (VPS)	~ 600	-	< 1200[a]
physical vapour deposition (PVD)	room temperature - ~ 300	-	< 1200[a]
chemical, electrochemical vapour deposition (CVD, ECVD)	1000 - 1200	-	1000 - 1200[a]

a. the sintering temperature as well as the operation temperature depend strongly on the materials used and on the desired microstructure (see Table 2).

Table 2: Selected SOFC components and typical fabrication temperatures

Component	Material	T_{sinter} [°C]	Micro-structure
Anode	Ni-cermet	1300	porous
Cathode	perovskite, ($La_{(1-x)}Sr_xMnO_3$, $La_{(1-x)}Sr_xCoO_3$)	1000-1200	porous
Electrolyte	ZrO_2, CeO_2 based	1400-1600	dense
Interconnector	$LaCrO_3$	1600	dense

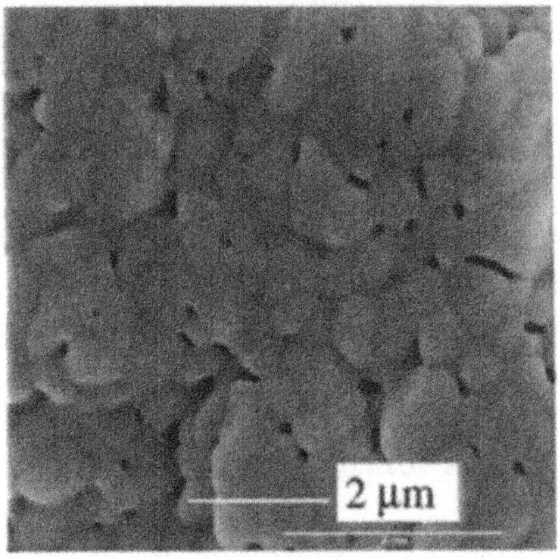

Fig. 1: SEM pictures of a thin ZrO_2(10 mol% Y_2O_3) electrolyte layer, deposited by PVD. Heat treated after sputtering at 700°C (left) and at 1300°C (right).

3 Limitation in processing sequence for different cell designs

Processing routes for three different cell designs will be outlined in the following for: a state-of-the-art SOFC based on a self supporting thick electrolyte (100-200 µm) with thin electrodes, for the tubular design based on a cathode support (both operating at a temperature T_{Oper} between 800 and 1000°C) and for an alternative planar cell design for Intermediate Temperature SOFC (IT-SOFC, T_{OP} = 600-800°C) based on an either cathodic or anodic support structure with a thin electrolyte (~5 µm).

3.1 ELECTROLYTE SUPPORTED PLANAR CELL DESIGN

In Figure 2 a sketch of the cell is shown. It consists of a self-supporting thick (100-200 µm) electrolyte with thin (5-20 µm) electrodes. This type of cell has to be operated at temperatures as high as 1000°C in order to decrease the ohmic losses due to the thick electrolyte.

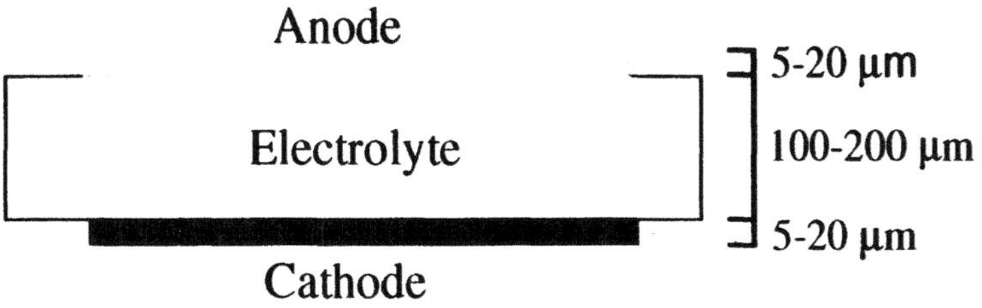

Fig. 2: State-of-the-art solid oxide fuel cell with self supporting, thick electrolyte (150 to 200 μm) and thin (5 to 20 μm) electrodes.

Typical processing steps to build this cell are shown in Figure 4. Following this route, the highest temperature (1400°C) is applied first to achieve a dense electrolyte. Second the anode is added and sintered at a lower temperature (1350°C). In a third step the cathode is added and sintered at a still lower temperature (1100°C). For both the anode and the cathode, the sintering temperature has been chosen to obtain a highly porous microstructure and a good adherence to the supporting electrolyte. In accordance with these requirements, enhanced sintering (i.e. coarsening of the microstructure) of either the electrolyte or the anode can be prevented during fabrication. It is worth to note that the operating temperature (900-1000°C) of such an electrolyte supported SOFC may lead to important changes in microstucture during operation e.g. Ni coarsening, cation diffusion.

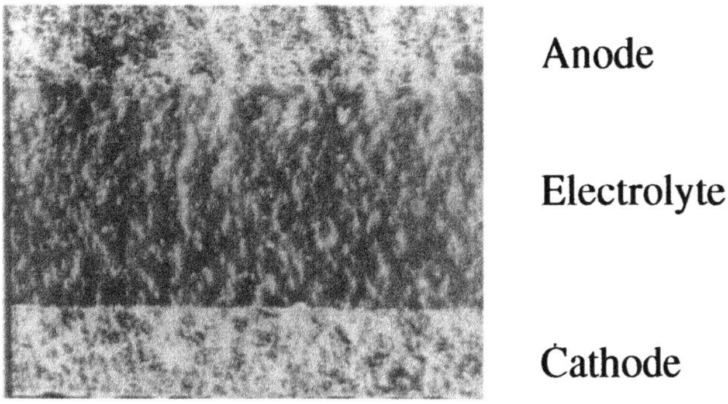

Fig. 3: Typical microstructure of a State-of-the-art solid oxide fuel cell with self supporting, thick electrolyte (150 to 200 μm) and thin (5 to 20 μm) electrodes.

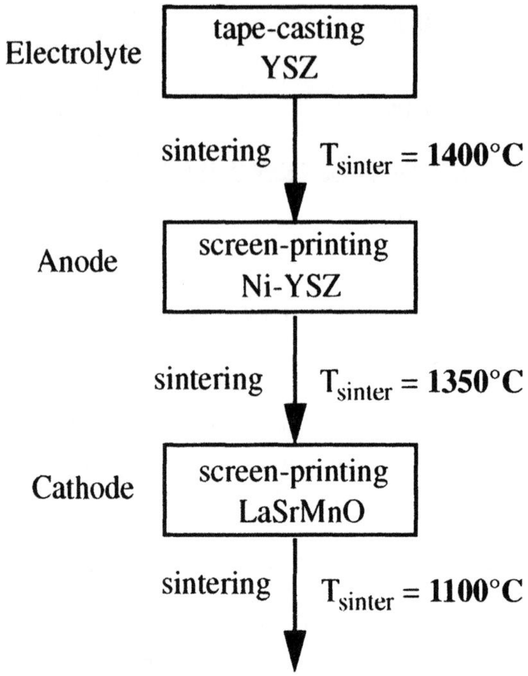

Fig. 4: Processing route for a electrolyte supported SOFC. The processing sequence is adapted to the typical sintering temperature of each component.

3.2 CATHODE SUPPORTED CELL DESIGN

This design is characterized by a thick and mechanically stable cathode structure which acts as support, as current collector and as substrate for electrolyte thin-film deposition. This support might be either a tubular or a planar [Tsai97][Klein97] base structure. The goal of the cathode supported design is to minimize the ohmic resistance of the electrolyte by thinning and to achieve a high fuel utilization by means of a thin-layer anode.

Tubular design

Figure 6 shows a sketch of the tubular cell design which consists of a porous support tube closed on one end, a porous air electrode which acts as substrate for the subsequent depositon of the zirconia electrolyte and of a Ni-cermet anode. In Table 3 the materials for each component are listed as well as the fabrication process. The first processing step is the sintering of the support tube which is carried out at 1550° to 1650°C. Next the cathode is slurry-coated on top of the tube and sintered typically at a temperature around 1400°C. Since the perovskite materials are generally sintered at about 1100°C in order to obtain the desired highly porous microstructure, carbon is added to prevent the perovskite from densification at a temperature this high. In the following processing steps, the electrolyte and the interconnect are deposited by chemical vapor deposition (CVD) at a temperature of about 1400°C. This technique was developed first by Westinghouse [Isenberg77] and it has demonstrated the ability to reproducibly deposit relatively thin films of e.g. Y_2O_3-stabilized ZrO_2 on porous substrates. The CVD process is at present a key technology for depositing uniform gas-tight fine-grained films of electrolytes and interconnects mainly in the tubular configuration. However, it is commonly considered as too costly due to the high reaction temperature, the presence of corrosive gases, the relatively low deposition rate, and the number of firing operations as far as the tubular design is concerned [Gellings96]. In a last step the anode is deposited by first dipping the tube in a Ni slurry and second to fix the anode by electrochemical vapor deposition (EVD) of Y_2O_3-stabilized ZrO_2. The processing route for the tubular cell design is shown in Figure 5. Figure 7 shows a micrograph of a fracture surface of a tubular solid oxide fuel cell.

Table 3: Cell components, materials, and fabrication processes for the tubular configuration of SOFC (Westinghouse)

Component	Material	Thickness	Fabrication process
Support tube	$ZrO_2(CaO)$	1.2 mm	Extrusion sintering
Air electrode (Cathode)	$La_{(1-x)}Sr_xMnO_3$	1.4 mm	Slurry coat-sintering
Electrolyte	$ZrO_2(Y_2O_3)$	40 μm	Electrochemical vapor deposition
Interconnector	$LaCr(Mg)O_3$	40 μm	Electrochemical vapor deposition
Fuel electrode (Anode)	$Ni\text{-}ZrO2(Y_2O_3)$	100 μm	Slurry coat-electrochemical vapor deposition

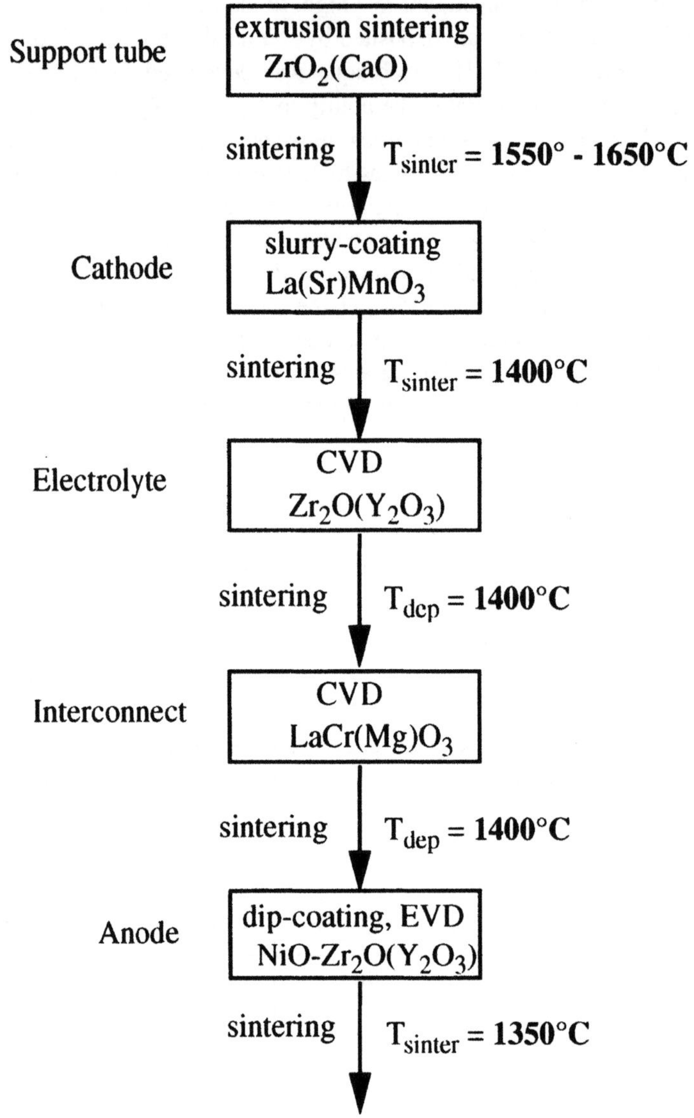

Fig. 5: Processing route for a tubular SOFC. The high processing temperature for the CVD electrolyte deposition requires special powder and slurry preparation for the air electrode (cathode).

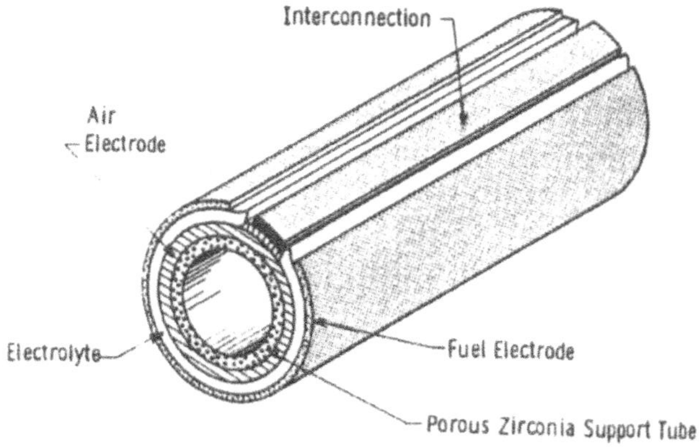

Fig. 6: Schematic representation of the tubular solid oxide fuel cell configuration (Westinghouse)[Singhal93].

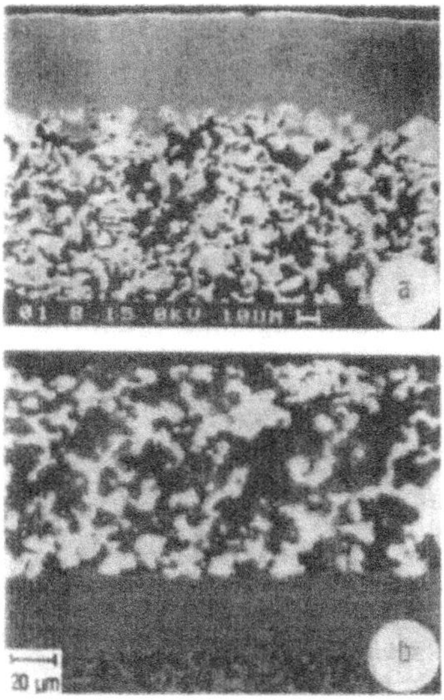

Fig. 7: Representative micrographs of the yttria-stabilized zirconia electrolyte electrochemically vapor deposited on the porous air electrode (a); nickel/yttria-stabilized zirconia fuel electrode on the electrolyte (b); (Westinghouse)[Singhal93].

In order to lower the air electrode losses the novel design of Westinghouse has eliminated the Ca-stabilized Zirconia support tube and replaced by a doped lanthanum manganite air electrode tube which acts as active cathode, as substrate for electrolyte deposition and as mechanical supporting element as well [Singhal97]. This decreases the resistance associated with oxygen diffusion towards the cathode, and significantly improves the power density and therefore the cell efficiency.

Planar design

As shown schematically in Figure 8 the cell consists of a highly porous ceramic foam as mechanical support with a co-fired cathode layer on top which serves as substrate for the subsequent thin film deposition of the electrolyte layer. The anode is applied as last step on top of the electrolyte. The purpose of this design is to minimize the ohmic resistance of the electrolyte by thinning and to achieve a high fuel utilization by means of a thin-layer anode. The processing route is shown in Figure 9. A photograph of the porous ceramic foam structure with the cathode layer on top is shown in Figure 10.

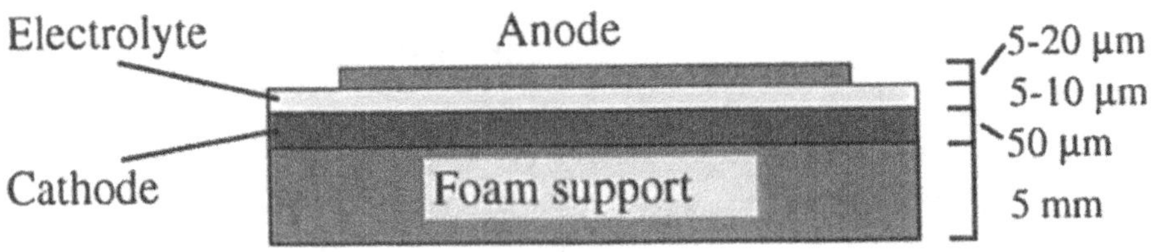

Fig. 8: Cathode supported solid oxide fuel cell with thin film (5-10 µm) electrolyte.

Figure 11 shows a SEM picture of the cross-section of the sputtered thin electrolyte layer on top of the cathode support.

The main drawback of this design are the low sintering temperatures of the anode layer on top of the thin-film electrolyte resulting in low in plane conductivity as well as in low adherence and contact areas to the electrolyte. Increasing the sintering temperature of the anode would increase the anode performance but also coarsen the cathode and the electrolyte microstructure. This implies that state-of-the-art anodes prepared by a conventional ceramic method as mentioned in Table 2 cannot be used and other processing routes e.g. PVD, VPS or ESD have to be used [Tsai97][Stelzer97].

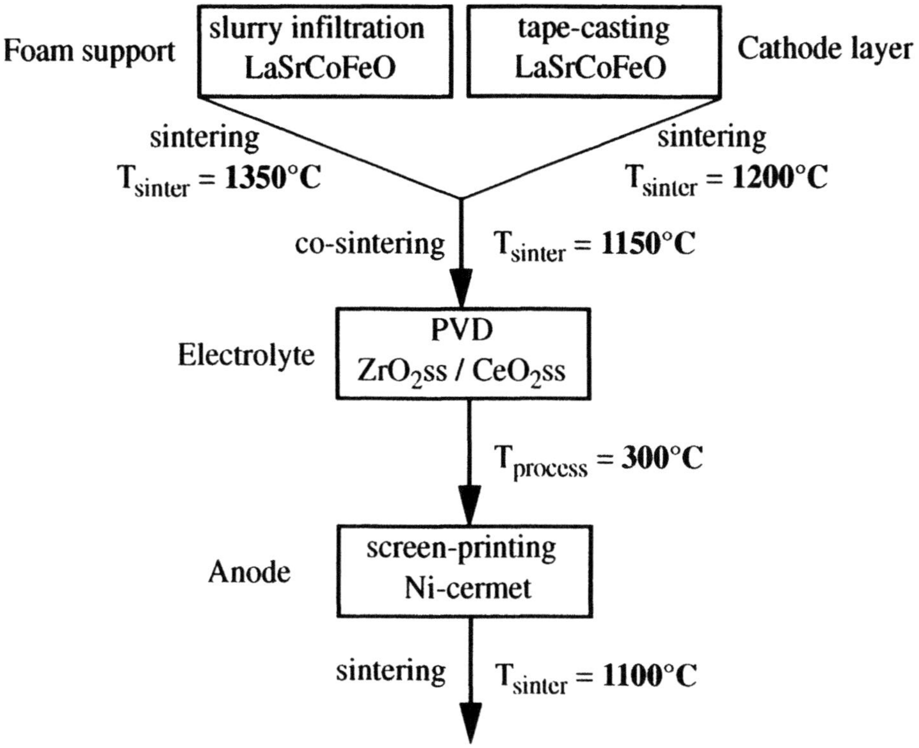

Fig. 9: Processing route for a cathode supported SOFC. Limitation in the sintering temperature of the anode due to the preceding processing steps means that a state-of-the-art Ni-YSZ anode with a typical sintering temperature of about 1350°C cannot be applied.

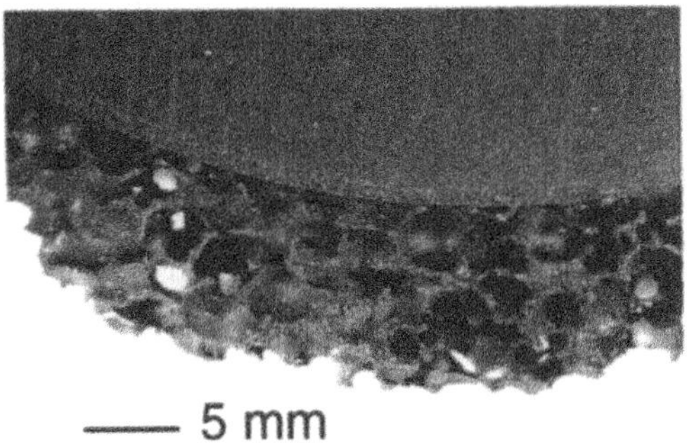

Fig. 10: Photograph of a cathode substrate which consists of a porous ceramic foam structure (bottom) and a 100 µm thin cathode layer on top.

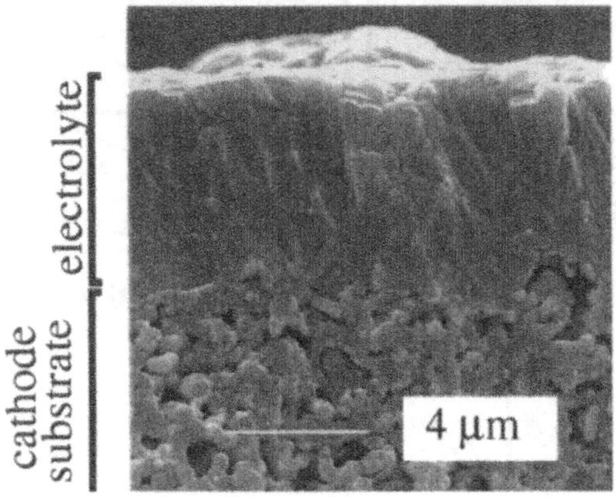

Fig. 11: SEM picture of the sputtered PVD electrolyte layer (5 μm) on top of the porous cathode layer.

3.3 ANODE SUPPORTED PLANAR CELL DESIGN

The anode support structure consists of a double layer structure composed of a microporous substrate on top of a macroporous ceramic foam. The ceramic foam acts as mechanical support, fuel distributor and current collector whereas the 100 μm thick microporous anode layer on top acts as active anode. A thin-film electrolyte is subsequently deposited on the anode support (Figure 12). It is important to use the same slurry compositions for tape casting of the active anode and for impregnation of the polymeric foam to prevent chemical reaction and delamination due to different thermal expansion coefficients.

The processing route is shown in Figure 13. The highest temperature is applied in the first step by co-sintering of the anode support structure with the ceramic foam. All following steps are processed at a lower temperature and no special care must be taken since the deposited electrolyte layer is stable up to temperatures where the $La(Sr)MnO_3$ cathode is typically sintered.

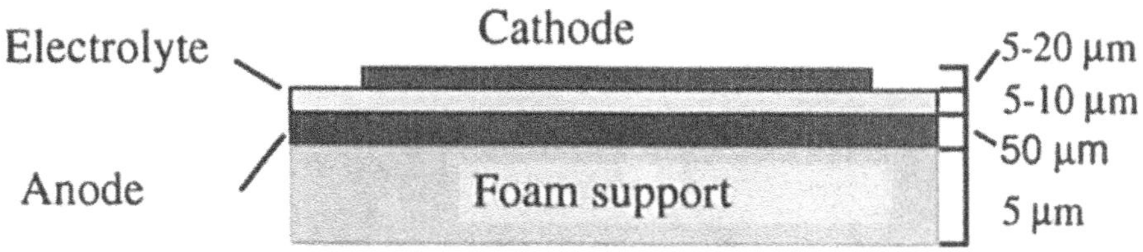

Fig. 12: Anode supported solid oxide fuel cell with thin film (5-10 µm) electrolyte.

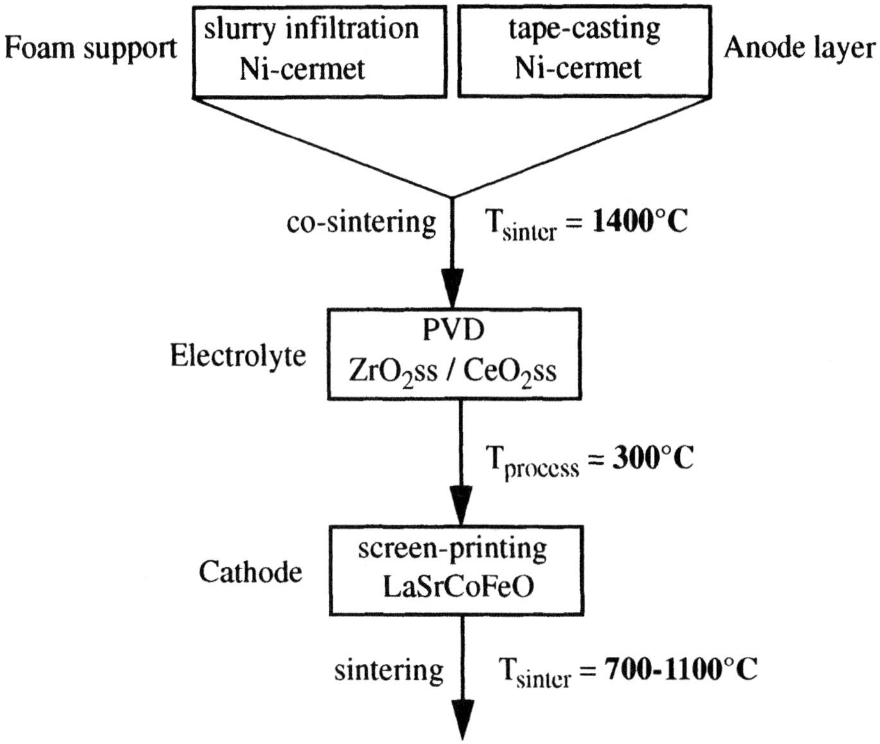

Fig. 13: Processing route for a anode supported SOFC. The processing sequence is adapted to the typical sintering temperature of each component.

Figure 14, shows a photograph of a ceramic foam with an anode layer on top and a fracture image (SEM) of the anode/electrolyte and cathode/electrolyte interface. The performance of an anode supported cell at different temperatures can be seen in Figure 15. The high power outputs were obtained at a low fuel utilisation.

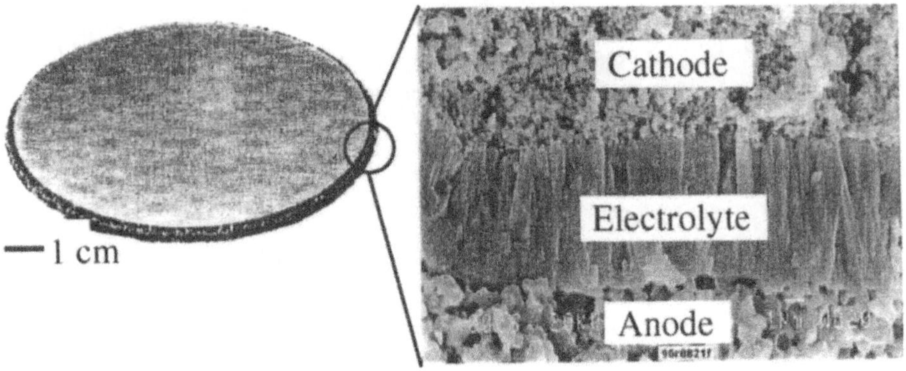

Fig. 14: Sintered anode substrate samples, diameter 120 mm, composed of foam and coverplate (left). Fracture cross section of PVD-cell after 500 hours of operation at 715°C (right) [Honegger97].

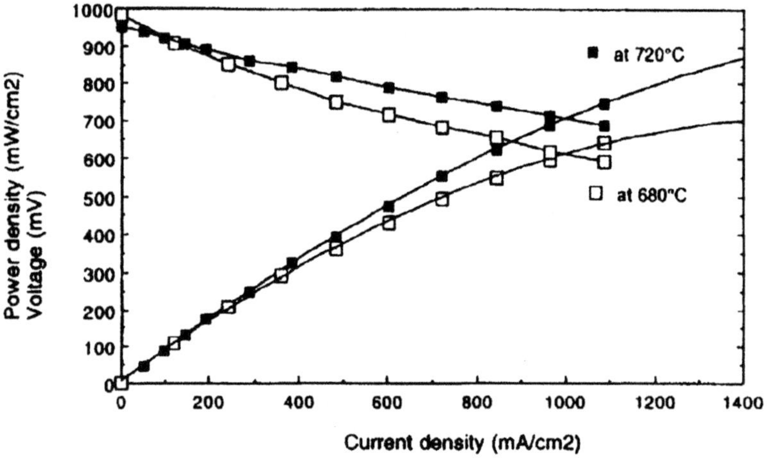

Fig. 15: Temperature dependence of PVD-cell discharge characteristic after 24 hours of service [Honegger97].

4 Summary

In the fabrication of solid oxide fuel cells the boundary conditions such as mechanical, chemical and thermal integrity of single processing steps have to be considered. Processing temperatures of single cell components are determined by the materials, by the microstructure and by the preparation process selected. For the state-of-the-art materials which are $ZrO_2(Y_2O_3)$ for the electrolyte, NiO-$ZrO_2(Y_2O_3)$ for the anode and La(Sr)MnO_3 for the cathode the sintering temperatures necessary to obtain the desired microstructure are well known. These are about 1400°C for the electrolyte, 1350°C for the anode and 1100°C for the cathode. Since the fabrication of a SOFC component usually starts with the highest sintering step, special material preparation or processing step selection is required for the tubular design and for the electrode supported designs.

5 References

[Mitterdorfer96] A. Mitterdorfer, M. Cantoni and L.J. Gauckler, in Proc. of the Second European Solid Oxide Fuel Cell Forum/1996, Oslo, B. Thorstensen, Ed., p. 373

[Kawada92] T. Kawada, N. Sakai, H. Yokokawa, M. Kokiya, Solid State Ionics 50 (1992), p. 189

[Visco93] S.J. Visco, C. Jacobson, L.C. De Jonghe, EPRI/GRI Workshop on Fuel Cell Technology Research and Development, Electric Power Research Institute, Palo Alto, CA, 1993

[Chen92] C.C. Chen, M.M Nasrallah, H. Anderson, in 1992 Fuel Cell Seminar Abstracts, Tucson, AZ, Courtesy Associates, Washington, DC, 1992, p. 515

[Christie93] G.M. Christie, P.H. Middleton, B.C.H. Steele, in Proc. of the third International Symposium on Solid Oxide Fuel Cell, 1993, Hawaii, S.C.Singhal, Ed., p. 315

[Honegger97] K. Honegger, E. Batawi, Ch. Sprecher and R. Diethelm, in Proc. of the fifth International Symposium on Solid Oxide Fuel Cell, 1997, Aachen, S.C. Singhal, Ed., p. 321

[Singhal93] S.C. Singhal, "Tubular Solid Oxide Fuel Cells", in Proc. of the third International Symposium on Solid Oxide Fuel Cell, 1993, Hawaii, S.C.Singhal, Ed., p. 665

[Barnett90] S.A. Barnett, "A new Solid Oxide Fuel Cell Design based on Thin Film Electrolytes", Energy, 15, 1 (1990)

[Tsai97] T. Tsai, S.A. Barnett, "Effect of LSM-YSZ Cathode on Thin-Electrolyte Solid Oxide Fuel Cell Performance", Solid State Ionics, 93, 207 (1997)

[Singhal97] S.C. Singhal, "Application of ionic and electronic conducting ceramics in solid oxide fuel cells", in Proc. of the 192nd meeting of the Electrochemical Society, 1997, Paris, No. 2136

[Isenberg77] A.O. Isenberg, Electrochemical Society Symposium Electrode Materials, Processes for Energy Conversion and Storage, 1977, 77(6), p. 572

[Gellings96] P.J. Gellings, H.J.M. Bouwmeester, Ed., "The CRC Handbook of Solid State Electrochemistry", 1996, p. 438.

[Stelzer97] N.H.J. Stelzer, C.H. Chen, L.N. van Rij, and J. Schoonman, "Electrostatic Spary Deposition of Doped YSZ Electrode Materials for a Monolithic Solid Oxide Fuel Cell Design", in 10th SOFC Workshop, A.J. McEvoy and K. Nisancioglu, Editor. Les Diablerets 1997, 236

[Will97] J. Will, L.J. Gauckler, " Ceramic Foams as Current Collectors in Solid Oxide Fuel Cells (SOFC): Electrical Conductivity and Mechanical Behaviour", in Proceedings of the Fifth International Symposium on Solid Oxide Fuel Cells, S.C. Singhal, Editor, Aachen 1997, 757

METALLIC INTERCONNECTOR

A. BIEBERLE, L.J. GAUCKLER
ETH Zürich
Department of Materials
Nonmetallic Materials
Sonneggstrasse 5
8092 Zürich, Switzerland

1 Introduction

In the planar design of a SOFC (= Solid Oxide Fuel Cell), the interconnector (Figure 1) has to satisfy two main purposes: first, it has to pass current from one cell to the next; second, it has to separate the fuel gas on the anode side (hydrogen, coal gas, or natural gas) from the oxidizing gas on the cathode side (air or oxygen). The interconnector is therefore directly in contact with two different reactive gas atmospheres. It must resist high temperatures up to 1000°C corresponding to the operating temperature of SOFCs.
Due to the environment and the high temperatures, solid-gas reactions are likely to occur during operation. The reactions may damage the metallic interconnector or may lead to a degradation of other SOFC elements. These two issues will be discussed in the following after a short paragraph on the materials selection.

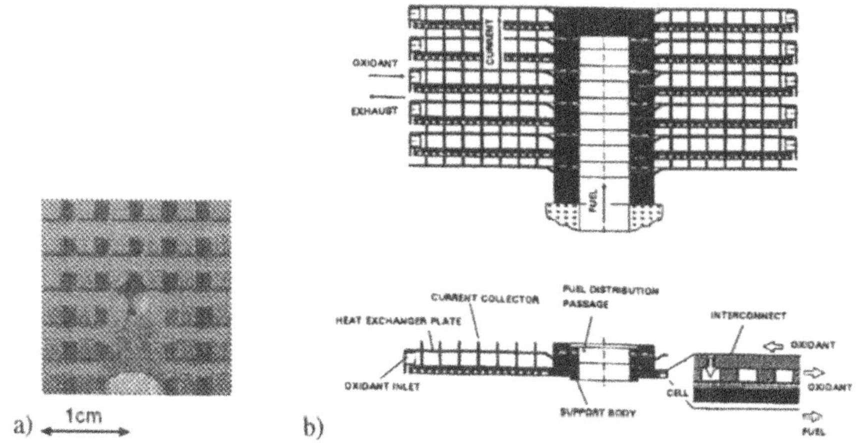

Figure 1: a) Interconnector for the planar SOFC design; b) schematic view of the interconnector in Sulzers HEXIS SOFC-system [1].

H.L. Tuller et al. (eds.), Oxygen Ion and Mixed Conductors and Their Technological Applications, 375–387.

2 Materials Selection

For the selection of a material for interconnector application it is important to consider not only the chemical compatibility but also electrical, mechanical, thermal, manufacturing, and economical requirements. According to studies carried out in the last years, only some oxide forming alloys are theoretically suitable metallic interconnector materials for high temperature and highly corrosive SOFC applications [2,3,4,5,6]. Results of an evaluation of alloys for interconnector application are listed in TABLE 1.

TABLE 1: Results of an evaluation for the selection of alloys for interconnector application [6].

	Physical Properties					Long Term Stability				Fabricability		
	Tensile Strength	Toughness	Creep Strength	Oxide Scale Conductivity	Thermal Expansion Compatibility	Air	Coal Gas	Natural Gas	Temperature-Induced Embrittlement	Forming	Welding	Brazing
AC 66	2	1	2	1	4	2	1	3	1	1	1	1
HA 230	2	1	2	1	2	2	2	2	1	1	1	1
HA 214	2	1	2	4	4	1	1	1	3	1	1	1
Incoloy MA 956	1	1	1	4	2	1	1	1	1	1	1	1
Cr-Co Alloys	1	3	1	2	1	4	2	4	1	3	4	1
Cr-Fe Alloys	1	2	1	2	1	2	4	4	1	2	2	1

1 = good; 2 = adequate; 3 = poor; 4 = inadequate

HA 230: Haynes 230: about 52 Ni, 20-24 Cr, 13-15 W
HA 214: Haynes 214: about 75 Ni, 16 Cr, 4.5 Al, 3 Fe
Incoloy MA 956: about 75 Fe, 20 Cr, 4.5 Al, 0.5 Y_2O_3

TABLE 2 contains a list of metallic interconnector materials that can be used at temperatures up to 1000°C. These alloys are, however, very expensive. In order to use less expensive materials, such as ferritic steel, the operating temperature must be reduced to an intermediate temperature range (about 700°C).

TABLE 2: Examples of metallic interconnector materials.

Material	**Ni**	**Cr**	**Fe**	**Al**	**Y**	**Oxide Scale**	**Reference**
Inconel 601	62.6	22.6	13.6	1.2	-	chromia	[3]
Hastelloy X	50.4	21	18.4	-	-	chromia	[3]
$CrFe5Y_2O_31$	-	94	5	-	1	chromia	[5,8,9]
NiCr25Fe10Al2Y		25	10	2	-	chromia	[5]

3 Corrosion Resistance

A good corrosion resistance of metallic interconnectors is obtained due to the formation of a dense protective oxide scale. In Al- and Cr-containing alloys, alumina and chromia scales are formed, respectively.

According to Figure 2, the corrosion resistance of alumina forming alloys is better than that of chromia forming alloys because of the smaller rate constant of oxidation. However, the electrical conductivity of the alumina scale is much lower than the electrical conductivity of the chromia scale ($\sigma_{alumina} = 10^{-6}$-$10^{-8}$S/cm at 900°C, $\sigma_{chromia} = 10^{-2}$-$10^{-1}$S/cm) [3]. Therefore, the electronic transport from the electrode to the interconnector in alumina forming alloys becomes much lower during operation than in chromia forming alloys. If the electronic transport is reduced too much, the cell performance breaks down. Therefore, alumina forming alloys are not considered as suitable metallic interconnector materials, even though they possess excellent corrosion resistance.

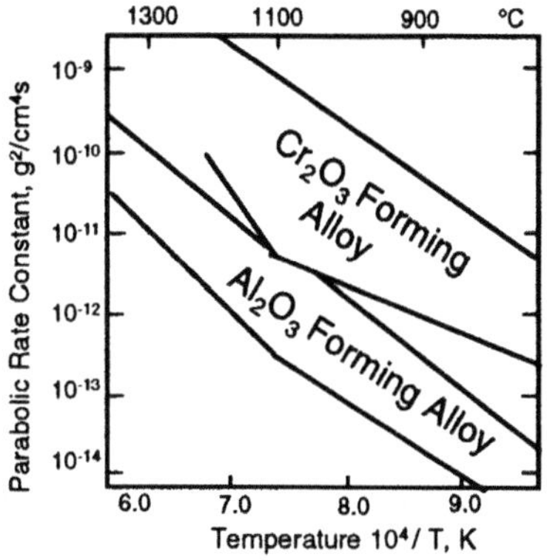

Figure 2: Scattering of literature values of the parabolic rate constants for the scale formation in chromia and alumina forming alloys [3].

The oxide scale in chromia forming alloys consists of Cr_2O_3. This oxide represents the only solid chromium oxide that is stable at high temperatures; higher oxides of chromium, such as CrO_2 and CrO_3, are not stable in the solid state at high temperatures.

The basic growth mechanism of the chromia scale in non-ODS materials (see “Improvement of Materials Properties“) can be summarized as follows (Figure 3a) [10]:

1. The scales grow by counter-current grain boundary diffusion of chromium and oxygen; the new oxide formation takes place at the grain boundaries.
2. Outward chromium diffusion is more important than inward oxygen diffusion, and the new oxide formation accordingly takes place in an outer region of the scale; inward oxygen diffusion becomes increasingly important the higher the oxygen prcssurc.
3. The counter-current chromium and oxygen diffusion and the formation of oxide within the scale result in lateral scale growth in addition to the growth normal to

the metal surface; this, in turn, results in large growth stresses so that the scale can plastically deform and often develops small cracks.

4. The ability of the scale to deform plastically increases with decreasing oxygen activity.
5. Due to the outward chromium diffusion, voids develop at or near the metal/scale interface; the chromia scales can easily be detached from the surface.

As a result of the above described reaction mechanism, the chromia scale is heavily deformed, buckled, and wrinkled [11].

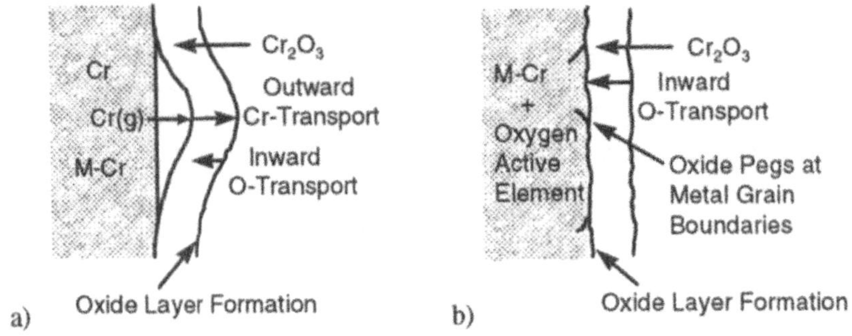

Figure 3: Schematic representation of the growth of chromia scale in a) non-ODS and b) ODS materials[10].

The formation of Cr_2O_3 on Cr-containing alloys follows a parabolic law at temperatures higher than 700°C (Figure 2) [10]. There is uncertainty in literature about the magnitude of the rate constant. The differences which are probably due to sample preparation (surface roughness, grain size, and microstructure) are found to be several orders of magnitude [10].

4 Cr-Evaporation

A major drawback of metallic interconnector materials containing chromium is the evaporation of chromium containing species during operation. It was proven that the degradation of a cell stack with a chromium containing metallic interconnector material is mainly caused by chemical contamination of the cathode due to chromium evaporation (Figure 4) [12].

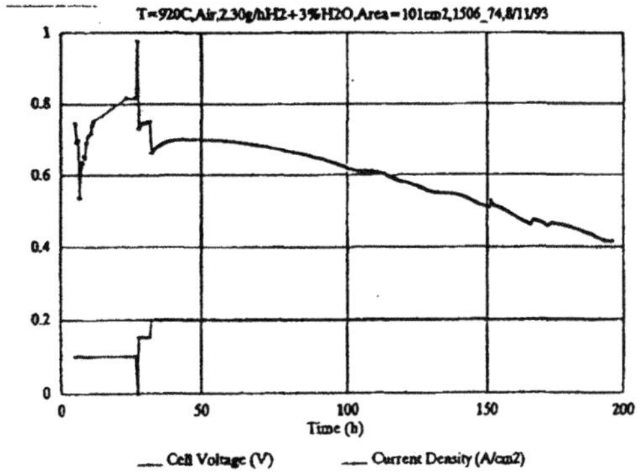

Figure 4: Degradation behavior of a single cell stack with metallic interconnectors on both electrodes [12].

As described in the previous paragraph, Cr_2O_3(s) formed during SOFC operation may react to various volatile chrome-oxides and chrome-oxyhydroxides at operating temperature [10]

$$\tfrac{1}{2}\, Cr_2O_3(s) + \tfrac{3}{4}\, O_2(g) \rightarrow CrO_3(g) \qquad (1)$$

Equation (1) indicates that a high partial pressure of oxygen enhances the evaporation of chromium as CrO_3 (Figure 5). This indicates that the evaporation of chromium is more critical at the cathode side than at the anode side of the SOFC.

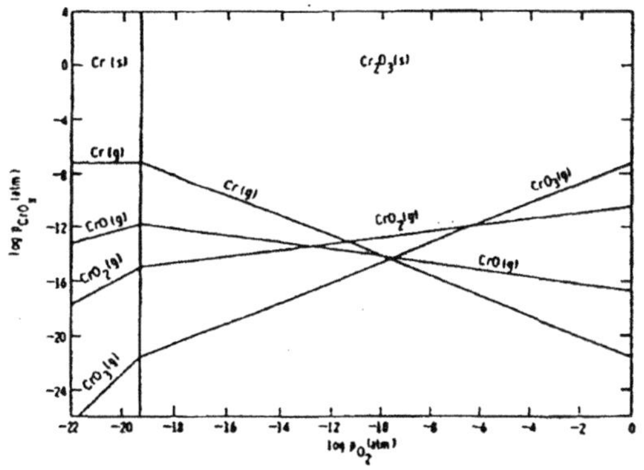

Figure 5: Thermochemical diagram for the Cr-O system at 1127°C [12].

The chromium volatility increases also, when the water content of the air at the cathode side is increased due to humidity or leakage between anode and cathode side [8]. Other volatile species which may be produced are CrO_2, CrO, Cr, CrOH, $Cr(OH)_2$, $Cr(OH)_3$, $Cr(OH)_4$, CrOOH, $CrO(OH)_4$, $CrO(OH)_3$, $CrO(OH)_2$, $CrO_2(OH)$, $CrO_2(OH)_2$ (Figure 6) [8,13].

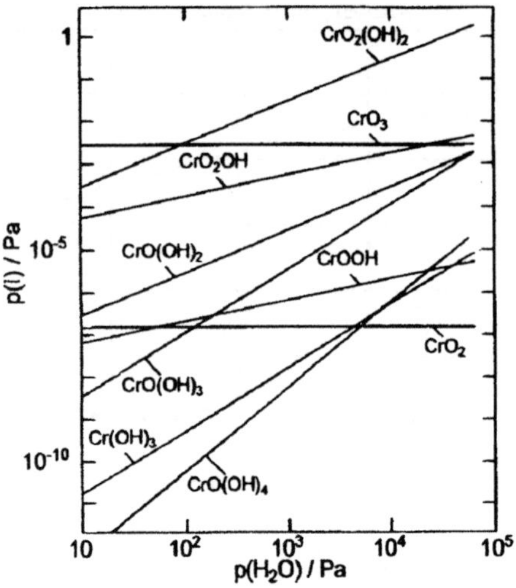

Figure 6: Partial pressures over $Cr_2O_3(s)$ at 1223K in humid air ($p(O_2) = 2.13 \cdot 10^4$ Pa) with different H_2O partial pressures [13].

The maximum amount of evaporated $CrO_3(g)$ can be calculated by Langmuir's equation [9]

$$K_i = \frac{a_i p_i}{(2p \cdot m_i kt)^{1/2}} \tag{2}$$

where a_i is the evaporation coefficient, p_i the vapor pressure, m_i the mass of the evaporating molecule, k the Boltzmann´s constant, and t the time.

Two mechanisms are discussed in literature concerning the degradation of the electrical properties of SOFCs by chromium evaporation [10,14]:

The first mechanism involves electrochemical reactions of the chromium containing gas (CrO_3, $CrO_2(OH)_2$, $CrO_2(OH)$). The chromium containing gas can be reduced, while O^{2-} ions are produced. The oxygen ions pass the electrolyte and react with the fuel at the anode side. The reaction results in the precipitation of $Cr_2O_3(s)$ at the cathode or the cathode/electrolyte phase boundary (Figure 7). The active sites for O_2 reduction are blocked by Cr_2O_3. Polarization losses increase.

$$2CrO_3(g) + 6e^- \rightarrow Cr_2O_3(s) + 3O^{2-} \qquad [8,9] \tag{3}$$

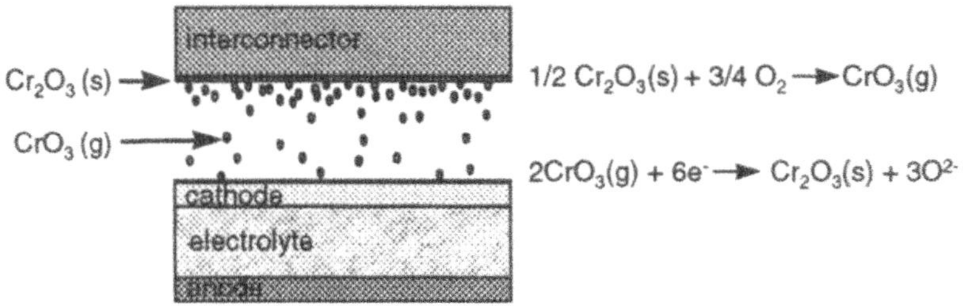

Figure 7: Cromium evaporation and formation of solid chromiumoxide at the cathode side of a SOFC [8].

The second mechanism involves chemical reactions of chromium with the cathode and the anode material, respectively. Equation (4) is an example for this mechanism [8]

$$La_{1-x}Sr_xMnO_3(s) + 1/2Cr_2O_3(s) \rightarrow La_{1-x}Sr_xMn_{1-y}Cr_yO_3(s) + (Cr_{1-y}Mn_y)O_{1.5-\delta}(s) + \delta/2\ O_2(g) \quad (4)$$

Chromia becomes dissolved in the cathode perovskite; spinel phases or $(Cr_{1-y}Mn_y)O_{1.5-\delta}$ solid solution can precipitate and deteriorate the electrical properties of the perovskite.

5 Improvement of Materials Properties

The literature proposes two ways to improve the properties of metallic interconnector materials.

5.1 OXIDE DISPERSION STRENGTHENING (= ODS)

Small amounts (0.4-1wt%) of oxide active elements, such as Y or Ce, are added to the metal. These elements are included in form of a finely dispersed oxide phase [15]. Apart from outstanding physical and mechanical properties, ODS Cr-alloys show improved corrosion resistance in oxidizing, nitriding, as well as carburizing hot gases up to 1000°C. Exposure to hydrogen containing atmospheres does not cause embrittlement of the material [6]. The rate of oxidation is illustrated by the square of the weight gain at 1000°C in dry air as a function of time (Figure 8). Fitting the data to the parabolic law

$$w^2 = 2k_p t + C \quad (5)$$

where k_p is the parabolic rate constant, t the exposure time, and C an integration constant, gives $k_p = 1 \cdot 10^{-12}\ g^2/cm^4s$ at 1000°C [16]. According to Figure 9, the oxidation increases rather slowly with temperature up to 1050°C; at higher temperatures the oxidation-rate increases rapidly [16].

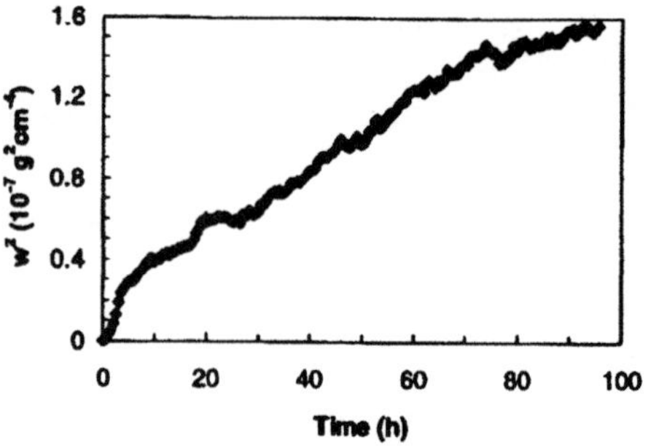

Figure 8: The square of the weight gain vs. time when oxidizing $Cr_{94}Fe_5(Y_2O_3)_1$ alloy in dry air at 1000°C [16].

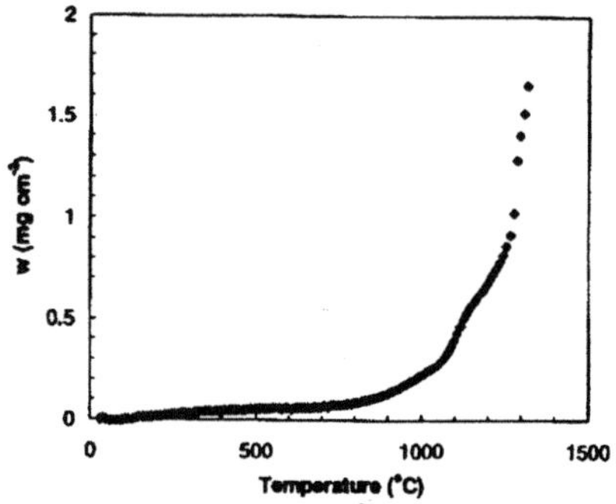

Figure 9: Weight gain of $Cr_{94}Fe_5(Y_2O_3)_1$ alloy in dry air vs. temperature (temperature scan rate of 100K/h) [16].

Oxygen active elements have the following effect on the oxidation behavior of chromia forming alloys [10]:

1. Selective formation of chromia scales is enhanced; i.e. continuous protective scales of Cr_2O_3 are obtained at lower chromium contents in the alloys.
2. The growth rate of the chromia scale is reduced.
3. The growth mechanism of the chromia scales is changed to involve a predominant inward oxygen diffusion.
4. The grain size of the chromia scale often remains small during oxidation and thus grain growth is impeded; for alloys without oxygen active elements a faster grain growth takes place in the scales during oxidation.
5. The adherence of chromia scales to the alloy substrate is improved.

The differences in the growth of the oxide scale with and without oxide active elements are demonstrated in Figure 3a and 3b.

A comparison of the microstructure of ODS and non-ODS materials is shown in (Figure 10). The non-ODS material NiCr25Fe10Al2Y has large macroscopic pores along the grain boundaries after operation in oxygen at 950°C for 1000h (Figure 10a). No pores were detected in the ODS material $CrFe5Y_2O_31$ (Figure 10b). Except of the oxide layer on top of the alloy, the alloy itself is not attacked.

Figure 11 shows the microstructure of the ODS material $CrFe5Y_2O_31$ after 1000h testing at 950°C in different environments: oxygen, air, and fuel gas. Fuel gas atmosphere produces less corrosive attack.

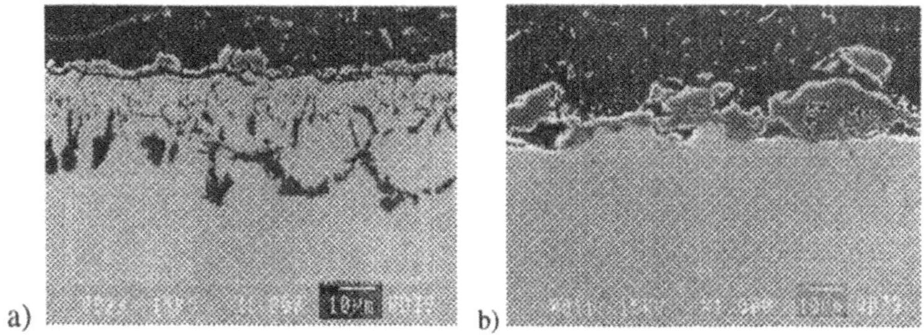

Figure 10: Microstructure of a) the non-ODS material NiCr25Fe10Al2Y and b) the ODS material $CrFe5Y_2O_31$ after 1000h testing at 950°C in oxygen [5].

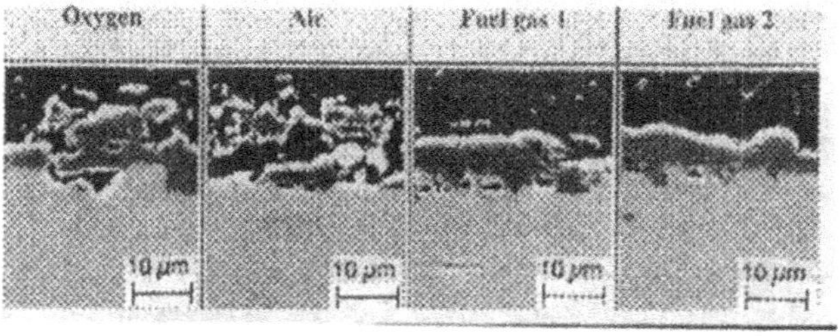

Figure 11: Microstructure of $CrFe5Y_2O_31$ after testing at 950°C [5].

The rate constants of corrosion are much smaller in the case of the ODS material $CrFe5Y_2O_31$ compared to the non-ODS material NiCr25Fe10Al2Y (Figure 12). The ODS material will therefore corrode much slower than the non-ODS material [5]. The same results were obtained in long term corrosion tests.

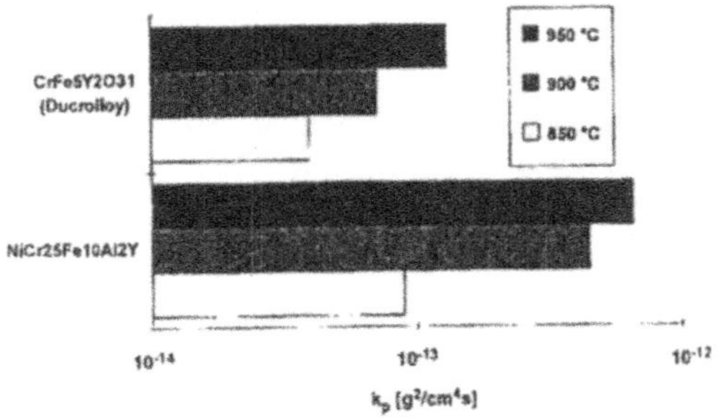

Figure 12: Rate constant of $CrFe5Y_2O_31$ and NiCr25Fe10Al2Y after 500h testing in oxygen [5].

Differences in the corrosion behavior of ODS and non-ODS materials concerning the weight changes as a function of time can be seen in Figure 13. The ODS material shows much lower weight changes meaning less corrosive attack than the non-ODS material which fails after about 400h of operation. Similar results were obtained with compositions where Cr was doped with 0.5wt.% Ce [11].

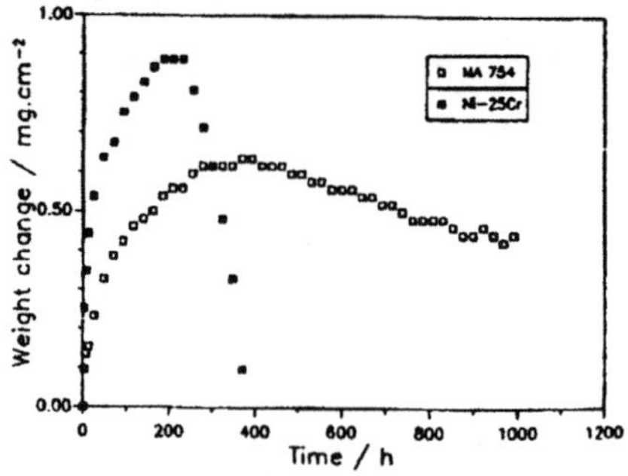

Figure 13: Cyclic oxidation of ODS-material (MA754) and non-ODS material (Ni-25Cr)[15].

5.2 COATING

The formation of a coating on the metallic interconnector material is another possibility to improve the corrosion resistance and to reduce chromium evaporation. Coatings for the air side of the SOFC are usually made of typical SOFC cathode materials such as e.g. $La_{0.85}Sr_{0.15}CoO_3$ [3, 7, 8, 17, 18]; the coating for the fuel side consists of nickel metal [3].

Low pressure plasma-spraying can e.g. be used to coat a metallic interconnector [3]. Plasma-spraying [21] belongs to the thermal spray deposition techniques. The powder is heated close its melting point, accelerated by a plasma, and directed to the substrate. The coating consists of many layers of overlapping, thin lamellar particles. Low pressure is used in order to obtain higher velocity, sputter cleaning, and less stress due to preheated parts.

According to [3], a coating on the metallic interconnector does not reduce the thickness of the oxide scale; it reduces, however, the electrical resistance. This is supposed to be due either to the formation of a highly conductive scale based on Cr_2O_3 or to the dissolution of Sr in Cr_2O_3 [3]. Figure 14 shows that the slope of the U-I-curve, i.e. the internal resistance of the SOFC, is lower for a coated interconnector compared to an uncoated one. The same result was obtained in [18] where the results were also confirmed for long term runs (400h).

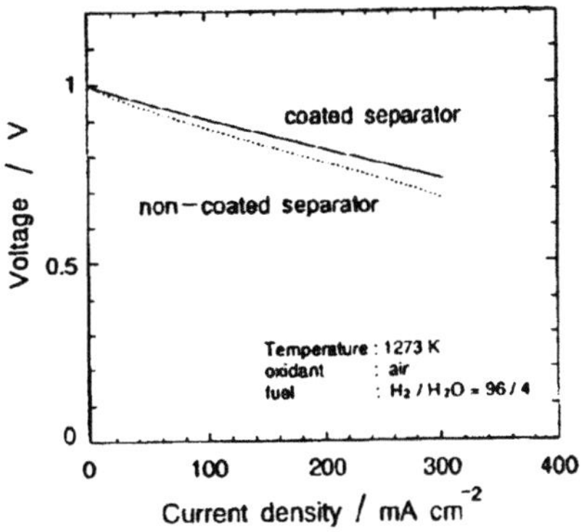

Figure 14: Current-voltage characteristics of a SOFC (144cm²) with an Inconel 601 interconnector at 1000°C [3].

6 Summary

Metallic interconnectors are made of chromia forming alloys. Al-containing alloys are not applicable, as they form an alumina scale that hinders electronic transport.
The corrosion resistance of metallic interconnectors is due to the formation of a protective scale which grows at the grain boundary.
The most serious problem concerning chromia forming alloys is the evaporation of chromium at elevated temperatures. The chromium evaporation contaminates the SOFC either electrochemically by the precipitation of Cr_2O_3 at the cathode or by the dissolution of chromia in the cathode and precipitation of a spinel phase. The evaporation results in a degradation of the electrical properties of the SOFC.
Two possibilities to improve the materials compatibility of metallic interconnectors are oxide dispersion strengthening or coating.

7 References

1. Diethelm, R., Honegger, K., (1993) Status of the Sulzer HEXIS Stack and System Development, in S.C. Singhal, H. Iwakara (eds.), *Proc. 3rd Int. Symp. SOFC*, The Electrochem. Soc., Pennington NJ, USA, 822-29

2. Ivers-Tiffée, E., Wersing, W., Schießl, M., Greiner, H. (1990) Ceramic and Metallic Components for a Planar SOFC, *Ber. Bunsenges. Phys. Chem.* **94**, 978-81.

3. Kadowaki, T., Shiomitsu, T., Matsuda, E., Nakagawa, H., Tsuneizumi, H., Maruyama, T. (1993) Applicability of Heat Resisting Alloys to the Separator of Planar Type Solid Oxide Fuel Cell, *Solid State Ionics* **67**, 65-69.

4. Quadakkers, W.J., Mallener, W., Grübmeier, H., Wallura, E. (1993) Corrosion and Compatibility Studies on Metallic Interconnector Materials for SOFCs, in P. Biedermann, B. Krahl-Urban (eds.), *Proc. 5th IEA Workshop*, Jülich, Germany, 87-99.

5. Greiner, H., Grögler, T., Köck, W., Singer, R.F. (1993) Chromium Based Alloys for High Temperature SOFC Applications, in S.C. Singhal, H. Iwakara (eds.), *Proc. 3rd Int. Symp. SOFC*, The Electrochem. Soc., Pennington NJ, USA, 879-88.

6. Köck, W., Martinz, H.P., Greiner, H., Janousek, M., (1995) Development and Processing of Metallic Cr Based Materials for SOFC Parts, in M. Dokiya, O. Yamamoto, H. Tagawa, S.C. Singhal (eds.), *Proc. 4th Int. Symp. SOFC*, The Electrochem. Soc., Pennington NJ, USA, 841-49.

7. Minh, N.Q., Takahashi, T. (1995) *Science and Technology of Ceramic Fuel Cells*, Elsevier, Amsterdam, Netherland.

8. Das, D., Miller, M., Nickel, H., Hilpert, K. (1994) Chromium Evaporation from SOFC Interconnector Alloys and Degradation Process by Chromium Transport, in U. Bossel (ed.), *Proc. 1st Europ. SOFC Forum*, Vol. 2, 703-13.

9. Quadakkers, W.J., Greiner, H., Köck, W. (1994) Metals and Alloys for High Temperature SOFC Application, in U. Bossel (ed.),*Proc. 1st Europ. SOFC Forum*, Vol. 2, 525-41.

10. Kofstad, P., (1988) *High Temperature Corrosion*, Elsevier Appl. Sci. Publ. LTD, England .

11. Bredesen, R., Kofstad, P. (1996) Formation of Cr_2O_3 Scales on Cr and Cr+0.5wt.%Ce in Oxidizing and Reducing Atmospheres, in F.W. Poulsen, N. Bonanos, S. Linderoth, M. Mogensen, B. Zachau-Christiansen (eds.),*Proc. 17th Riso Int. Sym. Mat. Sci.*, Riso Nat. Lab., Roskilde, Denmark, 187-92.

12. Batawi, E., Honegger, K., Diethelm, R. (1994) Factors Influencing the Degradation Performance of High Temperature Solid Oxide Fuel Cells Contacted with Metallic Bipolar Plates, in H. Nabielek, M. Brocco (eds.), *Proc. 6th IEA Workshop*, Rome, Italy, 175-81.

13. Hilpert, K., Das, D., Miller, M., Peck, D.H., Weiss, R. (1996) Chromium Vapor Species Over Solid Oxide Fuel Cell Interconnect Materials and Their Potential for Degradation Processes, *J. Electrochem. Soc.* **143**, 3642-47.

14. Quadakkers, W.J., Greiner, H., Köck, W., Buchkremer, H.P., Hilpert, K., Stöver, D. (1996) The Chromium Base Metallic Bipolar Plate - Fabrication, Corrosion and Cr Evaporation, in B. Thorstensen (ed.), *Proc. 2nd Europ. SOFC Forum*, Vol. 1, 297-306.

15. Quadakkers, W.J. (1990) Growth Mechanism of Oxide Scales on ODS Alloys in the Temperature Range 1000-1100°C, *Werkstoffe und Korrosion* **41**, 659-68.

16. Linderoth, S., Langvad, N., Mogensen, G. (1996) High-Temperature Oxidation Studies of $Cr_{94}Fe_5(Y_2O_3)_1$, in F.W. Poulsen, N. Bonanos, S. Linderoth, M. Mogensen, B. Zachau-Christiansen (eds.), *Proc. 17th Riso Int. Sym. Mat. Sci.*, Riso Nat. Lab., Roskilde, Denmark, 351-56.

17. Bata, E., Honegger, K., Diethelm, R., Wettstein, M., (1996) The Selection and Optimisation of Thermally Sprayed Perovskte-Based Coatings for Improved Long-Term Electrochemical Performance, in Proc. 2nd Europ. SOFC Forum, Vol. 1, Ed.: B. Thorstensen 307-14.

18. Fendler, E., Henne, R., (1996) Protecting Layers for the Bipolar Plates of Planar Solid Oxide Fuel Cells Produced by Vacuum Plasma Spraying, in B. Thorstensen (ed.), *Proc. 2nd Europ. SOFC Forum*, Vol. 1, 269-78.

19. Shiomitsu, T., Kadowaki, T. Ogawa, T., Maruyama, T. (1995) The Influence of $(LaSr)CoO_3$ Coatings on the Electrical Resistance of Ni-20Cr Alloys in High Temperature Oxidizing Atmosphere, in M. Dokiya, O. Yamamoto, H. Tagawa, S.C. Singhal (eds.), *Proc. 4th Int. Symp. SOFC*, The Electrochem. Soc., Pennington NJ, USA, 850-57.

20. Brückner, B., Landes, H., Rathjen, B. (1993) Improvement of the Electronic Contact Between the Bipolar Plate and Cathode of a SOFC Model Experiments and Screening of Interfacial Coatings, in S.C.

Singhal, H. Iwakara (eds.), *Proc. 3rd Int. Symp. SOFC*, The Electrochem. Soc., Pennington NJ, USA, 641-48.

21. Bunshah, R.F. (1994) *Handbook of Deposition Technologies for Films and Coartings*, 2nd Edition, Noyes Publicarions, Park Ridge, New Jersey, USA.

GLASS SEALS

A. BIEBERLE, L.J. GAUCKLER
ETH Zürich
Department of Materials
Nonmetallic Materials
Sonneggstrasse 5
8092 Zürich, Switzerland

1 Introduction

Glass seals have the purpose to join different components to a mechanically stable and gas-tight part. In planar SOFCs (Solid Oxide Fuel Cells), a glass seals the cell and the interconnector. Thus, the glass seal is in direct contact with the electrolyte (YSZ) on the one side and the interconnector ($LaCrO_3$ or chromia forming alloy ($CrFe5Y_2O_31$)) on the other side (Figure 1). Due to this situation, compatibility issues are of main interest for the design of a suitable glass seal.

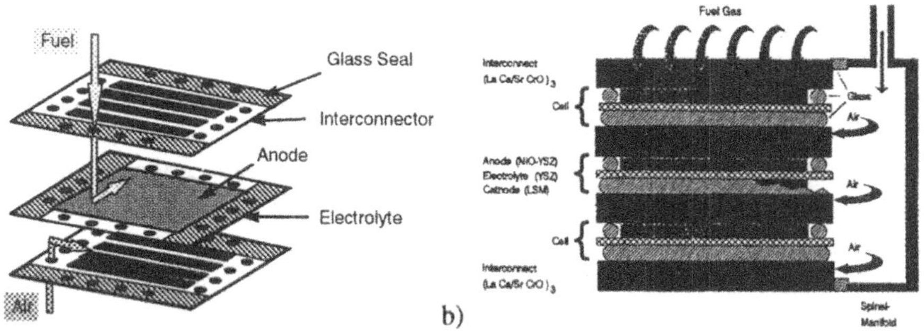

Figure 1: a) Schematic representation of a planar SOFC stack[1], b) cross-section of a flat plate design[2].

The glass seal is responsible for the gas-tight separation of the air and the fuel gas chambers and air-manifold from the fuel electrode and H_2-manifold from the air-electrode porosities [3]. Fuel gas losses have to be prevented because of detrimental effects on the energy efficiency of the SOFC. The sealant also has to be strong and stiff in order to achieve mechanically stable stacks that can be handled and which withstand pressure differences during operation. On the other hand, the sealant must be soft enough to reduce mechanical stresses occurring during production and in operation.
In addition, glass seals should be chemically compatible with the materials to be sealed. The glass seal should not react with other SOFC elements, such as the electrolyte (YSZ), the electrodes (Ni-YSZ cermet and LSM), or the interconnector

H.L. Tuller et al. (eds.), Oxygen Ion and Mixed Conductors and Their Technological Applications, 389–397.

($LaCrO_3$ or chromium alloy). The glass seal should be stable in oxidizing as well as reducing gas atmospheres at high temperatures between 950 and 1000°C.
In the following paragraphs, some examples of glass sealing materials are given and their behavior and properties in a SOFC stack are discussed.

2 Glass Sealing Materials

Many different glasses and glass ceramic materials were tested for sealing purposes in SOFCs. Due to the stringent requirements, most sealants were not practicable because of major drawbacks concerning either thermal expansion coefficient mismatch or due to reactions with electrodes, electrolyte, or interconnector materials [4]. Soda-lime glasses e.g. have suitable thermal expansion coefficients ($\sim 9.5 \cdot 10^{-6}$/K); however, they can not be used for sealing purposes in SOFCs, because they react excessively with state-of-the-art cathode materials [4].
Glass seals proposed for SOFC applications can be grouped into three categories: silicate, borate, and phosphate based glasses.

2.1 SILICATE BASED GLASSES

Several studies deal with silicate based glass seals [2,3,4,5,6]: Single phase silicate based glass seals gave unsatisfactory results [3]. A multiphase concept with separate functions for the different constituents (glassy and crystalline phase) was therefore proposed [7]. The glassy phase is necessary in order to obtain appropriate relaxation properties. The crystalline phase is required for long term stability. Therefore, alumina was added to the silicate glasses and the chemical stability was improved [7].
A similar glass seal was investigated in [6]. The mica glass-ceramics Macor and Photoveel [6] consist of micro-crystals of mica ($KMg_3AlSi_3O_{10}F_2$) in a SiO_2-B_2O_3-Al_2O_3-K_2O-MgO-F glass matrix. The product names Macor and Photoveel refer to similar mica glass-ceramics from different companies, Corning Glass Work Co., Ltd. and Sumitomo Photon Ceramics Co., Ltd., respectively.
The systems MgO-Al_2O_3-SiO_2 (e.g. 52.3% MgO, 15% Al_2O_3, 32.7% SiO_2) and LiO_2-Al_2O_3-SiO_2 were proposed in [5], however, no information is available on their chemical compatibility with the other SOFC components.
Alkali silicate glasses are in principle not suitable, as alkali cations react with the other fuel cell components. The alkaline-earth silicates have too low viscosities [4].

2.2 BORATE BASED GLASSES

In borate based glasses, the SiO_2 is replaced by B_2O_3 as primary glass forming constituent [2,4,8]. Other oxides are added to B_2O_3 in order to tailor special properties [4]: strontium oxide is added in order to increase the thermal expansion coefficient, lanthanum oxide to control the viscosity, alumina to retard the crystallization of strontium borate and other crystalline phases, and silica to enlarge the glass-forming composition range.

Alkali borosilicates, such as Pyrex [8], are not suitable for large area sealing since they have very low thermal expansion coefficients of about $3.2\cdot10^{-6}$/K [4]; in addition, the alkali cations tend to react with other fuel cell components.

Recently, AF 45 (SiO_2 - B_2O_3- BaO- Al_2O_3 - As_2O_3) was suggested as a successful glass for sealing SOFC stacks operating at temperatures between 850 and 950°C [9]. Part of the glass-forming agent, most likely B_2O_3, is lost at the high temperature and the glass crystallizes in a number of possible modifications of celsian and SiO_2.

2.3 PHOSPHATE BASED GLASSES

P_2O_5 based multicomponent glasses with and without small quantities of SiO_2 and stabilizers (e.g. alkaline metal earth oxides, Al_2O_3) were investigated [2]. La_2O_3 and Cr_2O_3 additions lower the reactivity with the $LaCrO_3$ interconnector.

The main problem in pure P_2O_5 based glasses is the volatilization of the phosphate component which leads to surface nucleated crystallization [2]. After evaporation of a certain amount of P_2O_5, the surface shifts to compositions that crystallize easily. Meta- or pyrophosphates are usually formed as crystalline phases. These phases show poor stability at high temperature in wet fuel gas atmosphere. The meta- or pyrophosphates will decompose further into stable orthophosphates, as exemplified with Mg-phosphates according to the following reaction:

$$\text{Mg-}P_2O_5\text{ glass} \rightarrow \underset{\text{meta}}{Mg(PO_3)_2} + P_2O_5\uparrow \rightarrow \underset{\text{pyro}}{Mg_2P_2O_7} + P_2O_5\uparrow \rightarrow \underset{\text{ortho}}{Mg_3(PO_4)_2} \qquad (1)$$

Oxides can be added in quantities of 2-10mol% in order to increase the stability. These oxides have to be stable in the entire $p(O_2)$ range. However, the drawback of such addition is that most oxides promote bulk crystallization (e.g. Cr_2O_3 and La_2O_3). Only SiO_2 and ZrO_2 were found to increase the chemical stability and to suppress crystallization tendencies.

The thermal expansion coefficients of phosphate based glasses do not match those of YSZ. In order to increase their thermal expansion coefficients, up to 10 mole% alkaline metal oxides may be added. This, however, is detrimental to the high temperature stability due to the crystallization of the glassy structure [2]. The viscosity of the glass seal increases thereby and stress can no longer be relieved by the glass seal.

3 Glass Seal in Contact with SOFC Elements

3.1 GLASS SEAL IN CONTACT WITH THE ELECTROLYTE

In the planar design of a SOFC, the glass sealant is in direct contact with the electrolyte. Interface reactions between the glass sealant and the electrolyte are therefore of main concern.

Alumosilicate based glass-ceramics in contact with YSZ do not show disadvantageous reaction zones and diffusion of specific elements when heat treated to temperatures up to 1000°C [6]. Even after 17,000 hours of operation, only a small zone which is formed predominately during soldering, can be detected [3]. The small reaction zone that arises from melting of the residual glass matrix around the glass-ceramic surface guarantees good adhesion of the sealant.

At temperatures higher than 1000°C, deformation and melting of the glass seal becomes possible (TABLE 1). Macor glass-ceramic completely joins the electrolyte without cracks and no reaction products at the interface up to a joining temperature of 1050°C (Figure 2a) [6].

Photoveel glass-ceramic joins without cracks up to 1150°C, but formation of a glassy layer and of two types of precipitates can be observed on the Photoveel side of the interface. One precipitate forms on the surface of the electrolyte and the other deposits in the glass layer (Figure 2b) [6].

The different behavior of the Photoveel and the Macor glass-ceramic is supposed to be due to the different joining temperatures. Mica crystals in glass-ceramics are stable up to 1050°C; additional mica crystals are formed at the surface at higher temperature [6].

TABLE 1: Results of joining tests of mica glass-ceramics[6].

Temperature / °C	1000	1050	1100	1150	1200
Macor					
Electrolyte	J	D	D	RM	
Interconnector	J	D	D	RM	
Photoveel					
Electrolyte	J	J	J	D	RM
Interconnector	J	J	J	D	RM

J: joined with no apparent deformation
D: joined with visible deformation
RM: strong reaction or melting

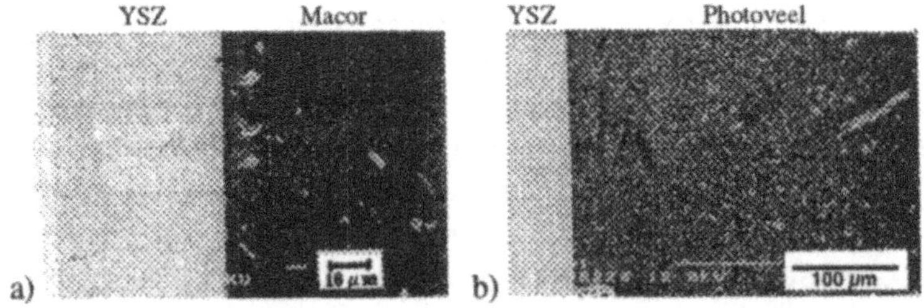

Figure 2: Interface between YSZ and a) mica glass-ceramic Macor at 1050°C, b) mica glass-ceramic Photoveel at 1150°C [6].

In the case of **borate based glass**, such as Pyrex, a new phase is formed at the interface between glass seal and electrolyte after 24h at 1200°C [8]. It consists mainly of $ZrSiO_4$. The distribution of the elements indicates that Y diffuses beyond the reaction layer into the glass phase, whereas relatively little zirconium can be observed in this phase (Figure 3). The main chemical processes are

$$ZrO_2 \text{ (in YSZ)} + SiO_2 \text{ (in glass)} \rightarrow ZrSiO_4 \quad (2)$$

with $\Delta G_1^0 = -44480 + 23.85\ T\ (\pm 1060)$ J/mol [10].

$$Y_2O_3 \text{ (in YSZ)} \rightarrow Y_2O_3 \text{ (in glass)} \quad (3)$$

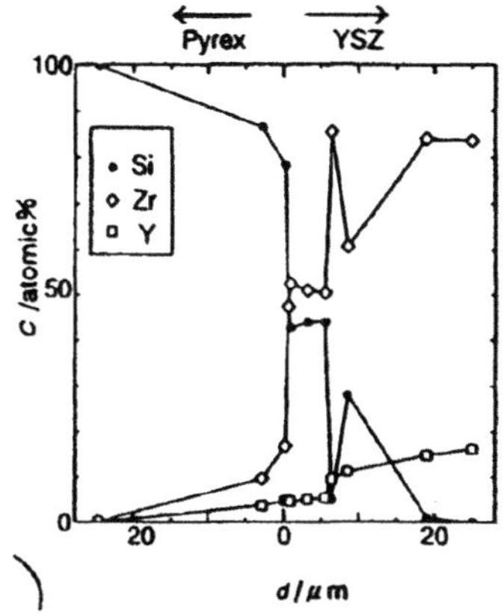

Figure 3: Element distribution between YSZ and Pyrex (1200°C, 48h)[8].

From thermodynamical estimation it can be expected that $ZrSiO_4$ already forms below 900°C. However, due to the low growth rate of the reaction layer at this temperature, it may not be observed easily. Above 1000°C, the growth of the reaction layer obeys a parabolic rate law which suggests that diffusion processes are rate-determining. The growth rates of the reaction layer for couples of $LaCrO_3$/glass and YSZ/glass are shown in Figure 4 for a temperature of 1000°C.

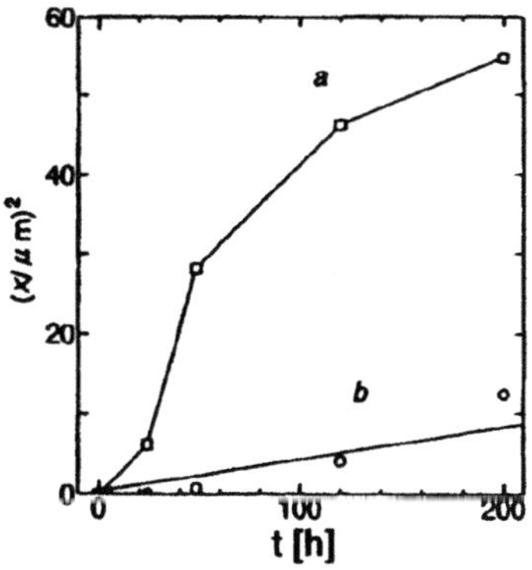

Figure 4: Growth rate of the reaction layers a) $LaCrO_3$ - Pyrex, b) YSZ - Pyrex (1000°C, air)[8].

3.2 GLASS SEAL IN CONTACT WITH THE ELECTRODES

Electrodes may be affected by a glass seal in two ways [2]: first, solid state (interface) reactions may occur, when the glass seal and the electrode are in direct contact. As these areas are quite limited, the overall electrode performance will not be influenced significantly. Second, volatile species of glass may be transported to the electrode and may react with the electrode material. Large areas of the anode may be affected in this case and severe performance losses may result.

Interface reactions between glass seal and electrode material (Ni-YSZ cermet) were investigated in [4]. Mechanical mixtures (1:1) of powdered anodes and silicate or phosphate based glasses, respectively, were heat treated at 1000°C for 100h in wet fuel gas. With **silicate based glasses** no reaction products between glass and anode can be found. The **phosphate based glass** shows extensive reaction with formation of nickel phosphide and zirconium-oxyphosphate.
In order to investigate the influence of vapor species from sealants on the anode performance, the glass seal is placed close to the anode, but without direct contact [2]. Little difference can be observed between a reference anode (Figure 5a) and the anode that was in an environment with **silicate glass** (the slightly enlarged Ni particles compared to the reference anode may be due to reproducibility problems) (Figure 5c). The anode that was in an environment with **phosphate glass** reacted extensively (Figure 5b) and the structure changed entirely compared to the reference anode. The isolated white phase was identified as NiP (EDS analyses) with a much larger particle size than the Ni particles in the reference anode. No elementary nickel remained in the sample after 360h [2].

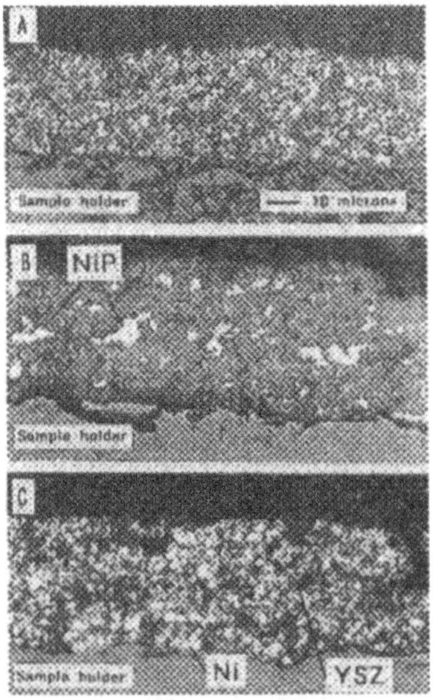

Figure 5: a) Reference anode, b) anode heat treated with phosphate based glass, c) anode heat treated with silicate based glass (magnification: 800x)[2].

3.3 GLASS SEAL IN CONTACT WITH THE INTERCONNECTOR

Alumosilicate based glass-ceramic sealants [3] in contact with the ceramic interconnector $LaCrO_3$ are less stable than the same glass-ceramics in contact with the YSZ electrolyte. However, the reaction zone is mostly formed during soldering and is not attributed to aging phenomena and therefore not critical.
The mica glass-ceramics Macor and Photoveel were studied in contact with the interconnector $La_{0.8}Ca_{0.22}CrO_3$ (TABLE 1) [6]. After heat treatment at 1000°C, both glass-ceramics joined the interconnector without any trace of chemical reaction; only some gaps were observed at the connected interface [6]. At higher temperatures, 1050°C for Macor and 1150°C for Photoveel, a glassy layer is formed at the mica glass-ceramic side of the interface and a trace of reaction layer can be observed at the interconnector side of the interface (Figure 6). In the glassy layer of the Macor glass ceramic, small spherical precipitates can be seen, whereas in the Photoveel glass ceramic large spherical and needle-like precipitates are present. The precipitates had almost the same chemical composition as the mica crystal [6].

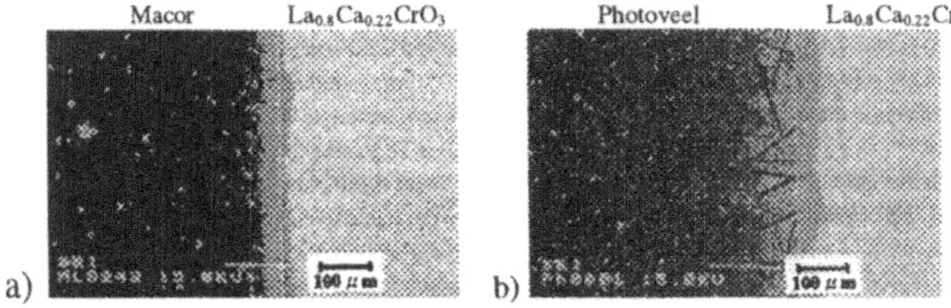

Figure 6: Interface between $La_{0.8}Ca_{0.22}CrO_3$ and a) mica glass-ceramic Macor at 1050°C, b) mica glass-ceramic Photoveel at 1150°C [5].

The diffusion of elements forming the interconnector ($La_{0.8}Ca_{0.2}CrO_3$) were studied in [6]. La and Ca diffuse widely into the glassy layer, whereas Cr does not diffuse into the glassy layer. Ca diffuses deeper into the glass than La. These results can be understood on the basis of phase diagrams: Ca dissolves more easily in alumosilicate glass than La. Cr_2O_3 has no solubility in SiO_2 [6].
On the other hand, the elements Mg, Al, and Si of the glass diffuse into the reaction layer at the interconnector side of the interface forming products such as $Mg(Cr,Al)_2O_4$ and $Ca_2La_8(SiO_4)_6O_2$ [6].

Borate based glasses react extensively with the $LaCrO_3$ interconnector both in air and wet fuel gas [2, 8]. A new phase forms between $LaCrO_3$ and Pyrex after 48h at 1200°C (Figure 7) [8]. Analysis of the interface indicates three zones: 1) the glass phase in which some components of the perovskite phase are dissolved, 2) the reaction zone, 3) the perovskite phase. The reaction mechanism which is mostly due to the migration of the calcium component can be summarized as follows:

Glass Phase	*Reaction Zone*		*Perovskite*
CaO dissolved	Ca-silicate	Ca-silicate $LaCrO_3$ La-silicate (Cr-oxides)	(La, Ca)CrO_3

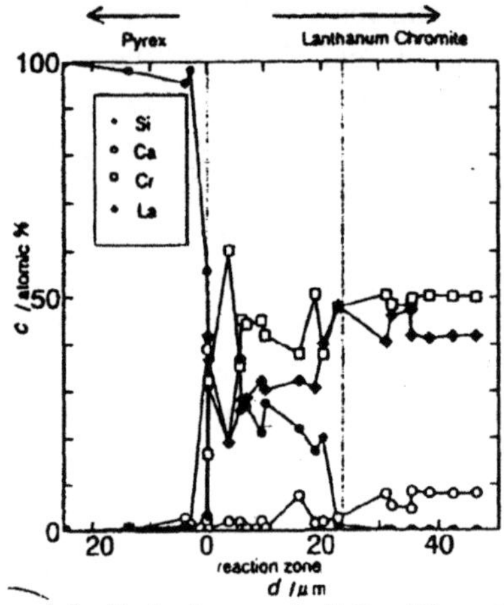

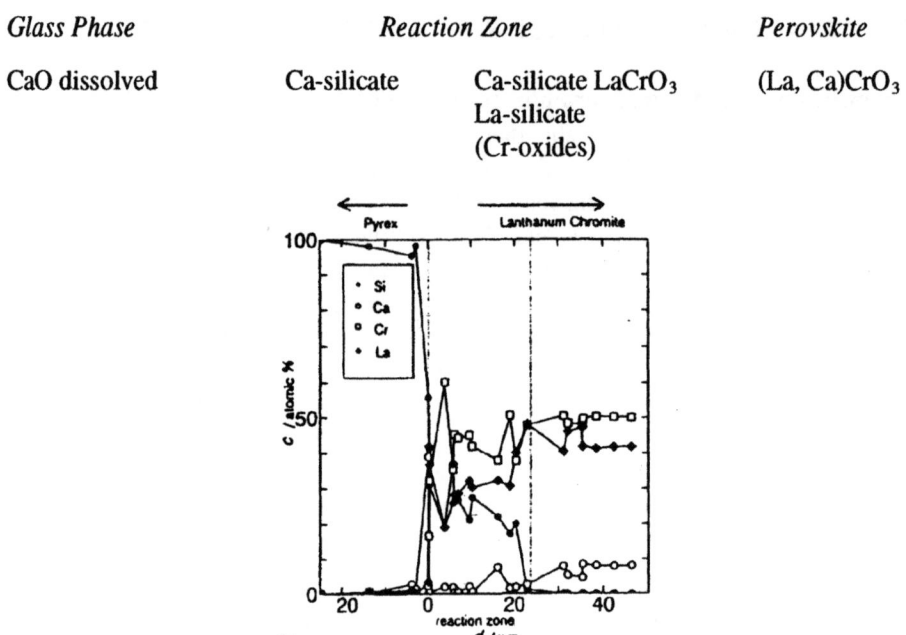

Figure 7: Elemental distribution between $LaCrO_3$ and Pyrex (1200°C, 48h) [8].

4 Summary

Three groups of possible glass sealants for SOFC application are discussed in literature: silicate, borate, and phosphate based glasses and glass-ceramics.
Silicate based glass seals do not react with the electrode and electrolyte materials at SOFC operating temperature. However, at temperatures higher than 1150°C, silicate based glass seals react with the electrolyte material. The interaction with interconnector materials is even more serious when using silicate based glass seals. Additions of alkali cations are detrimental, as alkali cations react with fuel cell components.
Borate based glass seals form a reaction layer when the glass sealant is in contact with the electrolyte as well as with the interconnector material. No information is available concerning interactions with electrodes.
Phosphate based glass seals react with the Ni/YSZ cermet anode. NiP is formed by interface reaction as well as by evaporation leading to undesired excessive coarsening of the Ni grains.
Summarizing, it is very difficult to propose the ideal glass for sealing purposes in SOFCs. The latest publication favors the glass AF45 (SiO_2 - B_2O_3- BaO- Al_2O_3 - As_2O_3) where part of the glass-forming agents are removed at operating temperature.

5 References

1. Yamamoto, T., Itho, H., Mori, M., Mori, N., Abe, T. (1995) Application of Mica Glass-Ceramics as Glass Sealing Materials for SOFC, in M. Dokiya, O. Yamamoto, H. Tagawa, S.C. Singhal (eds.), *Proc. 4th Int. Symp. SOFC*, The Electrochem. Soc., Pennington NJ, 245-54.

2. Larsen, P.H., Bagger, C., Mogensen, M., Larsen, J.G. (1995) Stacking of Planar SOFCs, in M. Dokiya, O. Yamamoto, H. Tagawa, S.C. Singhal (eds.), *Proc. 4th Int. Symp. SOFC*, The Electrochem. Soc., Pennington NJ, 69-75.

3. Stolten, D., Monreal, E., Schäfer, W. Soft (1994) Glass Ceramic Sealing for Gastight SOFC Stacks, in U. Bossel (eds.),*Proc. 1st Europ. SOFC Forum*, Vol. 2, 517-24.

4. Ley, K.L., Krumpelt, M., Kumar, R., Meiser, J.H., Bloom, I. (1996) Glass-Ceramic Sealants forSolid Oxide Fuel Cells: Part I. Physical Properties*J. Mater. Res.* 11, 1489-93.

5. Diekmann, U. (1995) Hochtemperaturbrennstoffzellen - eine Herausforderung an die Fügetechnik, *4th Int. Conf. on Brazing, high Temperature Brazing, and Diffusion Weding* , 126-29.

6. Yamamoto, T., Itoh, H., Mori, M., Mori, N., Watanabe, T. (1996) Compatibility of Mica Glass-Ceramics as Gas-Sealing Materials for SOFC,*Denki Kagaku* 64, 575-80.

7. Stolten, D., Soltész, U., Moreal, E. (1995) Properties of a Soft and Long-Term Stable Glass Ceramic Sealing for Solid Oxide Fuel Cells,*Glastech. Ber. Glass Sci. Technol.* 68 C1, 439-46.

8. Horita, T., Sakai, N., Kawada, T., Yokokawa, H., Dokiya, M. (1993) Reaction of SOFC Components with Sealing Materials,*Denki Kagaku* 61, 760-62.

9. Günther, C., Hofer, G., Kleinlein, W. (1997) The Stability of the Sealing Glass AF45 in H_2/H_2O - and O_2/N_2 - Atmospheres, in U. Stimming, S.C. Singhal, H. Tagawa, W. Lehnert (eds.), *Proc. 5th Int. Symp. SOFC*, The Electrochem. Soc., Pennington NJ, 746-56.

10. Pandit, S.S., Jacob, K.T. (1994) Phase Relations in the System $CaO-SiO_2-ZrO_2$, *Steel Research* 65, 410-13.

ELECTROCHEMICAL SENSORS

JOACHIM MAIER

Max-Planck-Institut für Festkörperforschung

Heisenbergstraße 1, 70569 Stuttgart, Germany

1. General Remarks

A sensor is a device which is constructed for the purpose of information gain. This is in contrast to batteries and actuators (see also the figure at the end of the paper). Certainly, in order to classify sensors it is important to know what aspects of information are needed and by which technique this is to be achieved. Information may be desired on parameters such as temperature, pressure, chemical composition while the technique itself may be optical, thermal, chemical, electrochemical and so on. For example, a hydrocarbon may be detected by catalytic oxidation and the resulting temperature change (thermal composition sensor).

The notation is not always unambiguous; the term thermal sensor, e. g., is not only used to characterise the measurement technique, but often also to characterise the purpose (temperature sensor). On one hand, chemical sensors usually denote devices used to analyse composition (composition sensors) while, on the other hand, electrochemical sensors refer to the fact that electrochemical principles are used. In view of the theme of the workshop, we will deal with electrochemical techniques for chemical analysis

H.L. Tuller et al. (eds.), Oxygen Ion and Mixed Conductors and Their Technological Applications, 399–421.

only (electrochemical composition sensors). For a comprehensive overview on techniques, the reader is referred to Ref. [1]. Also for conciseness, we will not elaborate on sensors making use of kinetic effects such as ampèrometric devices. Those usually rely on measuring the limiting current, a principle widely used in polarography.

With the focus of our interest on equilibrium sensors (including local and also partial electrochemical equilibrium), we consider essentially potentiometric (more precisely e.m.f.) and conductometric devices. We will also concentrate on gas sensors and solid state devices.

It is not at all exceptional to obtain a change in a given property of a material if the composition of the gas mixture to which it is exposed is varied. Obtaining a signal is only a necessary condition, a reasonable sensor signal must be unambiguous, sensitive and selective. It should also be preferably drift-free, long-time stable and easy to record. Other aspects such as size, cost and so on are of course also decisive for commercialisation.

Let us consider a signal S which is a function of the concentration of the species k to be detected. Generally the sensor will also respond to:

$$S = S(c_1, c_2, ...c_k, ...c_n). \tag{1}$$

The total change is given by

$$dS = \Sigma_{k' \neq k} \frac{\partial S}{\partial c_{k'}} dc_{k'} + \frac{\partial S}{\partial c_k} dc_k. \tag{2}$$

The term $\left|\frac{\partial S}{\partial c_k}\right|$ represents the sensitivity and should be as high as possible, whereas the other derivatives should be zero for an ideally selective sensor. Generally, if we have a set of signals $S_1, S_2...S_n$ which respond to the species $1, ...k, ...n$, the diagonal terms only should be non-zero for a selective sensing procedure. If only S_{kk} is non-zero, the procedure is said to be specific with respect to k. (One can also define the terms selectivity and specificity as continuous quantities, as done in Ref. [2]. The sensitivity may be defined in such cases by the determinant of the response matrix.)

An ideal sensor should not exhibit an explicit time dependence. If, however, c_k depends on t, it should immediately follow $c_k(t)$. If writing $S = S(c(t), t)$, we obtain

$$\frac{dS}{dt} = \left.\frac{\partial S}{\partial c}\right)_t \frac{dc}{dt} + \left.\frac{\partial S}{\partial t}\right)_c \tag{3}$$

which should reduce to $\frac{\partial S}{\partial c}\frac{dc}{dt}$ in the ideal case. If $S = A \cdot c^P$, as often fulfilled, $\left.\frac{\partial S}{\partial t}\right)_c = 0$ means of course that A does not depend on time. Response time and drift can be unambiguously distinguished if they occur at different time scales. Usually there is a fast response of the order of τ_R and then after having obtained a quasi steady state ("∞"), the signal degrades slowly. The response time can then be defined by the time needed to reach the steady state (e.g. $\left|\frac{S(\tau_R) - S(\text{“}\infty\text{”})}{S(\text{“}\infty\text{”})}\right| \simeq 1\%$) and the drift as the time change for $t \gg \tau_R$ $\left(\dot{S}\,(t \gg \tau_R)\right)$.

Let us first discuss the conventional sensor principles based on the use of electronic conductors, ionic conductors and mixed conductors which are widely used in particular for the detection of redox-active gases and then formulate the analogues for acid-base active gases. (The structure follows Ref. [3].) Only in the last case global equilibrium is established, whereas in the other ones we rely on contact (electrochemical) equilibria.

2. Redox-active Gases

Fig. 1a shows a mixed conducting oxide in which oxygen can be dissolved by ambipolar diffusion and both ionic and (usually more sensitively) the electronic carrier densities are homogeneously changed (mode 1). In Fig. 1b the oxide is an electronic conductor on the surface of which oxygen is adsorbed, trapping electrons to form negatively charged species at the

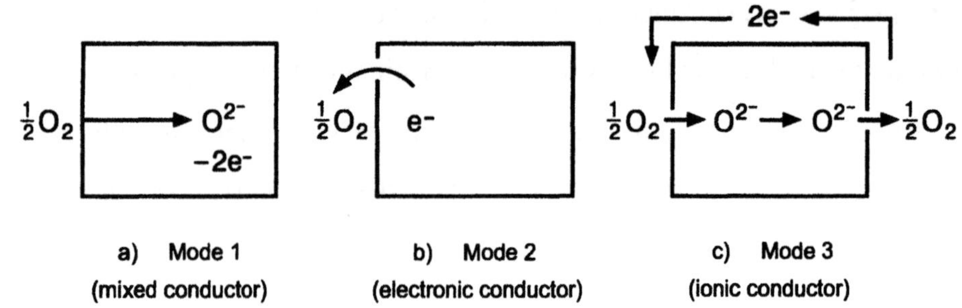

Figure 1. The use of mixed (a), pure electronic (b) and pure ionic conductors (c) in the three sensing modes for redox-active gases discussed in the text.

expense of space charges (depletion of electrons) at the conductor-gas interface. The surface electronic conductivity is thereby changed (mode 2). Figure 1c displays the other extreme case, viz. contact to an ionic conductor. In this case phase potential differences appear giving rise to the possibility of measuring an e.m.f. if the oxygen activities on both sides are different.

For the discussion of the detection of redox active gases, let us consider an elemental gas E_2 ($E_2 = Cl_2, O_2$, etc.) which forms anions according to

$$\frac{1}{2}E_2 + \varepsilon e^- \rightleftharpoons E^{\varepsilon-}. \tag{4}$$

In equilibrium

$$\frac{1}{2}\mu_{E_2} + \varepsilon\tilde{\mu}_{e^-} = \tilde{\mu}_{E^{\varepsilon-}} \tag{5}$$

holds where μ and $\tilde{\mu}$ are the chemical and electrochemical potential, respectively. If we refer to the same site ($\underline{r}$, same electrical potential, ϕ), we can also state that

$$\frac{1}{2}\mu_{E_2}(\underline{r}) + \varepsilon\mu_{e^-}(\underline{r}) = \mu_{E^{\varepsilon-}}(\underline{r})\,. \tag{6}$$

In the case of mode 1 (bulk conductivity sensor) overall equilibrium is reached after a certain time.

2.1. BULK CONDUCTIVITY SENSOR (MODE 1)

Let us formulate the introduction of E as

$$\text{(E)} \qquad \frac{1}{2}E_2(g) + V_E^{\varepsilon\cdot} + \varepsilon e' \rightleftharpoons E_E. \tag{7}$$

The mass action law leads to $[V_E^{\varepsilon\cdot}][e']^{\varepsilon} = K_E^{-1}P_{E_2}^{-1/2}$. The partial pressure dependence and the T-dependence may be derived from defect chemistry. Usually the electronic conductivity is measured. If we neglect ionic and electronic counter defects, the signal change $\delta \ln \sigma$ is proportional to $\delta \ln P_{E_2}$. In the intrinsic case the proportionality constant determining the sensitivity is $-\frac{1}{2(\varepsilon+1)}$, in the acceptor-doped case $-\frac{1}{2\varepsilon}$ (the same is valid if ionic disorder is overwhelming, e. g. $[V_E^{\varepsilon\cdot}] \simeq \left[E_i^{\varepsilon'}\right]$) and in the donor doped case 0. Thus sensitivity is highest in the acceptor doped case. In this case, however, the temperature sensitivity is also highest ($\Delta_E H^0/\varepsilon$ compared to $\Delta_E H^0/(\varepsilon+1)$ and 0 where $\Delta_E H^0$ is the enthalpy associated with reaction 7). (This T-dependence is a serious difficulty which may be overcome by comparison with a reference sample.) In addition, the response times are different. In the diffusion-controlled regime, they are (besides sample thickness) obviously determined by the chemical diffusion coefficients which should be as high as possible, for fastest response.

Let us consider $SrTiO_3$. It is usually acceptor doped. Ionic defects are oxygen vacancies, electronic defects are holes at higher partial pressures and conduction electrons at lower ones. In donor doped systems conduction electrons are compensating. The chemical diffusion coefficient is given by (4)

$$\tilde{D} = \frac{RT}{F^2}\tilde{\sigma}/\tilde{c} \tag{8}$$

and $\tilde{\sigma}$ and $\tilde{c}$ are the (weighted) harmonic means of the individual values ($\tilde{\sigma}^{-1} \simeq \sigma_{eon}^{-1} + \sigma_{ion}^{-1}; \tilde{c}^{-1} = \frac{1}{4}c_{ion}^{-1} + c_{eon}^{-1}; \sigma_{eon} = \sigma_{e'} + \sigma_{h^\cdot}, \sigma_{ion} = \sigma_{V_O^{\cdot\cdot}} + \sigma_{O_i''}; c_{ion} = c_{O_i''} + c_{V_O^{\cdot\cdot}}, c_{eon} = c_{e'} + c_{h^\cdot}$). For simplicity we exclude very low

P_{O_2}. For a sufficiently donor doped material, $\tilde{D}$ reduces to $D_{V_O^{\cdot\cdot}}$ ($\sigma_{eon} = \sigma_{e'} \gg \sigma_{ion}, c_{eon} = c_{e'} \gg c_{V_O^{\cdot\cdot}}$), to $4\frac{c_{V_O^{\cdot\cdot}}}{c_{h^\cdot}} D_{V_O^{\cdot\cdot}} = \frac{\sigma_{V_O^{\cdot\cdot}}}{\sigma_{h^\cdot}} D_{h^\cdot}$ for a moderately acceptor doped and finally to $D_{h^\cdot}$ for a heavily acceptor doped material (here, however, activity coefficients are important). One recognises the definite dependence of the response time $\tau \propto \tilde{D}^{-1}$ on the defect chemistry. Moreover, redox active impurities, even if negligible in the electroneutrality equation, have been shown to be extremely important for $\tilde{D}$ and, thus, the response time [4, 5]. Fe-doped $SrTiO_3$ possesses a much lower diffusivity than $SrTiO_3$ acceptor doped with a three-valent dopant (A^{3+}) of fixed valence and same active concentration ($[A'_{Ti}] = [Fe'_{Ti}]$). This effect stems from the reversible trapping or de-trapping of electronic carriers. In this way $\tilde{c}$ is tremendously increased (it is given by the value of the large "conservative ensemble", trapped plus free holes [4]), whereas $\tilde{\sigma}$ is the same in both cases. In contrast to the other sensor modes, this is obviously a detrimental effect for bulk conductivity sensors. In the same way the surface rate constant ($\tilde{k}_{eff}$) must be sufficiently high. If this process is rate limiting, the response time $\tau \propto \tilde{k}_{eff}^{-1}$. Here catalytic coatings (e.g. $YBa_2Cu_3O_{7-\delta}$ on $SrTiO_3$) or simply surface roughening can help [6]. The bulk conductivity sensor is obviously quite selective on principle. Even if the $\tilde{k}_{eff}$-values for competing reactions may be noticeable, this usually will not hold for the diffusion constants. An ionic species which can also diffuse in oxides is the proton. In this way H_2 and H_2O can interfere. Of course, the presence of redox active gases such as H_2, Cl_2, CO etc. will also change the oxygen partial pressure and so indirectly affect the sensor signal. This is a "regular phenomenon" and does not involve selectivity as such. Selectivity, however, is affected if in a metastable state the oxide acts catalytically and changes the composition in a undesired way.

2.2. SURFACE CONDUCTIVITY SENSORS (MODE 2)

The second mode to be described in detail, is the interaction with a purely electronically conducting oxide. The material of choice is SnO_2. It possesses a simple defect chemistry [7]. Oxygen vacancies are compensated by conduction electrons. At high temperatures both are mobile and the electronic bulk conductivity changes with oxygen partial pressure as -1/6 in pure SnO_2 (in doped SnO_2 this is only true for low P_{O_2}), at high P_{O_2} the characteristic exponent is -1/4 (acceptor doped) or 0 (donor doped). This simple defect chemistry is also reflected in the temperature dependence.

The surface conductivity sensor works at a comparably low temperature of the order of 100 °C . There the oxygen vacancy mobility is negligible. The high temperature defect density and the cooling conditions are important for the low temperature carrier conditions, especially if redox-active impurities are involved [8]. Oxygen is adsorbed at the surface and, according to its electronegativity, it traps conduction electrons out of the space charge region (presumably formation of O_2^-) and increases the surface resistance (Taguchi sensor). Thus, a depletion layer forms. If the majority defect being responsible for the space charge is immobile we end up with a Schottky barrier for which the resistance is roughly*

$$Z^{\perp} \propto c_{e'}(x=0)^{-1}\lambda^{*}. \tag{9}$$

The characteristic exponent at $x = 0$ (first layer of the space charge region) is not necessarily the same as in the bulk as discussed in the chapter on "Interfaces" [9]. λ^* is the width of the Schottky layer which depends on the interaction strength.

*Equation (9) refers to a step-like profile. More precisely, a correction factor has to be introduced which, however, does not significantly affect the dependencies to be discussed.

If measured along the interface, the overall parallel conductance is diminished by a volume fraction of λ^*/L ($L =$ width of the measured volume). Equation 9 refers to a measurement perpendicular to the layer.

On the other hand, if we contact SnO_2 with a reducing gas such as CO or CH_x, we increase [e′]. The enhanced conductivity if measured parallel is given by ($0, \infty$ denote surface and bulk)

$$Y^{\|} \simeq \Delta Y^{\|} \propto \lambda \left(c_{e'}(x=0)\, c_{e'}(x=\infty)\right)^{1/2}. \qquad (10)$$

If the electrons are majority carriers in the bulk, the bulk term cancels and $Y^{\|} \propto c_{e'}(x=0)^{1/2}$. Again a regime of thickness 2λ would fall out of the balance, if the conductance were measured in a perpendicular way. Again the dependence of $c_{e'}(x=0)$ on P_{O_2} may become complicated and only in special cases power laws may result, with exponents not necessarily equal to the bulk value. Even so, the exponent of the conductance signal would, if e′ are majority carriers in the bulk, be half of it [10]. These dependencies are crucial for the sensitivity of the signal. A sensitive signal also requires favourable measurement geometries, i. e. direction, area and thickness etc. have to be optimised. For this reason thin or thick films are used. The response time is determined by the rate constants of the rate limiting surface step. Unlike mode 1 the chemical diffusion coefficient must be as small as possible to keep drift phenomena as low as possible. Since many gases exhibit a redox active behaviour, the selectivity is rather low. Also complex gases such H_2O, CH_4, CH_x etc. NO_x interfere, the detailed mechanism being different.

Unlike the bulk conductivity sensor where the quite selective bulk process is in series, the decisive chemistry occurs at the boundary in the case of the surface conductivity sensor. As a consequence the decisive mechanisms may vary.

H_2O may induce a change in the surface chemistry of a material and thus show a cross effect relative to O_2 or H_2 performance. However, direct electronic effects have also been claimed to occur [11]. NO_x acts oxidising while CH_x usually gives a reducing signal. Depending on the conditions, oxidation products are alcohols, aldehydes and carbon acids. Attempts to increase the effective selectivity concern variation of the operating temperature (since the sorption rates are differently T-dependent), or the use of pattern recognition, by applying sensor arrays which have to be calibrated.

Variants of measuring the gas-oxide surface interaction are the Kelvin probe sensor and the CHEMFET. In the first case a vibrating capacitor measures the change of the work function, while in the second case the interaction is detected by the field effect transistor effect. In a usual field effect transistor an appropriate voltage between the gate over an insulating oxide such as SiO_2, to the underlying Silicon (p-doped) causes an inversion layer at the contact oxide/Si. This can be sensitively detected by measurement of a current between source and drain. In a CHEMFET the field is — so to speak — provided by a rudiment of a galvanic cell. The metal gate is replaced by an electrolyte which is in contact to a reference electrode. The process on the working side of the cell is the interaction with the insulating oxide. This arrangement provides the necessary field to cause the field effect on the source-drain current.

If we again refer to Eq. (4), in the surface sensor mode (unlike mode 1 in which $\nabla\tilde{\mu}_{e^-} = \nabla\tilde{\mu}_E = \nabla\tilde{\mu}_{E^{n-}} = 0$ and thus full equilibrium was achieved) the equilibrium is only established at the surface according to $\frac{1}{2}\mu_{E_2} + \tilde{\mu}_{e^-} = \tilde{\mu}_{E^{n-}}$, with $\nabla\tilde{\mu}_{e^-} = 0 \neq \nabla\tilde{\mu}_{E^{n-}}$ (and thus $\nabla\mu_E \neq 0$).

2.3. GALVANIC SENSORS (MODE 3)

In mode III, the galvanic sensor mode, again there is no global equilibrium but now $\nabla\tilde{\mu}_{E^{n-}} = 0 \neq \nabla\tilde{\mu}_{e^-}$ (thus $\nabla\mu_E \neq 0$). Owing to $\sigma_{eon} = 0$ in the electrolyte, no electronic current can be established and since $i = i_{ion} + i_{eon} = 0$, there is also no ionic current. Because $\sigma_{ion} \neq 0$, the gradient $\nabla\tilde{\mu}_{E^{n-}}$ disappears. In this way differences of the electrochemical potential, i.e. of the cell voltage, can be established over the sample. Consider first the well-known λ-sensor

$$\mathrm{Pt, O_2\,(P_L)\,|ZrO_2\,(Y_2O_3)|\,O_2\,(P_R)\,, Pt.}$$

On both sides the local heterogeneous equilibrium

$$\frac{1}{2}\mu_{O_2} + 2\tilde{\mu}_{e^-}(\mathrm{Pt}) = \tilde{\mu}_{O^{2-}}\,(\mathrm{ZrO_2}) \tag{11}$$

is fulfilled. Subtracting the two equilibria on both sides, we obtain Nernst's equation ($P = P_{O_2}$):

$$E = \frac{RT}{4F}\ln\frac{P_R}{P_L}. \tag{12}$$

($\Delta\tilde{\mu}_{O^{2-}} = 0, \Delta\tilde{\mu}_e = -FE,\ E =$ cell voltage.) With the help of this λ-cell P_{O_2} can be measured on one side, if the P_{O_2} on the other side is known. The signal can also be used to control the partial pressure.

Let us proceed to a formation cell, e.g.

$$\mathrm{(Pt)Ag\,|AgCl|\,Cl_2, Pt.}$$

Since silver chloride is an ionic conductor in itself, this is a minimum phase scheme. On the left hand side we may write

$$\mu^0_{Ag} = \tilde{\mu}^L_{Ag^+}(\mathrm{AgCl}) + \tilde{\mu}^L_{e^-}(\mathrm{Ag}) = \tilde{\mu}^L_{Ag^+}(\mathrm{AgCl}) + \tilde{\mu}^L_{e^-}(\mathrm{Pt})) \tag{13}$$

and on the right side hand

$$\frac{1}{2}\mu^R_{Cl_2} + \tilde{\mu}^R_{Ag^+}(\mathrm{AgCl}) + \tilde{\mu}^R_{e^-}(\mathrm{Pt}) = \mu^0_{AgCl}. \tag{14}$$

Again the difference in $\tilde{\mu}_{Ag^+}$ cancels, whereas the difference in $\tilde{\mu}_{e^-}$ constitutes the cell voltage, leaving

$$\begin{aligned} EF &= \mu^0_{Ag} + \frac{1}{2}\mu^0_{Cl_2} + RT\ln P^R_{Cl_2} - \mu^0_{AgCl} = -\Delta_f G_{AgCl} \\ &= -\Delta_f G^0_{AgCl} + RT\ln P^R_{Cl_2}. \end{aligned} \tag{15}$$

The equivalence between cell voltage and the reaction free enthalpy follows also from the laws of thermodynamics, since the reversible electrical work to be done by the cell is — under open circuit conditions — balanced by the reversible chemical work, which corresponds to the free enthalpy change during the cell reaction

$$Ag + \frac{1}{2}Cl_2 \rightleftharpoons AgCl \tag{16}$$

which may be split into the partial reactions

$$\begin{aligned} Ag &\rightleftharpoons Ag^+ + e^-(L) \\ Ag^+ + \tfrac{1}{2}Cl_2 + e^-(R) &\rightleftharpoons AgCl. \end{aligned} \tag{17}$$

Since on the chlorine side the contact $AgCl/Cl_2$ establishes a well defined silver activity, and since the contact Ag/AgCl provides a well defined Cl-partial pressure, the formation cell also serves as a chlorine or silver activity cell.

Let us view the situation from a more local standpoint. The difference in $\tilde{\mu}_{e^-}(Pt)$ equals the electric potential difference across the cell (cell voltage) and can thus be obtained by summing up all ϕ-changes. Across the Ag|AgCl boundary, e.g. $\tilde{\mu}_{Ag^+}$ is constant and thus $\Delta\phi$ is given by $-\Delta\mu_{Ag^+}/F$. The chemical potential of Ag^+ in Ag is equal to $\mu^0_{Ag} - \mu_{e^-}(Ag)$, the latter being related to $\mu_{e^-}(Pt)$ via the ϕ-jump at the boundary Pt|Ag. The total electrode potential is then given by

$$\Delta\phi^L \propto \mu^L_{Ag^+}(AgCl) + \mu^L_{e^-}(Pt) - \mu^L_{Ag}. \tag{18}$$

In the same way the other electrode potential is obtained as

$$\begin{aligned}\Delta\phi^{\mathrm{R}} &\propto -\mu^{\mathrm{R}}_{\mathrm{Cl}^-}(\mathrm{AgCl}) + \mu^{\mathrm{R}}_{\mathrm{e}^-}(\mathrm{Pt}) + \frac{1}{2}\mu^{\mathrm{R}}_{\mathrm{Cl}_2} \\ &\propto -\mu^{0}_{\mathrm{AgCl}} + \mu^{\mathrm{R}}_{\mathrm{Ag}^+}(\mathrm{AgCl}) + \mu^{\mathrm{R}}_{\mathrm{e}^-}(\mathrm{Pt}) + \frac{1}{2}\mu^{\mathrm{R}}_{\mathrm{Cl}_2}.\end{aligned} \tag{19}$$

In the difference, the $\mu_{\mathrm{e}^-}(\mathrm{Pt})$-values cancel, since the electronic concentration is essentially constant in Pt. The same applies to μ_{Ag^+} if the ionic conductor has a sufficiently high defect concentration which is usually fulfilled. Otherwise, the μ_{Ag^+} change across the electrolyte adds another $\Delta\phi$-value ($\Delta\tilde{\mu}_{\mathrm{Ag}^+} = 0$). Thus, $\mu^{\mathrm{R}}_{\mathrm{Ag}^+} - \mu^{\mathrm{L}}_{\mathrm{Ag}^+}$ cancels definitely if this $\Delta\phi$ contribution is taken into account. What happens if we refer to different silver halides such as in the combination

$$\mathrm{Pt}\,|\mathrm{Ag}|\,\mathrm{AgX}'\,|...|\,\mathrm{AgX}\,|\mathrm{X}_2, \mathrm{Pt}\,.$$

The difference $\mu^{\mathrm{L}}_{\mathrm{Ag}^+} - \mu^{\mathrm{R}}_{\mathrm{Ag}^+}$ is more delicate. It obviously only vanishes if we let both halides equilibrate. For X′ = Br and X = Cl a solid solution must be formed (Ag(Cl, Br)) and the phase boundary appears, for X′ = Br and X = I the equilibrated contact is $\mathrm{AgBr_{ss}}|\mathrm{AgI_{ss}}$. In both cases the steady state cell voltage refers to solid solutions rather than to the pure phases. Generally, we would not obtain the difference of the half cell potentials involving the pure halides.

We can also switch-in a solid electrolyte which only lets Ag^+ pass but blocks the anions such as for Ag^+-alumina. In this case $\mu^{\mathrm{L}}_{\mathrm{Ag}^+} - \mu^{\mathrm{R}}_{\mathrm{Ag}^+} = 0$ ($\mu^{\mathrm{L}}_{\mathrm{Ag}^+}(\mathrm{AgBr}) = \mu^{\mathrm{L}}_{\mathrm{Ag}^+}(\beta - \mathrm{alumina}) = \mu^{\mathrm{R}}_{\mathrm{Ag}^+}(\beta - \mathrm{alumina}) = \mu^{\mathrm{R}}_{\mathrm{Ag}^+}(\mathrm{AgCl})$) even though we refer to the pure phases. This allows the difference of the half cell potentials to be measured.

If we just allow the two pure phases to be contacted without such a membrane or equilibration, we have to add a (time dependent) diffusion potential.

The consideration so far was done so thoroughly to demonstrate that e.m.f. effects are ubiquitous but not always defined. There is no doubt that a manifold of processes can occur at an electrode and contribute to the electrode potential, especially if solids are involved. In this way the Ag|AgCl half cell may give a signal upon changes in the copper activity as the other half cell AgCl|Cl_2, Pt may respond to bromine. A sodium conductor may sense potassium activities or even H_2-partial pressures, as can be seen by formulating local exchange equilibria. Even if the diffusion potentials may be ill-defined they can be virtually constant (as in the case of the Daniell element Zn|$ZnSO_4$||$CuSO_4$|Cu). If there are different, otherwise independent processes (1 and 2) contributing to an electrode potential, a mixed potential occurs (see e.g. [12]). Let ε_1 be the potential for which $i_1 = 0$ (i.e. $|\vec{i}_1| = |\overleftarrow{i}_1|$) and equally $i_2(\varepsilon_2) = 0$, then the potential at which the total current ($|i_2| = |i_1|$) disappears, lies in between ε_1 and ε_2 ($i(\varepsilon_M) = i_1(\varepsilon_M) + i_2(\varepsilon_M) = 0$). (In this case $|\vec{i}_1| \neq |\overleftarrow{i}_1|$ and $|\vec{i}_2| \neq |\overleftarrow{i}_2|$, and a non-equilibrium situation is locally met.) Obviously and fortunately the mixed potential is closer to the process with the higher exchange current density (steeper $i(\varepsilon)$ curve), and thus processes occuring at low partial rates are not so important, they usually become even less important if the necessary measuring current is increased because of their higher effective resistance. Nevertheless this affects very much the selectivity of potentiometric sensors. (This point is less serious in the amperometric mode mentioned in the beginning. Such a device based on zirconia electrolytes has been used for oxygen sensing. It is essentially a gas pump which reaches a saturation current if the voltage is increased to a high enough value. For a given concentration in the gas phase, the current causes a depletion immediately at the boundary where the reaction takes place. Usually the width over which the gradient is established, is — for hydrodynamic reasons — roughly constant. If the concentration at the interface has essentially decreased to zero at the boundary, the diffusion controlled current cannot be increased any longer ($i_{diff} \propto \frac{c_\infty - c_0}{\delta}$, $i_{diff}(U \longrightarrow \infty) \propto c_\infty$). Hence i_{diff} is proportional to the concentration in the gas atmosphere.)

Potentiometric sensors are easily designed and often very sensitive. Also the T-dependence is moderate if the activities involved (at the reference side) do not change with temperature. In the λ-sensor a metal/metal oxide mixture (instead of using air) is an easy possibility to establish a defined reference partial pressure, but one which is strongly T-dependent due to the chemical equilibrium constant. Drift effects in galvanic cells occur due to irreversible processes either at the internal boundaries or due to chemical reactions with the gas phase.

A special irreversibility is met if the electrolyte has a measurable degree of electronic conduction. Then an internal mass flux occurs from the side of the higher to the lower P_{O_2}. If the partial pressures are maintained by gas flow, the e.m.f. nevertheless has a well-defined value. Let us consider an O^{2-}-conductor with some degree of mixed conduction, the result is, instead of Eq. (12) [13, 2]

$$E = \frac{RT}{4F} \int t_{ion} d\mu_{O_2} = \frac{RT}{4F} \langle t_{ion} \rangle \ln \frac{P_R}{P_L}. \tag{20}$$

as seen from the flux equations. The ionic current is given by $i_{ion} = \sigma_{O^{2-}} \nabla\tilde{\mu}_{O^{2-}}/2F$. Since $i_{ion} = -i_{eon} \neq 0$ the gradient $\nabla\tilde{\mu}_{O^{2-}}$ does not disappear. By splitting it into $\frac{1}{2}\nabla\mu_{O_2} - 2\nabla\tilde{\mu}_{e^-}$, and replacing i_{ion} by $-\sigma_{e^-}\nabla\tilde{\mu}_{e^-}/F$, we obtain for $\nabla\tilde{\mu}_e$

$$\nabla\tilde{\mu}_e \propto \frac{\sigma_{O^{2-}}}{\sigma_{O^{2-}} + \sigma_{e^-}} \nabla\mu_{O_2} \tag{21}$$

which yields Eq. (20) after integration.

It is more complicated if the ion can change its valence state. As shown in Ref. [14] the ionic transference number of the "conservative ensemble"

$t_{\{ion\}}$ appears in Eq. (20) instead of t_{ion}. An example of relevance may be a copper conductor with a Cu^{2+} and a Cu^{+} mobility, between which electronic equilibrium might be established. The ratio between E and the Nernst-value is found to be

$$t_{\{ion\}} = \frac{\sigma_{Cu^{2+}} + 2\sigma_{Cu^{+}}}{\sigma_{Cu^{2+}} + \sigma_{Cu^{+}} + \sigma_{e^{-}}}. \tag{22}$$

Note the factor of two in the numerator. This leads to a value different from one, even if $\sigma_{e^-} = 0$, which is simply explainable by the fact that a counter diffusion between Cu^{2+} and $2Cu^{+}$ can irreversibly occur (so-to-speak ion transport plus electron vehicle transport). The fact that $t_{\{ion\}}$ is greater than one may appear puzzling but it is simply caused by the use of a charge number of 2 in Eq. (20).

Another irreversible partial short-circuit appears in proton conducting oxides and thus for H_2O-permeation. We obtain:

$$\begin{aligned} E &= \frac{RT}{4F}\left[t_{O^{2-}} + 2t_{OH^-} - 2t_{H_3O^+}\right]\Delta \ln P_{O_2} + \\ &\quad -\frac{RT}{2F}\left[t_{H^+} - t_{OH} + t_{H_3O^+}\right]\Delta \ln P_{H_2} \\ &= \frac{RT}{4F}\left[t_{O^{2-}} + t_{H^+} + t_{OH^-} + t_{H_3O^+}\right]\Delta \ln P_{O_2} + \\ &\quad -\frac{RT}{2F}\left[t_{H^+} - t_{OH^-} + t_{H_3O^+}\right]\Delta \ln P_{H_2O}. \end{aligned} \tag{23}$$

For the derivation we have to consider the coupling relations $\nabla\tilde{\mu}_{OH^-} = \nabla\tilde{\mu}_{H_2O} - \nabla\tilde{\mu}_{H^+}$; $\nabla\tilde{\mu}_{H^+} = \frac{1}{2}\nabla\mu_{H_2} - \nabla\tilde{\mu}_{e'}$; $\nabla\tilde{\mu}_{H_3O^+} = \nabla\mu_{H_2O} + \nabla\tilde{\mu}_{H^+}$, $\nabla\tilde{\mu}_{O^{2-}} = \frac{1}{2}\nabla\mu_{O_2} + 2\nabla\tilde{\mu}_{e^-}$ and the flux equations $i_k = -\frac{\sigma_k}{z_k F}\nabla\tilde{\mu}_k$. Without restriction of generality we consider a slightly simplified case for which $\sigma_{H_3O^+} = \sigma_{O^{2-}} = 0$. The proof is as follows: Since $\Sigma_k i_k = 0$, we have $\frac{\sigma_{H^+}}{F}\nabla\tilde{\mu}_{H^+} - \frac{\sigma_{e^-}}{F}\nabla\tilde{\mu}_{e^-} - \frac{\sigma_{OH^-}}{F}\nabla\tilde{\mu}_{OH^-} = 0$. Using the coupling conditions between OH^-, H_2O, e^- and between H^+, H_2, e^-, we rewrite this as

$$\frac{\sigma_{OH^-} + \sigma_{H^+}}{2F}\nabla\mu_{H_2} - \frac{\sigma_{OH^-}}{F}\nabla\mu_{H_2O} - \frac{\sigma_{e^-} + \sigma_{H^+} + \sigma_{OH^-}}{F}\nabla\tilde{\mu}_{e^-} = 0 \tag{24}$$

from which after integration Eq. (23) follows.

The above equation (23), derived here in a straightforward way, has already been given by Norby et al. (see e.g. Ref. [15]). It is worth paying attention to the fact that the species OH^-, H_3O^+, H^+, O^{2-} appear differently, since it matters how much oxygen and hydrogen is transferred by the species under consideration.

3. Acid-base Active Gases

So far we concentrated on (redox-active) gases whose interactions with electrons are most relevant. On the other side, mostly complex gases such as NH_3, H_2O, CO_2 electrochemically react with ions, i.e. are acid-base active through

$$\begin{array}{lcl} NH_3 + H^+ & \rightleftharpoons & NH_4^+ \\ CO_2 + O^{2-} & \rightleftharpoons & CO_3^{2-} \\ H_2O + H^+ & \rightleftharpoons & H_3O^+. \end{array}$$

If they are detected via a redox change, this would most likely involve a non-characteristic, less selective sensing process. Let us construct the analogues to the 3 modes discussed above, in terms of acid-base interactions [3]. Let us start with mode 1.

Here we have to use a material in which the complex gas is soluble. A well-working example may be the proton and oxygen conducting perovskites mentioned above [16]. By an ambipolar diffusion of both carriers, water transport is established. Depending on the water content, the proton (and O^{2-}) ion conductivity changes and can be used as a sensor signal.

The chemical diffusion coefficient, $\tilde{D}_{H_2O} = f\left(c_{H_i^{\cdot}}, c_{V_O^{\cdot\cdot}}, \sigma_{H_i^{\cdot}}, \sigma_{V_O^{\cdot\cdot}}\right)$ which can be calculated as usually done for $\tilde{D}_{O_2} = f\left(c_{h^{\cdot}}, c_{V_O^{\cdot\cdot}}, \sigma_{h^{\cdot}}, \sigma_{V_O^{\cdot\cdot}}\right)$ if trapping effects are absent, as

$$\tilde{D}_{H_2O} = \frac{(2-x)D_{H_i^{\cdot}} \cdot D_{V_O^{\cdot\cdot}}}{xD_{H_i^{\cdot}} + 2(1-x)D_{V_O^{\cdot\cdot}}}, \tag{25}$$

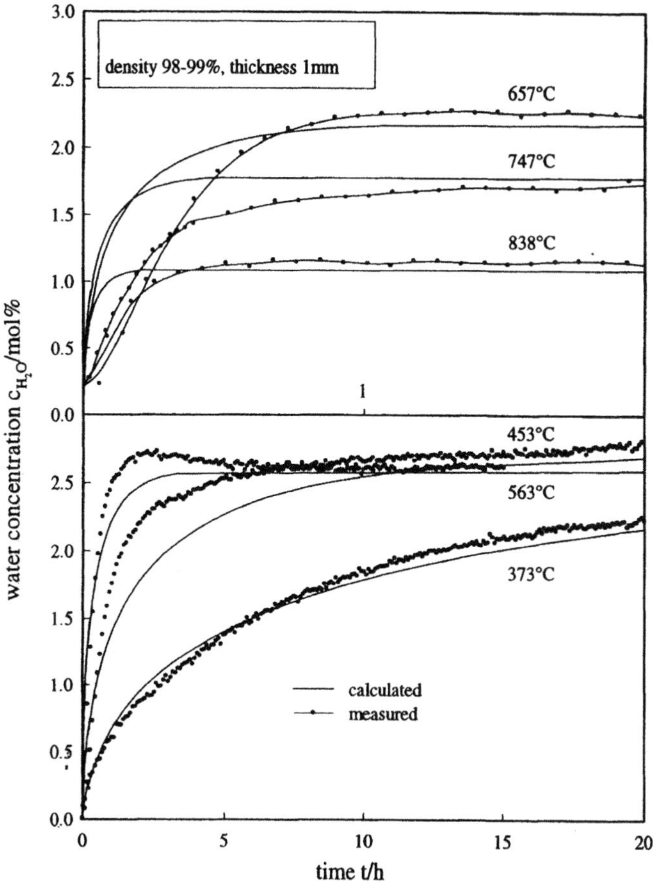

Figure 2. The water uptake of Gd-doped $BaCeO_3$ as a function of time is characterised by the chemical diffusion coefficient of H_2O [17].

x being short for $c_{H_2O}/c_{V_O^{\cdot\cdot}}$ [17]. $\tilde{D}$ turns out to be rather low in relevant cases and thin films are needed for a reasonable response time (see Fig. 2). This may be overcome by mode 2. We simply have to use an ionic conductor, the ionic carrier concentration of which we expect to be influenced by the acid-base active gas. The recently reported proton conductor $Ba_2YSnO_5(OH)$ seems [18] to be an appropriate candidate. Basic gases may trap protons and decrease the surface conductivity, whereas acidic gases may have the opposite effect. An already realised variant of this novel sensor principle introduced in Ref. [19], is the detection of NH_3 by measuring

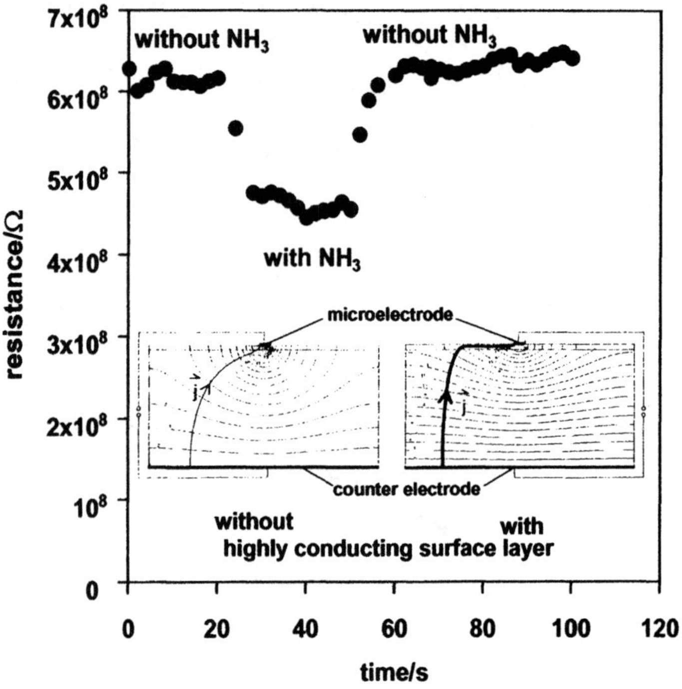

Figure 3. The conductivity response of AgCl on NH_3 (measured by microelectrodes) [20].

the surface conductivity of (highly porous) AgCl. According to its basic nature we expect NH_3 to trap Ag^+ (just as Al_2O_3 [19]) and to enhance the silver vacancy conduction as was recently confirmed. The correspondence with the Taguchi (SnO_2) sensor can be most closely expressed in terms of an energy level diagram (see [19]). The sluggish response time was overcome by the use of point electrodes which only probe their immediate surroundings (Fig. 3) [20]. In the same way the sensitivity can be increased. The last point refers to the measurement of acid-base active gases by galvanic cells (mode 3). The electrode reactions are necessarily redox reactions. By a skilful cell design one may nevertheless construct a cell, the overall reaction of which is only of the acid-base type.

Let us consider an electrochemical CO_2 sensor. The gas sensitive electrode usually used is an alkaline or alkaline earth carbonate. If we use Na_2CO_3, an appropriate electrolyte is Na-β''-alumina. If we use pairs of binary and a ternary oxide such as [21]

$$Na_2CO_3, CO_2 \left|Na^+ - \text{conductor}\right| Na_2ZrO_3, ZrO_2$$

we fix the chemical potential of sodium oxide on both sides, e.g. $\mu^L_{Na_2O} = \mu_{Na_2CO_3} - \mu^L_{CO_2}$ (so on an oxidic level the cell is similar to an oxygen concentration cell such as $O_2 \left|O^{2-}\text{-conductor}\right| Ni, NiO)$. However, the Na^+-conductor probes the sodium potential and not the sodium oxide potential. Thus we have to fix the sodium potential (cf. phase rule). On the reference side, this has been done in many previous galvanic cells by using e.g. Na or Na-alloys. Not to mention the materials problems, the fixing of μ_{Na} in such a way requires sealing. This is not only complicated but also disadvantageous. It leads to the fact that the oxygen partial pressure appears in the cell reaction. In our approach we simply use the oxygen of the atmosphere as the necessary third phase to fix μ_{Na} on both sides. The cell reactions then read

l.h.s.: $$Na_2CO_3 \rightleftharpoons Na^+ + e^- = +\frac{1}{2}O_2 + CO_2,$$

r.h.s.: $$Na^+ + e^- + \frac{1}{2}O_2 + ZrO_2 \rightleftharpoons Na_2ZrO_3.$$

In the overall reaction

$$Na_2CO_3 + ZrO_2 \rightleftharpoons Na_2ZrO_3 + CO_2$$

P_{O_2} cancels, and the e.m.f. is only dependent on CO_2!

$$FE = -\Delta_R G^0 - RT \ln P_{CO_2}. \qquad (26)$$

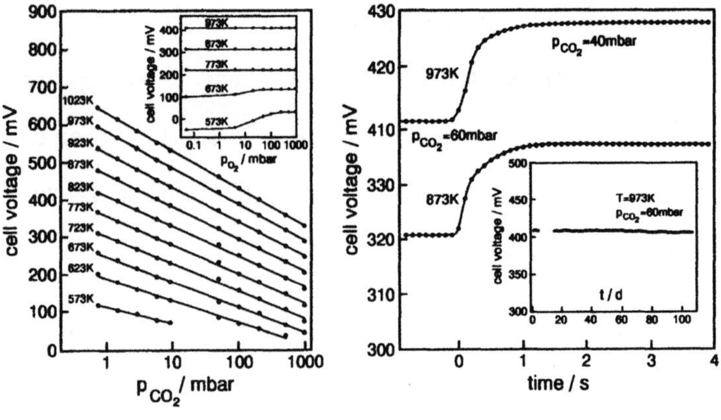

Figure 4. The performance of the open CO_2 sensor Au, O_2, CO_2 | Na_2CO_3 | Na-β-alumina| O_2, $Na_2Ti_6O_{13}$, TiO_2 [23].

This can also be derived by splitting the chemical potential of the ternaries into $\tilde{\mu}_{Na^+}$, $\tilde{\mu}_{e^-}$, μ_{O_2} and μ_{CO_2} or μ_{ZrO_2}, respectively. As usual $\Delta\tilde{\mu}_{Na^+}$ can be set to zero and $\Delta\tilde{\mu}_{e^-}$ transferred into the cell voltage. Since CO_2 does not harm the reference side in the condition window of interest ($\Delta_R G < 0$), the open sensor can be exposed to the O_2 and CO_2 containing atmosphere. Recently we used stannates and even more appropriately titanates instead of the sluggish zirconates [22]. The sensor

$$\mathrm{Au}\,|CO_2, O_2, Na_2CO_3|\,\mathrm{Na} - \beta'' - \mathrm{alumina}\,|(CO_2)O_2, Na_2Ti_6O_{13}, TiO_2|\,\mathrm{Au}$$

performs extremely well (Fig. 4). It is fast (response time of the order of 1 s at 500-600 °C), drift-free, thermodynamically well-defined (Nernst-equation is precisely fulfilled) at T $\gg$ 500 °C and, in that temperature range, independent of P_{O_2}. In addition, it is inexpensive to construct, is ecologically compatible and reflects all the advantages that mode 3 sensors can have. The only disadvantage which is common to the other competing e.m.f. solid state sensors, is the high operating temperature and the reactivity of the carbonates with acidic gases. Sensors based on the same

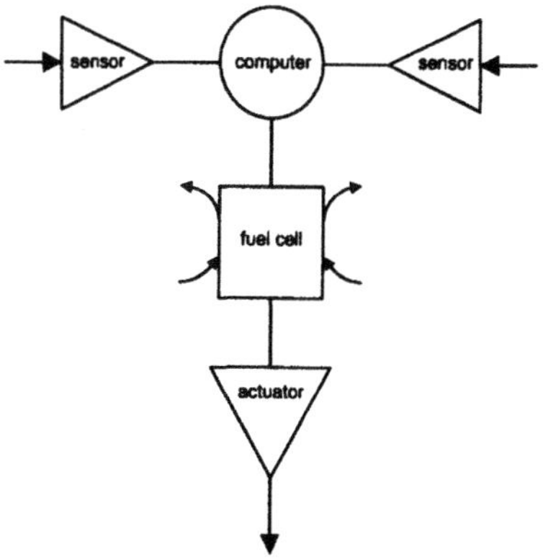

Figure 5. Autonomous system with solid state ionic/electronic "organs" [25].

principle, but using silicates, tungstates and molybdates are described by Möbius et al. [24].

Finally, two examples from liquid electrochemistry may be mentioned. One is the Severinghaus sensor. The heart of this consists of a membrane through which the gas has to permeate before it will undergo a hydrolysis in an appropriate solution. Acid-base active gases may be perceived by the induced pH-change. Such a pH-change may be detected e. g. by a glass electrode. The glass electrode, by the way, is a good example of an ion sensitive electrode and also a good example of the importance of exchange processes for the overall e.m.f., as discussed in Section 2.3. A decisive role is played by the exchange of H^+ ions in the neighbouring solution with the Na^+ ions in the glass membrane (for details see textbooks on electrochemistry).

Figure 5 may serve as a "paragon" of future autonomous systems which sense the environment (information gain), e. g. by artificial noses ("senses"), process and use the data by a Si-"brain" and in turn influence their environment by actuators ("legs and hands"). The necessary free energy is provided by a battery or fuel cell mechanism as a "metabolism". This picture is not necessarily meant as a realistic proposal, but rather to highlight the importance of solid state ionics and solid state electronics at a glance.

References

1. *Sensors, A Comprehensive Study* (1982), W. Göpel, J. Hesse, J. N. Zenel (eds.), VCH, Weinheim.
2. Dauter, K., Than, E., Molch, D. (1977) *Analytik*, Wissenschaftliche Verlagsgesellschaft, Stuttgart.
3. Maier, J. (1993) Electrical sensing of complex gaseous species by making use of acid-base properties, *Solid State Ionics* **62**, 105–111.
4. Maier, J. (1993) Mass transport in the presence of internal defect reactions – Concept of conservative ensembles. Parts I-IV, *J. Am. Ceram. Soc.* **76**, 1212–1217; 1218–1222; 1223–1227; 1223–1232; Kaiser, H. (1972) Zur Definition von Selektivität, Spezifität und Empfindlichkeit von Analysenverfahren, *Z. Anal. Chem.* **260**, 252–260.
5. Maier, J. and Münch, W. (1996) Chemical transport in mixed conductors: Application to the model materials $SrTiO_3$ and ZrO_2, *J. Chem. Soc., Faraday Trans.* **92**, 2143–2149; Denk, I., Traub, U., Noll, F., and Maier, J. (1995) In-situ optical investigation of oxygen diffusion profiles in $SrTiO_3$, *Ber. Bunsenges. Phys. Chem.* **99**, 798–801.
6. Denk, I., Claus, J., and Maier, J. (1997) Electrochemical Investigations of $SrTiO_3$ Boundaries, *J. Electrochem. Soc.*, 3526–3536; Leonhardt, M., Claus, J., Maier, J., in preparation.
7. Maier, J., Göpel, W. (1988) Investigations of the bulk defect chemistry of polycrystalline tin(IV) oxide, *J. Solid State Chem.* **72**, 293–302; Fonstad, C.G. and Rediker, R.H. (1971) Electrical properties of high-quality stannic oxide crystals, *J. Appl. Phys.* **42**, 2911–2918.
8. Sasaki, K., Haseidl, M., and Maier, J. (1997) The influence of redox-active impurity ions on defect chemistry and transport properties of solid electrolytes, *Proc. Eurosolid 4*, in press.
9. Sze, S.M. (1981) *Physics of Semiconductor Devices*, Wiley, New York.
10. Maier, J. (1998) *Interfaces*, this volume.
11. Janata, J. (1989) *Principles of Chemical Sensors*, Plenum Press, New York.
12. Bockris, J.O'M., Reddy, A.K.N. (1977) *Modern Electrochemistry*, Plenum Press, New York.
13. Wagner, C. (1972) The determination of small deviations from the ideal stoichiometric composition of ionic crystals and other binary compounds, in H. Reiss and J.O. McCaldin (eds.), *Progress in Solid State Chemistry*, **Vol. 6**, Pergamon Press, Oxford, pp. 1-15.

14. Maier, J. and Schwitzgebel, G. Theoretical treatment of the diffusion coupled with reaction, applied to the example of a binary solid compound MX, (1982) *phys. stat. sol. (b)* **113**, 535–547.

15. Sutija, D., Norby, T., Björnbom, P. (1994) Open-circuit voltages in oxides with mixed protonic, ionic and electronic conductivity, in T.A. Ramanarayanan, W.L. Worrell, H.L. Tuller (eds.), *Ionic and Mixed Conducting Ceramics*, **Vol. PV 94-12**, The Electrochemical Society, Pennington, pp. 27–38.

16. Iwahara, H., Uchida, H., and Kondo, J. (1983) Galvanic cell-type humidity sensor using high temperature-type proton conductive solid electrolyte, *J. Appl. Electrochem.* **13**, 365–370.

17. Kreuer, K.D., Schönherr, E., and Maier, J. (1994) Proton and oxygen diffusion in $BaCeO_3$ based compounds: A combined thermal gravimetric analysis and conductivity study, *Solid State Ionics* **70/71**, 278–284.

18. Murugaraj, P., Kreuer, K.D., He, T., Schober, T., Maier, J. (1997) High proton conductivity in barium yttrium stannate $Ba_2YSnO_{5.5}$, *Solid State Ionics* **98**, 1–6.

19. Maier, J. (1989) Heterogeneous solid electrolytes, in S. Chandra, A. Laskar (eds.), *Recent Trends in Superionic Solids and Solid Electrolytes*, Academic Press, New York, pp. 137–184.

20. Holzinger, M., Fleig, J., Maier, J., and Sitte, W. (1995) Chemical sensors for acid-base-active gases: Applications to CO_2 and NH_3, *Ber. Bunsenges. Phys. Chem.* **99**, 1427–1432.

21. Maier, J. and Warhus, U. (1986) Thermodynamic Investigations of Na_2ZrO_3 by Electrochemical Means, *J. Chem. Thermodynamics* **18**, 309-316.

22. Holzinger, M., Maier, J., and Sitte, W. (1996) Fast CO_2-selective potentiometric sensor with open reference Electrode, *Solid State Ionics* **86-88**, 1055–1062.

23. Spaeth, M., Kreuer, K.D., Dippel, Th., and Maier, J. (1997) Proton Transport Phenomena in Pure Alkaline Metal Hydroxides, *Solid State Ionics* **97**, 291–297.

24. Möbuis, H.-H., Shuk, P., Zastrow, W. (1996) Solid state systems for the potentiometric determination of CO_2, *Fresenius J. Anal. Chem.* **349**, 684–687.

25. Maier, J. (1997) Funktion durch Fehler: Zur Thermodynamik und Kinetik der Punktfehler in ionischen Festkörpern, in G. Ziegler, H. Cherdron, W. Hermel, J. Hirsch. H. Kolaska (eds.), *Werkstoff-Verfahrenstechnik*, DGM Informationsgesellschaft mbH, pp. 3–13.

SOLID OXIDE FUEL CELLS

B.C.H. STEELE
Imperial College, London, SW7 2BP, UK

1. Introduction

Unless technological breakthroughs occur in the cost of hydrogen and its storage then the concept of a hydrogen economy will remain a futuristic chimera. It is necessary, therefore, to assume that for the foreseeable future, except for certain niche markets, fuel cells will be supplied with a fossil fuel (eg. natural gas) which has to be converted to a hydrogen-rich fuel in an appropriate reformer. This requirement introduces problems for the polymeric, alkaline ,and phosphoric acid fuel cells which typically operate in the temperature range 60-200C. For these systems, due to the endothermic steam reaction, the reformer cannot be integrated into the fuel cell stack, and energy has to be supplied to an external reformer which reduces the overall system efficiency.For example operating on hydrogen the electrical conversion efficiency of a PAFC system can be around 60% whereas with natural gas fuel and an external reformer the overall system efficiency drops to around 40%.In contrast the relatively high operating temperature of the molten carbonate (MCFC) and solid oxide (SOFC) fuel cell systems allows the reforming reaction to be accomplished within the fuel cell stack ensuring high conversion efficiencies (55-60%).

The MCFC system has been under development for about twenty years, particularly in the US and Japan, and the selection of materials for the MCFC system (1a,1b) is dominated by the corrosive properties of the molten carbonate melt. This system has the merit of being constructed out of relatively cheap materials, and changes in processing routes have further reduced the overall cost without sacrificing performance. This is important because the power density of MCFC systems is relatively low (~0.15Wcm^{-2} at 0.7V) which implies relatively large areas of active components.The balance-of-plant also has to be designed to recycle CO_2 from anode to cathode to generate the electroactive CO_3^{2-} species. The low power density and system complexity suggest that the minimum size for MCFC units to be competitive in the market will be closer to 1MW than 100kW. To generate user confidence a variety of stack sizes have been or will

H.L. Tuller et al. (eds.), Oxygen Ion and Mixed Conductors and Their Technological Applications, 423–447.

be commissioned in the near future. These vary from the 2MW Santa Clara ERC demonstration unit to a number of 100kW stacks. Lifetime is still an issue for MCFC systems due to dissolution of the lithiated nickel oxide cathode, and corrosion of the stainless steel bi-polar plates under real operating conditions (ie high pressure and fuel conversion values around 85%), and lifetimes in excess of 1 year (8000 hrs) have still to be demonstrated.

In contrast SOFC systems are much more versatile (2a,2b,3) in that they can be designed to operate over a wide range of temperatures (500-1000C) with system sizes varying between 1kW and10MW. Moreover compared to other fuel cells, the SOFC system possesses other advantages in that it is a two phase gas/solid system.This overcomes many of the problems associated with liquid electrolytes such as corrosion, flooding, electrolyte distribution, and the maintenance of a stable Triple-Phase-Boundary (TPB) region.

2. Target Specifications for SOFC systems

At the conceptual stage it is desirable to adopt appropriate performance and cost criteria.Typical performance targets for single cells are about 0.5Wcm^{-2} at 0.7V and 0.7Acm^{-2} for 85% fuel conversion. It is recognised that these performance values will be lower in stacks due to additional resistive losses. Neither is it realistic to propose very high power density specifications as these are unlikely to be achieved in practice, and in any case would introduce additional complications for thermal management. The modified Ellingham diagram, Fig.1, incorporates a line which indicates how the equilibrium oxygen partial pressure varies with temperature for 85% utilisation at 1 bar for an H_2/CO fuel having a H/C ratio of 4:1.This line determines the value of the Nernst voltage of the exhaust gas and will be close to the operational voltage of the actual cell. It is immediately apparent that at high temperatures (~1000C) the total polarisation* ($IR+\eta_c+\eta_a$) cannot exceed about 0.15V if the target output voltage of 0.7V is to be attained. At 700C the Nernst voltage of the exhaust gas is about 0.95V thus allowing approximately 0.25V polarisation losses, and obviously higher losses can be tolerated at lower temperatures. For the preliminary evaluation in the present survey a total individual cell polarisation of 0.315V has been assumed. For current densities of 0.7Acm^{-2} this implies an area specific restivity (ASR) for the cell of 0.45Ωcm^2. This ASR value has been arbitrary divided equally between the cell components,ie. electrolyte(0.15Ωcm^2), cathode(0.15Ωcm^2), and anode(0.15Ωcm^2).These values

* At current densities of 0.7Acm^{-2} concentration polarisation is usually not important with well designed systems and electrodes, and so it is only necessary to consider electrolyte losses (IR), tgether with cathodic (ηc) and anodic (ηa) overpotentials.

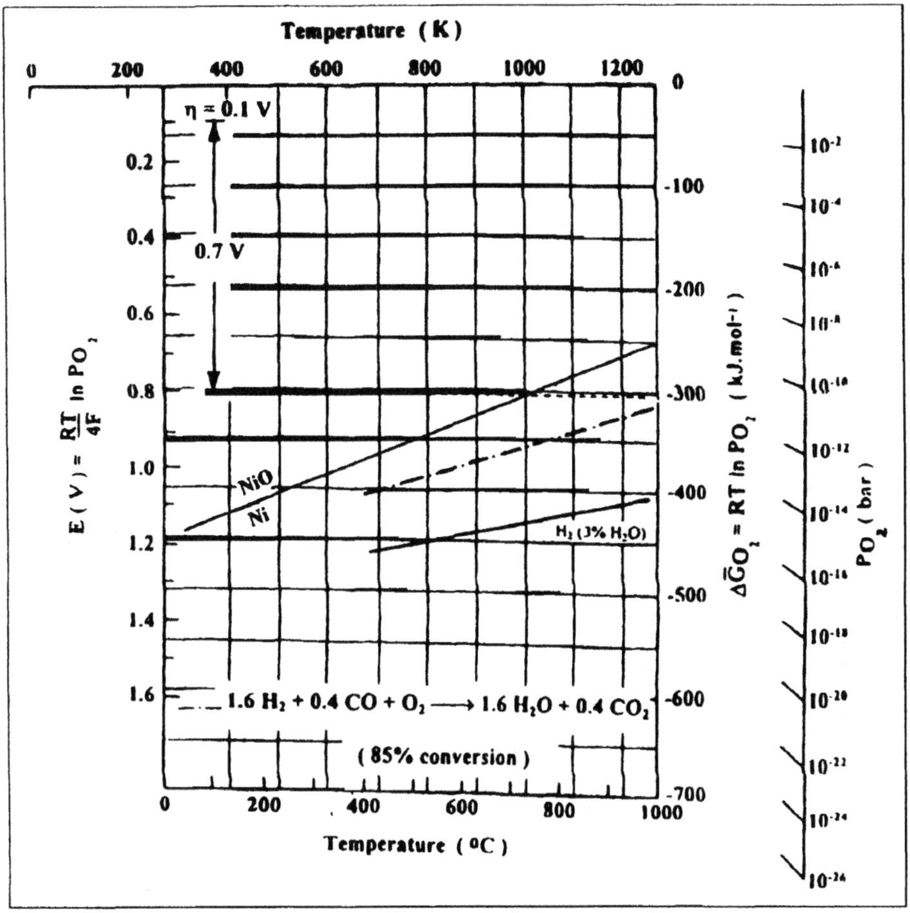

Figure.1 Modified Ellingham diagram depicting range of oxygen chemical potentials associated with operation of SOFC stacks

will be used for the various component selection exercises in the subsequent sections, but obviously other resistivity criteria could be used as appropriate. A number of papers (eg. 4,5) have described materials selection exercises for SOFC systems, and the Proceedings of the various International SOFC Symposia (eg. 6) should be consulted for detailed information about the various components incorporated in SOFC systems. It should be emphasised that materials selection parameters not only include transport properties but also thermochemical stability behaviour, thermophysical properties, thermomechanical behaviour, and of course, cost. However, as mechanical-microstructure relationships will be discussed in another contribution in the present volume, the present evaluation will focus on transport properties with comments on thermochemical behaviour as appropriate.

3. Operating temperature regimes

3.1 GENERAL

It is useful to designate three operating temperature regimes which are determined by the relevant properties of the the bi-polar plate (interconnect) material, and/or the behaviour of the associated ceramic electrolyte material. These regimes are shown in Fig 2 in which it has been assumed that the area specific resistance (ASR) of the electrolyte should be less than $0.15\Omega cm^2$ for the planar configuration. This ASR value is equal to L/σ where L (cm) is the thickness of an oxide ceramic exhibiting a specific ionic conductivity value of σ Scm^{-1}.

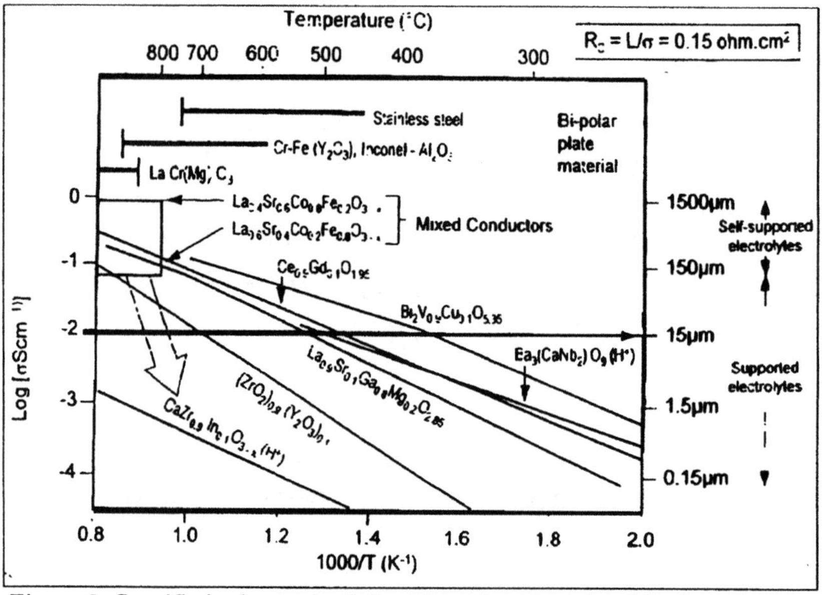

Figure 2. Specific ionic conductivity values for selected oxide electrolytes as a function of reciprocal temperature

Present day thick film ceramic technology can provide reliable material down to approximately 10-15μm which restricts the use of ZrO_2 based electrolytes to temperatures above approximately 700 C and CeO_2 based electrolytes to temperatures above 500 C. If the lower temperature limits for the electrolytes and $LaCrO_3$ based ceramic bi-polar plate are combined with the probable upper temperature limit for bi-polar plates incorporating metallic alloys (see Fig. 3) then three operating temperature regimes can be designated as follows:

(1) 850-1000 C: all ceramic fuel cells.

(2) 700-900 C : ZrO_2, CeO_2, or $LaGaO_3$ based electrolytes in association with special metallic(e.g. Siemens) or cermet (e.g.Tonen) bi-polar plate materials.

(3) 550-750 C : CeO_2 or $LaGaO_3$ based electrolytes in conjunction with ferritic stainless steel bi-polar plate materials.

3.2.OPERATION IN THE 900-1000C RANGE

For SOFC operation at 900-1000 C the following materials have been optimised over the past 25 years and shown to be compatible with little degradation after more than 70,000 hrs (~ 9 yrs) use, providing that processing conditions and compositions are selected to avoid the formation of the interfacial reaction products, $La_2Zr_2O_7$, and $SrZrO_3$.

Electrolyte	:	$Zr_{0.835}Y_{0.165}O_{1.917}$ (ZrO_2/Y_2O_3, 91/9)
Cathode	:	$La_{0.8}Sr_{0.2}MnO_{3+x}$
Anode :	:	Ni-ZrO_2 (Y_2O_3)
Interconnect	:	La $Cr_{0.8}Mg_{0.2}O_3$, $La_{0.8}Ca_{0.2}CrO_3$

Although there is a general consensus regarding the selection of these materials there is still a divergence of opinion about the best configuration to be adopted that will allow the economical mass production of reliable, strong tough ceramic components. Westinghouse have optimised a tubular arrangement which has been scaled up to 100 KW prototype units now under evaluation by gas and electric utilities. A porous cathode ($La_{0.8}Sr_{0.2}MnO_3$) support tube approximately 1cm diam. and 1.5m in length is first fabricated by conventional ceramic processing routes. Impermeable interconnect (~40μm) components are deposited using plasma-spraying techniques followed by the deposition of an impermeable electrolyte film(~35μm) using electrochemical vapour deposition (EVD) onto the porous cathode support tube. Finally a porous Ni-ZrO_2 anode coating is produced using a liquid spray process. The individual tubular cells which each generate 210W (active area 834cm2: ie 0.25Wcm^{-2}) are assembled into a bundle with nickel felt current collectors providing appropriate series-parallel conections. The Westinghouse design avoids the necessity of having high temperature seals, and the tubular module incorporates a ceramic heat exchanger so that the temperature of the exhaust gas is approximately 850 C which allows the use of conventional stainless steel ducting, pumps, etc. for the 'balance-of-plant' equipment. The size of the unit is being increased and Westinghouse expect to produce a 3MW prototype, SUREcell, for evaluation around 2001. This unit is pressurised (3 bar) and incorporates a 1.5MW SOFC system coupled to a 1.5MW gas turbine.It is projected that the overall electrical efficiency of this combined system will be about 60%, and the installed capital cost will in the region of \$1000/kW The principal queries associated with the

Westinghouse configuration are concerned with processing costs of the fabrication techniques employed and the low power density of the design. The relatively slow ECVD process, and associated high temperature masking procedures, introduce a major cost penalty. By eliminating two EVD steps originally used to fabricate the intercoact and anode structures, Westinghouse claim to be able to achieve the competitive figure of $1000/kW when scaled upto the 3MW size.However it should be noted that these cost reductions at present introduce some uncertainty about long term degradation trends. The excellent lifetimes (>70,000hrs) were achieved with components made by EVD techniques. However replacement of $LaCr_{0.8}Mg_{0.2}O_3$ by $La_{0.8}Ca_{0.2}CrO_3$ might allow degradation by stress corrosion mechanisms as the Ca stabilised $LaCrO_3$ is not thermodynamically stable (7) in the CO_2 partial pressures likely to be encountered in anodic environments. Moreover the relatively large lattice expansions associated with the loss of oxygen under reducing conditions can also create significant strains (8,9,10). Furthermore the employment of the EVD technique to create the Ni-ZrO_2 anode structure undoubtedly restricted the coarsening rate of the Ni particles and so excellent performance was observed over extended periods. Whether replacement of the EVD technique by a liquid spray process will result in such a stable anode structure remains to be seen.

Limited development work is proceeding (eg Dornier, MHI {Kobe}) using the planar configuration with doped $LaCrO_3$ ceramic bi-polar plates. However there remain major concerns over the long-term economic and structural aspects of this material when used in the planar design.

3.3. OPERATION IN THE 700-900C RANGE

In principle, the planar configuration, which is usually adopted for liquid electrolyte fuel cell systems, offers many advantages over the tubular design. A planar arrangement should develop higher power densities, and fabrication of the cell components can be accomplished by mass-production methods such as tape casting, calendering, or screen-printing. However the practical realisation of the planar configuration requires the individual cells to be sealed at the edges. Moreover it is doubtful whether the electrolyte component (150-200μm thick) can be reliably and cheaply fabricated at sizes much greater than 15 x15cm. Lowering the operating temperature to below 900C does allow, in principle at least, the use of metallic alloy bi-polar plates.Siemens/Plansee (11), for example,have developed an alloy composition, 94% Cr, 5%Fe, 1% Y_2O_3, which exhibits a thermal expansion very close to that of the zirconia ceramic electrolyte. A metallic 'window-frame' design allows the use of relatively small area (10x10cm) cell structures, which are in contact with large area metallic bi-polar plates. The introduction of a metallic framework and bi-polar plate introduces many benefits.For example , the superior thermal conduction of the metallic component reduces thermal gradients and improves the efficiency of the cooling fluid which is usually excess air in the cathode channels. Moreover the protective oxide has sufficient electronic conductivity to ensure that interfacial contact resistances are negligble. Long term tests are now underway to evaluate the stability of the metallic alloy and the metal/electrolyte seals used in the 'window-frame' arrangement. However it is already apparent (7) that the vapour pressures of gaseous species such as $CrO_2(OH)_2$ and CrO_3 are too high (Fig. 3) at 900C in the fuel cell environment.

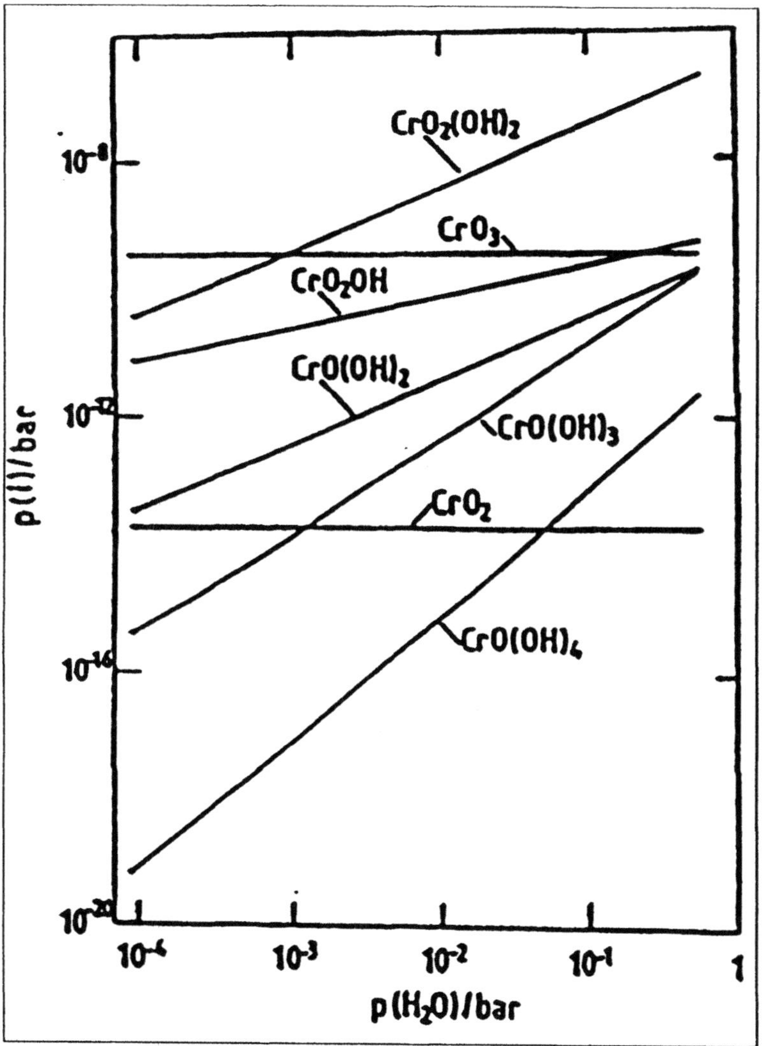

Figure 3. Equilibrium vapour pressures of gaseous species over Cr_2O_3 in air for different H_2O partial pressures

These chromium gaseous species distribute this element around the triple-phase-boundary producing a severe performance degradation. To ameliorate this situation steps are being taken to use protective coatings and to reduce the operating temperature (12). The Siemens planar configuration has been demonstrated at the 10kW size, and is being scaled upto 100kW units for evaluation in 1998/99.Smaller planar SOFC stacks incorporating self-supported $Zr(Y)O_{2-x}$ electrolytes and metallic alloy bi-polar plates have been demonstrated (13) by Z-Tek(USA), and Sulzer(Switzerland). Tonen (Japan) also claim promising results (14) using a cermet bi-polar plate with the composition, Inconel 600/Al_2O_3 (40/60 vol%).

The principal application for stacks operating around 850-900C is for integrated systems incorporating a gas turbine. The complexity of combined cycles together with the performance characteristics of gas turbines suggest that relatively large units (>1MW) will have to be manufactured to realise the economic benefits associated with these systems.

3.4 OPERATION IN 500-750C RANGE (IT-SOFC)

Within this intermediate temperature (IT) operating regime it becomes possible to specify ferritic stainless steel compositions for the bi-polar plate. These stainless steel compositions are relatively cheap and have excellent thermal expansion compatibility with CeO_2 based electrolytes. However more data are required about the long-term performance of these alloys in fuel cell environments, and the contact resistances associated with the interfacial electrode/bi-polar plate regions. At this temperature there is a choice of electrolyte materials as can be seen by examining Fig.2 and Table 1. Excellent power densities have been reported at 700C for electrode supported thick film YSZ electrolytes(15,16,17), self-supported $Ce_{0.9}Gd_{0.1}O_{1.95}$ (CGO) and $La_{0.9}Sr_{0.1}Ga_{0.8}Mg_{0.2}O_{2.85}$ (LSGM) electrolytes (18), and examples are provided in Fig.4. All these systems have advantages and disadvantages, and at present (mid-1997) it is not clear which electrolyte system will eventually emerge as the most appropriate for 700C operation. The results of long-term performance evaluation which is now beginning will help to clarify the situation but it will probably be another 2-3 years before confident predictions can be made. Some of the issues associated with the different electrolyte systems are summarised in subsequent sections.

Table 1
Selection of Electrolyte for 700C Operation

Electrolyte	Conductivity(Scm^{-1})	Thickness (μm)
YSZ	0.015	22
SSZ (scandia)	0.04	60
CGO	0.07	100
CSO	0.08	120
LSGM	0.06	90

Target ASR: $0.15\Omega cm^2$
ASR = L/σ

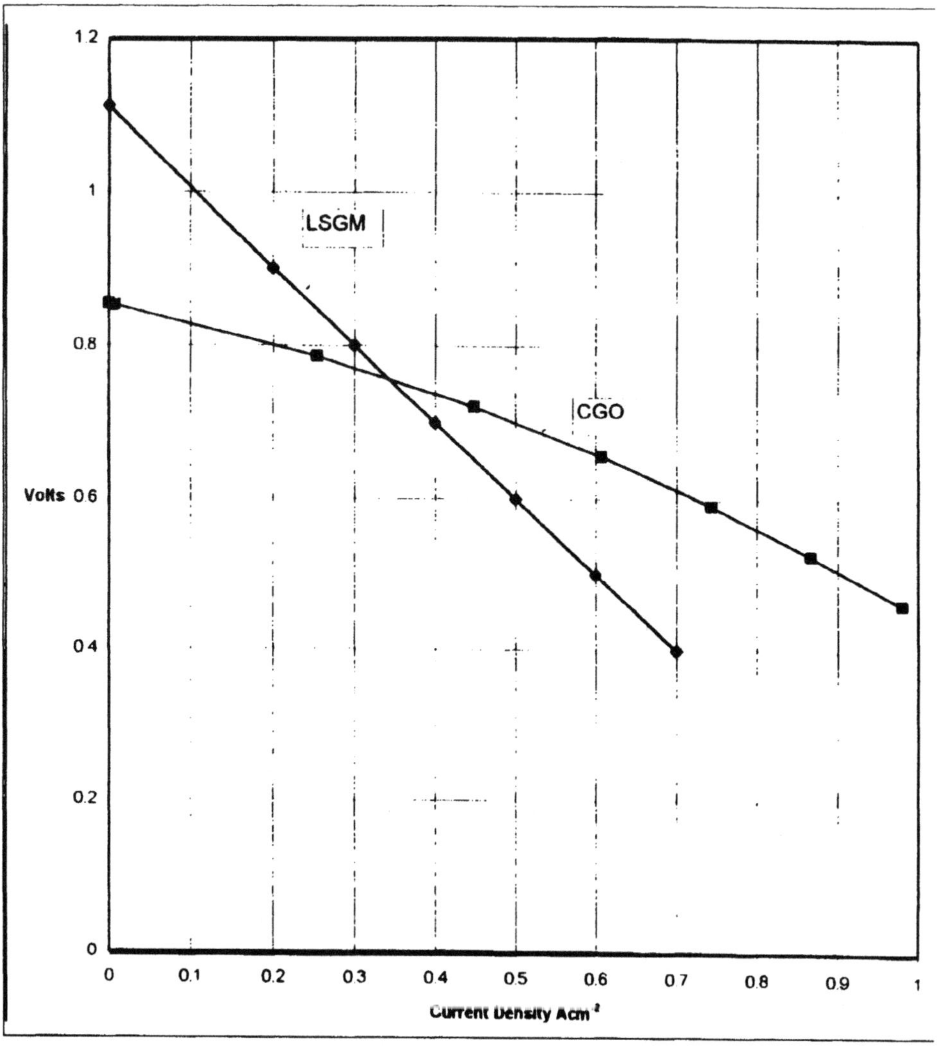

Figure4. I-V characteristics of CGO and LSGM 5x5cm cells operating under similar conditions at 973K

Lower temperature operation (~500C) is also possible with electrode supported CGO and LSGM electrolytes and applications such as micro-CHP (ie. <10kW output) and transport are being considered. A major advantage of SOFC systems is that CO is a fuel not an electrocatalytic poison, and so it is possible to envisage a direct methanol IT-SOFC system operating around 500C, and incorporating a thick film cell structure deposited on ferritic stainless steel (cf. zirconia based thermal barriers on gas turbine superalloys for aerospace applications). Such a configuration could withstand rapid heating/cooling cycles, and would form part of a SOFC/battery hybrid system for electric vehicles.

The performance of single IT-SOFC cells is already adequate at 600C, and recent developments in low cost processing technology (eg. electrostatic assisted vapour deposition) will enable IT-SOFC systems to be fabricated cheaply for operation at 500C. The main issues to be resolved in the short term are: scale-up to kW size, and optimisation of novel anode compositions for use with methanol. Partial oxidation reformers are being developed (eg A.D.Little, Argonne, Johnson Matthey) for gasoline which should make it possible to operate IT-SOFC systems using the existing liquid fuel infrastructure.

4. Selection of electrolytes

4.1. ZIRCONIA BASED ELECTROLYTES

Although operation at elevated temperatures is well documented (2), there is less information about the use of zirconia electrolytes at intermediate temperatures. A major issue concerns the cathodic overpotential around 700C and this is dicussed later. As already mentioned in the previous section, recent reports do indicate that good cathodic performances can be obtained using electrodes that are compatible with YSZ electrolytes.

A remaining uncertainty concerns the coefficient of thermal expansion (CTE) mismatch between YSZ (10.5 x $10^{-6}K^{-1}$) and ferritic stainless steel (12-13 x$10^{-6}K^{-1}$). Current scale-up activities will soon indicate whether this is a major problem or can be accommodated by appropriate design features.

In general ZrO_2/Y_2O_3 ratios of 91/9 mole% are favoured for operation in the range 900-1000C to ensure that the decrease in ionic conductivity due to vacancy ordering and formation of second phase is minimised. For operation at 700C there appears to be no long term operational evidence that favours either the 8 or 9 mole% Y_2O_3 dopant values as both 'cubic' compositions are metastable at this temperature.

It is known that ZrO_2-Sc_2O_3 solid solutions exhibit greater ionic conductivity values than ZrO_2- Y_2O_3 compositions. In thick film form the high cost of Sc_2O_3 would not be such a disadvantage and there may be designs/aplications where further consideration of this material might be appropriate, although examination of Table 1 suggests that that for thick film structures the conductivity of YSZ appears adequate.

Finally the concept of fabricating single phase p-i-n zirconia based electrolyte/electrode structures has been proposed (eg.19) using Tb_2O_3 and TiO_2 additions. However these designs are unlikely to perform well wthout the additions of further porous electrode layers to improve the relatively low oxygen surface exchange values of the doped zirconia surfaces.

4.2 CeO_2 SOLID SOLUTIONS

4.2.1 General comments

The superior ionic conductivity exhibited by CeO_2 based electrolytes was known over thirty years ago. Moreover Takahashi demonstrated the excellent electrode kinetics associated with CeO_2 electrolyte/electrode combinations, and these observations were confirmed by the oxygen self diffusion (D*) and surface exchange (k) values measured by Floyd and Steele (20). Because pure CeO_2 already exhibits the fluorite structure it is not necessary to add a dopant to stabilise the cubic structure, and so the composition can be optimised for maximum ionic conductivity which is usually attained for additions of around 10 mole% of $MO_{1.5}$ to yield oxygen vacancy concentrations of 2.5% (ie. $Ce_{0.9}$ $M_{0.1}$ $O_{1.95}$).In contrast the favoured zirconia composition is ZrO_2/Y_2O_3, 91/9, mole %, ie. $Zr_{0.835}Y_{0.165}O_{1.917}$ corresponds to an oxygen vacancy concentration of 4.1%, which is already in the region where vacancy ordering will occur.

The optimal dopant is Sm^{3+} ($\Delta E \sim 0.68eV$) closely followed by Gd^{3+} ($\Delta E \sim 0.7eV$) to ensure that V_o-M_{Ce} interactions are minimised (21,22). Although there are claims in the literature suggesting alternative dopants these conclusions are based on the interpretations of total conductivity measurements which include a grain boundary resistance. This parameter can be relatively large due to segregation of impurities and dopants but can be reduced to acceptable values (23). It is interesting to note that two recent papers on grain boundary impedances in $Ce(Gd)O_{2-x}$ (23) and $Zr(Y)O_{2-x}$ (24)confirm that even with 'pure' materials their appears to be an intrinsic grain boundary impedance.

The principal uncertainty arising from the selection of $Ce_{0.9}Gd_{0.1}O_{1.95}$ (or $Ce_{0.9}Sm_{0.1}O_{1.95}$) electrolytes for use at intermediate temperatures (500-700C) arises from the loss of oxygen at the oxygen partial pressures prevailing under anodic conditions according to the reaction,

$$O_o^x \Rightarrow V_o^{\cdot\cdot} + \frac{1}{2}O_2 + 2e^-$$

The increase in n-type conductivity which depends upon $\left(Po_2\right)^{-1/4}$, reduces the ionic transference number and associated cell performance. The formation of Ce^{3+} ions produces a lattice expansion which may have a deleterious effect on the relevant mechanical behaviour (see 25).

4.2.2. Electronic Transference Numbers

The influence of the electronic conductivity depends upon the value of P_θ, which is the oxygen partial pressure value at which $\sigma_{ion} = \sigma_e$, and the oxygen partial pressure imposed on the anode, which in turn is determined by the actual operating conditions of the stack. The role of electronic conductivity has been considered in recent publications (26, 27), and cthe I-V characteristics of a mixed conductor and an electrolyte exhibiting purely ionic conduction are compared in Fig.4.

Examination of literature values of P_θ indicate a considerable scatter of values for nominally the the same composition (eg. 28). Most investigators measure the total conductivity as a function of Po_2, and thus derive values of P_θ, although other techniques are available (29). The total conductivity measurements could include a grain boundary contribution which may complicate the interpretation, and collaborative studies are needed to establish definitive values of P_θ. The electronic conductivity values incorporate the product of two parameters, concentration of charge-carriers (n), and a mobilty term (μ),

ie. $\sigma_e = ne\mu$.

It is to be expected that the value of n, for small dopant levels, will be closely related to the thermodynamics of non-stoichiometric CeO_{2-x}, and so variations of σ_e probably reflect changes in μ. These variations in μ could arise due to trapping centres in the relevant CeO_{2-x} solid solutions, or due to Schottky barriers at the grain boundaries. Maricle et al (30) claimed that small additions of Pr^{3+} act as trapping centres thus decreasing the level of electronic conductivity and changing the value of P_θ. However Steele et al (23) were unable to confirm this observation although in principle the concept might work with other redox cations.

More recent data (27,31) has indicated that at low temperatures the situation is more complicated as the electronic conductivity does not follow the $(pO_2)^{-1/4}$ dependence due to the presence of disssolved water which appears to influence the grain boundary conductivity.

4.3 DOPEDLaGaO$_3$ ELECTROLYTES

The ionic conductivity of the composition $La_{0.9}Sr_{0.1}Ga_{0.8}Mg_{0.2}O_{2.85}$ (LSGM) is only slightly lower than that exhibited by CGO (Fig 2), and LSGM has the advantage of a much wider ionic domain (32,33) which is evident in Fig. 4. As commercial supplies of LSGM powder become available many groups are now hoping to exploit the excellent transport properties of this electrolyte and a number of outstanding issues will probably be resolved in the near future. For example, the thermomechanical properties (34) must be improved, and the stability of LSGM must be demonstrated in the CO_2 partial pressures likely to be ecountered in SOFC anodic environments to ensure that LaO_2CO_3 is not formed.Moreover it appears that LSGM reacts with anodes containing Ni. A comparison of the performance at 700C of PEN structures incorporating either CGO or LSGM is shown in Fig 4.

4.4 DUPLEX ELECTROLYTES INCORORATING CeO$_2$

Due to the onset of electronic conductivity in CeO_2 based electrolytes under anodic conditions it can be useful to incorporate an electronic blocking layer adjacent to the anode. This type of structure has been examined both with modelling calculations (35,36) and by experimental measurements (37, 38). Whilst the electronic conductivity is effectively blocked by thin layers of $Zr(Y)O_{2-x}$, the selection of $Zr(Y)O_{2-x}$ is not satisfactory due to the thermal expansion mismatch with $Ce_{0.9}Gd_{0.1}O_{1.95}$, and problems occur when the structures are scaled up in size. However LSGM should be better as the CTE value (~$11.0\times10^{-6}K^{-1}$) is closer to that for CGO ($12.3\times10^{-6}K^{-1}$).

5.Electrodes

5.1. CATHODES

Many papers have been devoted to interpreting the behaviour of porous oxide cathodes on solid electrolytes and it remains a controversial area (39,40). For LSM electrodes on YSZ it is generally agreed that at low overpotentials the ionic conductivity of LSM is too low to have a significant influence on the the distribution of the oxygen flux entering the electrolyte, and so most of the current passes through the triple-phase-bondary (TPB). Optimisation of the microstructural features of the LSM cathode, including fabrication of LSM-YSZ composite electrodes (41,42) can thus improve the performance of these porous cathodes.

Further improvement is dependant upon identifying the rate controlling step. It is known for example, that for $La_{0.8}Sr_{0.2}MnO_3$, the electrode resistivity decreases when the cathodic overpotential is around 0.4V at 1000C (43), which coincides with the change-over from oxygen excess to oxygen deficient stoichiometry(44). It is probable ,therefore, that most of the overpotential can be associated with the development of non-stoichiometry in the LSM cathode, and that the actual charge-transfer process across the electrode/electrolyte interfacial region is relatively facile resultng in small charge-transfer overpotentials in this area. Further support that most of the overpotential is associated with the development of nonstoichiometry is provided by the in-situ synchrotron X-ray diffraction experiments of Poulsen et al.(45). The kinetics associated with the stoichiometry changes will be controlled by a chemical reaction of the type:

$$\tfrac{1}{2}O_2 + V_o^{\cdot\cdot} + 2e^- \Rightarrow O_o^x$$

It should be emphasised that this is not a charge-transfer reaction as it involves the neutral combination of charged species.Steele (4, 46) has assumed that this is the rate determing process, and has provided a semi-quantitative approach to examine whether available surface exchange data can account for the observed resistivities for LSM/YSZ electrode structures.

Using this approach a surface exchange 'resistivity' can be defined:

$$R_s = RTV_m/(zF)^2k,$$

where V_m is the molar volume of the oxide cathode, and k ($cm.s^{-1}$) represents the oxygen surface exchange coefficient measured by the isoptopic exchange/diffusion profile (IEDP) technique (47).Taking into account microstructural features, w (grain size:µm), l_c (collection length:µm), and assuming tortuosity factors around 3-4 for 50% porous structures, then the electrode area specific resistivity, R_e, is given by:

$$R_e = \frac{RTV_m w}{(zF)^2 k2l_c}$$

Calculated values for R_e are provided in Table 2

Table 2

T(C)	T(K)	k $cm.s^{-1}$	R_s Ωcm^2	w µm	l_c µm	R_e Ωcm^2
1000	1273	1.1 E-7	85.28	1	100	0.43
900	1173	3.6 E-8	240.5	1	100	1.20
800	1073	9.5 E-9	829.6	1	100	4.15
700	973	2.6 E-9	3737	1	100	18.63

The calculated ASR values are in reasonable agreement with observed values in the temperature range 900-1000C and the associated activation energies (~1.4eV) also suggest surface exchange control.. If composite LSM/YSZ elecrodes are fabricated then the 800C value could be reduced to around $1\Omega cm^2$ but is clear that the calculated values cannot account for the high power densities observed at 700C (15,17) as the values are at least an order of magnitude too high. Alternative explanations are being sought for this dicrepancy between calculated and experimental LSM resisistivities.

For intermediate temperature operation most investigators select a mixed conducting oxide such as $La_{0.6}Sr_{0.4}Co_{0.2}Fe_{0.8}O_{3-x}$ (LSCF). Cell structures incorporating this cathode and either CGO or LSGM electrolytes can exhibit excellent power densities at 700C (18) due to both higher oxygen surface exchange (k) and self diffusion (D*) coefficients associated with this oxide. Even at 520C optimised LSCF cathodes have been fabricated with ASR values around $0.5\Omega cm^2$ (48). The behaviour of porous mixed conducting oxides has been modelled analytically (49) and using finite element methodology(50), and good agreement obtained between theory and experiment. It should be noted that non-charge transfer processes (solid state diffusion, surface exchange, and gas phase diffusion) cannot generally be resolved individually into separate equivalent circuits contributing to the total cell impedance, and so care must be taken in the interpretation of impedance measurements.

Unfortunately mixed conducting oxide cathodes such as LSCF cannot be used in direct contact with YSZ because of the formation of interfacial reaction products ($SrZrO_3$, $La_2Zr_2O_7$). One solution to this problem, which has been successfully demonstrated (51), is to use diffusion barriers such as CGO between the YSZ electrolyte and mixed conducting cathode. However this approach complicates the fabrication procedure and can be difficult to scale-up due to differential thermal expansivities.

5.2 ANODES

5.2.1 Ni - ZrO_2

The Ni-ZrO_2 cermet has been adopted as the anode material by almost all the SOFC teams following the pioneering work of Westinghouse in the 1970s. However the performance of the Ni - ZrO_2 cermet is very dependent upon the associated microstructure and because of the complexity of possible reaction mechanisms little progress has been made in our detailed understanding of the anode behaviour, a view often endorsed (e.g.52) at the recent SOFC-V symposium. Moreover, most of the investigations have used H_2/H_20 mixtures instead of the technological more important H_2O/CH_4 fuel gas compositions.

However phenomenological rate equations have been devised for CH_4 reforming on Ni - ZrO_2 anodes, and Odegard (53) concluded that the reforming rate values obtained at 1000C by different investigators are of the same order of magnitude, although the derived activation energies do exhibit a significant variation (0.6 - 1.0 eV). Using the rate equation suggested by Odegard (53) it is easy to demonstrate that the steam reforming reaction on Ni - ZrO_2 anodes at 1000C can provide sufficient H_2 for an average oxygen ion flux around 1A cm^{-2}. As these current densities are not usually attained in SOFC cells under normal operating conditions it would appear that the electrochemical reaction is rate - limiting at 1000C. Using the activation energy (0.6 eV) derived by Odegard then at 700C the hydrogen produced by the steam-reforming reaction could be consumed by an oxygen flux equivalent to 180 $mAcm^{-2}$. This value appears reasonable as molten carbonate fuel cells operating at 650C on H_2O/CH_4 fuel supplies rarely exceed 150 $mA.cm^{-2}$. It would appear therefore that alternative anode materials (and catalysts) need to be developed if the SOFC design/application requires either internal or external reforming of H_2O/CH_4 mixtures around 700C. One approach is to activate the Ni-oxide cermet by additions of dispersed noble metal electrocatalysts (54)

Before considering alternative materials a few comments are appropriate about the performance degradation often observed for Ni - ZrO_2 anodes. The rate of degradation appears relatively small for the Westinghouse Ni-ZrO_2 structure which involves the initial preparation of a Ni skeleton anode followed by an ECVD process which produce an intergrowth of $Zr(Y)O_{2-x}$ within the porous Ni framework, although this technology has now been replaced by a cheaper liquid spray process.. The morphology of the $Zr(Y)O_{2-x}$ intergrowth appears to restrict grain growth of the Ni particles and also provides an effective pathway for oxygen ions from the electrolyte to the adjacent Ni particle which is believed to be primarily responsible for the catalytic activation of methane particles. As it is very difficult to reproduce this specific microstructure starting with NiO-ZrO_2 particles it may be appropriate to investigate further the processing route developed by MIT (55) which also starts with Ni particles followed by an infiltration of $Zr(Y)O_{2-x}$ particles.

Another reason for replacing Ni-ZrO_2 cermet anodes is that at temperatures below about 700C the oxygen partial pressure in the anode compartment could be high enough to form NiO under typical operating conditions as indicated in Fig.1. Finally it should be noted that dry CH_4 can be oxidised on Ni-ZrO_2 cermets (56) provided that the Ni-ZrO_2 interface is supplied with sufficient oxygen ions.

5.3 CeO_2 BASED MATERIALS

Obvious contenders for the replacement of ZrO_2 in Ni-ZrO_2 cermet anodes are CeO_2 solid solutions.

More than thirty years ago Takahashi (57) had demonstrated the superior anodic kinetics associated with reduced CeO_2 ss due to the presence of relatively high electronic and oxygen ion conductivity. CeO_2 is incorporated into the Dornier anode (58), and Steele et al (59) reported that cells incorporating oxide anodes could directly electrochemically oxidise dry methane as the presence of mobile lattice oxygen reduced the rate of carbon deposition (60). British Gas (61) have examined the steam reforming rates of CH_4 or Ni-CeO_2 anodes and obtained comparable activation energies (0.52eV) to those reported by Odegard (53). However, using natural gas containing higher alkanes resulted in much higher activation energies (1.5 - 3.6eV). It has also been reported (62) that additions of precious metals to Ni-ZrO_2 based anodes can improve the electrode performance. However, anodes based on CeO_2 ss have not replaced Ni - ZrO_2 anodes because of the relatively low levels of electronic conductivity and the relatively large lattice expansion (28) associated with the formation of Ce^{3+} under anodic conditions which can eventually result in the anode spalling off the electrolyte. Fortunately this lattice expansion is much reduced at lower temperatures.

5.4 ALTERNATIVE ANODE MATERIALS

For fuel gases containing CH_4 the presence of Ni promotes the formation of carbon deposits and it is desirable to avoid the use of this material if at all possible, particularly as NiO is likely to be produced at temperatures below 700C (Fig.1). With oxides the calalytic oxidation of hydrocarbons appears to involve lattice oxygen (Mars van Krevelen mechanism), and there is also some evidence that the presence of protonic conductivity can also activate adsorbed CH_4 molecules (63). Accordingly Steele et al (64) proposed the following criteria for the selection of alternative oxide anode materials:

1) Good electronic conductivity, preferably > 10^2 Scm^{-1} at anode operating potentials (0.7-0.9V). Probably n-type behaviour preferable.
2) Predominant anion lattice disorder to enhance oxygen diffusion coefficients and possibly protonic conductivity.
3) High values for oxygen surface exchange kinetics
4) Fabrication of adherent films with minimal processing problems
5) Compatibility with solid electrolyte substrate (thermal expansion and interdiffusion values)

and interdiffusion values)

6) Stability in anode environment including gaseous species H_2, H_2O,CO,CO_2,etc

A number of oxide materials have been examined at Imperial College (64,65,66,67) including $La_{0.8}Ca_{0.2}CrO_3$, $Ti_{0.97}Nb_{0.03}O_2$, $Mg_{0.3}Nb_{0.1}Ti_{2.6}O_5$, $Sm_2Ti_{1.9}Nb_{0.1}O_7$, and $CrTi_2O_5$.

So far,however, it has not been possible to develop a single phase material that is stable over the oxygen partial pressure range associated with typical anode operation and exhibits sufficient electronic conductivity.Mixed valence (Ti^{4+}/Ti^{3+}) titanates such as $CrTi_2O_5$ exhibit excellent electronic conductivity at 954C at low Po_2 values ($<10^{-17}$ bar), but form new phases when the oxygen partial pressure is increased (Fig 1). Similar behaviour can be expected for the titanate compositions investigated by Irvine et al.(68). It is obviously going to be difficult to develop novel single phase oxide materials to satisfy all the criteria listed above, and so a more productive strategy would be to examine the anodic behaviour of oxides that are known to activate CH_4, and to rely on a second phase for the current collection function. Possible materials for the catalytic role include, Sm_2O_3 (65), doped $LaYO_3$ (63), together with doped $CaTiO_3$ (69,70), and doped $SrTiO_3$ (71).Some information about the possible ocurrence of protonic conductivity in such materials could be provided by relevant correlations (72).The electronic conducting phase could be an oxide, possibly a metal, and a preliminary evaluation should also include carbides, oxycarbides, silicides, borides, particularly for low temperature operation (~500C).

5.5 ANODES FOR METHANOL FUEL

The previous sections have assumed that the SOFC system would operate on CH_4 fuel. However for low temperature operation (~500C) there is interest in developing direct methanol SOFC systems for applications in electric vehicles and possibly for CHP in remote locations having access to bio-fuel. At 500C NiO is likely to form (Fig.1), and alternative materials such as stainless steel mesh and powder have been successfully used (23) in contact with thick film $Ce_{0.9}Gd_{0.1}O1_{.95}$ electrolytes. With these materials the ratio H_2O/CH_3OH should exceed 0.5 to prevent carbon build-up on the anode. Although the oxidation of CH_3OH can be rapid at 500C more work is required to optimise the electrode structure, and evaluate the relevant mechanistic pathways.

6. Strategies for commercialisation

6.1 GENERAL

It has already been mentioned that the efficiencyof SOFC systems can be high. Moreover the stack efficiency is not a strong function of size, and it should be emphasised that the efficiency remains high under part-load operating conditions. This is not the situation for rotating/reciprocating heat engines whose efficiency rapidly decreases away from the optimal design load.These considerations, together with low enviromental emission levels make SOFC units very appropriate for distributed power applications.At present, however, entry into the market is difficult due to the relatively high installation price associated with a new technology that does not have the benefit of the cost reductions associated with large volume production.

Like any developing technology, fuel cells face the typical 'Catch 22' of commercialisation: 'to enter the market, the production costs must come down, however, to lower these costs, the cumulative production must be greatly increased, ie. significant market penetration must occur'.Unless explicit steps are taken to address this question, fuel cell commercialisation will remain slow and require large subsidies for market entry. To address this commercialisation dilemma, it is necessary to follow a market -driven strategy and identify high-value entry markets that can support the current high costs of fuel cell systems, minimise the technical and financial risks of market entry, and sustain market penetration with market pull. In this regard small size (<10kW) stationary applications that have high operating duty cycles and reasonable competitive costs appear a low risk market entry strategy for SOFC systems. These systems can already compete with thermo-electric generators (~$25,000/kW) for remote applications, and Sulzer (73) are following this approach for their domestic CHP units.In addition there also appears to be opportunities for the leisure market including caravans, and yachts, where price considerations are relatively elastic. The increasing interest in intermediate temperature (500-700C) SOFC units for electric vehicles should also stimulate this sector.

It is appropriate, therefore, to compare the cost of electricity (COE) of fuel cells with other types of power producing systems and this has been done for 25kW (Fig.5) and 3MW (Fig.6) systems. Estimates of COE for emerging technologies can vary widely because of different assumptions regarding the 'learning curve', reliability, discount rate, and depreciation period, etc. Much of the data used in Figs 5 and 7 has been taken from the work of Mugerwa and Blomen (1a) who used the Technical Assessment Guide P-2410SR published by EPRI (1982).

The COE will depend upon the number of units being produced, and where appropriate, figures are are given for FS (first series: individual units), L (low volume: 10's of units per year), and H (high volume: 100's of units per year). It should be noted that mass production methods using robots, etc, have not been considered in the present study but obviously adoption of such manufacturing routes will result in a further reduction of costs. The size of unit will also influence the COE and estimates are provided for 25kW, and 3MW units.

The COE has usually been divided up into contributions associated with the stack (or equipment), balance-of-plant (important for phosphoric(PAFC) and molten carbonate(MCFC) fuel cells), operation and maintenance, and fuel. Only PAFC, MCFC, and SOFC have been considered because at present the polymer electrolyte fuel cell (PEFC) requires very pure hydrogen (<10 ppm CO), and with the external steam reformer/gas conditioner the overall efficiency is less than 35%. Morover the installation costs are very high at present due to the price of the polymer membrane, Pt catalyst loadings, and graphite bi-polar plate.

In the present study emphasis is placed on COE and therefore in general the systems with the highest efficiency will be the cheapest. However for combined heat and power (co-generation) applications the economic value of the waste heat can have a significant value, since the alternative is to generate the heat separately at an increased cost. Therefore in CHP applications it is necessary to examine the relative amounts of heat and electricity used which is specific to the particular application. When large amounts of heat and relatively low amounts of electricity are required, conventional technologies can be more appropriate than fuel cells unless the capital costs of this emerging technology can be significantly reduced.

6.2 COE OF 25kW SYSTEMS: FIGURE 5

The COE values for the PAFC (L) and PAFC (H) can be relied on as several PAFC systems in the range 20-40kW have been manufactured and evaluated in field tests. Likewise the values for the Gas Engine (GE) and Uninterruptable Battery Power Supply (UPS), as these are already commercially available. Electricity from the grid, priced at $0.11/kW-hr (~7p/kW-hr), was taken as the 'fuel' for the UPS system. The values for SOFC (FS) need to be treated with caution as prototypes (1-5kW) have only recently been started to be evaluated in field trials, principally by Sulzer for domestic applications.However many studies have indicated that small planar configuration SOFC systems are capable of attaining installation costs below $500/kW providing the operating temperature is around 700C which allows the use of stainless steel components for the bi-polar plate and balance-of-plant. The principal issues with small SOFC systems relate to ability to withstand repeated heating/cooling cycles, reliability, and performance degradation over long periods. Micro-turbines (turbo-generators) are just beginning to be field tested, particularly as power-trains in hybrid electric vehicles and and it is likely that the capital costs will

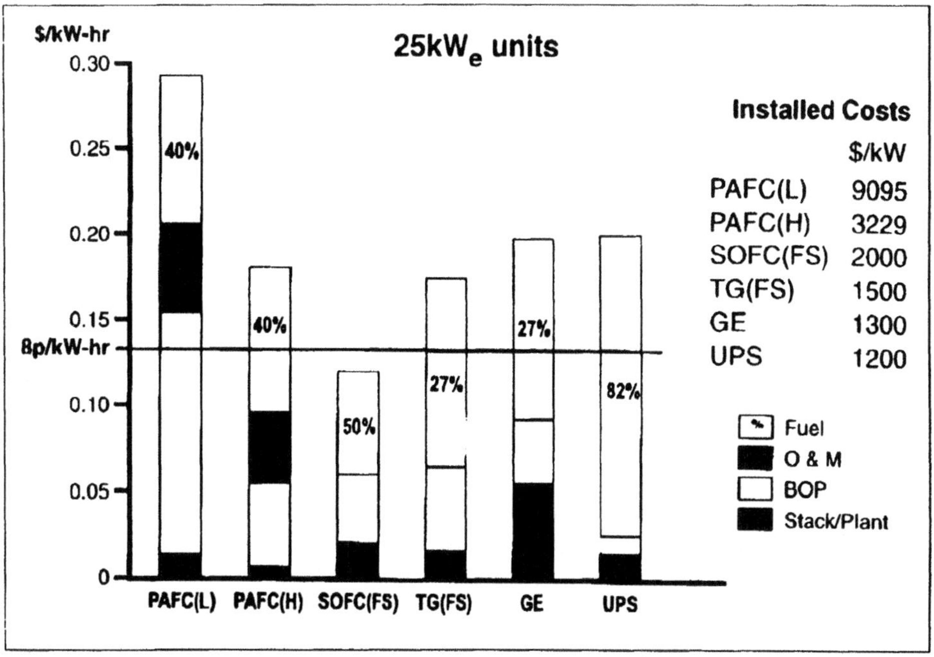

Figure 5. COE values for various 25kW technologies
% refers to LHV electrical efficiency

also fall below $500/kW. Small radial turbines have been used for many years as turbo-chargers in cars, and so are already available in high volume at relatively low cost. These are coupled with novel high speed (75-125,000 rpm) brushless alternators incorporating neodynium-boron- iron magnets on the rotor. Although high power densities (~10kW/dm^3) are available from the actual turbo-charger the system has to be contained in an enclosure incorporating sound insulation, heat exchangers, and exhaust gas clean-up devices which lower the initial power density. The effect of system efficiencies on fuel costs is very evident when comparing the COE value for the SOFC and TG units.

6.3 COE OF 3MW SYSTEMS: FIGURE 6

As the size of the MCFC systems is increased from 250kW to 3MW the installation costs are reduced from $4820-2810/kW for individual units (FS). It should be noted that a 2MW MCFC is at present being evaluated at Santa Clara, California, but most field tests are conducted on 100kW size systems. The values for the SOFC system are those provided by Westinghouse (74) for their 3MW pressurised (3 bar) SURE Cell system which incorporates a 1.5MW tubular SOFC system with the exhaust gas fed into a 1.5 MW combustion

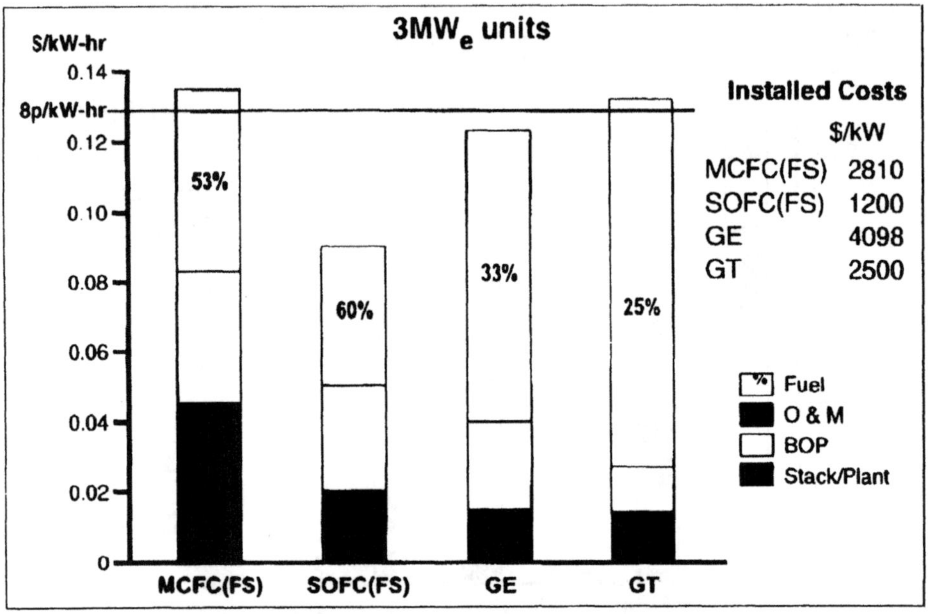

Figure 6. COE values for various 3MW technologies.
% refers to LHV electrical efficiency

turbine. A world-wide consortium is being formed to exploit this technology with prototypes available for evaluation around 2000.

7. Acknowledgements

Discussions with members of the Ceramic Ion Conducting Membrane Group (CICM) at Imperial College are gratefully acknowledged, together with financial support from the Leverhulme Trust for a Senior Research Fellowship.

8. References

1a..Blomen L.J.N and.Mugerwa M.N,(1993),Ed., *Fuel Cell Systems*, Plenum, New York.

1b. Kordesch K. and Simader G. (1996), *Fuel Cells and Their Applications*, VCH, Germany.

2a. Minh N.Q.and Takahashi T. (1995), *Science and Technology of Fuel Cells* Elsevier, Amsterdam.

2b. Hammou A.and Guindet J. (1997), Solid Oxide Fuel Cells, in P.J.Gellings and H.J.M.Bouwmeester (eds.),*Solid State Electrochemistry*,CRC Press, New York, p 407.

3. Steele B.C.H.(1996), *Phil.Trans.R.Soc.Lond. A*, **354**, 1695.

4. Steele B.C.H.(1996), *Solid State Ionics*, **86/88**, 1223.

5. Steele B.C.H.(1996), *Brit. Ceram. Proc.*, **56**, 151.

6. *Solid Oxide Fuel Cells-V* (1997), Proc.Vol. 97-40,Electrochem.Soc.,New Jersey,USA.

Jersey,USA.
7. Peck D.H,.Miller M, Nickel H., Das D., and.Hilpert K (1995), *Solid Oxide Fuel Cells- IV*, Proc.Vol 95-1, Electrochem.Soc., New Jersey, USA, p.858.
8. Hendriksen P.V.,.Carter J.D, and.Mogensen M. (1995), ibid, p.934
9. Yasuda I.and Hishinuma M. (1995), ibid, p.924
10..Mori K, Miyamoto H.,.Takenobu K, Matsudaira T.(1997), p.1301 in ref 6.
11. Janousek M.,.Kock W,.Baumgartner M, and.Greiner H. (1997) p.1225 in ref 6.
12. Urbanek J.,.Miller M., Schmidt H., and Hilpert K.(1996), in 2nd European Fuel Cell Forum, U.Bossel, Switzerland, ISBN 3-922 148-19-0, p 503.
13. Diethelm R., Brun J., Gamper Th., Keller R., Kruschwitz R.and Lenel D.(1997), p 79 in ref 6
14. Seto H., Mitayata T., Tsunoda A., Yoshida T., and Sakurada S.(1993), in *Solid Oxide Fuel Cells -III*, Proc.Vol. 93-4, Electrochem. Soc., New Jersey, USA, p 421
15..Minh N.Q and Montgomery K.(1997), p 153 in ref 6.
16. Buchkremer H.,Diekmann U., de Haart L.G.J.,Kabs H.(1997), p160 in ref 6
17. Visco S.J., Jacobsen C., and De Jonghe L.C.(1997), p 710 in ref 6.
18.Christie G.M,.van Berkel F.P.F, and Huijsmans J.P.P.(1998), to be published in Proc. of Grove Symp. V, (*J.Power Sources*,**71**)
19. Worrell W.L., Han P., Uchimoto Y., and.Davies P.K(1997), p.50 in ref 7
20. B.C.H.Steele and J.M.Floyd. (1971), *Brit. Ceram. Proc.***19**, 55.
21. Kilner J.A.and.Steele B.C.H. (1981), in *'Non Stoichiometric Oxides'* ed.O.Toft Sorensen (Academic ,New York), p.2336
22. Inaba H.and Tagawa H.(1996), *Solid State Ionics*, **83**, 1,
23. Steele B.C.H., Zheng K., Rudkin R.A., Kiratzis N., and.Christie M. (1995), p.1028 in ref 7
24. Aoki A., Chiang Y-M., Kosacki I., Lee J-R.,.Tuller H.L, and Liu Y.(1996), *J.Am.Ceram.Soc.* **79**,1169.
25. A.Atkinson,(1996), p 707 in ref 12
26. Godickemeier M., Sasaki K.,and Gauckler L.J.(1995), p.1072 in ref 7
27. Sahibzada M., Rudkin R.A., Steele B.C.H., Metcalfe I.S., and.Kilner J.A. (1997), p.244 in ref 6.
28. Mogensen M., Lindegaard T., Hansen U.R., and Mogensen G.(1994), in *'Ionic and Mixed Conducting Ceramics II'*, Proc.Vol. 94-12, Electrochem. Soc., New Jersey, USA, p.448.
29. Navarro L., Marques F.,and Frade J.(1997), *J.Electrochem.Soc.*,**144**,267.
30. Maricle D.L., Swarr T.E., and Karavolis S.(1992), *Solid State Ionics*, **52**, 173.
31. Kiratzis N., Steele B.C.H., Ralph J.M., Sahibzada M., Atkinson A.,and Kilner J(1997), to be published in *Proc. Ionic and Mixed Conducting*

Ceramics III, Proc. Vol.97-24, Electrochem.Soc.,New Jersey, USA.
32. Ishihara T., Matsuda H., Takita Y.(1994), *J.Am.Chem.Soc.*, **116**,3801.
33. Ishihara T.,.Honda M, Nishigushi H.,and Takita Y.(1997), p301 in ref 6.
34. Drennan J., Zelizco V., Hay D., Ciacchi F.T., Rajendran S., and Badwal S.P.S.(1997), *J.Mater.Chem*.7, 79.
35. Virkar A.V.(1991), *J.Electrochem.Soc.*, **138**,1481.
36. Yuan S.,and Pal U.(1996), *J.Electrochem.Soc.*, **143**, 3214.
37. Riess I.(1992), *Solid State Ionics*, **52**, 127.
38. Steele B.C.H.(1994), *J.Power Sources*, **49**, 1.
39. Comments(1997), *J.Electrochem.Soc.* **144**,1881.
40. van Heuveln F.H., Bouwmeester H.J.M., and van Berkel F.P.F.(1997), J.Electrochem.Soc. **144**, 126.
41. Ostergard M.J.L., Clausen C., Bagger C., and Mogensen M.(1995), *Electrochimica Acta,* **40**,1971.
42. Tanner C.W., Fung K-Z., and.Virkar A.V. (1997), *J.Electrochem.Soc.*, **144**, 21.
43..Hammouche A.et al (1991), *J.Electrochemical Soc.*,**138**, 1212.
44. Tagawa H., Mori N., Takai H., Yonemura Y., Minamiue H., Inaba H., Mizusaki J., and Hashimoto T.(1997), p 785 in ref 6.
45. Poulsen F-W., Sorby L., Poulsen H.F., and Garbi S.(1997), *10th SOFC Workshop, Materials and Processes*, IEA Annex VII, , Les Diablerets, Switzerland, p84
46. Steele B.C.H.(1997), *Solid State Ionics*, **94**, 239.
47. Kilner J.A.and De Souza R.A.(1996), in *Proc.17th Riso Intl. Symposium on Materials Science,* Riso National Laboratory, Roskilde, Denmark, p.41.
48. Bae J-M.(1996), *'Properties of Selected Oxide Cathodes for Solid Oxide Fuel Cells'*, Ph.D Thesis, (May, 1996), University of London. Sections accepted for publication in Solid State Ionics, 1997.
49 Adler S.B., Lane J.A., and.Steele B.C.H.(1996), *J.Electrochem.Soc.* **143**, 3554.
50. Bernier M., Herbin R., and Gehain E.(1996), p.203 in ref 12.
51. Tsai T.and.Barnett S.A. (1997), p.274 in ref 6.
52. Gubner A.,.Landes H, Metzger J., Seeg H., and Stubner R.(1997), p 844 in ref 6.
53. Odegard R., Johnsen E, and Karoliussen H.(1985), p 810 in ref 7.
54. Uchida H.,Mochizuki N.,and Watanabe M. (1996), *J.Electrochem.Soc.*,**143**, 1700
55. Chou K.C, Yuan S., and Pal U., in *'Solid Oxide Fuel Cells III'*, Eds. S.C.Singhal and H.Iwahara, Proc.Vol. 93-4, Electrochem. Soc., New Jersey, USA, p 431.
56.Aida T, Abudala A., Ihara M., Komiyama H.,and Yamada K.(1997), p.801

in ref. 6.
57. Takahashi T. (1972) in *'Physics of Electrolytes'*,Vol.2, Ed. J.Hladik, Academic Press.
58. Schafer W., Geier H., Lindemann G., and Stolten D.(1993), in *'High Temperature Electrochemical Behaviour of Fast Ion and Mixed Conductors'* Eds. F.W.Poulsen et al,14th Riso Symp.,Riso National Laboratory, Roskilde, Denmark, p.409.
59. Steele B.C.H., Kelly I.E., Middleton P.H.,and Rudkin R.A.(1998), *Solid State Ionics*, **28/30**, 1547.
60. I.S.Metcalfe (1992), *Solid State Ionics*, **57**,259.
61. British Gas (1995), *'Investigation of Internal Reforming Anodes on Solid Oxide Fuel Cell* ETSU Report No. F/01/00013/REP, DTI,UK ETSU, Harwell, Oxon., UK.
62. Watanabe M.,Uchida H.,Suzuki H.,and Tsuno A. (1995), *Solid Oxide Fuel Cells-IV*, Proc.Vol.95-1, Electrochemical Soc.,New Jersey,USA, p.750.
63. Alcock C.B. (1993), *J.Catal*. **140**, 557.
64. Steele B.C.H, Middleton P.H., and Rudkin R.A.(1990), *Solid State Ionics*, **40/41**,810
65. Middleton P.H.,Steiner H.J, Christie G.M., Baker R., Metcalfe I.S.,and B.C.H.Steele, (1993)in *'Solid Oxide Fuel Cells III'*, Eds.S.C.Singhal and H.Iwahara, Proc. Vol. 93-4.,Electrochem. Soc.,New Jersey, USA, p.542.
66. Steiner H.J., Middleton P.H., and Steele B.C.H,(1993), *J.Alloys and Cpds.*,**190**, 279
67. Baker R.T.and Metcalfe I.S., *in'Solid Oxide Fuel Cells IV'*,Eds.M. Dockiya et al.,Proc.Vol. 95-1, Electrochem.Soc., New Jersey,USA, p.781.
68. Fagg D.P., Fray S.M., and Irvine J.T.S.(1994), *Solid State Ionics*,**72**,235,
69. Iwahara H., Esaka T., and Mangahara T.(1988), *J.Appl. Electrochem.* **18**,173,
70. Selcuk A.and Steele B.C.H.,(1995) in Proc.of 4th Euro-Ceramics, eds. G.Gusmano and E.Traversa, Gruppo Editoriale Faenza Editrice,Italy, p 413.
71. Steinsvik S.,Norby T., and Kofstad P(1994), inProc.Vol II 'Electroceramics IV',Eds.R.Waser et al, Augustinus Buchhandlung, Aachen, 1994, p.691
72. Norby T. and Larring Y.(1997), *Current Opinion in Solid State and Materials Science*, **2**, 593.
73. Honegger K., Batawi E., Sprecher Ch., and Diethelm R.(1997), *Solid Oxide Fuel Cells-V*, Proc.Vol.97-40, Eds.U.Stimming et al., Electrochem.Soc., New Jersey, USA, p 321.
74. George R.(1998), Proc. Grove-V, *J.Power Sources*,**71.**

COMPARISON OF SOLID OXIDE FUEL CELLS WITH ALTERNATIVE FUEL CELLS AND COMPETITIVE TECHNOLOGIES

MOGENS MOGENSEN
Materials Research Department, Risø National Laboratory
DK-4000 Roskilde, Denmark

1. Introduction

The aim of this paper is to compare the solid oxide fuel cell (SOFC) with other types of fuel cells in order to put the special properties of the SOFC into perspective. Also the economic requirement of the SOFC is dealt with and compared to the costs of other types of electricity generators. The economy of fuel cells is naturally closely related to a number of technical aspects such as electrical efficiency and life time of the systems. An overview of the different types of fuel cells including their stage of development and first foreseen applications are given, the various types of effiency concepts are described, the costs of stacks and systems are discussed and finally some visions of possible future applications are sketched. For more comprehensive literature, the reader is referred to the series of books of the biannual meetings, Fuel Cell Seminar, Program and Abstracts, 1986, 1988, 1990, 1992, 1994, 1996 (Courtsey Associates, Inc., Washington D.C.), and references in these. Recent overviews are given by Appleby and Foulkes [1] and Minh and Takahashi [2].

2. The Principle of Fuel Cells

Fuel cells are electric batteries which are able to convert hydrogen and oxygen (usually from air) into electricity. Heat and steam are the only by-products, and they are products of value in many circumstances. There are several types of fuel cells. The high temperature fuel cells can also use natural gas and coal gas as fuel. Usually, all the different types of fuel cells are named according to their electrolyte. The five main types are: The alkaline fuel cell, AFC, having an aqueous KOH electrolyte, the phosphoric acid fuel cell, PAFC, the solid polymer proton conductor fuel cell, SPFC, the molten carbonate fuel cell, MCFC, and the solid oxide fuel cell, SOFC, of which the electrolyte is a ceramic oxide ion conductor.

Fig. 1 shows the basic principle of the structure of a fuel cell as well as the electrochemical reactions of the different cell types. In Table 1 some essential data for the five main types are listed. The individual cell only produces a voltage of about 0.5 to 1 volt depending on operation conditions. Consequently, it is necessary to build a stack or

H.L. Tuller et al. (eds.), Oxygen Ion and Mixed Conductors and Their Technological Applications, 449–469.

assembly of series connected cells to get a useful high voltage. The functionally right type of stack geometry, the so called bipolar flat plate design, is shown in Fig. 2. However, many other geometries have been suggested and some are produced, for instance tubular SOFC stacks which are more easily fabricated from a technical point of view. However, this and other geometries result in a longer current path than the one shown in Fig. 2 giving rise to larger power losses.

It is essential for the electrodes within a fuel cell to be good electron conductors. Furthermore, it is important that they are porous in order for the gas to penetrate and react, and that the shape of the porous structure enables the biggest possible contact area between the three phases: ion conducting electrolyte, electron conducting electrode and the reactants of the gas phase.

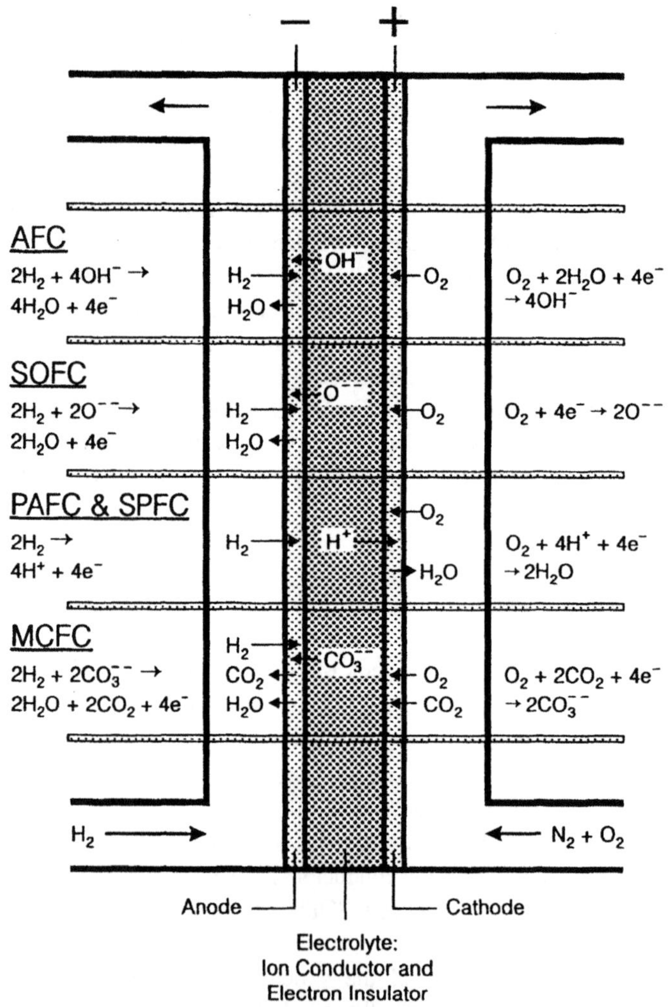

Figure 1. An illustration of the basic structure of a fuel cell: fuel gases are fed at the anode and cathode which are separated by an electrolyte. The electrochemical reactions arising in the five principal types of fuel cells are indicated.

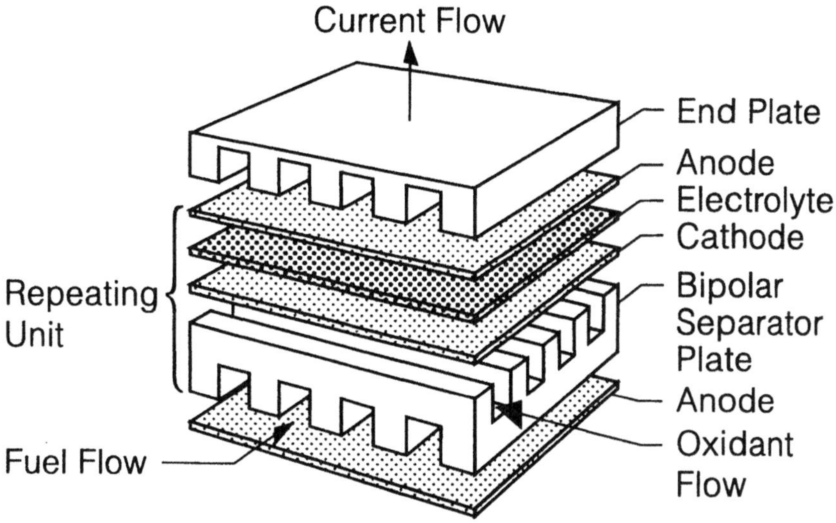

Figure 2. An expanded view of the basic fuel cell structure with the repeating unit of the fuel cell stack.

3. The Advantages of the Fuel Cell

The processes in a fuel cell are direct conversion of chemical energy to that of electricity without using thermal energy as an intermediate step. Consequently, they are not subject to the same thermodynamic limitations as are found for a Carnot engine (piston engine or turbine). Therefore, it is possible to achieve high efficiency. Furthermore, the operating temperature in fuel cells (even in SOFC, 1000°C) is low compared to that of a flame. This is the reason why no nitrogen oxide is evolved, and consequently, part of why fuel cells pollute less.

TABLE 1. Some typical data for the most important fuel cell types

Fuel Cell Type	Alkaline	Solid proton conductor	Phosphoric acid	Molten carbonate	Solid oxide
Abbreviation	AFC	SPFC	PAFC	MCFC	SOFC
Electrolyte	Aq. KOH	Polymer H^+-conduc.	H_3PO_4	$KLiCO_3$	$Zr(Y)O_2$
Anode material	Porous nickel	Grafite +Pt	Grafite +Pt	Porous nickel	Nickel $+Zr(Y)O_2$
Cathode material	Porous Ni or Ag	Grafite +Pt	Grafite +Pt	Porous NiO	$La(Sr)MnO_3$ $+Zr(Y)O_2$
Possible fuel	Pure H_2	H_2 (CO free)	H_2 (low CO)	H_2+ CO Nat. gas	H_2+ CO Nat. gas
App. operation temperature	100°C	100°C	200°C	650°C	1000°C
Approx efficiency, % *)	40	40	40	60	60

*) % of HHV of natural gas. To be taken as indicative only, see text.

Thus, a number of potential advantages are offered by fuel cells in general and solid oxide fuel cells (SOFC) in particular. According to the European Commission [3] fuel cells in the future can play a major role in increased decentralized electricity production due to:

- Low pollutant emission and low noise levels, which allows location in populated areas. The NOx emission of gas turbines and diesel engines is typically 10 and 100 times higher, respectively, than that of fuel cells. Generally, this makes combustion systems less attractive for location in urban areas;
- Modularity which allows optimum adaption to the energy use requirements and cheap mass production;
- Fuel cells such as MCFC and SOFC produce high temperature waste heat and are suitable for cogeneration in both buildings and industry;
- Low maintenance costs due to low number of moving parts and to autonomous operation;

Consequently, many industrialized countries support considerable SOFC R&D efforts in spite of the fact, that it may take more than 10 years before the SOFC technology is commercialized.

4. The Types of Fuel Cells

4.1. THE LOW TEMPERATURE FUEL CELLS

The different fuel cell materials and operation temperatures make them somewhat difficult to compare. However, as they operate according to the same basic principles, a formal comparison may be made. Some values for the performance of various types are given in Table 2.

In comparison to other electric power generating technologies, the differences become so large that only values like the investment costs per kW_e (kW electrical output), lifetime and electric efficiency may be compared. The final comparison should (and will eventually) be based on the cost of kWh (of electricity). The cost of a kWh is, however, very much dependent on a number of fluctuating (in time and geographic position) variables such as fuel price and energy taxes (energy policy) and therefore not easy to use.

Below a brief description of the various types of fuel cells are given. The comparison with the possible efficiency and price of other technologies is given in a subsequent section.

TABLE 2. State of the art information for various fuel cell types. R_i is the area specific internal resistance and P is the area power density at a cell voltage of 0.7. The values should be taken as indicative only because each developer has his own sets of data. SOFC/W and SOFC/ASAC means Westinghouse and Allied Signal Airo-space corporation type respectively.

Fuel cell type	Demonstrated size in kW_e	R_i Ωcm^2	P W/cm^2	Demonstrated stack lifetime, h	Foreseen near term application	Refs.
AFC	~200 kW	2.5	0.1	4000	Already used in space crafts and military	[4,5,6]
SPFC/H_2	200	0.4	0.5	4000	Used in space crafts , automobiles	[7,8,9]
SPFC/DMFC	Lab. size	~7	0.04 *)		Automobiles	[5,10]
PAFC	11000	2	0.1	30000	Combined heat and power	[5,11,12, 13]
MCFC	2000	2	0.1	7000	Combined heat and power	[14,15]
SOFC/W (1000°C)	25	0.8	0.2	15000	Topping cycle for gas turbine	[16,17]
SOFC/ASAC (800°C)	0.25	0.4	0.4	100	?	[18]

*) 100 mA/cm^2 at 0.37 V/cell

4.1.1. *The Alkaline Fuel Cell*

The alkaline fuel cell has been developed to a quite high standard and has been produced commercially for special niches (space and military which are insensitive to cost). The reported problems are [6]: 1) Degradation of electrodes especially at high polarisations when the cheaper electrocatalysts like Ni and Ag are used. 2) Costs of electrode fabrication especially in case durable noble metal catalysts are used. 3) The classical diaphragm which separates the electrodes from touching each other is made of asbestos, the use of which has become banned in several countries. 4) Intolerance to CO_2, both in the air and in the hydrogen.

The main reason why the alkaline fuel cell AFC has not yet experienced a broader commercial success is that it does not tolerate CO_2. This reacts with OH^- and CO_3^{2-} is formed and the electrolyte is gradually destroyed. The carbonate was also previously believed to precipitate in the electrodes blocking them and thus be responsible for the AFC degradation. This is recently reported not to be the case in all circumstances [6]. Nevertheless, the CO_2 content in the hydrogen as well as the oxygen must be minimized to a few ppm. Naturally, it is quite expensive to produce such clean gasses. The recognition of this led in the 1960s and 1970s to the development of CO_2-tolerant fuel cells (see below). In spite of this many AFC development programs continued into the 1990s. Now only very few AFC R&D programs remain active [6].

4.1.2. *The Solid Proton Conductor Fuel Cell*

One of the CO_2 tolerant cells, the solid proton conductor fuel cell (SPFC), has a proton conductor ion exchanger membrane as the electrolyte. The proton conductor polymer is quite expensive [4]. And the Pt-catalyst in the anode of the cell is poisoned by CO at the relatively low operation temperature which is the reason why hydrogen made by reforming hydrocarbons cannot be used without a purification having a price-raising effect. Therefore, SPFC has not yet been put to use except in the area of space crafts (see below).

SPFC has gained a rapidly increasing attention during the latest decade. It is believed widely that the SPFC will become the fuel cell for cars. Sofar, no convincing cost calculations have been presented even though the Canadian company Ballard claims to have tested "a new low cost membrane electrolyte" [9,19]. From a purely economic point of view the passenger car market is the most difficult (se below). In spite of this a number of car producers e.g. Daimler-Benz has recently invested larger sums (in the 100 M$ class) in developing SPFC-systems. The main reason seems to be the expectation that extra cost will be allowed in the future if the vehicles are polluting less and this is probably especially so for city busses.

The solid polymer fuel cell is also often referred to as the polymer electrolyte fuel cell (PEFC) or the polymer electrolyte membrane fuel cell (PEMFC).

A special type of SPFC is being developed for direct oxidation of methanol (DMFC = direct methanol fuel cell). Main technical problems are the high anode polarization resistance and the diffusion of CH_3OH through the polymer electrolyte to the cathode side.

4.1.3. *The Phosphoric Acid Fuel Cell*

The other CO_2-tolerant low temperature fuel cell is the phosphoric acid fuel cell, PAFC. The PAFC is the fuel cell type which has been developed furthest towards commercialization. About 100 units of 200 kW (the PC25) have been manufactured by IFC (International Fuel Cells) and sold to various utilities world wide for demonstration [13]. Still the price has not come down to a really commercial level. A unit of 11 MW was demonstrated in Tokyo during the early 1990s [12]. It is anticipated that PAFC will probably gain a certain market share, but it may not become extensively used because its electric efficiency is bound to be relatively low (as for all low temperature fuel cells). For small (< 200 kW) combined heat and power units an efficiency of 40% may be quite acceptable if the investment costs are low enough. And small PAFC units have a cost problem because of a relatively complicated system (balance of plant) is necessary, e.g. the graphite based cathode can only withstand open circuit conditions for a very short period. Otherwise the graphite will oxidises. Therefore, the PAFC unit system must automatically change the cathode gas from air to inactive nitrogen in case the load suddenly is taken off the PAFC. Such arrangements may not be of great economical importance for multi-MW plants in terms of cost per kW but for small systems it may be prohibitive. And large stand alone fuel cell systems will have difficulties in competing with gas turbines (see later).

The low temperature fuel cells have a common problem: they cannot convert hydrocarbons or carbon monoxides directly. Thus, it is necessary to feed them with hydrogen which usually has to be produced from fossil fuel. Furthermore, the hydrogen must not contain CO (PAFC tolerates low CO-concentrations up to about 100 ppm). The reforming of the fossil fuel into hydrogen involves a considerable energy loss, so that the resulting efficiency becomes about 40% of the upper heating value for methane (natural gas, see table 1). The main argument for using fuel cells should be a higher efficiency, but this argument is lost as the gas turbines used for electricity production today have an efficiency of about 40% for medium size units (10 -100MW) and about 50% for large gas turbines above 100MW.

4.2. HIGH TEMPERATURE FUEL CELLS

The high temperature fuel cells MCFC and SOFC can convert natural gas as well as CO either by internal reforming or by direct electrochemical reaction. This means that they potentially have a considerably higher efficiency than that of the low temperature fuel cells.

4.2.1. *The Molten Carbonate Fuel Cell*

In the MCFC process it is necessary to add CO_2 to the cathode together with air, as it is the CO_3^{2-} ion which takes care of the electrolytic charge transport, Fig. 1. If CH_4 is converted in the cell, CO_2 is developed at the anode. This means that it is necessary to extract CO_2 from the anode exhaust and add it to the cathode inlet.

The molten salt saturated with O_2 and CO_2 in the MCFC is extremely corrosive. This is the most important problem for the MCFC today, as it limits the life time of the cell. Both the NiO-cathode and the stainless steel interconnects (or separator plates as they are called in the MCFC literature) are corroded especially if the MCFC is pressurized. The higher CO_2 pressure the more corrosive is the molten carbonate electrolyte. Another problem associated with the MCFC is the evaporation and migration of the molten electrolyte. This means that lost electrolyte must be replaced regularly.

4.2.2. *The Solid Oxide Fuel Cell*

The international SOFC R&D has resulted in a continuous improvement of cell and stack performance during the last decade in contrast to all other types of fuel cells[5]. And the improvements seem to continue.

The SOFC which is totally produced from solid oxides (see table 1) seems to have the biggest potential of the fuel cells. High efficiency (50%) and long life time (>80,000 h for single cells) have been demonstrated. A further increase in efficiency to ~60% seems possible.

Two main types of SOFC designs are developed at present. The one is the Westinghouse tubular type which has been demonstrated on the 25 kW level [16], and the other is the bipolar flate plate which is persued by a number of companies and research laboratories [10,17,20].

The main SOFC main problem is to develop a sufficiently cheap and reproducible fabrication technology for the thin ceramic layers, but also serious materials problems still remain for the bipolar flate plate type. A pO_2-related dimensional instability exhibited by the doped $LaCrO_3$ interconnect cannot be tolerated in high performance stacks [20]. Thus it is necessary to intensify the research for ceramic interconnect materials with much better dimensional stability in reducing atmosphere. Metallic interconnect is an alternative to ceramics, but then reduced operating temperatures are required. The increase in electrode overpotential and electrolyte resistance with decreasing temperatures, however, necessitates development of improved electrodes on thin electrolytes (~25μm) in order to reach the necessary low R_i achieved at higher temperatures.

Presently, only few materials have been identified as potential candidates for metallic interconnect, all with high chromium contents. Oxidation of Cr and volatilisation of Cr_2O_3 which influences R_i and cathode performance adversely with time necessitates further development. The presently most advanced metallic interconnect material uses dispersed yttria to reduce the oxidation rate [21]. Fabrication of Cr-based alloys usually requires relatively expensive powder metallurgic processes. Additionally the presence of yttria makes post-fabrication component shaping very difficult and boosts the cost far beyond the possibilities in commercialisation.

Probably, a commercialized SOFC will be fuelled with natural gas. If todays standard anode, the Ni-YSZ-cermet, is exposed to natural gas, carbon precipitation will occur and the anode will be destroyed. Therefore, steam reforming is necessary. e Economic calculations show that it is very important to avoid external reforming units which will add to the cost of the SOFC-system (balance of plant). As Ni is a good steam reforming catalyst, internal reforming is in principle possible. However, at 1000°C the very endothermic reforming process will be too fast resulting in unacceptable steep temperature gradients at the fuel inlet. Consequently, it is necessary to develop anodes which are either more suitable for internal reforming than todays standard, or preferably able to facilitate direct oxidation of the methane in prereformed natural gas which seems possible using CeO_2-based anodes [22]. Prereforming of natural gas means to reform the hydrocarbons containing two or more carbon atoms. This will always be necessary in order to avoid carbon deposition in the gas preheating tubes. Natural gas usually contains 5-10% of C_2's or higher hydrocarbons.

5. Electric Efficiency

5.1. THEORETICAL EFFICIENCY

People working within power stations often calculate the efficiency by taking the Higher Heating Value (% HHV) of the prime fuel as their starting point. The Higher Heating Value is numerically the same as the change in standard enthalpy, ΔH^{o}_{298} (298 K, 1 atm.) but with an opposite sign. The maximum amount of work obtainable from a chemical is reaction is equal to the change og Gibb's free energy, ΔG. The connection between the changes in free energy, ΔG, and enthalpy, ΔH, and the entropy, ΔS, is given by the well-known formula:

$$\Delta G = \Delta H - T\Delta S \tag{1}$$

The theoretical efficiency, η_t, which can be achieved for a fuel cell at a given working temperature is, therefore,

$$\eta_t = \frac{\Delta G}{\Delta H^{o}_{298}} = \frac{\Delta H - T\Delta S}{\Delta H^{o}_{298}} \tag{2}$$

where ΔG, ΔH and ΔS refer to the temperature, T (in Kelvin), and the actual pressure condition in the fuel cell.
For Carnot engines

$$\eta_t = \frac{T - T_o}{T} \tag{3}$$

where T again is the working temperature and T_0 is the temperature in the exit gas.

When arguing for the development of fuel cells it has often been claimed that by comparing the very equations (2) and (3) it is seen that the fuel cells have a possibility of achieving higher efficiencies than those of the Carnot engines.

This is, however, a very simplified argumentation. Figure 3 shows a comparison of theoretical efficiencies for fuel cells for three different fuel gases (H_2, CO and CH_4) together with the Carnot curve. It is seen that only regarding CH_4-fuel the efficiency of the fuel cells is higher than that of a Carnot cycle in the whole temperature area of 25-1000°C.

The explanation of the difference in temperature dependence of the efficiency for the different fuel gases is a consequence of the number of gas molecules on the left and right side of the overall equations of reaction:

$$2H_2 + O_2 \rightarrow 2H_2O \qquad \text{(I)}$$
$$2CO + O_2 \rightarrow 2CO_2 \qquad \text{(II)}$$
$$CH_4 + 2O_2 \rightarrow CO_2 + 2H_2O \qquad \text{(III)}$$

In reactions (I) and (II) the number of gas molecules is changed from 3 to 2 during the reaction, which results in a temperature dependent entropy loss. In reaction (III) there is no change in the number of molecules.

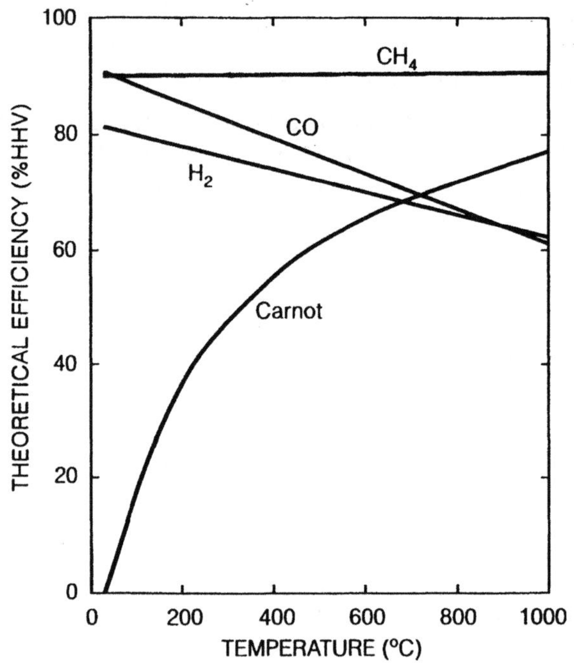

Figure 3. Comparison between the theoretical efficiency of fuel cells for CH_4, CO and H_2 fuel gases and the Carnot efficiency.

For a fuel cell operated on hydrogen it is seen (Figure 3) that it is most advantageous to operate at low temperatures. However, this only proves true in practice in special situations where a direct hydrogen source is available. Normally, hydrogen is made from reaction of coal, oil or gas with water. During the reforming process a large amount of heat has to be added at a rather high temperature (600-700°C). This heat, however, is regained in the fuel cell but at a too low temperature in the low temperature fuel cells. Practically, this means that the low temperature cells, which can only operate on hydrogen, have efficiencies which are not much above 40% HHV of the fossil fuel.

Regarding the high temperature fuel cells, however, it is really possible to increase the efficiency using methane (natural gas) as fuel. Naturally, the theoretical efficiency will not be reached. In practice there will always be losses.

5.2. PRACTICAL EFFICIENCIES

The resistance of the materials and the polarization of the electrodes determine the size of the loss developed in the form of heat resulting from the internal resistance of the fuel cell stack. To some degree the internal resistance can be decreased by choosing a design having the shortest possible way in the thin electrodes for the current to flow. This is obtainable by a plate design (Fig. 2) in which the current only has to flow maximum half of the width of the gas channel in the electrodes to reach the thicker cell interconnect. A half gas channel width will typically be in the order of 1 mm.

For a fuel cell the efficiency and the power are dependent on the cell load. These correlations will be elucidated in more detail.

The area specific internal resistance, R_i (incl. electrode polarization), is in the first approximation taken as independent of the current density. This means that Ohm's law is applicable and the following simple correlation is found:

$$P = EI - R_i I^2 \tag{4}$$

where P is the power density (W/cm^2), E is the open circuit voltage and I is the current density. The cell voltage, U, at the current density, I (in A/cm^2), is:

$$U = E - R_i I \tag{5}$$

The voltage efficiency, η_v, is defined as:

$$\eta_v = \frac{U}{E} \tag{6}$$

The open circuit voltage, E, also called the electromotive force, is about 1 volt for fuel cells.

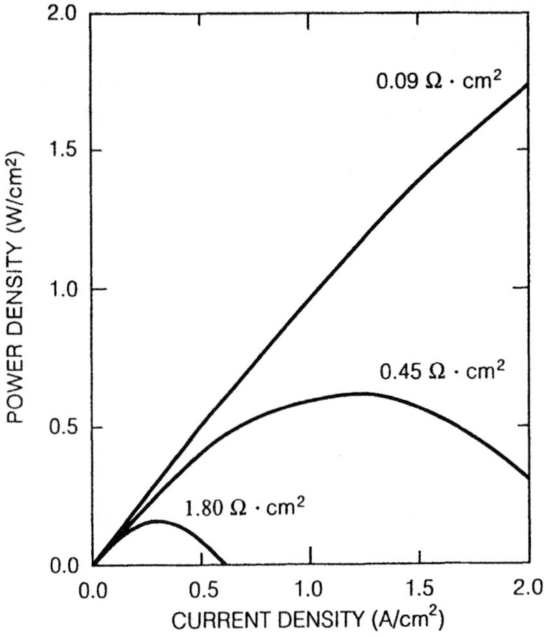

Figure 4. The power density of a solid oxide fuel cell as a function of current density for three different values of internal resistance.

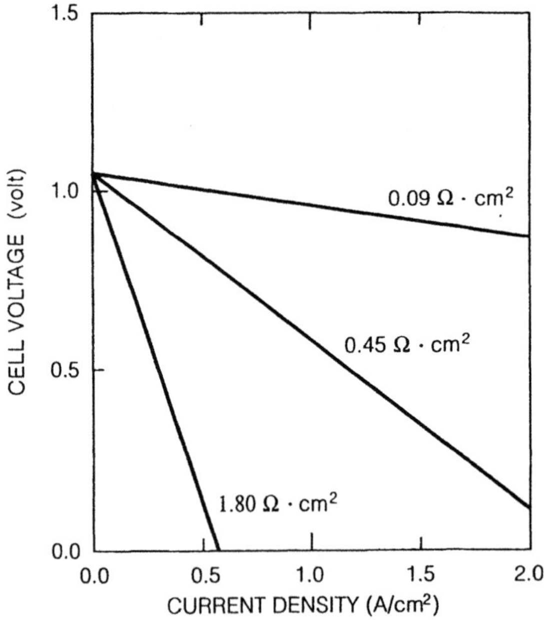

Figure 5. Cell voltage as a function of current density for a solid oxide fuel cell for the three internal resistances of Fig. 4. As the electromotive force E in the chosen case is close to 1 V, the cell voltage values are here approximately equal to the voltage efficiency of eq. (6).

In Figs. 4 and 5, it is illustrated what happens to the power density and the cell voltage when real values for R_i are put into the formulas. The three values for R_i apply to SOFC at three development levels. The highest R_i-value 1.8 Ωcm^2, applies to stacks of the early development stage. The medium value, 0.45 Ωcm^2, applies to present days best SOFC stacks. The lowest value, 0.09 Ωcm^2, is the resistance value calculated from the measured conductivities of the classic SOFC electrolyte and electrode materials (150 μm 8YSZ, 80 μm Ni-YSZ cermet, 80 μm LSM, 2.5 mm LSC). This applies on the assumption that the electrodes are not polarized, or in other words: the electrodes have an unlimited electrocatalytic effect and there are no mass transport limitations.

The electromotive force, E, is given by:

$$E = \frac{\Delta G}{nF} \tag{7}$$

where n is the number of electrons per mol and F is Faraday's number. The real efficiency, η, is:

$$\eta = \eta_t \eta_v \alpha \tag{8}$$

where α is the fuel utilization. It equals the amount of converted fuel divided by the amount of supplied fuel. By combining the formulas (2), (6), (7) and (8) the result is:

$$\eta = \frac{\alpha nFU}{\Delta H^o_{298}} \tag{9}$$

In the formulas (8) and (9) η is the efficiency in fraction of the higher heating value of the actual fuel gas put into the cell. For instance, loss from reforming coal or natural gas into hydrogen is not included in (8) and (9).

5.3. THE MAXIMUM PRACTICAL EFFICIENCY

If a pressurized high temperature fuel cell system (SOFC or MCFC) is put in front of a gas turbine, and a part, say 50%, of the fuel is converted in the fuel cell and the rest is burnt in the gas turbine, very high real electric efficiencies around 70% (for the combined system) may be reached. In fact the highest alowable stack costs are obtained in such a combination. Thus, in contrast to the low temperature fuel cells, the high temperature ones may in the futue get a share of the large multi-MW power plant market.

6. Cost Analysis

The problems related to fuel cell costs are not well described in the open literature, probably because this information is essential for the competition in the further developments. Therefore, these problems are illustrated here by an extract from an economic analysis [23] done as part of the Danish Solid Oxide Fuel cell program, DK-SOFC.

The problems which must be solved before SOFC-systems are competitive with todays power production technology are of both technical (as described above) and economical nature. The cost of SOFC stacks at the 25 kW level of today is about 30,000 ECU/kW [3] and it is bound to come down to about 500 ECU/kW. The allowable cost of a SOFC system is anticipated to be around 1500 ECU/kW.

As part of the Danish SOFC program (DK-SOFC) a 0.5 kW stack was built and tested during the second half of 1995 [20]. Based upon the experience gained, an economic analysis has been made. And the tools required to approach an economically acceptable solution have been outlined below.

6.1. TECHNOLOGICAL STARTING POINT

The Danish 0.5 kW SOFC stack was built according to the Bipolar Flat Plate SOFC (BFP-SOFC) concept. 50 cells, each with an active area of 50 cm^2, produced the targeted 0.5 kW with an overall area specific internal resistance of 1 Ωcm^2 at 1000°C. Cheap ceramic techniques with a potential for upscaling were selected initially for the technological development, which led to the first stack production.

Each cell was built from a tape cast and sintered yttria stabilized (8 mole%) zirconia electrolyte plate. The 8x8 cm plate was made from high purity zirconia with a thickness of 160μm.

Thin (5-10μm) electrochemically active electrode layers were mounted by spray painting of ceramic slurries and sintered individually. The composite cathode layer was based upon yttria stabilized zirconia and $La_{0.85}Sr_{0.15}Mn_{1.1}O_3$ (LSM) [24]. The latter was synthesized from pure metal nitrates with an organic complexing agent, which additionally acted as internal fuel in a drip pyrolysis process. The composite anode layer was made from yttria stabilized zirconia and NiO [25].

A cathode current collecting layer was made from high purity LSM by tape casting, mounted on the sintered composite cathode by hand followed by sintering. Anode current collection was accomplished by spray painted layers of yttria stabilized zirconia with an increasing NiO concentration gradient in the direction orthogonal to the anode and one sintering.

Ceramic interconnects were fabricated from $La_{0.8}Sr_{0.2}Cr_{0.97}V_{0.03}O_3$ [26] made from an organically complexed drip pyrolysis process with pure materials, similar to the afore mentioned production of LSM. Fractions of the powder were thermally treated at various temperatures, mixed and uniaxially pressed prior to sintering at high temperature in air. The sintered plates were machined with diamonds to give suitable surfaces, well defined border lines and cross flow gas channels.

Substacks of 10 cells each were assembled from approved interconnect plates and cells using spray painted (anode) and tape cast (cathode) contact layers. Subsequent to assembling electrode, seals were established with strings of developed glass types.

6.2. CALCULATED COSTS

The fabrication cost of a stack made according to the above guidelines has been calculated to 3,000-10,000 ECU/kW, depending on the internal stack resistance. The lower limit of the fabrication costs corresponds to an internal area specific resistance, R_i, of 0.4 Ωcm^2. The direct materials cost constitutes ~50% (Fig. 6). The materials cost includes actual losses during the various fabrication processes and components rejected by the quality controls. The remaining 50% are the costs of capital and manpower in a pilot plant with a total production capacity of 2 MW SOFC stacks per year. In spite of the fundamentally cheap ceramic techniques originally selected, the stack cost nevertheless requires a reduction of 6-20 times before it becomes of general *(non-niche related)* commercial interest (500 ECU/kW).

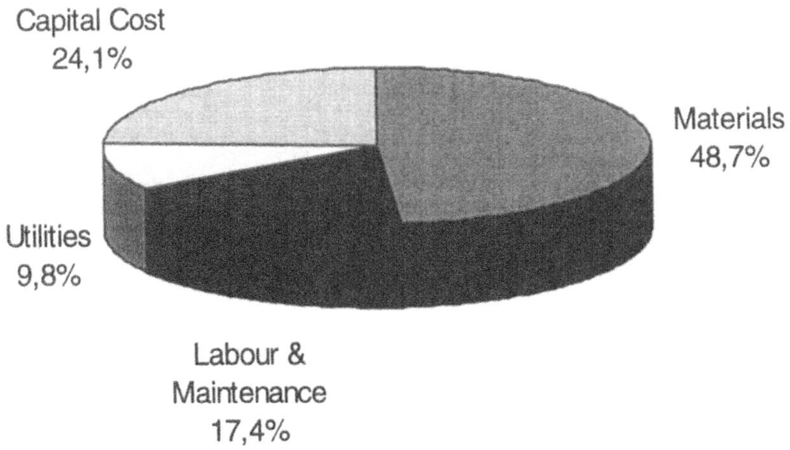

Figure 6. Distribution of present stack fabrication cost.

A number of tools are available for reduction of the stack cost:
1. Redesign of identified high-cost elements
2. Cheaper materials where acceptable
3. Reduction of the number of components in the repetitive unit of the stack
4. Reduction of losses during fabrication
5. Automation of the production

The interconnect plate is by far the most expensive individual stack element. The requirement of gas tightness implicates low materials porosity and the presence of gas channels makes the stack element the most voluminous one, altogether maximizing the consumption of the relatively costly lanthanum chromite. The thickness and the layout with channels in cross flow configuration limit the range of applicable ceramic shaping techniques. A requirement for planarity minimizes in-plane current conduction in the adjacent electrodes (part of R_i) but introduces high cost mechanical grinding after sintering. Reduction of the interconnect element cost clearly requires reconsideration of the design, leading to lower complexity of the element.

Materials of high purity are usually used for the development of new materials and components in laboratory scale. Commercialisation, however, will require determination of the importance of purity to enable substitution of frequently used materials with cheaper ones where such substitution has no detrimental effects. The cost of pure lanthanum compounds is high. The relatively high content in a stack (Fig. 7) makes the use of much cheaper lanthanide mixtures desirable, and a substitution seems to be acceptable in current collecting elements [17]. The cost of yttrium stabilized zirconia is high, too, when the content of silica which decrease the ion conductivity, is to be kept at a low level. Additions to YSZ which may neutralize the effect of Si in cheaper zirconia are desired.

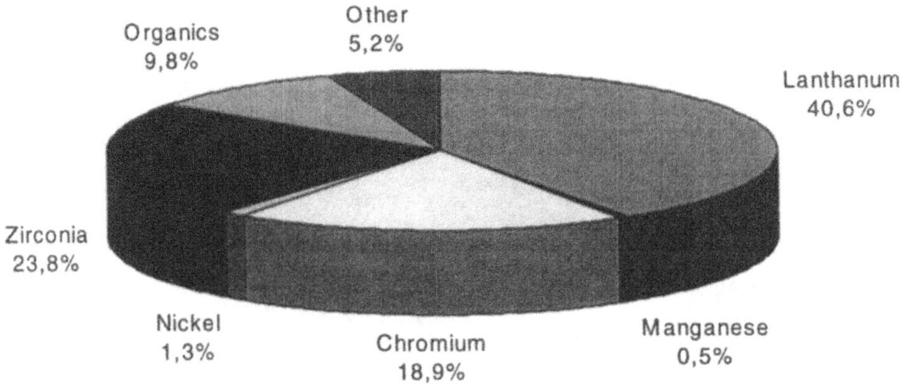

Figure 7. Distribution of present materials cost

Eight different stack components were used to build the repetitive stack segment in the first Danish 1/2 kW stack and fabrication involved 7 sintering procedures. A cost reduction will require a cutdown of the number of components as well as the number of sinterings.

Generally, losses were high during the laboratory scale fabrication of the stack elements. Mass production and automation will of course reduce the waste fraction somewhat, but losses are to a significant extent inherent to the individual ceramic techniques (cut-away green tape, slurry remnants). It is therefore of importance to establish recycling of waste in the fabrication processes.

Cofiring - in the sense of simultaneous sintering of adjacent layers - is often mentioned as a means of reducing the number of sintering steps. It is doubted, if the balance between gain due to the reduced number of sintering processes and increase of component losses due to less than perfect parameter control and materials reactions actually is in favour of cofiring. Cofiring which includes the electrolyte material is difficult, because sintering temperatures much below 1300°C of YSZ will result in incomplete sintering. Cofiring of YSZ with NiO-based anodes at temperatures above 1300°C may affect the bend strength of the electrolyte adversely assumably due to NiO dissolution in the electrolyte layer, the effect increasing with temperature and time. Cofiring of YSZ with LSM based cathodes at temperatures even below 1300°C may result in the formation of zirconates, preventing full densification of the electrolyte and may affect the cathode performance adversely. Cofiring of Ni-based anodes and Cr-alloy based as well as $LaCrO_3$-based interconnect materials at relevant temperatures will also lead to undesired reactions. However, cofiring of non-adjacent layers, e.g. anode and cathode sintering on the two sides of an already sintered electrolyte is a means for reduction of the overall number of sinterings.

Application of the above principles indicates that a reduction of the stack cost by a factor of 3 may be realised as a result of the identified possible improvements of the technology. The relative cost distribution in Fig. 8 shows that the materials cost is reduced below 25% of the total, while capital investment cost increases to above 40%.

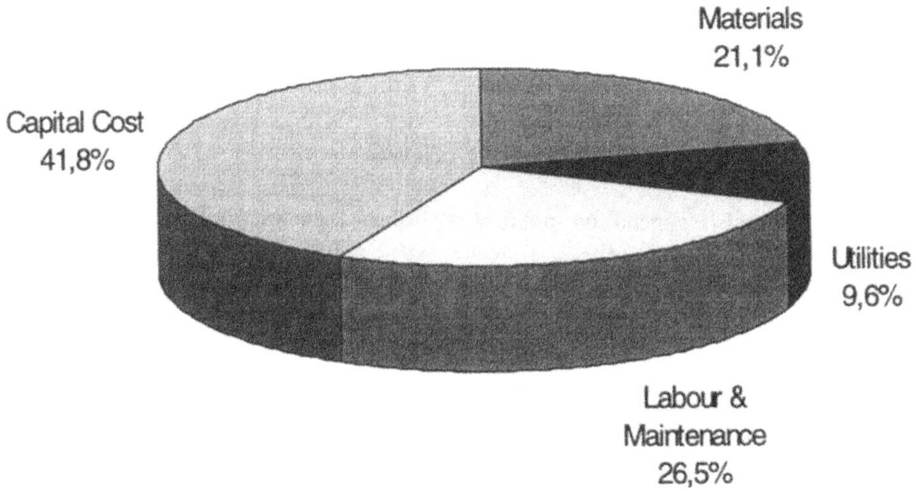

Figure 8. Assessed distribution of stack fabrication cost after planned optimizations

The only realistic way of further reduction of the stack cost is though further decrease of R_i because the cost of the entire fuel cell system is in the first approximation proportional to R_i.

7. Applications of Fuel Cells

7.1. THE ENERGY SUPPLY OF SPACE CRAFTS

During the 1960'ies, alkaline cells were developed for use in the Apollo space craft bound for the moon. Fuel cells have high energy as well as power densities, and contrary to most other energy sources they do not evolve a lot of waste products. The only reaction product of the oxygen and hydrogen is pure water which in fact was part of the astronauts' water supply.

In this connection, it is interesting to note that the first fuel cell used in practice was a polymer proton conductor fuel cell developed by General Electric for the Gemini programme. This type of fuel cell had not been of any particular interest before. However, the fact that space travel required lightness, resulted in the development of this particular lightweight type. For the later Apollo programme the "old" alkaline fuel cell was chosen because, in a special space craft design, its power per kilogramme was bigger than that of the SPFC. Today a special space version of the alkaline cell is used in the space shuttle. The weight of a 18 kW unit is 120 kg.

The best electrode materials for low temperature fuel cells are precious metals. As regards the space version, the air electrode is made from a mixture of gold and platinum black on gold coated nickel, and the hydrogen electrode is made from a platinum-palladium alloy. Although NASA has not published the price of fuel cells for space crafts, figures in the order of 1 to 2 mill. US$ per kW installed have been mentioned.

Thus, the choice of materials for fuel cells for space travel has been rather exotic. However, much experience has been gained which is applicable in the development of cheaper types for use on earth.

7.2. ELECTRIC CARS AND POWER STATIONS

It will very much depend on political decisions regarding the protection of the environment whether fuel cells to a greater extent will be used in electric cars in the future. A dc electric engine costs about as much as a petrol engine. An ac engine is significantly cheaper but in this case a dc to ac inverter is required, and the price of this seems to be in the range of the price of the petrol engine (about 50-150 $/kW). This means that the price of fuel cells in cars has to compete with the price of a petrol tank, which means compete with a bent and welded metal sheet. This is the reason why electric cars can probably never be competitive on equal economic terms with the petrol car but must be treated economically different. This may well be justified by the fact that the fuel cells do not pollute [2].

Whereas fuel cells and electric engines cannot compete as to prices with petrol and diesel engines for cars, the situation is totally different in the case of production of electric power for the grid. As the background for comparison with conventional technology it should be mentioned that the electric efficiency is up towards 40% HHV for advanced coal fired boilers/steam turbines multi-MW power plants, and the cost is about 1000 US$/kW. For advanced gas turbine power plants bigger than about 50 MW the efficiency is about 50% HHV and the cost about 150 US$/kW [28]. In the case of the Otto engine the efficiency is in the range of 20-25% and the cost is also about 150 US$ [6].

When it comes to the question of competition, there are three main points to consider: cost of construction, life time and efficiency. The fact that the price of car engines is normally below 200 US$ per kW power illustrates that it is not the cost of construction alone which is especially decisive. The fact is that the life time of a car engine is normally below 3000 hours of operation (300,000 km with 100 km/h) corresponding to 4 months of continous operation. Such a short life time is of course unacceptable as the power stations require a life time for the plants of many years.

8. Visions

Eventually, the possibilities represented by fuel cells can influence the everyday life of man to a considerable extent. One example is that most types of fuel cells can be used as electrolysers. Thus, a fuel cell can be connected with e.g. a windmill. When there is a surplus of wind this surplus is used to electrolyze water into oxygen and hydrogen. When the weather is calm, the hydrogen is transmitted into the fuel cell which consequently produces electricity. SOFC is especially qualified for water electrolysis.

Fuel cells need not take up much room and, as for instance the oxide fuel cell can be operated by natural gas, it seems reasonable to put fuel cell systems in every house with natural gas at its disposal. A 10 kW SOFC unit will probably take no more room up than a central heater unit does today. Such a fuel cell system would be able to supply a house with both electricity and heat. In fact only few kW per house is needed in average over night and day. The Swiss company, Sulzer, is in fact trying to realize this vision using a special design, HEXIS [29], but many years of further development is probably necessary before all problems are solved in an economic feasible manner.

Apart from a large-scale power production there are of course a large number of special applications for fuel cells: For instance a combined electricity and chemical production is possible; examples are: selective oxidation of methanol to formaldehyde and oxidation of ammonia to nitric acid for the fertilizer industry. Research is being made in the area of implanting small fuel cells into the human body for the operation of pacemakers, pumps etc. These fuel cells may be able to operate by using the blood sugar as fuel. Finally, research is being carried out in the coupling of biological systems (algae etc.) with fuel cells. In these bio-fuel cells the micro organisms produce convertible substance from CO_2 and solar energy in a chamber directly connected to the fuel cell anode.

9. Acknowledgement

This work was supported by the Danish Energy Agency and ELSAM.

10. References

1. Appleby, A. J., Foulkes,F. R. (1989). Fuel Cell Handbook. Van Nostrand Reinhold, New York.
2. Minh, N. Q., Takahashi, T. (1995). Science and Technology of Ceramic Fuel cells. Elsevier, Amsterdam.
3. European Commission (1995). A ten year fuel cell research, development and demonstration strategy for Europe. Separate document from the European Commission's Directorate General XII and XVII.
4. Lindstrøm, O. (1988). Muscles, engines and fuel cells, Chemtech 17, 686-693.
5. Lindstrøm, O. (1994). A critical assessment of fuel cell technology, Program and Abstracts of 1994 Fuel Cell Seminar, 227-298.
6. Gülzow, E. (1996). Alkaline fuel cells: a critical view, *J.Power Sources*, **61**, 99-104.
7. Komaki, H., Txuchiyama, S. (1996). Performance of a 1 kW PEFC, Program and Abstracts of 1996 Fuel Cell Seminar, 307-310.
8. Prater, K.B. (1992). Solid polymer fuel cell development at Ballard, J. Power Sources (1992) 181-188.
9. Prater, K.B. (1996). Solid polymer fuel cells for transport and stationary applications, J. Power Sources, 61, 105-109.
10. Waidhas, M., Drenckhahn, W., Preidel, W., Landes, H. (1996). Direct-fueld fuel cells, J. Power Sources, 61, 91-97.
11. Sjunnesson, L. (1994). The experiences from the demonstration of fuel cells in Europe, - The user's prespectives and prospects for the future, Program and Abstracts of 1994 Fuel Cell Seminar, 15-20.
12. Homma, T. Mori, S. Nakaoka, A. (1994). Current status of fuel cells in Japan, Program and Abstracts of 1994 Fuel Cell Seminar, 675-678.
13. Bonvill, L.J., Scheffler, G.W. and Smith, M.J. (1996). Progress and prospects for phosphoric acid fuel cell power plants, Program and Abstracts of 1996 Fuel Cell Seminar, 24-27.
14. Baker, B.S., Maru, H.C. (1997) Carbonate fuel cells, a decade of progress, in: Proc. 4th International Symp. Carbonate Fuel Cell Technology, Proc. Vol. 97-4, The Electrochemical Society Inc., Pennington, pp. 14-27.
15. Ding, J., Patel, P.S., Farooque, M., Maru, H.C. (1997). A computer model for direct carbonate fuel cells, in: Proc. 4th International Symp. Carbonate Fuel Cell Technology, Proc. Vol. 97-4, The Electrochemical Society Inc., Pennington, pp. 127-138.
16. Singhal, S.C. (1997). Recent progress in tubular solid oxide fuel cell technology, in: Solid Oxide Fuel Cells V, Proc. vol. 97-40, The Electrochemical Society Inc., Pennington, pp. 37-50.
17. Singhal, S.C. (1996). Status of Solid Oxide Fuel Cell Technology, in: Proc. 17th Risø International Symposium on Materials Science, pp. 123-138.
18. Minh, N.Q. Montgomery, K. (1997). Performance of reduced-temperature SOFC stacks, in: Solid Oxide Fuel Cells V, Proc. vol. 97-40, The Electrochemical Society Inc., Pennington, pp.153-159.

19. Pow, R., Reindl, M., Tillmetz, W. (1996). High power density fuel cell stack development for automotive applications, Program and Abstracts of 1996 Fuel Cell Seminar, 276-279.
20. Bagger, C., Juhl, M., Hendriksen, P.V., Larsen, P.H., Mogensen, M., Larsen, J.G. and Pehrson, S. (1996). Development and testing of a Danish 0.5 kW SOFC stack. In: Proc. 2nd European Solid Oxide Fuel Cell Forum, 6-10 May 1996, Oslo, Norway. Edited by Bernt Thorstensen, p. 175-184.
21. Köck, W., Martinz, H.-P., Greiner, H. and Janousek, M. (1995). Development and processing of metallic Cr based materials for SOFC parts. In: Solid oxide fuel cells IV. Edited by M. Dokiya, O. Yamamoto, T. Tagawa and S.C. Singhal, p. 841-849.
22. Mogensen, M., Kindl, B. (1994). Solid state fuel cell and process for the production thereof. European Patent 0571494 and US Patent 5, 350, 641.
23. Bagger, C.,Christiansen, N., Hendriksen, P.V., Jensen, E.J., Larsen, S.S., Mogensen, M. (1996) Techno-economic problems of SOFC commercialisation. In: Proc. 17th Risø International Symp. Materials Science, pp 167 - 174.
24. Bagger, C., Kindl, B. and Mogensen, M. (1995). Solid Oxide Fuel Cell, European patent, pending, no. 94908982.5
25. Bagger, C. (1992). Improved production methods for YSZ electrolyte and Ni-YSZ anode for SOFC. In: Program and Abstracts of 1992 Fuel Cell Seminar, November 29 - December 2, Tucson, Arizona, 241-244.
26. Carter, J.D. (1996). A sintering acid for doped lanthanum chromite. European patent, pending, no. 94923682.2.
27. Lloyd, A.C., Leonard, J.H., George, R. (1994). Fuel cells and air quality: a California perspective, J. Power Sources, 49, 209-233.
28. Krist, K., Wright, J.D., Romero, C., Chen, T.P. (1996) Cost projections for planar solid oxide fuel cell systems, Program and Abstracts of 1996 Fuel Cell Seminar, 497 - 500.
29. Diethelm, R., Brun, J. Gamper, Th., Keller, M., Kruschwitz, R. and Lenel, D., Status of the Sulzer Hexis solid oxide fuel cell (SOFC) system development. In: Solid Oxide Fuel Cells V. Edited by U. Stimming, S.C. Singhal, H. Tagawa, W. Lehnert, The Electrochemical Society Proceedings, Vol. 97-40, p.79-87.

INDEX

GPSR Compliance

The European Union's (EU) General Product Safety Regulation (GPSR) is a set of rules that requires consumer products to be safe and our obligations to ensure this.

If you have any concerns about our products, you can contact us on ProductSafety@springernature.com

In case Publisher is established outside the EU, the EU authorized representative is:

Springer Nature Customer Service Center GmbH
Europaplatz 3
69115 Heidelberg, Germany

Batch number: 09624545

Printed by Printforce, the Netherlands